London Mathematical Society Lecture Note Series: 348

Transcendental Dynamics and Complex Analysis

A Tribute to Noel Baker

Edited by

PHILIP J. RIPPON
The Open University

GWYNETH M. STALLARD
The Open University

CAMBRIDGE
UNIVERSITY PRESS

CAMBRIDGE UNIVERSITY PRESS
Cambridge, New York, Melbourne, Madrid, Cape Town, Singapore, São Paulo, Delhi

Cambridge University Press
32 Avenue of the Americas, New York, NY 10013-2473, USA

www.cambridge.org
Information on this title: www.cambridge.org/9780521683722

© Cambridge University Press 2008

This publication is in copyright. Subject to statutory exception
and to the provisions of relevant collective licensing agreements,
no reproduction of any part may take place without
the written permission of Cambridge University Press.

First published 2008

Printed in the United Kingdom at the University Press, Cambridge

A catalog record for this publication is available from the British Library.

Library of Congress Cataloging in Publication data

Transcendental dynamics and complex analysis / edited by Philip J. Rippon,
Gwyneth M. Stallard.
  p. cm. – (London Mathematical Society lecture note series; 348)
Includes bibliographical references and index.
ISBN 978-0-521-68372-2 (pbk.)
1. Functions of complex variables. 2. Differentiable dynamical systems.
3. Mathematical analysis. I. Rippon, P. J. II. Stallard, Gwyneth M.
QA331.7.T73   2008
515'.9–dc22   2007050517

ISBN 978-0-521-68372-2 (paperback)

Cambridge University Press has no responsibility for the persistence or accuracy of URLs for
external or third-party Internet Web sites referred to in this publication and does not guarantee
that any content on such Web sites is, or will remain, accurate or appropriate.

QA 614.8 TRA

LONDON MATHEMATICAL SOCIETY LECT

Managing Editor: Professor N. J. Hitchin, Mathematical I
University of Oxford, 24–29 St Giles, Oxford OX1 3LB, U

The titles below are available from booksellers, or from C:
at www.cambridge.org/mathematics

159 Groups St Andrews 1989 volume 1, C.M. CAMPBE
160 Groups St Andrews 1989 volume 2, C.M. CAMPBELL & E.F. ROBERTSON (eds)
161 Lectures on block theory, B. KÜLSHAMMER
163 Topics in varieties of group representations, S.M. VOVSI
164 Quasi-symmetric designs, M.S. SHRIKANDE & S.S. SANE
166 Surveys in combinatorics, 1991, A.D. KEEDWELL (ed)
168 Representations of algebras, H. TACHIKAWA & S. BRENNER (eds)
169 Boolean function complexity, M.S. PATERSON (ed)
170 Manifolds with singularities and the Adams-Novikov spectral sequence, B. BOTVINNIK
171 Squares, A.R. RAJWADE
172 Algebraic varieties, G.R. KEMPF

251 Gröbner bases and applications, B. BUCHBERGER & F. WINKLER (eds)
252 Geometry and cohomology in group theory, P. KROPHOLLER, G. NIBLO, R. STÖHR (eds)
253 The q-Schur algebra, S. DONKIN
254 Galois representations in arithmetic algebraic geometry, A.J. SCHOLL & R.L. TAYLOR (eds)

255	Symmetries and integrability of difference equations,	P.A. CLARKSON & F.W. NIJHOFF (eds)
256	Aspects of Galois theory,	H. VÖLKLEIN et al
257	An introduction to noncommutative differential geometry and its physical applications 2ed,	J. MADORE
258	Sets and proofs,	S.B. COOPER & J. TRUSS (eds)
259	Models and computability,	S.B. COOPER & J. TRUSS (eds)
260	Groups St Andrews 1997 in Bath, I,	C.M. CAMPBELL et al
261	Groups St Andrews 1997 in Bath, II,	C.M. CAMPBELL et al
262	Analysis and logic,	C.W. HENSON, J. IOVINO, A.S. KECHRIS & E. ODELL
263	Singularity theory,	B. BRUCE & D. MOND (eds)
264	New trends in algebraic geometry,	K. HULEK, F. CATANESE, C. PETERS & M. REID (eds)
265	Elliptic curves in cryptography,	I. BLAKE, G. SEROUSSI & N. SMART
267	Surveys in combinatorics, 1999,	J.D. LAMB & D.A. PREECE (eds)
268	Spectral asymptotics in the semi-classical limit,	M. DIMASSI & J. SJÖSTRAND
269	Ergodic theory and topological dynamics,	M.B. BEKKA & M. MAYER
270	Analysis on Lie Groups,	N.T. VAROPOULOS & S. MUSTAPHA
271	Singular perturbations of differential operators,	S. ALBEVERIO & P. KURASOV
272	Character theory for the odd order function,	T. PETERFALVI
273	Spectral theory and geometry,	E.B. DAVIES & Y. SAFAROV (eds)
274	The Mandelbrot set, theme and variations,	T. LEI (ed)
275	Descriptive set theory and dynamical systems,	M. FOREMAN et al
276	Singularities of plane curves,	E. CASAS-ALVERO
277	Computational and geometric aspects of modern algebra,	M.D. ATKINSON et al
278	Global attractors in abstract parabolic problems,	J.W. CHOLEWA & T. DLOTKO
279	Topics in symbolic dynamics and applications,	F. BLANCHARD, A. MAASS & A. NOGUEIRA (eds)
280	Characters and automorphism groups of compact riemann surfaces,	T. BREUER
281	Explicit birational geometry of 3-folds,	A. CORTI & M. REID (eds)
282	Auslander-Buchweitz approximations of equivariant modules,	M. HASHIMOTO
283	Nonlinear elasticity,	Y. FU & R.W. OGDEN (eds)
284	Foundations of computational mathematics,	R. DEVORE, A. ISERLES & E. SÜLI (eds)
285	Rational points on curves over finite fields,	H. NIEDERREITER & C. XING
286	Clifford algebras and spinors 2ed,	P. LOUNESTO
287	Topics on Riemann surfaces and Fuchsian groups,	E. BUJALANCE et al
288	Surveys in combinatorics, 2001,	J. HIRSCHFELD (ed)
289	Aspects of Sobolev-type inequalities,	L. SALOFF-COSTE
290	Quantum groups and Lie theory,	A. PRESSLEY (ed)
291	Tits buildings and the model theory of groups,	K. TENT (ed)
292	A quantum groups primer,	S. MAJID
293	Second order partial differential equations in Hilbert spaces,	G. DA PRATO & J. ZABCZYK
294	Introduction to the theory of operator spaces,	G. PISIER
295	Geometry and integrability,	L. MASON & Y. NUTKU (eds)
296	Lectures on invariant theory,	I. DOLGACHEV
297	The homotopy category of simply connected 4-manifolds,	H.-J. BAUES
298	Higher operads, higher categories,	T. LEINSTER
299	Kleinian groups and hyperbolic 3-manifolds,	Y. KOMORI, V. MARKOVIC & C. SERIES (eds)
300	Introduction to Möbius differential geometry,	U. HERTRICH-JEROMIN
301	Stable modules and the D(2)-problem,	F.E.A. JOHNSON
302	Discrete and continuous nonlinear Schrödinger systems,	M.J. ABLORWITZ, B. PRINARI & A.D. TRUBATCH
303	Number theory and algebraic geometry,	M. REID & A. SKOROBOGATOV (eds)
304	Groups St Andrews 2001 in Oxford Vol. 1,	C.M. CAMPBELL, E.F. ROBERTSON & G.C. SMITH (eds)
305	Groups St Andrews 2001 in Oxford Vol. 2,	C.M. CAMPBELL, E.F. ROBERTSON & G.C. SMITH (eds)
306	Peyresq lectures on geometric mechanics and symmetry,	J. MONTALDI & T. RATIU (eds)
307	Surveys in combinatorics 2003,	C.D. WENSLEY (ed)
308	Topology, geometry and quantum field theory,	U.L. TILLMANN (ed)
309	Corings and comdules,	T. BRZEZINSKI & R. WISBAUER
310	Topics in dynamics and ergodic theory,	S. BEZUGLYI & S. KOLYADA (eds)
311	Groups: topological, combinatorial and arithmetic aspects,	T.W. MÜLLER (ed)
312	Foundations of computational mathematics, Minneapolis 2002,	F. CUCKER et al (eds)
313	Transcendantal aspects of algebraic cycles,	S. MÜLLER-STACH & C. PETERS (eds)
314	Spectral generalizations of line graphs,	D. CVETKOVIC, P. ROWLINSON & S. SIMIC
315	Structured ring spectra,	A. BAKER & B. RICHTER (eds)
316	Linear logic in computer science,	T. EHRHARD et al (eds)
317	Advances in elliptic curve cryptography,	I.F. BLAKE, G. SEROUSSI, N. SMART
318	Perturbation of the boundary in boundary-value problems of partial differential equations,	D. HENRY
319	Double affine Hecke algebras,	I. CHEREDNIK
320	L-functions and Galois representations,	D. BURNS, K. BUZZARD & J. NEKOVÁŘ (eds)
321	Surveys in modern mathematics,	V. PRASOLOV & Y. ILYASHENKO (eds)
322	Recent perspectives in random matrix theory and number theory,	F. MEZZADRI & N.C. SNAITH (eds)
323	Poisson geometry, deformation quantisation and group representations,	S. GUTT et al (eds)
324	Singularities and computer algebra,	C. LOSSEN & G. PFISTER (eds)
325	Lectures on the Ricci flow,	P. TOPPING
326	Modular representations of finite groups of Lie type,	J.E. HUMPHREYS
328	Fundamentals of hyperbolic manifolds,	R.D. CANARY, A. MARDEN, & D.B.A. EPSTEIN (eds)
329	Spaces of Kleinian groups,	Y. MINSKY, M. SAKUMA & C. SERIES (eds)
330	Noncommutative localization in algebra and topology,	A. RANICKI (ed)
331	Foundations of computational mathematics, Santander 2005,	L. PARDO, A. PINKUS, E. SULI & M. TODD (eds)
332	Handbooks of tilting theory,	L. ANGELERI HÜGEL, D. HAPPEL & H. KRAUSE (eds)
333	Synthetic differential geometry 2ed,	A. KOCK
334	The Navier-Stokes equations,	P.G. DRAZIN & N. RILEY
335	Lectures on the combinatorics of free probability,	A. NICA & R. SPEICHER
336	Integral closure of ideals, rings, and modules,	I. SWANSON & C. HUNEKE
337	Methods in Banach space theory,	J.M.F. CASTILLO & W.B. JOHNSON (eds)
338	Surveys in geometry and number theory,	N. YOUNG (ed)
339	Groups St Andrews 2005 Vol. 1,	C.M. CAMPBELL, M.R. QUICK, E.F. ROBERTSON & G.C. SMITH (eds)
340	Groups St Andrews 2005 Vol. 2,	C.M. CAMPBELL, M.R. QUICK, E.F. ROBERTSON & G.C. SMITH (eds)
341	Ranks of elliptic curves and random matrix theory,	J.B. CONREY, D.W. FARMER, F. MEZZADRI & N.C. SNAITH (eds)
342	Elliptic cohomology,	H.R. MILLER & D.C. RAVENEL (eds)
343	Algebraic cycles and motives Vol. 1,	J. NAGEL & C. PETERS (eds)
344	Algebraic cycles and motives Vol. 2,	J. NAGEL & C. PETERS (eds)
345	Algebraic and analytic geometry,	A. NEEMAN
346	Surveys in combinatorics, 2007,	A. HILTON & J. TALBOT (eds)
347	Surveys in contemporary mathematics,	N. YOUNG & Y. CHOI (eds)

Contents

Preface		vii
Introduction		ix
1	Iteration of inner functions and boundaries of components of the Fatou set *D. Bargmann*	1
2	Conformal automorphisms of finitely connected regions *A. F. Beardon and D. Minda*	37
3	Meromorphic functions with two completely invariant domains *W. Bergweiler and A. Eremenko*	74
4	A family of matings between transcendental entire functions and a Fuchsian group *S. Bullett and M. Freiberger*	90
5	Singular perturbations of z^n *R. L. Devaney, M. Holzer, D. M. Look, M. Moreno Rocha and D. Uminsky*	111
6	Residual Julia sets of rational and transcendental functions *P. Domínguez and N. Fagella*	138
7	Bank-Laine functions via quasiconformal surgery *D. Drasin and J. K. Langley*	165
8	Generalisations of uniformly normal families *W. K. Hayman and A. Hinkkanen*	179
9	Entire functions with bounded Fatou components *A. Hinkkanen*	187
10	On multiply connected wandering domains of entire functions *M. Kisaka and M. Shishikura*	217
11	Fractal measures and ergodic theory of transcendental meromorphic functions *J. Kotus and M. Urbański*	251
12	Combinatorics of bifurcations in exponential parameter space *L. Rempe and D. Schleicher*	317
13	Baker domains *P. J. Rippon*	371
14	Escaping points of the cosine family *G. Rottenfusser and D. Schleicher*	396
15	Dimensions of Julia sets of transcendental meromorphic functions *G. M. Stallard*	425
16	Abel's functional equation and its role in the problem of *croissance régulière* *G. Szekeres*	447

Professor Noel Baker (1932–2001)

PREFACE

This book was written in honour of Noel Baker following his sudden death in 2001. It comprises a collection of articles written by friends, colleagues and former students of Noel. In particular, we are delighted that Noel's MSc supervisor and long-time friend, George Szekeres, was able to contribute a paper to this volume — he sadly died before the book was published.

All of these articles deal with topics that interested Noel and, in most cases, they are in areas where Noel's own work has been very influential. Several of the papers are survey articles that we hope will be a valuable addition to the literature. There are also new results that Noel would have been delighted to have seen. Most of the papers deal with the iteration of transcendental meromorphic functions — the field in which Noel was pre-eminent and in which he carried out much of the pioneering work — and there are also some papers in closely related topics that he would have enjoyed. As this volume shows, much of the recent work in complex dynamics (as the subject of iteration theory is now called) builds on ideas and techniques that Noel introduced and that will continue to be used by all those who work in this field. We hope that this book will be a fitting memorial to a man who inspired so many of us.

Phil Rippon and Gwyneth Stallard

Department of Mathematics and Statistics
The Open University
Milton Keynes MK7 6AA

INTRODUCTION

In this introduction, we summarise the mathematical career of Noel Baker and indicate how the papers in this volume relate to his work. Much of the material is taken from the obituary of Noel Baker that appeared in the *Bulletin of the London Mathematical Society* [17].

Noel Baker was born on 10 August 1932 and died, of a heart attack, on 20 May 2001. He grew up in Australia and was first introduced to the theory of iteration by his MSc supervisor, George Szekeres, who suggested that he work on the functional equation

$$f(f(z)) = F(z),$$

where f and F are analytic functions. In his first mathematical paper (1), Noel used the theory of iteration of analytic functions, which had been developed principally by Fatou and Julia and which was not well known at that time. He used this theory to show, amongst other things, that if F belongs to a certain class of entire functions, which includes the exponential function, then the above equation has no entire solution. This first paper also contains examples that were constructed using Wiman-Valiron theory. Throughout his career Noel was to find ever more techniques from classical complex analysis that can usefully be applied to iteration theory.

In 1955 Noel won a German government scholarship to the University of Tübingen, where he worked under Hellmuth Kneser. Noel's doctoral thesis, published in (2), continued his study of functional equations. From 1957 to 1959, Noel taught mathematics at the University of Alberta in Edmonton, Canada. In 1959 he moved to Imperial College London, where he remained until retirement in 1997.

In his research, Noel worked on many problems in complex analysis and had a wide range of collaborators, but iteration theory, his great love, was for many years a lone interest. However, when the subject was reborn around 1980, partly as a result of the advent of accessible computer graphics, it became clear to the new adherents that Noel had for many years been quietly and carefully completing the foundations begun earlier in the century by the French mathematicians Pierre Fatou and Gaston Julia. He had also pointed the way towards many future developments, both by proving new results and by posing challenging problems. In the explosion of research on iteration theory that took place in the subsequent years, many of the papers published on iteration made reference to Noel's work and he received many invitations to speak at international conferences on iteration. At these he would often appear reserved, much preferring to let others speak about the latest work, even though he was the acknowledged authority on many matters, and the person whose judgement about the validity of a new proof was always sought. Noel continued his research after his retirement and one of his last papers was dedicated to George Szekeres on the occasion of the latter's 90th birthday.

Noel's early work on functional equations led him to consider problems about periodic points, which play a very important role in complex dynamics. It was already known that for an entire function there must be infinitely many periodic points of period p, for all $p \geq 2$, but Noel considered the unsolved problem of the existence of periodic points of a given *exact* period. He showed in (6) that for all non-linear entire functions there exist periodic points of exact period p, for all p with at most one exception; for example, $f(z) = z + e^z$ has no fixed points. In a later paper (13) Noel showed that for a polynomial the only possible exceptional value in this result is $p = 2$, the corresponding exceptional functions being $f(z) = z^2 - z$ and other quadratics 'similar' to this one. He also conjectured that for a transcendental entire function the only possible exceptional value is $p = 1$, and this was proved by Bergweiler [2].

We now describe the origins of complex dynamics. Let f be a rational function of degree at least 2 or a transcendental entire function. The set of points near which the sequence of iterates f^n forms a normal family is called the *Fatou set* $F(f)$ and its complement is called the *Julia set* $J(f)$. Roughly speaking, the dynamics are stable on the Fatou set and chaotic on the Julia set. Also, the Julia set often exhibits great topological complexity as well as 'self-similarity'; for example, the paper in this volume by Devaney *et al* discusses a family of rational functions whose Julia sets in some cases contain Cantor sets of curves and in other cases contain Sierpinski curves.

The fundamental properties of the sets $F(f)$ and $J(f)$ were first established for rational functions in [13] and [8], and for transcendental entire functions in [9]. In the last paper, Fatou studied the iteration of transcendental entire functions in some detail, giving examples that pointed to significant differences to the theory that had been developed for rational functions. He asked the following fundamental questions about a transcendental entire function f:

1. Are the repelling periodic points of f dense in $J(f)$?
2. Are there examples where $J(f) = \mathbb{C}$? In particular, is this true for $f(z) = e^z$?
3. Can $J(f)$ be totally disconnected?
4. Must $J(f)$ contain infinitely many unbounded analytic curves, at each point of which $f^n \to \infty$?

Question 1 is of great theoretical importance, and it had been answered 'yes' for rational functions by both Fatou and Julia. Fatou had also given an example of a rational function f for which $J(f)$ is totally disconnected, and Lattès [14] an example for which $J(f) = \mathbb{C}$. Most of Fatou's questions were solved by Noel during the decade 1965–1975, as we now indicate.

The first question was answered in the affirmative in the paper (22), which is of fundamental importance in complex dynamics and appropriately dedicated to Hellmuth Kneser. Here, Noel called on a deep covering theorem due to Ahlfors (see [11, page 148]) to show that arbitrarily close to each point of $J(f)$ there is a repelling periodic point of f. From this, he deduced the general result that if f is any non-linear entire function, then the set of entire functions that commute with f is countable. Many authors have tried to simplify the proof in (22) that the repelling periodic points are dense in $J(f)$, in order to avoid the deep theorem of Ahlfors. Eventually, more elementary proofs based on a renormalisation technique were given by Schwick [18], Bargmann [1], and Berteloot and Duval [4].

Two years later, in (25), Noel answered the first part of Fatou's second question by showing that there is a function of the form $f(z) = kze^z$, where $k > 0$, such that $J(f) = \mathbb{C}$. A proof that if $f(z) = e^z$, then we have $J(f) = \mathbb{C}$ was given ten years later by Misiurewicz [15].

Noel answered Fatou's third question in the negative in (32). If $J(f)$ is totally disconnected, then $F(f)$ must have a single unbounded multiply connected component. Noel had already constructed in (9) an example of a transcendental entire function for which $F(f)$ has at least one multiply connected component. This function was of the form

$$f(z) = Cz^2 \prod_{n=1}^{\infty} \left(1 + \frac{z}{r_n}\right),$$

in which the positive constants $r_1 < r_2 < \ldots$ have the property that

$$f(A_n) \subset A_{n+1}, \quad \text{where } A_n = \{z : r_n^2 < |z| < r_{n+1}^{1/2}\}.$$

However, Noel did not determine in (9) whether $F(f)$ has a single unbounded multiply connected component or a sequence of bounded multiply connected components. In (33) he used Schottky's theorem [11, page 169], yet another result from classical complex analysis, to show that the latter must be the case. This solved another important problem in complex dynamics, open since the work of Fatou and Julia, by showing that the above function has a sequence of *wandering domains*, that is, distinct components U_n of $F(f)$ such that $f(U_n) \subset U_{n+1}$, for $n = 1, 2, \ldots$. In contrast, Sullivan [16] showed that rational functions do not have wandering domains. The paper (32), written later than but published earlier than (33), used Schottky's theorem once again to show that a transcendental entire function cannot have an unbounded multiply connected component of $F(f)$, thus proving that $J(f)$ can never be totally disconnected.

The results in (32) and (33) led to much further work. In (53), Noel showed that wandering domains for transcendental entire functions may be infinitely connected. For many years it was not known whether such wandering domains could be finitely connected. In this volume, Kisaka and Shishikura show that they can in fact have any given finite connectivity.

The result in (32) shows that if f is a transcendental entire function, then $J(f)$ must contain a continuum, so its Hausdorff dimension $\dim_H J(f)$ is at least 1. It remains an open question whether $\dim_H J(f) = 1$ is possible. In this volume, there is a survey article on dimensions of Julia sets by Stallard, complemented by a survey article on fractal measures and ergodic theory by Kotus and Urbański.

Noel's wandering domains example mentioned earlier shows that the answer to Fatou's fourth question (as stated here) is 'no'. However, the structure of the 'escaping set', where $f^n \to \infty$, continues to stimulate much work, including the paper by Rottenfusser and Schleicher that appears in this volume.

Sullivan's remarkable result [16] that rational functions do not have wandering domains was proved using new techniques based on quasiconformal conjugacy. Noel quickly saw that these new techniques would also apply to various families of transcendental entire functions, and a proof that exponential functions have no wandering domains appeared in (49). This was one of a number of papers at that time that established many of the basic dynamical properties of the exponential family and began the description of the corresponding parameter space, the 'exponential Mandelbrot set', which has since been the subject of much study — see, for example, the paper by Rempe and Schleicher in this volume.

In (41), Noel initiated another major development by showing that if a transcendental entire function f has order of growth at most $1/2$, minimal type, then $F(f)$ has no unbounded invariant components, and he also gave a more restrictive condition on the maximum modulus of f that forces every component of $F(f)$ to be bounded. The question of whether the latter conclusion follows from order at most $1/2$, minimal type, remains open, though many authors have obtained partial results in this direction; this volume contains a survey article on this problem by Hinkkanen.

A key step in Noel's proof in (41) is to exclude unbounded invariant components of $F(f)$ in which $f^n \to \infty$. He did this by establishing estimates for the growth of iterates in such components, which he later refined in (57). In recognition of his work on Fatou components of this type, Eremenko and Lyubich introduced the name *Baker domain* for such components in [7]. In this fundamental paper, Eremenko and Lyubich showed that if the set $S(f)$ of inverse function singularities of a transcendental entire function f is bounded, then f has no Baker domains and if $S(f)$ is finite, then f has no wandering domains; see also [10]. A survey article on Baker domains by Rippon appears in this volume.

Yet another fundamental contribution to the iteration of transcendental entire functions came in the papers (65), (73) and (74). Once again an unbounded invariant component U of $F(f)$ was considered, but now the aim was to describe the nature of the boundary of U. Some special cases had been investigated by other authors, following the appearance of computer pictures of Julia sets, but Noel and his students Weinreich and Domínguez attacked the general case. In (65), it was shown that

- if U is not a Baker domain (that is, U is an attracting basin, a parabolic basin, or a Siegel disc), then ∂U is sufficiently complicated that ∞ belongs to the impression of every prime end of U;
- if ∂U is a Jordan curve in the extended complex plane $\hat{\mathbb{C}}$ (and such U do exist), then not only must U be a Baker domain, but f must be univalent in U.

The key tool introduced in this work arises from the fact that if Ψ is a conformal map from the unit disc D onto U, then $\Psi^{-1} \circ f \circ \Psi$ is an *inner function*, that is, an analytic self-map of D whose angular limits have modulus 1 almost everywhere on ∂D. The paper (65) initiated a version of Fatou-Julia theory for inner functions, a topic now of interest in its own right, and this theory was taken further in (73). Further results on this theory are given in the paper by Bargmann in this volume.

Many of Noel's final papers are joint papers with his last student, Domínguez, and concern the connectedness properties of the Julia set. Many of these results are described and extended in the paper by Domínguez and Fagella in this volume.

Fatou-Julia theory of the iteration of *general* transcendental meromorphic functions was established in the fundamental papers (62), (63), (64) and (66) by Baker, Kotus and Lü. The Fatou set $F(f)$ is here taken to be the set of points near which the iterates f^n are defined and form a normal family, and then $J(f) = \hat{\mathbb{C}} \setminus F(f)$. Many of the basic results turn out to be similar to those for rational and entire functions, but there are some striking differences. For example, in (62) the authors showed that $J(f)$ is once again the closure of the repelling periodic points of f, and this fact is used to give a complete classification of those transcendental meromorphic functions, such as $f(z) = \tan z$, for which $J(f)$ is a subset of the real line; there are no transcendental entire functions for which the Julia set is contained in the real line. Then, in (63), they used techniques from approximation theory, pioneered by Eremenko and Lyubich [6], to construct transcendental meromorphic functions with wandering domains of all possible connectivities.

The question of periodic components was taken up in (64), where the authors showed that precisely five possible types can arise for a transcendental meromorphic function, namely, attracting basins, parabolic basins, Siegel discs, Herman rings and Baker domains. Moreover, any invariant components of $F(f)$ must be simply connected, doubly connected, or infinitely connected.

But perhaps the most striking result here was the construction of a transcendental meromorphic function f with a preperiodic component of $F(f)$ of any given finite connectivity. This construction used the powerful technique of quasiconformal surgery, introduced by Shishikura [19], which also appears in many of the papers in this volume — namely, those by Drasin and Langley, Domínguez and Fagella, and Kisaka and Shishikura. Finally, in (66), Sullivan's method of quasiconformal conjugacy was adapted to show that a transcendental meromorphic function of finite type has no wandering domains. These four papers opened a new and fruitful area of research, made even more accessible by the excellent survey article [3], which appeared soon after.

One of the differences between the iteration of entire functions and meromorphic functions is the number of completely invariant components of the Fatou set that can occur. In (24) Noel proved that a transcendental entire function can have at most one completely invariant component of the Fatou set. In (64) Baker, Kotus and Lü proved that a transcendental meromorphic function of finite type can have at most two completely invariant Fatou components, and in this volume it is shown by Bergweiler and Eremenko that, in these circumstances, the Julia set must be a Jordan curve. (An example of a function with these properties is $f(z) = \tan z$.) It is an open question whether a general transcendental meromorphic function can have at most two completely invariant Fatou components.

Fatou-Julia theory can be developed in many further directions. For a transcendental meromorphic function f, the iterates f^n need not be meromorphic. It is desirable, however, to have a closed system of iterates, so that we can consider, for example, the Fatou set of f^n, for $n \geq 2$. To obtain such a system, Noel's student Herring [12], and independently Bolsch [5], developed Fatou-Julia theory for functions, such as $f(z) = e^{\tan z}$, which are meromorphic outside certain compact totally disconnected subsets of $\hat{\mathbb{C}}$. Much of this theory, and its subsequent developments, is expounded in Noel's last papers (75), (77), (78) and (79).

This volume also contains papers that, while not explicitly about complex dynamics, are on closely related topics. The paper by Hayman and Hinkkanen is concerned with the growth of meromorphic functions that belong to certain normal families, the paper by Beardon and Minda classifies conformal automorphisms of finitely connected regions of the plane, and the paper by Szekeres is on possible connections between 'regular growth' and Abel's functional equation, a topic in which Noel had a great interest. Finally, the paper by Bullett and Freiburger is on the theory of holomorphic correspondences, a generalisation of complex dynamics. Here they investigate, for the first time, holomorphic correspondences that involve transcendental entire functions.

Publications of I. N. Baker

(1) The iteration of entire transcendental functions and the solution of the functional equation $f(f(z)) = F(z)$, *Math. Ann.* 129 (1955) 174–180.
(2) Zusammensetzungen ganzer Funktionen, *Math. Z.* 69 (1958) 121–163.
(3) Fixpoints and iterates of entire functions, *Math. Z.* 71 (1959) 146–153.
(4) Solutions of the functional equation $(f(x))^2 - f(x^2) = h(x)$, *Canad. Math. Bull.* 3 (1960) 113–120.
(5) Some entire functions with fixpoints of every order, *J. Austral. Math. Soc.* 1 (1959/61) 203–209.
(6) The existence of fixpoints of entire functions, *Math. Z.* 73 (1960) 280–284.
(7) Permutable entire functions, *Math. Z.* 79 (1962) 243–249.
(8) Permutable power series and regular iteration, *J. Austral. Math. Soc.* 2 (1961/62) 265–294.
(9) Multiply-connected domains of normality in iteration theory, *Math. Z.* 81 (1963) 206–214.
(10) Length of a graph, *Solution in Amer. Math. Monthly* 71 (1964) 217–218.
(11) Partition of a domain, *Solution in Amer. Math. Monthly* 71 (1964) 219–220.
(12) Fractional iteration near a fixpoint of multiplier 1, *J. Austral. Math. Soc.* 4 (1964) 143–148.
(13) Fixpoints of polynomials and rational functions, *J. London Math. Soc.* 39 (1964) 615–622.
(14) Entire functions with linearly distributed values, *Math. Z.* 86 (1964) 263–267.
(15) Sets of non-normality in iteration theory, *J. London Math. Soc.* 40 (1965) 499–502.
(16) The distribution of fixpoints of entire functions, *Proc. London Math. Soc.* 16 (1966) 493–506.
(17) On a class of meromorphic functions, *Proc. Amer. Math. Soc.* 17 (1966) 819–822.
(18) On some results of A. Rényi and C. Rényi concerning periodic entire functions, *Acta Sci. Math. (Szeged)* 27 (1966) 197–200.
(19) A series associated with the logarithmic function, *J. London Math. Soc.* 42 (1967) 336–338.
(20) Non-embeddable functions with a fixpoint of multiplier 1, *Math. Z.* 99 (1967) 377–384.
(21) (with F. GROSS) On factorizing entire functions, *Proc. London Math. Soc.* 18 (1968) 69–76.
(22) Repulsive fixpoints of entire functions, *Math. Z.* 104 (1968) 252–256.

(23) (with F. GROSS) Further results on factorization of entire functions, *Entire Functions and Related Parts of Analysis (Proc. Symp. Pure Math. La Jolla, Calif., 1996)* (Amer. Math. Soc., 1968) 30–35.
(24) Completely invariant domains of entire functions, *Mathematical Essays Dedicated to A. J. Macintyre* (ed. H. Shankar, Ohio Univ. Press. 1970) 33–35.
(25) Limit functions and sets of non-normality in iteration theory, *Ann. Acad. Sci. Fenn. Ser. A I Math.* 467 (1970) 11 pp.
(26) The value distribution of composite entire functions, *Acta. Sci. Math. (Szeged)* 32 (1971) 87–90.
(27) (with L. S. O. LIVERPOOL) Picard sets for entire functions, *Math. Z.* 126 (1972) 230–238.
(28) (with L. S. O. LIVERPOOL) Further results on Picard sets of entire functions, *Proc. London Math. Soc.* 26 (1973) 82–98.
(29) Linear Picard sets for entire functions, *Math. Nachr.* 64 (1974) 263–276.
(30) (with J. A. DEDDENS and J. L. ULLMAN) A theorem on entire functions with applications to Toeplitz operators, *Duke Math. J.* 41 (1974) 739–745.
(31) (with E. MUES) Zur Faktorisierung endlicher Blaschkeproductke, *Arch. Math. (Basel)* 26 (1975) 388–390.
(32) The domains of normality of an entire function, *Ann. Acad. Sci. Fenn. Ser. A I Math.* 1 (1975) 277–283.
(33) An entire function which has wandering domains, *J. Austral. Math. Soc. Ser. A* 22 (1976) 173–176.
(34) Analytic mappings between two ultrahyperelliptic surfaces, *Aequationes Math.* 14 (1976) 461–472.
(35) (with C. C. YANG) An infinite order periodic entire function which is prime, *Complex Analysis,* Lecture Notes in Math. Vol 599 (Springer, 1977) 7–10.
(36) (with L. S. O. LIVERPOOL) Sylvester series and normal families, *Solution in Amer. Math. Monthly* 85 (1978) 290–291.
(37) (with L. S. O. LIVERPOOL) The value distribution of entire functions of order at most one, *Acta Sci. Math. (Szeged)* 41 (1979) 3–14.
(38) (with Ch. POMMERENKE) On the iteration of analytic functions in a halfplane II, *J. London Math. Soc.* 20 (1979) 255–258.
(39) Condition for a composite of polynomials, *Solution in Amer. Math. Monthly* 87 (1980) 228.
(40) Entire functions with two linearly distributed values, *Ann. Acad. Sci. Fenn. Ser. A I Math.* 5 (1980) 381–386.
(41) The iteration of polynomials and transcendental entire functions, *J. Austral. Math. Soc. Ser. A* 30 (1980/81) 483–495.

(42) (with J. M. ANDERSON and J. G. CLUNIE) The distribution of values of certain entire and meromorphic functions, *Math. Z.* 178 (1981) 509–525.

(43) Entire functions whose a-points lie on systems of lines, *Factorization theory of meromorphic functions*, Lecture Notes in Pure and Appl. Math. 78 (ed. C. C. Yang, Marcel Dekker, 1982) 1–18.

(44) Complex function theory: a sequence of entire functions converging pointwise, *James Cook Math. Notes*, Townsville, Qld, Australia, Issue 29, Vol. 3 (August 1982) 3112–3114.

(45) (with Z. RUBINSTEIN) Simultaneous iteration by entire or rational functions and their inverses, *J. Austral. Math. Soc. Ser. A* 34 (1983) 364–367.

(46) (with P. J. RIPPON) Convergence of infinite exponentials, *Ann. Acad. Sci. Fenn. Ser. A I Math.* 8 (1983) 179–186.

(47) (with L. S. O. LIVERPOOL) The entire solutions of a polynomial difference equation, *Aequationes Math.* 27 (1984) 97–113.

(48) Composition of polynomials, *Solution in Amer. Math. Monthly* 91 (1984) 317.

(49) (with P. J. RIPPON) Iteration of exponential functions, *Ann. Acad. Sci. Fenn. Ser. A I Math.* 9 (1984) 49–77.

(50) Wandering domains in the iteration of entire functions, *Proc. London Math. Soc.* 49 (1984) 563–576.

(51) (with P. J. RIPPON) A note on infinite exponentials, *Fibonacci Quart.* 23 (1985) 106–112.

(52) (with P. J. RIPPON) A note on complex iteration, *Amer. Math. Monthly* 92 (1985) 501–504.

(53) Some entire functions with multiply-connected wandering domains, *Ergodic Theory Dynam. Systems* 5 (1985) 163–169.

(54) Wandering domains for maps of the punctured plane, *Ann. Acad. Sci. Fenn. Ser. A I Math.* 12 (1987) 191–198.

(55) (with A. EREMENKO), A problem on Julia sets, *Ann. Acad. Sci. Fenn. Ser. A I Math.* 12 (1987) 229–236.

(56) Iteration of entire functions: an introductory survey, Proceeding of the Symposium on Complex Analysis, 21–22 May 1987, Xian, China, *Lectures on Complex Analysis* (World Sci. Publishing, 1988) 1–17.

(57) Infinite limits in the iteration of entire functions, *Ergodic Theory Dynam. Systems* 8 (1988) 503–507.

(58) (with P. BHATTACHARRYA) On a class of non-embeddable entire functions, *J. Ramanujan Math. Soc.* 3 (1988) 151–159.

(59) (with P. J. RIPPON) Iterating exponential functions with cyclic exponents, *Math. Proc. Cambridge Phil. Soc.* 105 (1989) 357–375.

(60) (with P. J. RIPPON) Towers of exponents and other composite maps, *Complex Variables Theory Appl.*, Volume in honour of Albert Edrei and Wolfgang Fuchs, 12 (1989) 181–200.

(61) (with P. J. RIPPON) On compositions of analytic self-mappings of a convex domain, *Arch. Math. (Basel)* 55 (1990) 380–386.

(62) (with J. KOTUS and LÜ YINIAN) Iterates of meromorphic functions II: Examples of wandering domains, *J. London Math. Soc.* 42 (1990) 267–278.

(63) (with J. KOTUS and LÜ YINIAN) Iterates of meromorphic functions I, *Ergodic Theory Dynam. Systems* 11 (1991) 241–248.

(64) (with J. KOTUS and LÜ YINIAN) Iterates of meromorphic functions III: Preperiodic domains, *Ergodic Theory Dynam. Systems* 11 (1991) 603–618.

(65) (with J. WEINREICH) Boundaries which arise in the dynamics of entire functions, Analyse complexe (Bucharest, 1989), *Rev. Roumaine Math. Pures Appl.* 36 (1991) 413–420.

(66) (with J. KOTUS and LÜ YINIAN) Iterates of meromorphic functions IV: Critically finite functions, *Results Math.* 22 (1992) 651–656.

(67) (with R. N. MAALOUF) Convergence of a modified iteration process, *Computational Methods and Function Theory, 1994 (Penang)* (ed. S. Ruscheweyh, World Sci. Publishing, 1995) 49–55.

(68) (with A. P. SINGH) Wandering domains in the iteration of compositions of entire functions, *Ann. Acad. Sci. Fenn. Ser. A I Math.* 20 (1995) 149–153.

(69) (with A. P. SINGH) A note on differential polynomials, *Bull. Calcutta Math. Soc.* 87 (1995) 63–66.

(70) (with G. M. STALLARD) Error estimates in a calculation of Ruelle, *Complex Variables Theory Appl.* 29 (1996) 141–159.

(71) On factorizing meromorphic functions, *Aequationes Math.* 54 (1997) 87–101.

(72) (with P. DOMÍNGUEZ) Analytic self-maps of the punctured plane, *Complex Variables Theory Appl.* 37 (1998) 67–91.

(73) (with P. DOMÍNGUEZ) Boundaries of unbounded Fatou components of entire functions, *Ann. Acad. Sci. Fenn. Math.* 24 (1999) 437–464.

(74) (with P. DOMÍNGUEZ) Some connectedness properties of Julia sets, *Complex Variables Theory Appl.* 41 (2000) 371–389.

(75) (with P. DOMÍNGUEZ) Residual Julia sets, *J. Anal.* 8 (2000) 121–137.

(76) Dynamics of slowly growing entire functions, *Bull. Austral. Math. Soc.* 63 (2001) 367–377.

(77) (with P. DOMÍNGUEZ and M. HERRING) Dynamics of functions meromorphic outside a small set, *Ergodic Theory Dynam. Systems* 21 (2001) 647–672.

(78) Limit functions in wandering domains of meromorphic functions, *Ann. Acad. Sci. Fenn. Math.* 27 (2002) 499–505.

(79) (with P. DOMÍNGUEZ and M. HERRING) Functions meromorphic outside a small set: completely invariant domains, *Complex Variables Theory Appl.* 49 (2004) 95–100.

References

[1] D. Bargmann, Simple proofs of some fundamental properties of the Julia set, *Ergodic Theory Dynam. Systems* 19 (1999) 553–558.
[2] W. Bergweiler, Periodic points of entire functions: proof of a conjecture of Baker, *Complex Variables Theory Appl.* 17 (1991) 57–72.
[3] W. Bergweiler, Iteration of meromorphic functions, *Bull. Amer. Math. Soc.* 29 (1993) 151–188.
[4] F. Berteloot and J. Duval, Une démonstration directe de la densité de cycles répulsif dans l'ensemble de Julia, *Complex analysis and geometry (Paris 1997)*, Progr. Math. 188 (Birkhäuser, Basel, 2000) 221–222.
[5] A. Bolsch, Repulsive periodic points of meromorphic functions, *Complex Variables Theory Appl.* 31 (1996) 75–79.
[6] A. E. Eremenko and M. Yu. Lyubich, Examples of entire functions with pathological dynamics, *J. London Math. Soc.* 36 (1987) 458–468.
[7] A. E. Eremenko and M. Yu. Lyubich, Dynamical properties of some classes of entire functions, *Ann. Inst. Fourier (Grenoble)* 42 (1992) 989–1020.
[8] P. Fatou, Sur les équations fonctionelles, *Bull. Soc. Math. France* 47 (1919) 161–271; 48 (1920) 33–94, 208–314.
[9] P. Fatou, Sur l'itération des fonctions transcendantes entières, *Acta Math.* 47 (1926) 337–370.
[10] L. Goldberg and L. Keen, A finiteness theorem for a dynamical class of entire functions, *Ergodic Theory Dynam. Systems* 6 (1986) 183–192.
[11] W. K. Hayman, *Meromorphic functions* (Clarendon Press, Oxford, 1964).
[12] M. Herring, An extension of the Julia–Fatou theory of iteration, Ph.D. thesis, University of London (1994).
[13] G. Julia, Mémoire sur l'itération des fonctions rationelles, *J. Math. Pure Appl.* 1 (1918) 47–245.
[14] S. Lattès, Sur l'itération des substitutions rationelles et les fonctions de Poincaré, *C.R. Acad. Sci. Paris Sér. I Math.* 166 (1918) 26–28 (Errata: p. 88).
[15] M. Misiurewicz, On iterates of e^z, *Ergodic Theory Dynam. Systems* 1 (1981) 103–106.
[16] D. Sullivan, Quasiconformal homeomorphisms and dynamics I, *Ann. Math.* 122 (2) (1985) 401–418.
[17] P. J. Rippon, Obituary: Irvine Noel Baker 1932–2001, *Bull. London Math. Soc.* 37 (2005) 301–315.
[18] W. Schwick, Repelling periodic points in the Julia set, *Bull. London Math. Soc.* 29 (1997) 314–316.
[19] M. Shishikura, On the quasiconformal surgery of rational functions, *Ann. Scient. Ec. Norm. Sup.* (4) 20 (1987) 1–29.

ITERATION OF INNER FUNCTIONS AND BOUNDARIES OF COMPONENTS OF THE FATOU SET

DETLEF BARGMANN

ABSTRACT. Let D be an unbounded invariant component of the Fatou set of a transcendental entire function f. Let $\phi : \mathbb{D} \to D$ be a Riemann map. Then the set $\Theta := \{\theta \in \partial \mathbb{D} : \lim_{r \to 1} \phi(r\theta) = \infty\}$ is closely related to the Julia set of the corresponding inner function $g := \phi^{-1} \circ f \circ \phi$. In the first part of the paper we further develop the theory of Julia sets of inner functions and the dynamical behaviour on their Fatou sets. In the second part we apply these results to iteration of entire functions by using the above relation and obtain some new results about the boundaries of components of the Fatou set of an entire function.

1. INTRODUCTION

Dynamics of inner functions have turned out to be a very useful tool to study the boundary structure of unbounded invariant components of the Fatou set of a transcendental entire function. The key method in this area has been developed by I.N. Baker and P. Domínguez [3]. We recall some of their techniques and results in Subsection 1.1 and then state the aims and results of this paper in Subsection 1.2

1.1. The method of Baker-Domínguez. Let f be a transcendental entire function, with Fatou set $\mathcal{F}(f)$ and Julia set $\mathcal{J}(f)$; see [7] for background information on these concepts. Suppose that D is an unbounded invariant component of $\mathcal{F}(f)$. Then D is simply connected [2, Theorem 1], which implies that there exists a biholomorphic (Riemann) map $\phi : \mathbb{D} \to D$. Then

$$g := \phi^{-1} \circ f \circ \phi$$

is an inner function, i.e. a holomorphic self-map of the unit disk such that

$$\lim_{r \to 1} |g(r \exp(2\pi i \alpha))| = 1,$$

for almost every $\alpha \in [0, 1]$. According to Fatou's theorem [18, p. 139] this implies that $\lim_{r \to 1} g(r \exp(2\pi i \alpha))$ exists and is contained in $\partial \mathbb{D}$, for almost every $\alpha \in [0, 1]$.

Keywords: inner function, Julia set, Fatou set, boundary, radial limit, Baker domain.
AMS subject classification: 30D05, 58F08.

If $f|_D$ is a proper self-map of D, then g is a (finite) Blaschke product, i.e. there exist $m, n \in \mathbb{N}_0$, $\lambda \in \partial \mathbb{D}$, and $a_1, \ldots, a_n \in \mathbb{D} \setminus \{0\}$ such that

$$g(z) = \lambda z^m \prod_{j=1}^{n} \frac{z - a_j}{1 - \overline{a_j} z},$$

for each $z \in \mathbb{D}$.

Since D is unbounded $f|_D$ need not be a proper self-map of D. In this case g has at least one singularity on the boundary of the unit disk.

Definition 1.1. *Let g be an inner function of $\mathbb{D}$. A point $\zeta \in \partial \mathbb{D}$ is called a singularity of g if g cannot be continued analytically to a neighbourhood of ζ. Denote the set of all singularities of g by* $\operatorname{sing}(g)$.

Throughout this paper we assume that an inner function is always continued to $\hat{\mathbb{C}} \setminus \overline{\mathbb{D}}$ by the reflection principle, where $\hat{\mathbb{C}}$ denotes the complex sphere, and to $\partial \mathbb{D} \setminus \operatorname{sing}(g)$ by analytic continuation.

It follows from the theory of inner functions that the composition of two inner functions is again an inner function; see [3, Lemma 4]. In particular, the n-th iterate g^n of an inner function g is an inner function. Now, the Julia set of an inner function can be defined in the following way.

Definition 1.2. *Let g be an inner function of the unit disk $\mathbb{D}$. The Fatou set $\mathcal{F}(g)$ of g is the set of all points $z \in \hat{\mathbb{C}}$ for which there is an open neighbourhood $U \subset \hat{\mathbb{C}}$ of z such that $U \cap \operatorname{sing}(g^n) = \emptyset$, for each $n \in \mathbb{N}$, and $\{g^n|_U : n \in \mathbb{N}\}$ is normal. The Julia set $\mathcal{J}(g)$ of g is the complement of $\mathcal{F}(g)$ in $\hat{\mathbb{C}}$.*

Remark 1.3. *It follows from Montel's theorem that $\mathcal{J}(g) \subset \partial \mathbb{D}$. Moreover, for the case of a finite Blaschke product this definition coincides with the usual definition of the Julia set of a rational function.*

Baker and Domínguez initiated the study of Julia sets of inner functions by proving the following result [3, Lemma 8].

Theorem 1.4 (Baker-Domínguez). *Let g be an inner function of the unit disk $\mathbb{D}$. Then the following properties hold.*
(1) $g(\mathcal{F}(g)) \subset \mathcal{F}(g)$.
(2) *If g is non-Möbius, then $\mathcal{J}(g)$ is a perfect set.*
(3) *If g is non-rational, then $\mathcal{J}(g) = \overline{\bigcup_{n \in \mathbb{N}} \operatorname{sing}(g^n)}$.*

The main tool in the proof of Theorem 1.4 is the the following lemma on inner functions [3, Lemma 5], which we use later. For the definition of a Stolz angle we refer the reader to [18, p. 6].

Lemma 1.5 (Baker-Domínguez). *Let g be an inner function of the unit disk. Suppose that $\zeta \in \operatorname{sing}(g)$. Then, for each $\theta \in \partial \mathbb{D}$ and each neighbourhood U of ζ, there exist $\eta \in U \setminus \{\zeta\}$ and a path $\gamma : [0,1) \to \mathbb{D}$ such that $\lim_{t \to 1} \gamma(t) = \eta$ and $g(\gamma(t)) \to \theta$ in a Stolz angle as $t \to 1$.*

Following Baker and Domínguez [3], we relate the boundary behaviour of the Riemann map ϕ to the dynamical behaviour of the corresponding inner function g. This process involves the sets

$$\Xi := \{\zeta \in \partial \mathbb{D} : \infty \in C(\phi; \zeta)\},$$

and

$$\Theta := \{\theta \in \partial \mathbb{D} : \lim_{r \to 1} \phi(r\theta) = \infty\}.$$

Here $C(\phi; \zeta)$ denotes the cluster set of ϕ at ζ, i.e. the set of all values $w \in \hat{\mathbb{C}}$ for which there is sequence $(z_n)_{n \in \mathbb{N}}$ in $\mathbb{D}$ such that $z_n \to \zeta$ and $\phi(z_n) \to w$ as $n \to \infty$. Since D is unbounded the set Ξ is always non-empty. In general, it is not known whether the set Θ is always non-empty. However, for the case when D is a Baker domain, i.e. $f^n|_D \to \infty$ locally uniformly, it is easy to see that $\Theta \neq \emptyset$. Throughout this paper the sets Ξ and Θ will relate to a Riemann mapping $\phi : \mathbb{D} \to D$ of the invariant Fatou component D of the function f under consideration.

There is a close connection between Ξ and $\mathcal{J}(g)$.

Lemma 1.6. *If $f|_D$ is not an automorphism of D, then $\mathcal{J}(g) \subset \Xi$.*

Proof. First, it is easy to see that Ξ is closed.

Case 1. Suppose that g is rational. Then g is a finite Blaschke product and it is easy to see that Ξ is backward invariant under g. Since g is locally injective on $\partial \mathbb{D}$ (see Remark 2.19) we conclude that Ξ is an infinite set. Hence Ξ is a closed, backward invariant set which contains at least three points. This implies that Ξ is a superset of $\mathcal{J}(g)$.

Case 2. Suppose that g is not rational. By Theorem 1.4 we need only show that the singularities of the iterates of g are contained in Ξ. Let $n \in \mathbb{N}$ and ζ be a singularity of g^n. Then $C(g^n; \zeta) = \overline{\mathbb{D}}$ (see for instance [13, Theorem 5.4]), which implies that

$$\infty \in \overline{\mathbb{D}} \subset C(\phi \circ g^n; \zeta) = C(f^n \circ \phi; \zeta).$$

Thus we conclude that $\infty \in C(\phi; \zeta)$. $\square$

Using Lemma 1.5, Baker and Domínguez obtained a similar result for the set Θ; see [3, Lemma 13].

Lemma 1.7 (Baker-Domínguez). *Suppose that $f|_D$ is not an automorphism of D and $\Theta \neq \emptyset$. Then $\mathcal{J}(g) \subset \overline{\Theta}$.*

Hence the Julia set of the corresponding inner function g is a lower bound for the size of the set Ξ and, provided that $\Theta \neq \emptyset$, for the size of the set $\overline{\Theta}$. This provides a strategy to show that the sets Ξ and $\overline{\Theta}$ are equal to the unit circle by showing that the Julia sets of the corresponding inner functions are the unit circle. Baker and Domínguez gave two cases when $\mathcal{J}(g) = \partial \mathbb{D}$; see [3, Lemma 9 and Lemma 10].

Theorem 1.8 (Baker-Domínguez). *Let g be a non-Möbius inner function of $\mathbb{D}$ with a fixed point $p \in \mathbb{D}$. Then $\mathcal{J}(g) = \partial \mathbb{D}$.*

Theorem 1.9 (Baker-Domínguez). *Let g be an inner function of $\mathbb{D}$. Suppose that there exists $p \in \partial \mathbb{D}$ such that, for each $z \in \mathbb{D}$, $g^n(z) \to p$ in an arbitrarily small Stolz angle as $n \to \infty$. Then g is non-Möbius and $\mathcal{J}(g) = \partial \mathbb{D}$.*

The assumptions of Theorem 1.8 are satisfied for the corresponding inner function g when D is an attracting domain of f. Theorem 1.9 can be applied to the corresponding inner function g when D is a parabolic domain of f. Taken together, these lead to the following result; see [4, Theorem 1] and [3, Theorem 1.1].

Theorem 1.10 (Baker-Weinreich, Baker-Domínguez). *Suppose that D is an attracting domain, a parabolic domain, or a Siegel disk of f. Then*

$$\Xi = \partial \mathbb{D} \quad \text{and} \quad \Theta \neq \emptyset \Rightarrow \overline{\Theta} = \partial \mathbb{D}.$$

This result does not carry over in general to Baker domains, since there are examples of Baker domains (due to Baker-Weinreich and Bergweiler, see Examples 3.4 and 3.5 in Subsection 3.1) whose boundaries are Jordan curves. On the other hand, Baker and Weinreich proved [4, Theorem 4] that if D is a Baker domain whose boundary is a Jordan curve, then $f|_D$ has to be univalent. Thus it is natural to ask whether Theorem 1.10 holds for those Baker domains for which $f|_D$ is not univalent. In this case, Baker and Domínguez proved that Θ contains a perfect set and hence is infinite; see [3, Theorem 1.2].

1.2. **Aims and results of this paper.** The aim of this paper is twofold. In Section 2 we further develop the theory of Julia sets of inner functions, independently of the application to iteration of entire functions. Then, in Section 3, we apply these results to iteration of entire functions. Here, we use the method of Baker-Domínguez described in Subsection 1.1 and further extend it.

In Section 2, the main results are in Subsections 2.3 and 2.4. In Subsection 2.3 we prove the following theorem, which is a generalization of Theorems 1.8 and 1.9. For a hyperbolic domain G in the complex sphere $\hat{\mathbb{C}}$, λ_G always denotes the hyperbolic metric on G.

Theorem 2.24 *Let g be an inner function such that $\lambda_{\mathbb{D}}(g^n(z), g^{n+1}(z)) \to 0$ as $n \to \infty$ for some $z \in \mathbb{D}$. Then $\mathcal{J}(g) = \partial \mathbb{D}$.*

Theorem 2.24 will be an easy consequence of two more general theorems which give necessary and sufficient conditions for an inner function to be eventually conjugated to a certain Möbius transformation on the Fatou set and which also classify the possible eventual conjugacies which can arise. See Subsection 2.1 for the meaning of 'eventual conjugacy'.

The different types of components of $\mathcal{F}(g) \cap \partial \mathbb{D}$ will be introduced and classified in Subsection 2.2.

In Subsection 2.4 we prove that the Julia set of an inner function coincides with the closure of the set of repelling periodic points (with a suitable definition of repelling periodic point, see Definition 2.30). Analogous results are known for rational and entire functions and it is an interesting fact that inner functions have this property, too.

Theorem 2.34 *Let g be a non-Möbius inner function. Then $\mathcal{J}(g)$ is the closure of the set of the repelling periodic points of g.*

At the end of Section 2, we give some examples of inner functions; in particular, we show that all the possible types of eventual conjugacy can occur.

In Section 3, we apply the results of Section 2 to the iteration of entire functions. For instance, using Lemma 1.7 and our Theorem 2.24 we can prove the following generalization of Theorem 1.10.

Theorem 3.2 *Let f be a transcendental entire function. Suppose that D is an unbounded invariant component of the Fatou set of f such that*

$$\lambda_D(f^n(z), f^{n+1}(z)) \to 0 \quad \text{as } n \to \infty,$$

for some $z \in D$. If $\Theta \neq \emptyset$, then $\overline{\Theta} = \partial \mathbb{D}$.

We can use Theorem 3.2 to extend Theorem 1.10 to a certain class of Baker domains. Here we use the symbol $\sim$ to indicate an eventual conjugacy.

Theorem 3.1 *Let f be a transcendental entire function. Suppose that D is a Baker domain of f such that $f|_D \sim \operatorname{id}_{\mathbb{C}} + 1$. Then $\overline{\Theta} = \partial \mathbb{D}$.*

Moreover, we see in Subsection 3.1 that, for a whole class of examples, the set Θ is dense in $\partial \mathbb{D}$ whenever f is not univalent on the Baker domain D; see Lemma 3.3.

In Subsection 3.2 we improve Lemma 1.7, at least for the case when D is a completely invariant component of the Fatou set.

Theorem 3.8 *Let f be a transcendental entire function. Suppose that D is a completely invariant component of the Fatou set of f. Let $\phi : \mathbb{D} \to D$ be a Riemann map and let $g := \phi^{-1} \circ f \circ \phi$ be the corresponding inner function. If $\Theta \neq \emptyset$, then $\mathcal{J}(g)$ is equal to the set of accumulation points of Θ.*

As a consequence of this result, we shall be able to prove the following results about boundaries of components of the Fatou set; see Subsection 3.3.

Theorem 3.11 *Let f be a transcendental entire function. Suppose that D is a completely invariant component of the Fatou set of f. Let $G \subset \mathbb{C}$ be a bounded Jordan domain such that $G \cap \mathcal{J}(f) \neq \emptyset$. Then $\partial G \cap D$ has infinitely many components.*

Theorem 3.12 *Let f be a transcendental entire function. Suppose that there exists an unbounded component of the Fatou set of f. Let $G \subset \mathbb{C}$ be a bounded Jordan domain such that $G \cap \mathcal{J}(f) \neq \emptyset$. Then $\partial G \cap \mathcal{F}(f)$ has infinitely many components.*

Theorem 3.14 *Let f be an entire function. Let D be a component of the Fatou set of f. Suppose that at least one of the following conditions is satisfied:*

(1) *f is transcendental and there exists an unbounded component of the Fatou set of f, or*
(2) *$\bigcup_{n \in \mathbb{N}} f^n(D)$ is bounded.*

Let $\phi : \mathbb{D} \to D$ be a Riemann map. Let $\mathrm{Acc}(D)$ be the set of finite accessible boundary points of D, and let Z be the set of all $\zeta \in \partial \mathbb{D}$ such that $\phi(\zeta) := \lim_{r \to 1} \phi(r\zeta)$ exists and is finite. Then the map

$$Z \to \mathrm{Acc}(D), \ \zeta \mapsto \phi(\zeta)$$

is a bijection.

Corollary 3.15 *Let f be an entire function. Suppose that D is a Siegel disk for f. Then*

(1) *There is no periodic point of f in ∂D which is an accessible boundary point of D.*
(2) *f is univalent on the set of finite accessible boundary points of D.*

Theorem 3.16 *Let f be an entire function. Let $\mathcal{E}$ be a finite set of components of the Fatou set of f. Suppose that at least one of the following conditions is satisfied:*

(1) *f is transcendental and there exists an unbounded component of the Fatou set of f, or*
(2) *$\bigcup_{n \in \mathbb{N}} f^n(D)$ is bounded, for each $D \in \mathcal{E}$.*

Then there are at most $\mathrm{card}(\mathcal{E}) - 1$ points in $\mathbb{C}$ which are common accessible boundary points of at least two components in $\mathcal{E}$.

Acknowledgement. I would like to thank the late Professor I.N. Baker and Professor W. Bergweiler for their support and helpful discussions.

2. Iteration of inner functions

2.1. Holomorphic self-maps of hyperbolic domains.
In this subsection we recall some facts about the dynamical behaviour of a holomorphic self-map

of the unit disk or, more generally, of a hyperbolic domain in the complex sphere. We start with an old theorem of A. Denjoy [15] and J. Wolff [22]; see also [12, p. 79] or [21, p. 43]. For the notion of the angular limit we refer the reader to [18, p. 6].

Theorem 2.1 (Denjoy-Wolff). *Let h be a non-Möbius holomorphic self-map of $\mathbb{D}$. Then there is a point $p \in \overline{\mathbb{D}}$ such that $h^n \to p$ locally uniformly on $\mathbb{D}$. Moreover, if $p \in \partial \mathbb{D}$, then h has the angular limit p at p.*

The point p referred to in this theorem is often called the Denjoy-Wolff point of h. It may appear that the dynamical behaviour of a holomorphic self-map depends only on the question whether its Denjoy-Wolff point p is inside the disk or on its boundary. But, in fact, the case $p \in \partial \mathbb{D}$ can be further subdivided. Here we make use of a classification due to C. Cowen [14] who has shown that a holomorphic self-map of the unit disk is eventually conjugated to a certain Möbius transformation. Roughly speaking, eventually conjugated means that the function is semi-conjugated to a Möbius transformation and, starting at an arbitrary point and iterating, one eventually lands in a simply connected region where the function is even conjugated to this Möbius transformation. More precisely, we have the following definition.

Definition 2.2. *Let h be a holomorphic self-map of a hyperbolic domain $G \subset \hat{\mathbb{C}}$. Let T be a biholomorphic self-map of a simply connected domain $\Omega \subset \mathbb{C}$. Then we say that*
$$h \sim T \quad (h \text{ is eventually conjugated to } T)$$
if there exist a holomorphic function $\Phi : G \to \Omega$ and a simply connected domain $V \subset G$ such that the following conditions are satisfied:
(1) *$\Phi \circ h = T \circ \Phi$.*
(2) *Φ is univalent on V.*
(3) *V is a fundamental set for h on G, i.e.*
$$h(V) \subset V \quad \text{and} \quad \forall\, z \in G\, \exists\, n \in \mathbb{N} : h^n(z) \in V.$$
(4) *$\Phi(V)$ is a fundamental set for T on Ω.*
In this case, (Ω, T, Φ, V) is called an eventual conjugacy of h on G.

Eventual conjugacies are unique in the following sense; see [14, p. 79-80].

Lemma 2.3 (Cowen). *Let h be a holomorphic self-map of a hyperbolic domain $G \subset \hat{\mathbb{C}}$. Let $(\Omega_1, T_1, \Phi_1, V_1)$ be an eventual conjugacy of h on G.*
(1) *Let $\tau : \Omega_1 \to \mathbb{C}$ be an injective holomorphic function. Then $(\tau(\Omega_1), \tau \circ T_1 \circ \tau^{-1}, \tau \circ \Phi_1, V_1)$ is an eventual conjugacy of h on G.*
(2) *Let $(\Omega_2, T_2, \Phi_2, V_2)$ be another eventual conjugacy of h on G. Then there exists a component W of $V_1 \cap V_2$ such that W is a fundamental set for h on G and, for each $j \in \{1, 2\}$, $\Phi_j(W)$ is a fundamental set for T_j on Ω_j. Moreover, there exists a biholomorphic map $\tau : \Omega_1 \to \Omega_2$ such that $\Phi_2 = \tau \circ \Phi_1$ and $T_2 = \tau \circ T_1 \circ \tau^{-1}$.*

Cowen's result can be stated as follows. Here, $\mathbb{H} := \{z \in \mathbb{C} : \operatorname{Im}(z) > 0\}$.

Theorem 2.4 (Cowen). *Let h be a holomorphic self-map of the unit disk $\mathbb{D}$ without a fixed point. Then exactly one of the following statements holds.*
(1) $h \sim \operatorname{id}_{\mathbb{C}} + 1$.
(2) *There exists exactly one $\sigma \in \{-1, 1\}$ such that $h \sim \operatorname{id}_{\mathbb{H}} + \sigma$.*
(3) *There exists exactly one $\lambda > 1$ such that $h \sim \lambda \operatorname{id}_{\mathbb{H}}$.*

For the proof of Theorem 2.4, see [14, Theorem 3.2] and the following remark.

Remark 2.5. *By applying Lemma 2.3 it is easy to see that, for each $\sigma \in \{-1, 1\}$,*
$$h \sim \operatorname{id}_{\mathbb{H}} + \sigma \iff h \sim \operatorname{id}_{\sigma\mathbb{H}} + 1 \iff h \sim \operatorname{id}_{-i\mathbb{H}} - i\sigma$$
and, for each $\lambda > 1$,
$$h \sim \lambda \operatorname{id}_{\mathbb{H}} \iff h \sim \lambda \operatorname{id}_{-\mathbb{H}} \iff h \sim \lambda \operatorname{id}_{-i\mathbb{H}}.$$

H. König [17] has given geometrical conditions to determine which Möbius transformation the function h is eventually conjugated to. Expressing these conditions in terms of the hyperbolic metric we obtain the following lemma. Recall that we denote the hyperbolic metric on a hyperbolic domain $G \subset \hat{\mathbb{C}}$ by λ_G.

Lemma 2.6. *Let h be a holomorphic self-map of a hyperbolic domain $G \subset \hat{\mathbb{C}}$ without a fixed point. For each $z \in G$, let*
$$\rho(z) := \inf_{n \in \mathbb{N}} \lambda_G(h^n(z), h^{n+1}(z)) = \lim_{n \to \infty} \lambda_G(h^n(z), h^{n+1}(z)).$$
Then we have that
(1) $h \sim \operatorname{id}_{\mathbb{C}} + 1 \Rightarrow \rho = 0$,
(2) $h \sim \operatorname{id}_{\pm\mathbb{H}} + 1 \Rightarrow \rho > 0$ *and* $\inf \rho(G) = 0$,
(3) $\exists \lambda > 1 : h \sim \lambda \operatorname{id}_{\mathbb{H}} \Rightarrow \inf \rho(G) > 0$.

Remark 2.7. *If additionally G is simply connected, then Cowen's result implies that all these implications are equivalences. In general, a holomorphic self-map of a hyperbolic domain need not be semi-conjugated to a Möbius transformation at all.*

To prove Lemma 2.6 we make use of the following lemma.

Lemma 2.8. *Let $\Omega \in \{\mathbb{C}, \mathbb{H}, -\mathbb{H}\}$ and let $T := \operatorname{id}_\Omega + 1$. Let $W \subset \Omega$ be a fundamental set for T on Ω. Let $r > 1$ and let $w \in \Omega$ be the center of an open disk Q with radius r such that $\overline{Q} \subset \Omega$. Then*
$$\lim_{n \to \infty} \lambda_W(w + n, w + n + 1) \leq \frac{1}{2} \log\left(\frac{r+1}{r-1}\right).$$

Proof. Since W is a fundamental set for $\mathrm{id}_\Omega + 1$ on Ω there exists $n \in \mathbb{N}$ such that $\overline{Q} + n \subset W$. By the Schwarz-Pick lemma we have that

$$\lambda_W(w+n, w+n+1) \le \lambda_{Q+n}(w+n, w+n+1) = \frac{1}{2}\log\left(1 + \frac{2}{r-1}\right). \quad \square$$

Proof of Lemma 2.6. Let (Ω, T, Φ, V) be an eventual conjugacy of h on G, where $\Omega \in \{\mathbb{C}, \mathbb{H}, -\mathbb{H}\}$. Let $W := \Phi(V)$. For each $z \in G$ and $n \in \mathbb{N}$ such that $h^n(z) \in V$, the Schwarz-Pick lemma implies that

$$\begin{aligned}
(*) \quad \lambda_\Omega(\Phi(z), T(\Phi(z))) &= \lambda_\Omega(T^n(\Phi(z)), T^{n+1}(\Phi(z))) \\
&= \lambda_\Omega(\Phi(h^n(z)), \Phi(h^{n+1}(z))) \\
&\le \lambda_G(h^n(z), h^{n+1}(z)) \\
&\le \lambda_V(h^n(z), h^{n+1}(z)) \\
&= \lambda_W(\Phi(h^n(z)), \Phi(h^{n+1}(z))) \\
&= \lambda_W(T^n(\Phi(z)), T^{n+1}(\Phi(z))),
\end{aligned}$$

where the first inequality makes sense only for the case when $\Omega \in \{\mathbb{H}, -\mathbb{H}\}$.

To prove (3) we observe that, for each $\lambda > 1$ and $w \in \mathbb{H}$,

$$\lambda_\mathbb{H}(\lambda w, w) = \log \frac{1 + (\lambda w - w)/(\lambda w - \overline{w})}{1 - (\lambda w - w)/(\lambda w - \overline{w})} \ge \log \lambda,$$

which together with $(*)$ implies that $\inf \rho(G) > 0$ if $h \sim \lambda \mathrm{id}_\mathbb{H}$.

To prove (2) suppose that there exists $\sigma \in \{-1, 1\}$ such that $T = \mathrm{id}_{\sigma \mathbb{H}} + 1$. Then, for each $w \in \sigma \mathbb{H}$, we have that

$$\lambda_{\sigma\mathbb{H}}(w+1, w) = \log \frac{1 + |\sigma/(\sigma + 2i\mathrm{Im}(w))|}{1 - |\sigma/(\sigma + 2i\mathrm{Im}(w))|} \ge \log\left(1 + \frac{1}{|\mathrm{Im}(w)|}\right),$$

which together with $(*)$ implies that $\rho > 0$. Because of $(*)$ it remains to show that

$$\inf_{w \in W} \inf_{n \in \mathbb{N}} \lambda_W(w+n, w+n+1) = 0.$$

This is an easy consequence of Lemma 2.8 because $\sigma \mathbb{H}$ contains an open disk with an arbitrarily large radius.

To prove (1) suppose that $T = \mathrm{id}_\mathbb{C} + 1$. Because of $(*)$ it remains to show that, for each $w \in \mathbb{C}$,

$$\inf_{n \in \mathbb{N}} \lambda_W(w+n, w+n+1) = 0.$$

This is an easy consequence of Lemma 2.8 because each $w \in \mathbb{C}$ is the center of an open disk in $\mathbb{C}$ with arbitrarily large radius. $\quad \square$

The case when the Denjoy-Wolff point of a holomorphic self-map of $\mathbb{D}$ is inside the unit disk also leads to an eventual conjugacy. More generally, we have the following result. Although this lemma might be folklore, for the sake of completeness we give a short proof.

Lemma 2.9. *Let h be a holomorphic self-map of a hyperbolic domain $G \subset \hat{\mathbb{C}}$ such that h is not an automorphism of G and there is a fixed point p of h in G such that $\lambda := h'(p) \neq 0$. Then $|\lambda| < 1$ and $h \sim \lambda \mathrm{id}_{\mathbb{C}}$.*

Proof. By Montel's theorem, the family $\{h^n : n \in \mathbb{N}\}$ is normal. Since h is not an automorphism of G we conclude that no subsequence of $(h^n)_{n \in \mathbb{N}}$ converges to a non-constant limit function (see for instance [6, Theorem 7.2.4]). Hence $h^n \to p$ locally uniformly on G, which implies that $|\lambda| < 1$. Hence there is an open and connected neighbourhood V of p in G, a number $r > 0$, and a biholomorphic map $\phi : V \to D(0, r)$ such that $h(V) \subset V$, $\phi(p) = 0$, $\phi'(p) = 1$ and $\phi \circ h|_V = \lambda \phi$. It is easy to see that

$$\Phi(z) := \frac{1}{\lambda^n} \phi(h^n(z)) \quad \text{if} \quad h^n(z) \in V$$

is a well-defined holomorphic function on G such that $(\mathbb{C}, \lambda \mathrm{id}_{\mathbb{C}}, \Phi, V)$ forms an eventual conjugacy of h on G. □

Moreover, we make use of the following theorems due to P. Bonfert [11, Theorem 5.7 and Theorem 6.1].

Theorem 2.10 (Bonfert). *Let h be a holomorphic self-map of a hyperbolic domain $G \subset \mathbb{C}$ without a fixed point, and without an isolated boundary fixed point, i.e. there is no isolated boundary point $a \in \partial G$ such that h extends holomorphically to a and fixes a. Suppose that $\lambda_G(h^n(z), h^{n+1}(z)) \to 0$ as $n \to \infty$ for some $z \in G$. Let $z_0 \in G$. Define*

$$\phi_n : G \to \mathbb{C}, z \mapsto \frac{h^n(z) - h^n(z_0)}{h^{n+1}(z_0) - h^n(z_0)}.$$

Then the sequence $(\phi_n)_{n \in \mathbb{N}}$ converges locally uniformly in G to a holomorphic function $\phi : G \to \mathbb{C}$ such that $\phi(h(z)) = \phi(z) + 1$ for all $z \in G$.

Theorem 2.11 (Bonfert). *Let T be a Möbius transformation and $G \subset \mathbb{C}$ be a hyperbolic domain such that $T(G) \subset G$, $T(\infty) = \infty$, and T has no fixed point in G. Then*

$$\lambda_G(T^n(z), T^{n+1}(z)) \to 0 \quad \text{as} \quad n \to \infty \quad \text{for (any)} \quad z \in G$$

if and only if

$$\bigcup_{n \in \mathbb{N}} T^{-n}(G) = \mathbb{C} \quad \text{or} \quad \bigcup_{n \in \mathbb{N}} T^{-n}(G) = \mathbb{C} \setminus \{b\},$$

where $b \in \mathbb{C} \setminus G$ is a fixed point of T.

2.2. The components of $\mathcal{F}(g) \cap \partial \mathbb{D}$.

We distinguish between the following types of components of $\mathcal{F}(g) \cap \partial \mathbb{D}$.

Definition 2.12. *Let g be a non-Möbius inner function such that $\mathcal{J}(g) \neq \partial \mathbb{D}$. A component I of $\mathcal{F}(g) \cap \partial \mathbb{D}$ is called*

(1) *absorbing, if, for each component J of $\mathcal{F}(g) \cap \partial \mathbb{D}$, there exists $n \in \mathbb{N}$ such that $g^n(J) \subset I$,*
(2) *invariant, if $g(I) \subset I$,*
(3) *eventually invariant, if there exists $n \in \mathbb{N}$ such that $g^n(I)$ is contained in an invariant component of $\mathcal{F}(g) \cap \partial \mathbb{D}$,*
(4) *wandering, if, for all $m \neq n \in \mathbb{N}$, $g^m(I)$ and $g^n(I)$ are contained in different components of $\mathcal{F}(g) \cap \partial \mathbb{D}$.*

As a consequence of Theorem 2.1 we can make the following observation.

Lemma 2.13. *Let g be a non-Möbius inner function such that $\mathcal{J}(g) \neq \partial \mathbb{D}$. Then there is at most one invariant component of $\mathcal{F}(g) \cap \partial \mathbb{D}$.*

Proof. Let $p \in \overline{\mathbb{D}}$ be the Denjoy-Wolff point of g. By Theorem 1.8 we have that $p \in \partial \mathbb{D}$. Since $g^n|_{\mathcal{F}(g)} \to p$ we conclude that $p \in \overline{I}$ for each invariant component I of $\mathcal{F}(g) \cap \partial \mathbb{D}$. Since $\mathcal{J}(g)$ is perfect there is at most one component J of $\mathcal{F}(g) \cap \partial \mathbb{D}$ such that $p \in \overline{J}$. Hence the conclusion follows. □

This leads to the following dichotomy of the set of components of $\mathcal{F}(g) \cap \partial \mathbb{D}$.

Lemma 2.14. *Let g be a non-Möbius inner function such that $\mathcal{J}(g) \neq \partial \mathbb{D}$. Then each component of $\mathcal{F}(g) \cap \partial \mathbb{D}$ is either eventually invariant or wandering. Moreover, each absorbing component of $\mathcal{F}(g) \cap \partial \mathbb{D}$ is invariant.*

Proof. It remains to show that each component I of $\mathcal{F}(g) \cap \partial \mathbb{D}$ for which there exists $n \in \mathbb{N}$ such that $g^n(I) \subset I$ is invariant. Let I be a component of $\mathcal{F}(g) \cap \partial \mathbb{D}$ and $n \in \mathbb{N}$ such that $g^n(I) \subset I$. Let J be the component of $\mathcal{F}(g) \cap \partial \mathbb{D}$ which contains $g(I)$. Then I and J are both invariant components of $\mathcal{F}(g^n) \cap \partial \mathbb{D}$ which by Lemma 2.13 implies that $I = J$. □

Remark 2.15. *The converse of the second statement in Lemma 2.14 is not true. There may be an invariant component which is not absorbing (see Example 2.38 in Subsection 2.5).*

The next theorem gives some information about the mapping behaviour of an inner function g on a component of $\mathcal{F}(g) \cap \partial \mathbb{D}$.

Definition 2.16. *Let $\zeta, \theta \in \partial \mathbb{D}$ such that $\zeta \neq \theta$. Let $\alpha \in (0, 2\pi)$ such that $\theta = \zeta \exp(i\alpha)$. Then define*

$$(\zeta, \theta) := \zeta \exp(i(0, \alpha)).$$

Theorem 2.17. *Let g be a non-Möbius inner function. Let $I = (\zeta_1, \zeta_2)$ be a component of $\mathcal{F}(g) \cap \partial \mathbb{D}$. Then*

(1) $g|_I$ is injective.
(2) $\eta_j := \lim_{z \to \zeta_j} g|_I(z)$ exists, for each $j \in \{1, 2\}$, and is equal to the angular limit of g at ζ_j. Moreover, $\eta_1 \neq \eta_2$, and $g(I) = (\eta_1, \eta_2)$ or $g(I) = (\eta_2, \eta_1)$.

We make use of the following lemmata to prove Theorem 2.17.

Lemma 2.18. *Let g be a non-Möbius inner function. Let $\zeta, \theta \in \partial \mathbb{D}$ be distinct and let g be holomorphic on $I := (\zeta, \theta)$. Suppose that*

$$\eta := \lim_{z \to \zeta} g|_I(z) \quad and \quad \lim_{z \to \theta} g|_I(z)$$

exist and are equal. Then $\partial \mathbb{D} \setminus \{\eta\} \subset g(I)$.

Proof. The set $Z := g(I) \setminus \{\eta\}$ is a non-empty open subset of $\partial \mathbb{D} \setminus \{\eta\}$. Let $w \in \overline{Z} \setminus \{\eta\}$. Then there exists a sequence $(x_n)_{n \in \mathbb{N}}$ in I such that $\lim_{n \to \infty} g(x_n) = w \neq \eta$. We may assume that $x_n \to x \in \overline{I}$. Since $\lim_{z \to \zeta} g|_I(z) = \eta = \lim_{z \to \theta} g|_I(z)$ we conclude that $x \in I$ and $w \in Z$. Since $\partial \mathbb{D} \setminus \{\eta\}$ is connected we have that $Z = \partial \mathbb{D} \setminus \{\eta\}$. □

Remark 2.19. *As a consequence of Lemma 2.18 we obtain that an inner function g is locally injective at any point $\zeta \in \partial \mathbb{D}$ where it is holomorphic. This can also be concluded from the Julia-Wolff lemma (see Theorem 2.29).*

Lemma 2.20. *Let g be a non-Möbius inner function. Let $\zeta, \theta \in \partial \mathbb{D}$ such that $\zeta \neq \theta$ and g is holomorphic on $I := (\zeta, \theta)$. Then either $g((\zeta, \eta)) = \partial \mathbb{D}$, for each $\eta \in I$, or $\lim_{z \to \zeta} g|_I(z)$ exists.*

Proof. Suppose that there exists $\eta \in I$ such that $g((\zeta, \eta)) \neq \partial \mathbb{D}$. From Lemma 2.18 we conclude that g is injective on (ζ, η). Hence g is strictly orientation preserving or strictly orientation reversing on (ζ, η). In both cases it is easy to see that $\lim_{z \to \zeta} g|_I(z)$ exists. □.

The next result is an easy consequence of the Lehto-Virtanen theorem.

Lemma 2.21. *Let g be a non-Möbius inner function. Let $\zeta, \theta \in \partial \mathbb{D}$ such that $\zeta \neq \theta$ and g is holomorphic on $I := (\zeta, \theta)$. Suppose that $\eta := \lim_{z \to \zeta} g|_I(z)$ exists. Then the angular limit of g at ζ exists and equals η.*

Proof. Choose a disk D which has its center at a point $\alpha \in I$ such that $\zeta \in \partial D$, $D \cap \partial \mathbb{D} \subset I$, and $g(D \cap \partial \mathbb{D}) \subset D(\eta; 1)$. Then $D(-\eta; 1) \cap \partial \mathbb{D} \subset \hat{\mathbb{C}} \setminus g(D)$ which implies that g is a normal function on D. Thus from the Lehto-Virtanen theorem (see [18], p. 71) we have that there is a path $\gamma : [0, 1) \to \mathbb{D} \cap D$ such that $\gamma(t) \to \zeta$ and $\gamma(t) \to \eta$ as $t \to 1$. Again we apply the Lehto-Virtanen theorem and conclude that the angular limit of g at ζ exists and equals η. □

Proof of Theorem 2.17. Since $g(I) \subset \mathcal{F}(g) \cap \partial \mathbb{D}$ and $\mathcal{J}(g) \neq \emptyset$ we conclude from Lemma 2.18 and Lemma 2.20 that g is injective on I, $\eta_j := \lim_{z \to \zeta_j} g|_I(z)$ exists, for each $j \in \{1, 2\}$, $\eta_1 \neq \eta_2$, and $\{\eta_1, \eta_2\} \cap g(I) = \emptyset$. From this it can be easily deduced that $g(I) = (\eta_1, \eta_2)$ or $g(I) = (\eta_2, \eta_1)$. Moreover, it follows from Lemma 2.21 that the angular limit of g at ζ_j exists, for each $j \in \{1, 2\}$, and equals η_j. □

2.3. Eventual conjugacies of inner functions.
The following theorem gives necessary and sufficient conditions for an inner function to be eventually conjugated to a Möbius transformation on the whole Fatou set. Moreover, it restricts the eventual conjugacies that can occur.

Theorem 2.22. *Let g be a non-Möbius inner function such that $\mathcal{J}(g) \neq \partial \mathbb{D}$. Then the following statements are equivalent.*

(1) *There exist a simply connected domain $\Omega \subset \mathbb{C}$ and a biholomorphic self-map T of Ω such that $g|_{\mathcal{F}(g)} \sim T$.*
(2) *There exists an absorbing component I of $\mathcal{F}(g) \cap \partial \mathbb{D}$.*
(3) *There is no wandering component of $\mathcal{F}(g) \cap \partial \mathbb{D}$.*

Let $p \in \partial \mathbb{D}$ be the Denjoy-Wolff point of g. If one (and hence all) of these equivalent conditions (1), (2), (3) is satisfied, then exactly one of the following statements holds.

(4) $p \in \mathcal{F}(g)$ *and* $g|_{\mathcal{F}(g)} \sim g'(p) \mathrm{id}_{\mathbb{C}}$ *and* $g|_{\mathbb{D}} \sim g'(p) \mathrm{id}_{\mathbb{H}}$.
(5) $p \in \mathcal{J}(g)$ *and* $g|_{\mathcal{F}(g)} \sim \mathrm{id}_{\mathbb{C}} + 1$ *and* $g|_{\mathbb{D}} \sim \mathrm{id}_{\mathbb{H}} \pm 1$.
(6) $p \in \mathcal{J}(g)$ *and there exists $\lambda > 1$ such that* $g|_{\mathcal{F}(g)} \sim \lambda \mathrm{id}_{\mathbb{H}}$ *and* $g|_{\mathbb{D}} \sim \lambda^2 \mathrm{id}_{\mathbb{H}}$.

We see in Subsection 2.5 that all the listed cases (4), (5), and (6) may occur. For the case when $g|_{\mathcal{F}(g)} \sim \mathrm{id}_{\mathbb{C}} + 1$ we get two more equivalences.

Theorem 2.23. *Let g be a non-Möbius inner function such that $\mathcal{J}(g) \neq \partial \mathbb{D}$. Let $p \in \partial \mathbb{D}$ be the Denjoy-Wolff point of g. Then the following conditions are equivalent:*

(1) $g|_{\mathcal{F}(g)} \sim \mathrm{id}_{\mathbb{C}} + 1$.
(2) $p \in \mathcal{J}(g)$ *and* $\lambda_{\mathcal{F}(g)}(g^n(z), g^{n+1}(z)) \to 0$ *as* $n \to \infty$ *for some (all)* $z \in \mathcal{F}(g)$.
(3) $g|_{\mathbb{D}} \sim \mathrm{id}_{\pm \mathbb{H}} + 1$ *and there exists an absorbing component of $\mathcal{F}(g) \cap \partial \mathbb{D}$.*
(4) $g|_{\mathbb{D}} \sim \mathrm{id}_{\pm \mathbb{H}} + 1$ *and there exists an invariant component of $\mathcal{F}(g) \cap \partial \mathbb{D}$.*

As a consequence of Theorem 2.22 and 2.23 we obtain the following result which is a generalization of Theorems 1.8 and 1.9.

Theorem 2.24. *Let g be an inner function such that $\lambda_{\mathbb{D}}(g^n(z), g^{n+1}(z)) \to 0$ as $n \to \infty$ for some $z \in \mathbb{D}$. Then $\mathcal{J}(g) = \partial \mathbb{D}$.*

Proof. Let $z \in \mathbb{D}$ such that $\lambda_{\mathbb{D}}(g^n(z), g^{n+1}(z)) \to 0$ as $n \to \infty$. Then g is not a Möbius transformation. Let p be the Denjoy-Wolff point of g. If $p \in \mathbb{D}$, then the conclusion follows from Theorem 1.8. Hence we may assume that $p \in \partial \mathbb{D}$. From Lemma 2.6 we see that $g|_{\mathbb{D}} \sim \mathrm{id}_{\mathbb{C}} + 1$.

Assume that $p \in \mathcal{F}(g)$. Then $\lambda := g'(p) \in (0,1]$ and from Lemma 2.9 we conclude that $g|_{\mathcal{F}(g)} \sim \lambda \mathrm{id}_{\mathbb{C}}$. By Theorem 2.22 this implies that $g|_{\mathbb{D}} \sim \lambda \mathrm{id}_{\mathbb{H}}$. This is a contradiction to $g|_{\mathbb{D}} \sim \mathrm{id}_{\mathbb{C}} + 1$. Hence $p \in \mathcal{J}(g)$.

Assume that $\mathcal{F}(g) \cap \partial \mathbb{D} \neq \emptyset$. From the Schwarz-Pick lemma we conclude that $\lambda_{\mathcal{F}(g)}(g^n(z), g^{n+1}(z)) \to 0$ as $n \to \infty$. By Theorem 2.23 this implies that $g|_{\mathbb{D}} \sim \mathrm{id}_{\mathbb{H}} \pm 1$. This contradicts $g|_{\mathbb{D}} \sim \mathrm{id}_{\mathbb{C}} + 1$. Hence $\mathcal{J}(g) = \partial \mathbb{D}$. □

This leads to the following theorem.

Theorem 2.25. *Let g be a non-Möbius inner function such that $\mathcal{J}(g) \neq \partial \mathbb{D}$. Suppose that there exists an invariant component of $\mathcal{F}(g) \cap \partial \mathbb{D}$ which is not absorbing. Then there exists $\lambda > 1$ such that $g|_{\mathbb{D}} \sim \lambda \mathrm{id}_{\mathbb{H}}$.*

Proof. From Theorem 2.24 we know that $g|_{\mathbb{D}} \not\sim \mathrm{id}_{\mathbb{C}}+1$. From Theorem 2.23 we conclude that $g|_{\mathbb{D}} \not\sim \mathrm{id}_{\mathbb{H}} \pm 1$. From Theorem 1.8 we see that g does not have a fixed point in $\mathbb{D}$. Hence the conclusion follows from Theorem 2.4. □

Remark. An example of a non-Möbius inner function which has an invariant but not absorbing component of $\mathcal{F}(g) \cap \partial \mathbb{D}$ is given in Subsection 2.5.

The rest of this subsection is devoted to the proofs of Theorem 2.22 and 2.23. The main part of the proofs is contained in the following lemma.

Lemma 2.26. *Let g be a non-Möbius inner function such that $\mathcal{J}(g) \neq \partial \mathbb{D}$. Let $p \in \partial \mathbb{D}$ be the Denjoy-Wolff point of g. Suppose that there exists an invariant component I of $\mathcal{F}(g) \cap \partial \mathbb{D}$. Let $G := \mathbb{D} \cup I \cup (\hat{\mathbb{C}} \setminus \overline{\mathbb{D}})$. Then exactly one of the following statements holds.*

(1) $p \in \mathcal{F}(g)$ and $g|_G \sim g'(p)\mathrm{id}_{\mathbb{C}}$ and $g|_{\mathbb{D}} \sim g'(p)\mathrm{id}_{\mathbb{H}}$.

(2) $p \in \mathcal{J}(g)$ and $g|_G \sim \mathrm{id}_{\mathbb{C}} + 1$ and $g|_{\mathbb{D}} \sim \mathrm{id}_{\mathbb{H}} \pm 1$.

(3) $p \in \mathcal{J}(g)$ and there exists $\lambda > 1$ such that $g|_G \sim \lambda \mathrm{id}_{\mathbb{H}}$ and $g|_{\mathbb{D}} \sim \lambda^2 \mathrm{id}_{\mathbb{H}}$.

Proof. By conjugating with an appropriate transformation we may assume that g is an inner function of the upper half plane $\mathbb{H}$ with Denjoy-Wolff point $p = \infty$ and that I is an invariant component of $\mathcal{F}(g) \cap (\mathbb{R} \cup \{\infty\})$. Since $\mathcal{J}(g)$ is a perfect set the simply connected domain $G := \mathbb{H} \cup I \cup -\mathbb{H}$ is hyperbolic. By Theorem 2.4 and Lemma 2.9 there exist

$$\begin{aligned}(\Omega, T) \in \ &\{(\mathbb{C}, \mathrm{id}_{\mathbb{C}} + 1)\} \\ \cup\ &\{(-i\mathbb{H}, \mathrm{id}_{-i\mathbb{H}} + i\sigma) : \sigma \in \{-1, 1\}\} \\ \cup\ &\{(-i\mathbb{H}, \lambda \mathrm{id}_{-i\mathbb{H}}) : \lambda > 1\} \\ \cup\ &\{(\mathbb{C}, g'(p)\mathrm{id}_{\mathbb{C}}\},\end{aligned}$$

a simply connected domain $V \subset G$, and a holomorphic function $\Phi : G \to \Omega$ such that (Ω, T, Φ, V) forms an eventual conjugacy of g on G. Here $(\Omega, T) = (\mathbb{C}, g'(p)\mathrm{id}_{\mathbb{C}})$ if and only if $\infty = p \in I$. By Lemma 2.13 this is the case if and only if $p \in \mathcal{F}(g)$.

By the reflection principle $\{\overline{z} : z \in V\}$ is also a fundamental set for g on G and $\Psi(z) := \overline{\Phi(\overline{z})}$ is holomorphic such that, for each $z \in G$,

$$\Psi(g(z)) = \overline{\Phi(\overline{g(z)})} = \overline{\Phi(g(\overline{z}))} = \overline{T(\Phi(\overline{z}))} = S(\Psi(z)),$$

where $S(w) := \overline{T(\overline{w})}$, for each $w \in \Omega$. Thus we have that $g|_G \sim T$ and $g|_G \sim S$.

Assume that there exists $\sigma \in \{-1, 1\}$ such that $(\Omega, T) = (-i\mathbb{H}, \text{id}_{-i\mathbb{H}} + i\sigma)$. Then $S = \text{id}_{-i\mathbb{H}} - i\sigma$ and hence $g|_G \sim \text{id}_{-i\mathbb{H}} + i$ and $g|_G \sim \text{id}_{-i\mathbb{H}} - i$. This is a contradiction to Theorem 2.4. Thus we conclude that $T \neq \text{id}_{-i\mathbb{H}} + i$ and $T \neq \text{id}_{-i\mathbb{H}} - i$.

For the remaining cases we have that $S = T$. From Lemma 2.3 we conclude that there exists a biholomorphic map $\tau : \Omega \to \Omega$ such that $T(\tau(z)) = \tau(T(z))$ and $\Psi(z) = \tau(\Phi(z))$.

For the case when $T = \text{id}_\mathbb{C} + 1$ we obtain that there exists $b \in \mathbb{C}$ such that $\tau(z) = z + b$ for every $z \in \mathbb{C}$. Since, for each $z \in G \cap \mathbb{R}$,

$$2i\text{Im}\Phi(x) = \Phi(x) - \overline{\Phi(x)} = \Phi(x) - \Psi(x) = \Phi(x) - \tau(\Phi(x)) = -b \in i\mathbb{R},$$

we see that, by passing over to $\Phi + b/2$ instead of Φ, we may assume that $\Phi(x) \in \mathbb{R}$ for each $x \in G \cap \mathbb{R}$.

For the case when $p \in \mathcal{F}(g)$ and $T = g'(p)\text{id}_\mathbb{C}$ we obtain that there exists $a \in \mathbb{C}\setminus\{0\}$ such that $\tau(z) = az$ for each $z \in \mathbb{C}$. Since, for each $z \in G \cap \mathbb{R}$,

$$\overline{\Phi(x)} = \Psi(x) = a\Phi(x),$$

we see that, by passing to $\sqrt{a}\Phi$ instead of Φ, we may assume that $\Phi(x) \in \mathbb{R}$ for each $x \in G \cap \mathbb{R}$.

For the case when there exists $\lambda > 1$ such that $T = \lambda\text{id}_{i\mathbb{H}}$ we obtain that there exists $c > 0$ such that $\tau(z) = cz$ for each $z \in -i\mathbb{H}$. Since

$$|\Phi(x)| = |\overline{\Phi(x)}| = |\Psi(x)| = c|\Phi(x)|$$

for each $x \in G \cap \mathbb{R}$ we conclude that $c = 1$. Hence we have that $\Phi(x) \in \mathbb{R}$ for each $x \in G \cap \mathbb{R}$.

Thus in any case we have that $\Phi(x) \in \mathbb{R}$ for each $x \in G \cap \mathbb{R}$. We can now show that

(*) there exists $\sigma \in \{1, -1\}$ such that $\Phi(\mathbb{H}) \subset \sigma\mathbb{H}$ and $\Phi(-\mathbb{H}) \subset -\sigma\mathbb{H}$.

Proof of (*). First, we prove that there exists an unbounded component J of $I \cap V$ such that $g(J) \subset J$. This is clear for the case when $p \in I$. Hence we may assume that $I \subset \mathbb{R}$. Let $x \in I$. Let K be the compact interval in $\mathbb{R}$ with endpoints x and $g(x)$. Since $K \subset I$ there exists $n \in \mathbb{N}$ such that $g^n(K) \subset I \cap V$. Since $g(I \cap V) \subset I \cap V$ it is easy to see that the component J of $g^n(K)$ in $I \cap V$ is unbounded such that $g(J) \subset J$.

Now let $x \in G$ such that $\Phi(x) \in \mathbb{R}$. We have that $\Phi(J)$ is an open interval in $\mathbb{R}$ such that $T(\Phi(J)) \subset \Phi(J)$ and $0 \in \Phi(J)$ if $T = g'(p)\text{id}_\mathbb{C}$. In any case this implies that there exists $n \in \mathbb{N}$ such that $\Phi(g^n(x)) = T^n(\Phi(x)) \in \Phi(J)$

and $g^n(x) \in V$. Let $y \in J$ such that $\Phi(g^n(x)) = \Phi(y)$. Since g is injective on V we conclude that $g^n(x) = y$ which implies that $x \in \mathbb{R}$ because g is an inner function. Hence we have that $\Phi^{-1}(\mathbb{R}) \subset \mathbb{R}$. This implies $(*)$.

Next, we prove that there exists a fundamental set D for g on $\mathbb{H}$ such that $\Phi(D)$ is a fundamental set for T on $\Omega \cap \sigma\mathbb{H}$. To this end let $z \in \mathbb{H}$ and let K be a compact connected subset of $\mathbb{H}$ such that $\{z, g(z)\} \subset K$. Then there exists $n \in \mathbb{N}$ such that $g^n(K) \subset V \cap \mathbb{H}$. Let D be the component of $V \cap \mathbb{H}$ which contains $g^n(K)$. Since $g(V \cap \mathbb{H}) \subset V \cap \mathbb{H}$ we conclude that $g(D) \subset D$. Since V is simply connected we have that D is simply connected. Let $y \in \mathbb{H}$. Then there exists a compact connected set $L \subset \mathbb{H}$ such that $\{z, y\} \subset L$. Choose $m \geq n$ such that $g^m(L) \subset V \cap \mathbb{H}$. Since $g^m(z) \in D$ we conclude that $g^m(L) \subset D$ and hence $g^m(x) \in D$. Thus D is a fundamental set for g on $\mathbb{H}$.

Let $u \in \Omega \cap \sigma\mathbb{H}$. Since $\Phi(V)$ is a fundamental set for T on Ω we find $k \in \mathbb{N}$ such that $T^k(u) \in \Phi(V) \cap \sigma\mathbb{H}$. According to $(*)$ we find $v \in V \cap \mathbb{H}$ such that $\Phi(v) = T^k(u)$. Choose $l \in \mathbb{N}$ such that $g^l(v) \in D$. Then we have that

$$T^{k+l}(u) = T^l(\Phi(v)) = \Phi(g^l(v)) \in \Phi(D).$$

Hence $\Phi(D)$ is a fundamental set for T on $\Omega \cap \sigma\mathbb{H}$.

Let $E := \Omega \cap \sigma\mathbb{H}$. We obtain that $(E, T|_E, \Phi|_\mathbb{H}, D)$ is an eventual conjugacy for g on $\mathbb{H}$. For the case when $g|_G \sim \mathrm{id}_\mathbb{C} + 1$ we conclude that $g|_\mathbb{H} \sim \mathrm{id}_{\sigma\mathbb{H}} + 1$. For the case when $p \in I$ and $g|_G \sim g'(p)\mathrm{id}_\mathbb{C}$ we conclude that $g|_\mathbb{H} \sim g'(p)\mathrm{id}_\mathbb{H}$. Finally, suppose that there exists $\lambda > 1$ such that $g|_G \sim \lambda \mathrm{id}_{-i\mathbb{H}}$. Let $\phi(z) := \Phi(z)^2$ for each $z \in \mathbb{H}$. Then we have that

$$\phi(g(z)) = \Phi(g(z))^2 = \lambda^2 \Phi(z)^2 = \lambda^2 \phi(z)$$

for each $z \in \mathbb{H}$. Moreover, it is easy to see that ϕ is injective on D and $\phi(D)$ is a fundamental set for $\lambda^2 \mathrm{id}_{\sigma\mathbb{H}}$ on $\sigma\mathbb{H}$. Thus $(\sigma\mathbb{H}, \lambda^2 \mathrm{id}_{\sigma\mathbb{H}}, \phi, D)$ is an eventual conjugacy for g on $\mathbb{H}$. Hence $g|_\mathbb{H} \sim \lambda^2 \mathrm{id}_\mathbb{H}$. □

Moreover, we make use of the following two lemmata.

Lemma 2.27. *Let g be a non-Möbius inner function such that $\mathcal{J}(g) \neq \partial\mathbb{D}$. Suppose that there exists an invariant component I of $\mathcal{F}(g) \cap \partial\mathbb{D}$, and let $G := \mathbb{D} \cup I \cup (\hat{\mathbb{C}} \setminus \overline{\mathbb{D}})$. Let $\Omega \subset \mathbb{C}$ be a simply connected domain and T be a biholomorphic self-map of Ω such that $g|_G \sim T$. If I is an absorbing component of $\mathcal{F}(g) \cap \partial\mathbb{D}$, then $g|_{\mathcal{F}(g)} \sim T$.*

Proof. Let $V \subset G$ be a simply connected domain and let $\phi : G \to \Omega$ be a holomorphic function such that (Ω, T, Φ, V) is an eventual conguagacy for g on G. Since I is an absorbing component of $\mathcal{F}(g) \cap \partial\mathbb{D}$ it is clear that V is a fundamental set for g on $\mathcal{F}(g)$. It is easy to see that

$$\Phi(z) := T^{-n}(\phi(g^n(z))) \quad \text{if} \quad g^n(z) \in V$$

is a well defined holomorphic function on $\mathcal{F}(g)$ such that $\Phi|_G = \phi$. By the identity theorem we have that $\Phi \circ g|_{\mathcal{F}(g)} = T \circ \Phi$. Hence (Ω, T, Φ, V) is an eventual conjugacy for g on $\mathcal{F}(g)$. □

The next lemma is proved in [1, p. 116, Lemma 2].

Lemma 2.28. *Let γ be a closed rectifiable curve in $\mathbb{C}$ such that $\gamma^{-1}(\mathbb{H}) \neq \emptyset \neq \gamma^{-1}(-\mathbb{H})$ and $\gamma^{-1}(\mathbb{R})$ consists of exactly two points a and b. Then the open interval with endpoints $\gamma(a)$ and $\gamma(b)$ is contained in $\mathrm{int}(\gamma)$.*

Now, we are able to prove Theorem 2.22 and Theorem 2.23.

Proof of Theorem 2.22.

(2) $\Rightarrow$ (1) and ((4) or (5) or (6)). Suppose that there exists an absorbing component I of $\mathcal{F}(g) \cap \partial \mathbb{D}$. By Lemma 2.14, the set I is invariant under g. Let $G := \mathbb{D} \cup I \cup (\hat{\mathbb{C}} \setminus \overline{\mathbb{D}})$. By Lemma 2.26, exactly one of the following statements holds:

(4') $p \in \mathcal{F}(g)$ and $g|_G \sim g'(p)\mathrm{id}_{\mathbb{C}}$ and $g|_{\mathbb{D}} \sim g'(p)\mathrm{id}_{\mathbb{H}}$,

(5') $p \in \mathcal{J}(g)$ and $g|_G \sim \mathrm{id}_{\mathbb{C}} + 1$ and $g|_{\mathbb{D}} \sim \mathrm{id}_{\mathbb{H}} \pm 1$,

(6') $p \in \mathcal{J}(g)$ and there exists $\lambda > 1$ such that $g|_G \sim \lambda \mathrm{id}_{\mathbb{H}}$ and $g|_{\mathbb{D}} \sim \lambda^2 \mathrm{id}_{\mathbb{H}}$.

Since I is absorbing we conclude from Lemma 2.27 that (4') implies (4), (5') implies (5), and (6') implies (6).

(3) $\Rightarrow$ (2). Suppose that there are no wandering components of $\mathcal{F}(g) \cap \partial \mathbb{D}$. Since $\mathcal{F}(g) \cap \partial \mathbb{D} \neq \emptyset$ we conclude from Lemma 2.14 that there exists an invariant component I of $\mathcal{F}(g) \cap \partial \mathbb{D}$. Let J be a component of $\mathcal{F}(g) \cap \partial \mathbb{D}$. By Lemma 2.14 there exists $n \in \mathbb{N}$ such that $g^n(J)$ is contained in an invariant component of $\mathcal{F}(g) \cap \partial \mathbb{D}$. By Lemma 2.13 there is at most one invariant component which implies that $g^n(J) \subset I$. Hence I is an absorbing component of $\mathcal{F}(g) \cap \partial \mathbb{D}$.

(1) $\Rightarrow$ (3). Suppose that there exists an eventual conjugacy (Ω, T, Φ, V) of g on $\mathcal{F}(g)$. We may assume that $p \in \mathcal{J}(g)$ because otherwise the component I of p in $\mathcal{F}(g) \cap \partial \mathbb{D}$ is absorbing and the conclusion follows. By conjugating with an appropriate transformation we may further assume that g is an inner function of $\mathbb{H}$ and $p = \infty$. Let I be a component of $\mathcal{F}(g) \cap \mathbb{R}$. Let $x \in I$ and let γ be the positively oriented circle centered at $(x + g(x))/2$ which contains x and $g(x)$. Then γ is a compact subset of $\mathcal{F}(g)$ and there exists $n \in \mathbb{N}$ such that $g^n(\gamma) \subset V$. Let J be the component of $\mathcal{F}(g) \cap \mathbb{R}$ which contains $g^n(x)$. Since V is simply connected and $g^n \circ \gamma$ is a closed analytic curve in V we conclude that $\mathrm{int}(g^n \circ \gamma) \subset V$. From Lemma 2.28 we see that the open interval K with endpoints $g^n(x)$ and $g^{n+1}(x)$ is contained in $\mathrm{int}(g^n \circ \gamma) \subset V \subset \mathcal{F}(g)$. Hence $K \subset J$ and $\{g^n(x), g^{n+1}(x)\} \subset J$. This implies that I is not wandering. □

Proof of Theorem 2.23.

(1) $\Rightarrow$ (2). Suppose that $g|_{\mathcal{F}(g)} \sim \mathrm{id}_{\mathbb{C}} + 1$. Then g does not have a fixed point in $\mathcal{F}(g)$ which implies that $p \in \mathcal{J}(g)$. From Lemma 2.6 we know that $\lambda_{\mathcal{F}(g)}(g^n(z), g^{n+1}(z)) \to 0$ as $n \to \infty$ for every $z \in \mathcal{F}(g)$.

(2) $\Rightarrow$ (3). Suppose that $p \in \mathcal{J}(g)$ and $\lambda_{\mathcal{F}(g)}(g^n(z), g^{n+1}(z)) \to 0$ as $n \to \infty$ for some $z \in \mathcal{F}(g)$. By conjugating appropriately we may assume

that g is an inner function of $\mathbb{H}$ and $p = \infty$. Since $\mathcal{J}(g)$ is perfect, $\mathcal{F}(g)$ is a hyperbolic domain in $\mathbb{C}$ such that $\mathcal{F}(g)$ has no isolated boundary points. Let $x \in \mathcal{F}(g) \cap \mathbb{R}$ and define

$$\phi_n : \mathcal{F}(g) \to \mathbb{C}, \ z \mapsto \frac{g^n(z) - g^n(x)}{g^{n+1}(x) - g^n(x)} = T_n(g^n(z)),$$

where $T_n(w) := (w - g^n(x))/(g^{n+1}(x) - g^n(x))$. From Theorem 2.10 we know that the sequence $(\phi_n)_{n \in \mathbb{N}}$ converges locally uniformly on $\mathcal{F}(g)$ to a holomorphic function $\phi : \mathcal{F}(g) \to \mathbb{C}$ such that $\phi(g(z)) = \phi(z) + 1$ for all $z \in \mathcal{F}(g)$. Let $G := \phi(\mathcal{F}(g))$. We first prove that

$$(*) \quad \bigcup_{n \in \mathbb{N}} (G - n) = \mathbb{C}.$$

Proof of $(*)$. We may assume that G is hyperbolic because otherwise $\mathbb{C} \setminus G$ contains at most one point and $(*)$ follows immediately. Since $\phi \circ g|_{\mathcal{F}(g)} = \phi + 1$ we have that $(\mathrm{id}_\mathbb{C} + 1)(G) \subset G$. Moreover, from the Schwarz-Pick lemma we conclude that

$$\begin{aligned}
\lambda_G(\phi(z) + n, \phi(z) + n + 1) &= \lambda_G(\phi(g^n(z)), \phi(g^{n+1}(z))) \\
&\leq \lambda_{\mathcal{F}(g)}(g^n(z), g^{n+1}(z)) \\
&\to 0
\end{aligned}$$

as $n \to \infty$ for some $z \in \mathcal{F}(g)$. Thus we can apply Theorem 2.11 and obtain that $(*)$ holds.

Since $G + 1 \subset G$, the sequence $(G - n)_{n \in \mathbb{N}}$ is increasing, which implies that there exists $k \in \mathbb{N}$ such that $[0, 1] \subset G - k$, and hence $[\phi(g^k(x)), \phi(g^{k+1}(x))] = [k, k+1] \subset G \cap \mathbb{R} = \phi(\mathcal{F}(g)) \cap \mathbb{R}$. Since $\phi_n \to \phi$ locally uniformly in $\mathcal{F}(g)$ we find $n_0 \in \mathbb{N}$ such that, for each $n \geq n_0$,

$$[\phi_n(g^k(x)), \phi_n(g^{k+1}(x))] \subset \phi_n(\mathcal{F}(g)) \cap \mathbb{R} = \phi_n(\mathcal{F}(g) \cap \mathbb{R}).$$

Hence, for each $n \geq n_0$,

$$[T_n(g^{n+k}(x)), T^n(g^{n+k+1}(x))] \subset T_n(\mathcal{F}(g) \cap \mathbb{R}),$$

which implies that the closed interval with endpoints $g^{k+n}(x)$ and $g^{k+n+1}(x)$ is contained in $\mathcal{F}(g) \cap \mathbb{R}$ for every $n \geq n_0$. Let I be the component of $\mathcal{F}(g) \cap \mathbb{R}$ which contains $u := g^{k+n_0}(x)$. Then $g(u) \in I$ and hence $g(I) \subset I$ which implies that I is unbounded because $(g|_{\mathcal{F}(g)})^n \to \infty$ as $n \to \infty$. Moreover, $g|_I$ does not have any fixed point which implies that either $g(z) > z$ for each $z \in I$ and $\sup I = \infty$ or $g(z) < z$ for each $z \in I$ and $\inf I = -\infty$.

We show that I is an absorbing component of $\mathcal{F}(g) \cap \mathbb{R}$. To this end let $y \in \mathcal{F}(g) \cap \mathbb{R}$. Then $\phi_n(y) \in \mathbb{R}$, for each $n \in \mathbb{N}$, and hence $\phi(y) \in \mathbb{R}$. Thus we find $m \in \mathbb{N}$ such that $\phi(g^m(y)) = \phi(y) + m > 1$. Hence there exists $n \geq n_0 + k$ such that

$$\frac{g^n(g^m(y)) - g^n(x)}{g^{n+1}(x) - g^n(x)} = \phi_n(g^m(y)) \geq 1.$$

Let $l := n - n_0 - k$. If now $g(z) > z$, for each $z \in I$, then
$$g^{n+1}(x) = g^{l+1}(u) > g^l(u) = g^n(x),$$
which implies that $g^n(g^m(y)) > g^n(x) = g^l(u)$ and hence $g^{n+m}(y) \in I$ because $\sup I = \infty$. We can argue analogously for the case when $g(z) < z$ for each $z \in I$. Thus $g^{n+m}(y) \in I$ and I is absorbing.

It remains to show that $g_\mathbb{H} \sim \text{id}_{\pm\mathbb{H}} + 1$. To this end, first note that $\phi_n(\mathbb{H}) \subset \mathbb{H}$ or $\phi_n(\mathbb{H}) \subset -\mathbb{H}$ for each $n \in \mathbb{N}$. Since ϕ is not constant we conclude that there exists $\sigma \in \{-1, 1\}$ such that $\phi(\mathbb{H}) \subset \sigma\mathbb{H}$. According to Theorem 2.4 there exists a fundamental set $V \subset \mathbb{H}$ for g on $\mathbb{H}$. Since $g(V) \subset V$ and $g|_V$ is injective we have that $\phi_n|_V$ is injective, for each $n \in \mathbb{N}$, and hence ϕ is injective on V. Since
$$\phi(g(z)) = \phi(z) + 1 \quad \text{for every} \quad z \in \mathbb{H}$$
it remains to show that $W := \phi(V)$ is a fundamental set for $\text{id}_{\sigma\mathbb{H}} + 1$. Let $w \in \sigma\mathbb{H}$. According to (*) there exists $q \in \mathbb{N}$ such that $w + q \in G \cap \sigma\mathbb{H} = \phi(\mathbb{H})$. Choose $z \in \mathbb{H}$ such that $w + q = \phi(z)$. Since V is a fundamental set for g on $\mathbb{H}$ there exists $r \in \mathbb{N}$ such that $g^r(z) \in V$ and hence
$$w + q + r = \phi(z) + r = \phi(g^r(z)) \in \phi(V) = W.$$
Thus W is a fundamental set for $\text{id}_{\sigma\mathbb{H}} + 1$ and $g|_\mathbb{H} \sim \text{id}_{\sigma\mathbb{H}} + 1$. This proves **(3)**.

The implication **(3)** $\Rightarrow$ **(4)** follows immediately from Lemma 2.14. To prove **(4)** $\Rightarrow$ **(1)** suppose that $g|_\mathbb{D} \sim \text{id}_\mathbb{H} \pm 1$ and there exists an invariant component I of $\mathcal{F}(g) \cap \partial\mathbb{D}$. Let $G := \mathbb{D} \cup I \cup (\hat{\mathbb{C}} \setminus \overline{\mathbb{D}})$. From Lemma 2.26 we conclude that $p \in \mathcal{J}(g)$ and $g|_G \sim \text{id}_\mathbb{C} + 1$. By Lemma 2.6 we have that $\lambda_G(g^n(z), g^{n+1}(z)) \to 0$ as $n \to \infty$, for each $z \in G$, which by the Schwarz-Pick lemma implies that $\lambda_{\mathcal{F}(g)}(g^n(z), g^{n+1}(z)) \to 0$ as $n \to \infty$ for every $z \in G$. Thus (2) holds and we conclude from the implication (2) $\Rightarrow$ (3) that there exists an absorbing component of $\mathcal{F}(g) \cap \partial\mathbb{D}$. Since each absorbing component is invariant and there is at most one invariant component of $\mathcal{F}(g) \cap \partial\mathbb{D}$ we have that I is absorbing. Thus we can apply Lemma 2.27 and obtain that $g|_{\mathcal{F}(g)} \sim \text{id}_\mathbb{C} + 1$. $\square$

2.4. Periodic points of an inner function.

It is well-known that the Julia set of a non-injective rational, entire, or meromorphic function f is equal to the closure of the set of repelling periodic points of f. In this subsection we see that the same is true for inner functions if the definition of a repelling periodic point is relaxed in a certain sense.

We make use of the following theorem which is known as the Julia-Wolff lemma [18, Proposition 4.13]. For the definition and the main properties of the angular derivative we refer the reader to [18].

Theorem 2.29 (Julia-Wolff). *Let h be a holomorphic self-map of the unit disk. Let $\zeta, \eta \in \partial \mathbb{D}$ such that h has the angular limit η at ζ. Then the angular derivative $h'(\zeta)$ exists in $(0, \infty]$ and satisfies*
$$h'(\zeta) = \frac{\eta}{\zeta} \sup_{z \in \mathbb{D}} \frac{1 - |z|^2}{|\zeta - z|^2} \frac{|\eta - h(z)|^2}{1 - |h(z)|^2}.$$

Definition 2.30. *Let h be a holomorphic self-map of the unit disk. Let $\zeta \in \partial \mathbb{D}$.*
(1) *We call ζ a fixed point of h if h has the angular limit ζ at ζ. In this case the fixed point ζ of h is called repelling if the angular derivative $h'(\zeta)$ is contained in $(1, \infty]$.*
(2) *We call ζ a (repelling) periodic point of h if there exists $n \in \mathbb{N}$ such that ζ is a (repelling) fixed point of h^n.*

Theorem 2.1 represents only one part of the theorem of Denjoy and Wolff. The second part of the Denjoy-Wolff theorem (see [15] and [22]) is a statement about the angular derivatives at the fixed points of a holomorphic self-map of $\mathbb{D}$:

Theorem 2.31 (Denjoy-Wolff). *Let h be a non-Möbius holomorphic self-map of the unit disk. Let $p \in \overline{D}$ be the Denjoy-Wolff point of h. Then p is the unique fixed point of h which is not repelling.*

Hence there is at most one periodic point of a non-Möbius inner function g which is not repelling. Since $\mathcal{J}(g)$ is perfect it remains to show that the periodic points of g are dense in the Julia set. This is a consequence of the following theorem.

Theorem 2.32. *Let g be an inner function. Suppose that $\zeta \in \partial \mathbb{D}$ is a singularity of g. Then ζ is an accumulation point of fixed points of g.*

We use the following lemma to prove Theorem 2.32.

Lemma 2.33. *Suppose that g is a non-Möbius inner function of the upper half plane $\mathbb{H}$. Then the following statements are equivalent.*
(1) *∞ is the Denjoy-Wolff point of g.*
(2) *$\operatorname{Im}(g(z)) \geq \operatorname{Im}(z)$, for each $z \in \mathbb{H}$.*
(3) *$G(z) := g(z) - z$ is an inner function of the upper half plane.*
(4) *$\operatorname{Im}(g(z)) > \operatorname{Im}(z)$, for each $z \in \mathbb{H}$.*

Proof. Suppose that (1) holds. Let $T(z) := i(1+z)/(1-z)$. Then $h := T^{-1} \circ g \circ T$ is an inner function of the unit disk with Denjoy-Wolff point 1. An easy calculation shows that
$$\operatorname{Im}(T(z)) = \frac{(1 - |z|)^2}{|1 - z|^2}$$

for each $z \in \mathbb{D}$. Since the angular derivative $h'(1)$ is contained in $(0, 1]$ we conclude from Theorem 2.29 that $\mathrm{Im}(g(z)) \geq \mathrm{Im}(z)$, for each $z \in \mathbb{H}$.

Suppose that (2) holds. Let $G(z) := g(z) - z$. Then $\mathrm{Im}(G(z)) \geq 0$, for each $z \in \mathbb{H}$, which implies that G is an inner function of $\mathbb{H}$ or G is constant. Since g is non-Möbius we conclude that G is non-constant and hence an inner function. The implication (3) $\Rightarrow$ (4) is trivial. Suppose that (4) holds. It is easy to see that, for each $p \in \mathbb{H} \cup \mathbb{R}$, there exists $z \in \mathbb{H}$ such that $(g^n(z))_{n \in \mathbb{N}}$ does not converge to p. Hence we conclude from the Denjoy-Wolff theorem that ∞ is the Denjoy-Wolff point of g. □

Proof of Theorem 2.32. Let $p \in \overline{\mathbb{D}}$ be the Denjoy-Wolff point of g.

Case 1. Suppose that $p \in \mathbb{D}$. By conjugating with a Möbius transformation if necessary we may assume that $p = 0$. It is easy to see that

$$h(z) := \frac{g(z)}{z}$$

is an inner function with a singularity at ζ. Let U be a neighbourhood of ζ. By Lemma 1.5 there exists $\theta \in (U \cap \partial \mathbb{D}) \setminus \{\zeta\}$ and a path $\gamma : [0, 1) \to \mathbb{D}$ such that $\lim_{r \to 1} \gamma(r) = \theta$ and $h(\gamma(r)) \to 1$ as $r \to 1$ in a Stolz angle. Hence we have that $g(\gamma(r)) \to \theta$ as $r \to 1$. By the Lehto-Virtanen theorem [18, p. 71], we conclude that g has angular limit θ at θ.

Case 2. Suppose that $p \in \partial \mathbb{D}$. We may assume that $p = 1$. Let $T(z) := i(1 + z)/(1 - z)$. Then $G := T \circ g \circ T^{-1}$ is an inner function of the upper half plane with Denjoy-Wolff point ∞ and a singularity at $T(\zeta)$. From Lemma 2.33 we conclude that $H(z) := G(z) - z$ is an inner function of the upper half plane which has a singularity at $T(\zeta)$. Let U be a neighbourhood of $T(\zeta)$. By Lemma 1.5 there exists $\theta \in (U \cap \mathbb{R}) \setminus \{\zeta\}$ and a path $\gamma : [0, 1) \to \mathbb{H}$ such that $\lim_{r \to 1} \gamma(r) = \theta$ and $H(\gamma(r)) \to 0$ as $r \to 1$ in a Stolz angle. Hence we have that $G(\gamma(r)) \to \theta$ as $r \to 1$. By the Lehto-Virtanen theorem again, we conclude that G has angular limit θ at θ. □

Theorem 2.34. *Let g be a non-Möbius inner function. Then $\mathcal{J}(g)$ is the closure of the set of the repelling periodic points of g.*

Proof. For the case when g is a finite Blaschke product this is an old theorem of G. Julia and P. Fatou; see for instance [6, Theorem 6.9.1] or [20], [5] and [10] for simpler proofs.

For the case when g has a singularity on $\partial \mathbb{D}$ it follows from Theorem 1.4 and Theorem 2.32 that $\mathcal{J}(g)$ is contained in the closure of the set of the periodic points of g. Since there is at most one periodic point of g which is not repelling and $\mathcal{J}(g)$ is perfect we conclude that even the repelling periodic points of g are dense in $\mathcal{J}(g)$.

On the other hand, the only periodic point of g which might be in $\mathcal{F}(g)$ is the Denjoy-Wolff point. Hence all repelling periodic points are contained in $\mathcal{J}(g)$. □

2.5. **Examples.** In this subsection we give examples to see that all cases listed in Theorem 2.22 may occur (see Examples 2.36, 2.37, and 2.39). Moreover, we give an example of an inner function g of the upper half plane having an invariant component of $\mathcal{F}(g) \cap (\mathbb{R} \cup \{\infty\})$ which is not absorbing (see Example 2.38).

Most of the examples of inner functions in this subsection are of the following type. Let Q be an at most countable subset of $\mathbb{R}$. Let $w : Q \to (0, \infty)$ be such that $\sum_{q \in Q} w(q)$ converges. Let $a \in [0, \infty)$ and let $b \in \mathbb{R}$. Then it is easy to see that

$$g(z) := az + b - \sum_{q \in Q} \frac{w(q)}{z - q}$$

is an inner function of the upper half plane. If Q is an infinite set, then the singularities of g are the accumulation points of Q. If $\mathrm{dist}(z, Q)$ is large, then $g(z)$ is nearly $az + b$. In particular, this is the case when $|\mathrm{Im}(z)|$ is large. Moreover, it follows from Lemma 2.33 that ∞ is the Denjoy-Wolff point of g if and only if $a \geq 1$.

First, we discuss some examples of inner functions g on the upper half plane which have the property that $g|_\mathbb{H} \sim \lambda \mathrm{id}_\mathbb{H}$, for some $\lambda > 1$. We make use of the following lemma.

Lemma 2.35. *Let g be a non-Möbius inner function of the upper half plane. Suppose that there exists $r > 1$ such that $\mathrm{Im}(g(z)) > r\mathrm{Im}(z)$, for each $z \in \mathbb{H}$. Then ∞ is the Denjoy-Wolff point of g and there exists $\lambda > 1$ such that $g|_\mathbb{H} \sim \lambda \mathrm{id}_\mathbb{H}$.*

Proof. Since $\mathrm{Im}(g(z)) \geq \mathrm{Im}(z)$, for each $z \in \mathbb{H}$, we conclude from Lemma 2.33 that the Denjoy-Wolff point of g must be ∞. Moreover, for each $z, w \in \mathbb{H}$, we have that

$$\lambda_\mathbb{H}(z, w) = \log \frac{1 + |(z-w)/(z-\overline{w})|}{1 - |(z-w)/(z-\overline{w})|},$$

which in particular implies that $\lambda_\mathbb{H}(z, w) \geq \lambda_\mathbb{H}(i\mathrm{Im}(z), i\mathrm{Im}(w))$. Now, for each $z \in \mathbb{H}$,

$$\begin{aligned}\frac{|\mathrm{Im}(g(z)) - \mathrm{Im}(z)|}{|\mathrm{Im}(g(z)) - \mathrm{Im}(\overline{z})|} &= \frac{\mathrm{Im}(g(z)) - \mathrm{Im}(z)}{\mathrm{Im}(g(z)) + \mathrm{Im}(z)} \\ &= 1 - \frac{2\mathrm{Im}(z)}{\mathrm{Im}(g(z)) + \mathrm{Im}(z)} \\ &\geq 1 - \frac{2\mathrm{Im}(z)}{(r+1)\mathrm{Im}(z)} \\ &= 1 - \frac{2}{r+1},\end{aligned}$$

which implies that

$$\inf\{\lambda_\mathbb{H}(z, g(z)) : z \in \mathbb{H}\} > 0.$$

From Theorem 2.4 and Lemma 2.6 we conclude that there exists $\lambda > 1$ such that $g|_\mathbb{H} \sim \lambda \mathrm{id}_\mathbb{H}$. □

Example 2.36. *Let Q be a countable subset of $[0,\infty)$ which is dense in $[0,\infty)$. Let $w : Q \to (0,\infty)$ be such that $\sum_{q \in Q} w(q)$ converges. Let $a > 1$ and $b \in \mathbb{R}$. Then*
$$g(z) := az + b - \sum_{q \in Q} \frac{w(q)}{z-q}$$
is an inner function of the upper half plane with the following properties.
  (1) *There exists an absorbing component of $\mathcal{F}(g) \cap (\mathbb{R} \cup \{\infty\})$.*
  (2) *The Denjoy-Wolff point ∞ of g is contained in $\mathcal{J}(g)$.*
  (3) *There exists $\lambda > 1$ such that $g|_{\mathcal{F}(g)} \sim \lambda \mathrm{id}_\mathbb{H}$ and $g|_\mathbb{H} \sim \lambda^2 \mathrm{id}_\mathbb{H}$.*

Proof. Since Q is bounded from the left we conclude that there exists $c > 0$ such that $\mathrm{Re}(g(z)) < -c$ whenever $\mathrm{Re}(z) < -c$. Hence $(-c,\infty)$ is contained in an invariant component I of the Fatou set of g. Since Q is dense in $[0,\infty)$ we conclude that $[0,\infty) \cup \{\infty\} \subset \mathcal{J}(g)$. Let J be another component of $\mathcal{F}(g) \cap \mathbb{R}$. Then $f^n(J) \subset \mathcal{F}(g) \cap \mathbb{R} \subset (-\infty, 0)$, for each $n \in \mathbb{N}$, and $f^n \to \infty$ locally uniformly on J as $n \to \infty$. This implies that there exists $n \in \mathbb{N}$ such that $f^n(J) \subset I$. Thus I is absorbing. From Lemma 2.35 we conclude that there exists $\mu > 1$ such that $g|_\mathbb{H} \sim \mu \mathrm{id}_\mathbb{H}$. Now, the conclusion follows from Theorem 2.22 . □

Example 2.37. *Let Q be a bounded subset of $\mathbb{R}$ which is not empty and at most countable. Let $w : Q \to (0,\infty)$ be such that $\sum_{q \in Q} w(q)$ converges. Let $a > 1$ and $b \in \mathbb{R}$. Then*
$$g(z) := az + b - \sum_{q \in Q} \frac{w(q)}{z-q}$$
is an inner function of the upper half plane with the following properties.
  (1) *There exists an absorbing component of $\mathcal{F}(g) \cap (\mathbb{R} \cup \{\infty\})$.*
  (2) *The Denjoy-Wolff point ∞ of g is contained in $\mathcal{F}(g)$.*
  (3) $g|_{\mathcal{F}(g)} \sim \frac{1}{a}\mathrm{id}_\mathbb{C}$ *and* $g|_\mathbb{H} \sim \frac{1}{a}\mathrm{id}_\mathbb{H}$.

Proof. Since Q is bounded there exists $c > 0$ such that $\mathrm{Re}(g(z)) < -c$ whenever $\mathrm{Re}(z) < -c$, and $\mathrm{Re}(g(z)) > c$ whenever $\mathrm{Re}(z) > c$. This implies that there exists an invariant component I of $\mathcal{F}(g) \cap (\mathbb{R} \cup \{\infty\})$ which contains the Denjoy-Wolff point ∞. Hence I is absorbing and the conclusion follows from Theorem 2.22 and the fact that the eigenvalue of the fixed point ∞ of g is equal to $1/a$. □

Example 2.38. *The function*
$$g(z) := 2z - \sum_{n \in \mathbb{N}} \frac{2^{-n}}{z - 2^n}$$
is an inner function of the upper half plane with the following property: g has a wandering and an invariant component of $\mathcal{F}(g) \cap (\mathbb{R} \cup \{\infty\})$. In particular, the invariant component is not absorbing.

Proof. Since $Q := \{2^n : n \in \mathbb{N}\}$ is bounded from the left there exists $c \in (0, \infty)$ such that $\operatorname{Re}(g(z)) < -c$ whenever $\operatorname{Re}(z) < -c$. This implies that there exists an invariant component I of $\mathcal{F}(g) \cap (\mathbb{R} \cup \{\infty\})$. Since ∞ is a accumulation point of Q we conclude that $\infty \in \mathcal{J}(g)$.

Choose $1 < a < b < 2$ such that
$$x(x-1) \geq \frac{1}{2} \quad \text{and} \quad x(2-x) \geq \frac{1}{2},$$
for each $x \in [a, b]$. We prove:

(∗) $\forall a \leq \alpha < \beta \leq b, \ \forall m \in \mathbb{N}, \ \forall x \in [2^m \alpha, 2^m \beta]$:
$$2^{m+1}\alpha(1 - 2^{-m}) \leq g(x) \leq 2^{m+1}\beta(1 + 2^{-m}).$$

Proof of (∗). Let $a \leq \alpha < \beta \leq b$, $m \in \mathbb{N}$, and $x \in [2^m \alpha, 2^m \beta]$. Then
$$-\sum_{n \in \mathbb{N}} \frac{2^{-n}}{x - 2^n} \leq \sum_{n \geq m+1} \frac{2^{-n}}{2^n - x} \leq \sum_{n \geq m+1} \frac{2^{-n}}{2^n - 2^m \beta}$$
$$= 4^{-m} \sum_{n \in \mathbb{N}} \frac{2^{-n}}{2^n - \beta} \leq 4^{-m}(2 - \beta)^{-1}.$$

This implies that
$$g(x) \leq 2x + 4^{-m}(2-\beta)^{-1} \leq 2^{m+1}\beta + \beta 4^{-m}(\beta(2-\beta))^{-1}$$
$$\leq 2^{m+1}\beta(1 + 2^{-m}).$$

On the other hand,
$$-\sum_{n \in \mathbb{N}} \frac{2^{-n}}{x - 2^n} \geq -\sum_{n \leq m} \frac{2^{-n}}{x - 2^n} \geq -\sum_{n \leq m} \frac{2^{-n}}{2^m \alpha - 2^n}$$
$$\geq -\sum_{n \leq m} \frac{2^{-n}}{2^m \alpha - 2^m} \geq -2^{-m}(\alpha - 1)^{-1},$$

which implies that
$$g(x) \geq 2x - 2^{-m}(\alpha-1)^{-1} \geq 2^{m+1}\alpha - \alpha 2^{-m}(\alpha(\alpha-1))^{-1}$$
$$\geq 2^{m+1}\alpha(1 - 2^{-m}).$$

Thus we have proved (∗). Since $\prod_{m \in \mathbb{N}}(1 - 2^{-m})$ and $\prod_{m \in \mathbb{N}}(1 + 2^{-m})$ are convergent, we can find $a < \alpha_0 < \beta_0 < b$ and $m_0 \in \mathbb{N}$ such that
$$\alpha_0 \prod_{m \geq m_0}(1 - 2^{-m}) \geq a \quad \text{and} \quad \beta_0 \prod_{m \geq m_0}(1 + 2^{-m}) \leq b.$$

Applying (∗) we conclude that, for each $x \in J := [2^{m_0}\alpha_0, 2^{m_0}\beta_0]$ and $n \in \mathbb{N}$,

$$g^n(x) \in [2^{m_0+n}\alpha_0 \prod_{k=m_0}^{m_0+n-1}(1-2^{-m}), 2^{m_0+n}\beta_0 \prod_{k=m_0}^{m_0+n-1}(1+2^{-m})]$$
$$\subset [2^{m_0+n}a, 2^{m_0+n}b] \subset (2^{m_0+n}, 2^{m_0+n+1}).$$

Hence all iterates are defined and holomorphic on J, and this implies that $J \subset \mathcal{F}(g)$. On the other hand, all elements of Q are singularities of g^2 which implies that $Q \subset \mathcal{J}(g)$. Hence $f^n(J)$ and $f^m(J)$ are contained in different components of $\mathcal{F}(g) \cap \mathbb{R}$ whenever $m \neq n \in \mathbb{N}$. Thus J is contained in a wandering component of $\mathcal{F}(g) \cap \mathbb{R}$. □

Example 2.39. *Let Q be a non-empty and at most countable subset of $[0, \infty)$. Let $w : Q \to (0, \infty)$ such that $\sum_{q \in Q} w(q)$ converges. Then*

$$g(z) := z - 1 - \sum_{q \in Q} \frac{w(q)}{z - q}$$

is an inner function of the upper half plane with the following properties.

(1) *There exists an absorbing component of $\mathcal{F}(g) \cap (\mathbb{R} \cup \{\infty\})$.*
(2) *The Denjoy-Wolff point ∞ of g is contained in $\mathcal{J}(g)$.*
(3) $g|_\mathbb{H} \sim \mathrm{id}_\mathbb{H} \pm 1$ *and* $g|_{\mathcal{F}(g)} \sim \mathrm{id}_\mathbb{C} + 1$.

Proof. Since Q is bounded from the left we can find $c \in (0, \infty)$ such that $\mathrm{Re}(g(z)) < -c$ whenever $\mathrm{Re}(z) < -c$. This implies that there exists an invariant component I of $\mathcal{F}(g) \cap (\mathbb{R} \cup \{\infty\})$ that contains $(-\infty, -c)$. Define

$$h(z) := -\sum_{q \in Q} \frac{w(q)}{z - q}.$$

Then $h(z) \to 0$ as $|\mathrm{Im}(z)| \to \infty$. Hence

$$\frac{|g(z) - z|}{|g(z) - \overline{z}|} = \frac{|h(z) - 1|}{|h(z) - 1 + 2i\mathrm{Im}(z)|} \to 0 \quad \text{as} \quad |\mathrm{Im}(z)| \to \infty.$$

This implies that $\inf\{\lambda_\mathbb{H}(z, g(z)) : z \in \mathbb{H}\} = 0$. Since $\mathcal{F}(g) \cap \mathbb{R} \neq \emptyset$ we conclude from Theorem 2.24 that $g|_\mathbb{H} \not\sim \mathrm{id}_\mathbb{C} + 1$. Thus from Lemma 2.6 and Theorem 2.4 we conclude that $g|_\mathbb{H} \sim \mathrm{id}_\mathbb{H} \pm 1$. Now, Theorem 2.23 yields that I is absorbing and $g|_{\mathcal{F}(g)} \sim \mathrm{id}_\mathbb{C} + 1$. In particular, the Denjoy-Wolff point ∞ of g is contained in $\mathcal{J}(g)$. □

Finally, we discuss a whole class of examples of inner functions that will be of interest in Subsection 3.1. These functions are lifts of certain other inner functions under a universal covering map. We make use of the following lemma.

Lemma 2.40. *Let h be an inner function of the upper half plane. Suppose that there exists $c \in \mathbb{H}$ such that $h(x+iy) \to c$ as $y \to \infty$ uniformly in x. Let $g(z) := z + h(z)$. Then g is an inner function of the upper half plane such that $g|_\mathbb{H} \sim \mathrm{id}_\mathbb{C} + 1$.*

Proof. There exists $r > 0$ such that $\text{Im}(h(z)) > \text{Im}(c)/2$ whenever $\text{Im}(z) > r$. Thus $\text{Im}(g(z)) > \text{Im}(z) + \text{Im}(c)/2$ whenever $\text{Im}(z) > r$, which implies that $\text{Im}(g^n(z)) \to \infty$ as $n \to \infty$ locally uniformly on $G := \{z \in \mathbb{H} : \text{Im}(z) > r\}$. For each $z \in G$, we conclude that

$$\frac{|g^{n+1}(z) - g^n(z)|}{|g^{n+1}(z) - \overline{g^n(z)}|} = \frac{|h(g^n(z))|}{|h(g^n(z)) + 2i\text{Im}(g^n(z))|} \to 0 \quad \text{as } n \to \infty,$$

which implies that $\lambda_{\mathbb{H}}(g^{n+1}(z), g^n(z)) \to 0$ as $n \to \infty$. From Lemma 2.6 and Theorem 2.4 we see that $g|_{\mathbb{H}} \sim \text{id}_{\mathbb{C}} + 1$. □

Lemma 2.41. *Let $g \neq \text{id}_{\mathbb{D}}$ be an inner function such that $g(0) = 0$ and $g^{-1}(0) = \{0\}$. Let $\phi : \mathbb{D} \to \mathbb{D} \setminus \{0\}$ be a universal covering map, and let $G : \mathbb{D} \to \mathbb{D}$ be holomorphic such that $\phi \circ G = g \circ \phi$. Then G is an inner function with the following properties.*

(1) *The Denjoy-Wolff point of G is the unique point on $\partial \mathbb{D}$ which is a singularity of ϕ.*
(2) *If $g'(0) = 0$, then there exists $\lambda > 1$ such that $G|_{\mathbb{D}} \sim \lambda \text{id}_{\mathbb{H}}$.*
(3) *If $0 < |g'(0)| < 1$, then $G \sim \text{id}_{\mathbb{C}} + 1$.*
(4) *If $|g'(0)| = 1$, then G is conjugated to $\text{id}_{\mathbb{H}} \pm 1$.*
(5) *If g is rational, then G is a Möbius transformation.*
(6) *If g is non-rational, then $\mathcal{J}(G) = \partial \mathbb{D}$.*

Proof. Choose $n \in \mathbb{N}$ and a holomorphic function $h : \mathbb{D} \to \mathbb{C}$ such that $h(0) \neq 0$ and $g(z) = z^n h(z)$, for each $z \in \mathbb{D}$. If g is rational then $h \equiv \mu \in \partial \mathbb{D}$. If g is non-rational, then h is a non-rational inner function. We can find a holomorphic function $H : \mathbb{H} \to \mathbb{C}$ such that

$$h(\exp(iz)) = \exp(iH(z)),$$

for each $z \in \mathbb{H}$. If g is rational, then $H \equiv x \in \mathbb{R}$. If g is non-rational, then H is an inner function of the upper half plane. In this case we have that $0 < |h(0)| < 1$ and we conclude that there exists $c \in \mathbb{H}$ such that $H(x + iy) \to c$ as $y \to \infty$ uniformly in x. Let $f(z) := nz + H(z)$. Then f is an inner function of the upper half plane such that

$$(*) \quad \text{Im}(f(z)) \geq n\text{Im}(z) \quad \text{and} \quad \exp(if(z)) = g(\exp(iz)).$$

If f had a fixed point $p \in \mathbb{H}$, then $\exp(ip)$ would be a fixed point of g in $\mathbb{D} \setminus \{0\}$, which is not the case. Hence the Denjoy-Wolff point of f is contained in $\mathbb{R} \cup \{\infty\}$. Since $\text{Im}(f(z)) \geq \text{Im}(z)$, for each $z \in \mathbb{H}$, we conclude that ∞ is the Denjoy-Wolff point of f. Note that ∞ is the unique singularity of $z \mapsto \exp(iz)$.

Case 1. Suppose that $g'(0) = 0$. Then $n \geq 2$ and we conclude from $(*)$ and Lemma 2.35 that there exists $\lambda > 1$ such that $f|_{\mathbb{H}} \sim \lambda \text{id}_{\mathbb{H}}$. If g is rational, then H is constant and hence f is injective. In this case f is even conjugated to $\lambda \text{id}_{\mathbb{H}}$.

Case 2. Suppose that $0 < |g'(0)| < 1$. Then g is non-Möbius and hence not a finite Blaschke product because $g^{-1}(0) = \{0\}$. Thus g is non-rational which implies that h and H are inner functions. Since $0 < |h(0)| < 1$ and $\exp(iH(x+iy)) \to h(0)$ as $y \to \infty$ uniformly in x we conclude that there exists $c \in \mathbb{H}$ such that $H(x+iy) \to c$ as $y \to \infty$ uniformly in x. Hence Lemma 2.40 yields that $f|_{\mathbb{H}} \sim \mathrm{id}_\mathbb{C} + 1$.

Case 3. Suppose that $|g'(0)| = 1$. Then the Schwarz lemma gives that $h \equiv \mu := g'(0)$. Since $g \neq \mathrm{id}_\mathbb{D}$ we see that there exists $t \in \mathbb{R} \setminus \{0\}$ such that $H \equiv t$. Hence f is a Möbius transformation and conjugated to $\mathrm{id}_\mathbb{H} \pm 1$.

Finally, suppose that g is non-rational. Assume that $\mathcal{F}(f) \cap (\mathbb{R} \cup \{\infty\}) \neq \emptyset$. Then we can find a non-empty open interval $I \subset \mathcal{F}(f) \cap \mathbb{R}$ such that $f^n(I) \subset \mathbb{R}$, for each $n \in \mathbb{N}$. This implies that all iterates of g are analytic on $\exp(iI)$. Since g is non-rational we conclude that $\exp(iI) \subset \mathcal{F}(g)$. On the other hand, we know from Theorem 1.8 that $\mathcal{J}(g) = \partial \mathbb{D}$. This is a contradiction. Thus we have that $\mathcal{J}(f) = \mathbb{R} \cup \{\infty\}$.

Since universal covering maps are uniquely determined up to isomorphy it follows that ϕ is of the form $\exp \circ (iM)$, where M is a Möbius transformation such that $M(\mathbb{D}) = \mathbb{H}$. By passing over to $f + 2\pi k$, $k \in \mathbb{Z}$, if necessary we can achieve that $G = M^{-1} \circ f \circ M$. Thus the conclusion follows. □

3. Boundaries of components of the Fatou set

3.1. Boundaries of Baker domains.
Cowen's result (Theorem 2.4) implies that if D is a Baker domain of a transcendental entire function f, then $f|_D$ is eventually conjugated to either $\mathrm{id}_\mathbb{C} + 1$, or $\mathrm{id}_\mathbb{H} \pm 1$, or $\lambda \mathrm{id}_\mathbb{H}$, where $\lambda > 1$. Indeed, all types of conjugacies can occur on Baker domains (see Lemma 3.3 below).

Using our results of Subsection 2.3 we are able to extend Theorem 1.10 to those Baker domains where $f|_D$ is eventually conjugated to $\mathrm{id}_\mathbb{C} + 1$.

Theorem 3.1. *Let f be a transcendental entire function. Suppose that D is a Baker domain of f such that $f|_D \sim \mathrm{id}_\mathbb{C} + 1$. Then $\overline{\Theta} = \partial \mathbb{D}$.*

More generally, we can prove the following result, which is a generalization of Theorem 1.10.

Theorem 3.2. *Let f be a transcendental entire function. Suppose that D is an unbounded invariant component of the Fatou set such that $\lambda_D(f^n(z), f^{n+1}(z)) \to 0$ as $n \to \infty$, for some $z \in D$. If $\Theta \neq \emptyset$, then $\overline{\Theta} = \partial \mathbb{D}$.*

Theorem 3.2 is more general than Theorem 3.1 because it also covers the cases of an unbounded attracting domain and an unbounded parabolic domain, because on parabolic domains the maps are always eventually conjugated to $\mathrm{id}_\mathbb{C} + 1$, which by Lemma 2.6 implies that $\lambda_D(f^n(z), f^{n+1}(z)) \to 0$ as $n \to \infty$, for each $z \in D$. For the attracting and parabolic case this result is already included in Theorem 1.10 but using our Theorem 2.24 we can give an alternative proof:

Proof of Theorem 3.2. Since $\lambda_D(f^n(z), f^{n+1}(z)) \to 0$ as $n \to \infty$, for some $z \in D$, $f|_D$ is not an automorphism of D. Since D is a Baker domain we have that $\Theta \neq \emptyset$. Thus we can apply Lemma 1.7 and obtain that $\mathcal{J}(g) \subset \overline{\Theta}$. From Theorem 2.24 we conclude that $\overline{\Theta} = \partial \mathbb{D}$. □

Remark. A statement related to Theorem 3.1 is presented in [16]. In the situation of Theorem 3.1 the author states that the set

$$\Xi := \{\xi \in \partial \mathbb{D} : \infty \in C(\phi; \xi)\}$$

is equal to $\partial \mathbb{D}$ whenever $f|_D$ is semi-conjugated to $\mathrm{id}_\mathbb{C} + 1$.

Here one has to be careful because there is an important difference between semi-conjugacies and eventual conjugacies. In particular, semi-conjugacies are not unique. For instance, each function which is semi-conjugated to $\mathrm{id}_\mathbb{H}+1$ is also semi-conjugated to $\mathrm{id}_\mathbb{C} + 1$. Presumably in the above statement the author means eventually conjugated instead of semi-conjugated.

A wide class of examples of Baker domains (including those of Baker-Weinreich and Bergweiler where the boundary is a Jordan curve) is given by the following method of construction:

Suppose that F is a transcendental entire function such that $F(0) = 0$, $F^{-1}(0) = \{0\}$, and $0 \in \mathcal{F}(F)$. Let V be the component of $\mathcal{F}(F)$ which contains 0. Since $F \circ \exp$ does not have any zeros there exists an entire function f such that $\exp \circ f = F \circ \exp$. It is easy to see that $D := \exp^{-1}(V)$ is connected. In [9] it is proved that $\mathcal{F}(f) = \exp^{-1}(\mathcal{F}(F))$. Thus D is a component of the Fatou set of f and it is easy to see that $f^n \to \infty$ locally uniformly in D as $n \to \infty$.

The following lemma deals with Baker domains constructed in this way. First, it gives criteria to determine which transformation $f|_D$ is eventually conjugated to. Second, we see that this method of construction leads to Baker domains D such that either $f|_D$ is an automorphism of D or $\overline{\Theta} = \partial \mathbb{D}$.

Lemma 3.3. *Let F be a transcendental entire function such that $F(0) = 0$, $F^{-1}(0) = \{0\}$, and $0 \in \mathcal{F}(f)$. Let V be the component of $\mathcal{F}(F)$ which contains 0. Let f be an entire function such that $\exp \circ f = F \circ \exp$. Then $D := \exp^{-1}(V)$ is a Baker domain of f. Then the following statements hold.*

(1) *If $|F'(0)| = 1$, then $f|_D \sim \mathrm{id}_\mathbb{H} \pm 1$ and $f|_D$ is an automorphism of D.*
(2) *If $0 < |F'(0)| < 1$, then $f|_D \sim \mathrm{id}_\mathbb{C} + 1$. In particular, $\overline{\Theta} = \partial \mathbb{D}$.*
(3) *If $F'(0) = 0$, then there exists $\lambda > 1$ such that $f|_D \sim \lambda \mathrm{id}_\mathbb{H}$. Moreover, either $F|_V$ is a proper self-map of V and $f|_D$ is an automorphism of D, or $F|_V$ is not a proper self-map of V and $\overline{\Theta} = \partial \mathbb{D}$.*

Proof. Let $\psi : \mathbb{D} \to V$ be a biholomorphic map such that $\psi(0) = 0$. Let $g := \psi^{-1} \circ F \circ \psi$ be the corresponding inner function of F. Let $G := \phi^{-1} \circ f \circ \phi$ be the corresponding inner function of f. We have that $p := \psi^{-1} \circ \exp \circ \phi$ is

a universal covering map of $\mathbb{D}$ onto $\mathbb{D} \setminus \{0\}$. Moreover, for each $z \in \mathbb{D}$,
$$\begin{aligned} g(p(z)) &= (g \circ \psi^{-1})(\exp(\phi(z))) \\ &= \psi^{-1} \circ (F \circ \exp) \circ \phi(z) \\ &= \psi^{-1} \circ \exp \circ (f \circ \phi)(z) \\ &= p(G(z)). \end{aligned}$$
Now statements (1) and (2) follow immediately from Lemma 2.41 and Theorem 3.1. Since $F|_V$ is a proper self-map of V if and only if g is rational and $\mathcal{J}(G) \subset \overline{\Theta}$ we also conclude from Lemma 2.41 that (3) holds. □

Condition (1) holds if and only if V is a Siegel disk of F. Using results of Herman and Ghys about boundaries of Siegel disks, Baker and Weinreich gave the following example of a Baker domain whose boundary is a Jordan curve.

Example 3.4 (Baker-Weinreich). *There exists $\alpha \in [0,1]$ such that the function $F(z) := \exp(2\pi i \alpha) z \exp(z)$ has a Siegel disk V around 0 which is bounded by a quasicircle. Hence $D := \exp^{-1}(V)$ is a Baker domain of the function $f(z) := z + 2\pi i \alpha + \exp(z)$ such that $f|_D \sim \mathrm{id}_{\mathbb{H}} \pm 1$, $f|_D$ is an automorphism of D, and ∂D is a Jordan curve in $\hat{\mathbb{C}}$.*

Proof. See [4, Theorem 3] and use Lemma 3.3. □

Example 3.4 shows that, in general, Theorem 3.1 does not carry over to those Baker domains where $f|_D \sim \mathrm{id}_{\mathbb{H}} \pm 1$.

Case (3) of Lemma 3.3 holds if and only if V is a Böttcher domain of F. In this case we have to distinguish the cases where $F|_V$ is a proper self-map of V or not. Both cases are possible.

Example 3.5 (Bergweiler). *The function $F(z) := \frac{1}{2} z^2 \exp(2 - z)$ has a Böttcher domain V around 0 such that $F|_V$ is a proper self-map of V and ∂V is a Jordan curve. Hence $D := \exp^{-1}(V)$ is a Baker domain of the function $f(z) := 2 - \log 2 + 2z - \exp(z)$ such that $f|_D \sim \lambda \mathrm{id}_{\mathbb{H}}$, for some $\lambda > 1$, $f|_D$ is an automorphism of D, and ∂D is a Jordan curve.*

Proof. See [8, Theorem 2] and use Lemma 3.3. □

Example 3.5 shows that, in general, Theorem 3.1 does not carry over to those Baker domains where $f|_D \sim \lambda \mathrm{id}_{\mathbb{H}}$, for some $\lambda > 1$.

Example 3.6. *The function $F(z) := z^2 \exp(z - 3)$ has a Böttcher domain V around 0 such that $F|_V$ is not a proper self-map of V. Hence $D := \exp^{-1}(V)$ is a Baker domain of the function $f(z) := 2z - 3 + \exp(z)$ such that $f|_D \sim \lambda \mathrm{id}_{\mathbb{H}}$, for some $\lambda > 1$, $f|_D$ is not an automorphism of D, and $\overline{\Theta} = \partial \mathbb{D}$.*

Proof. Since 0 is a superattracting fixed point of F the component V of 0 in $\mathcal{F}(f)$ is a Böttcher domain. It is easy to see that $F((-\infty, 0)) \subset \mathbb{D} \subset V$

which implies that $(-\infty, 0) \subset V$. Since $\lim_{x \to -\infty} F(x) = 0$ we have that $F|_V$ is not a proper self-map of V, so the conclusion follows from Lemma 3.3. □

3.2. An improvement of Lemma 1.7 for the case of a completely invariant component.
In this subsection we see that the lower bound for the size of the set $\overline{\Theta}$ given in Lemma 1.7 is sharp at least for the case of a completely invariant component of the Fatou set. This is an easy consequence of the following result.

Lemma 3.7. *Let f be a transcendental entire function. Suppose that D is a completely invariant component of the Fatou set of f. Let $\phi : \mathbb{D} \to D$ be a Riemann map and let $g := \phi^{-1} \circ f \circ \phi$ be the corresponding inner function. Let $\zeta, \theta \in \partial \mathbb{D}$ such that $\zeta \neq \theta$ and $\lim_{r \to 1} \phi(r\zeta)$ and $\lim_{r \to 1} \phi(r\theta)$ exist and are equal. Then $(\zeta, \theta) \cap \mathcal{J}(g) \neq \emptyset$.*

Theorem 3.8. *Let f be a transcendental entire function. Suppose that D is a completely invariant component of the Fatou set of f. Let $\phi : \mathbb{D} \to D$ be a Riemann map and let $g := \phi^{-1} \circ f \circ \phi$ be the corresponding inner function. If $\Theta \neq \emptyset$, then $\mathcal{J}(g)$ is equal to the set of accumulation points of Θ.*

Proof. Since D is a completely invariant component of the Fatou set of f it follows that $f|_D$ is not an automorphism of D and g is not a Möbius transformation. From Lemma 1.7 we know that $\mathcal{J}(g) \subset \overline{\Theta}$. Since $\mathcal{J}(g)$ is a perfect set we conclude that $\mathcal{J}(g)$ is contained in the set of accumulation points of Θ.

Conversely, suppose that $x \in \partial \mathbb{D}$ is an accumulation point of Θ. Let U be a neighbourhood of x. Then there exist $\zeta, \theta \in U \cap \Theta$ such that $\zeta \neq \theta$ and $(\zeta, \theta) \subset U$. From Lemma 3.7 we conclude that $U \cap \mathcal{J}(g) \supset (\zeta, \theta) \cap \mathcal{J}(g) \neq \emptyset$. Since $\mathcal{J}(g)$ is closed it follows that $x \in \mathcal{J}(g)$. □

We now prove a more elaborate version of Lemma 3.7 which, in the next subsection, will lead to more general results about the boundaries of components of the Fatou set of a transcendental entire function. For the definition of a crosscut and its elementary properties we refer the reader to [18].

Lemma 3.9. *Let f be a transcendental entire function. Suppose that D is a completely invariant component of the Fatou set of f. Let $\phi : \mathbb{D} \to D$ be a Riemann map and let $g := \phi^{-1} \circ f \circ \phi$ be the corresponding inner function. Let Q be an open subset of $\mathbb{D}$ such that there exists a finite set $\mathcal{C}$ of pairwise disjoint crosscuts of $\mathbb{D}$ with the following properties.*
  (1) $\partial Q \cap \mathbb{D} = \bigcup_{C \in \mathcal{C}} C.$
  (2) *For each $C \in \mathcal{C}$ and each endpoint θ of C, $\lim_{z \to \theta} \phi|_C(z)$ exists if the angular limit of ϕ at θ exists.*
Suppose that $\partial \mathbb{D} \setminus \overline{Q} \subset \mathcal{F}(g)$. Then $\partial D \subset \overline{\phi(Q)}$.

First, let us see how Lemma 3.7 can be deduced from Lemma 3.9.

Proof of Lemma 3.7. Let $w := \lim_{r \to 1} \phi(r\zeta) = \lim_{r \to 1} \phi(r\theta) \in \hat{\mathbb{C}}$. Let $C := \zeta[0,1) \cup \theta[0,1)$. Then C is a crosscut of $\mathbb{D}$ which satisfies condition (2) of Lemma 3.9. Moreover, $\gamma := \{w\} \cup \phi(C)$ is a Jordan curve in $\hat{\mathbb{C}}$. Let $Q := \{r\eta : r \in (0,1), \eta \in (\theta, \zeta)\}$ and let E be the component of $\hat{\mathbb{C}} \setminus \gamma$ which contains $\phi(Q)$. Now, $\lim_{r \to 1} \phi(r\eta)$ exists almost everywhere on $\partial \mathbb{D}$ and is $\neq w$ almost everywhere; see [18, Theorem 1.7 and Corollary 6.14]. This implies that there exists $\eta \in (\zeta, \theta)$ such that $y := \lim_{r \to 1} \phi(r\eta) \in \hat{\mathbb{C}} \setminus \gamma$. Let W be the component of $\hat{\mathbb{C}} \setminus \gamma$ which contains y. Since $\eta \in (\zeta, \theta)$ and γ is a Jordan curve we conclude that $W \neq E$ which implies that $y \notin \overline{\phi(Q)}$. From Lemma 3.9 we conclude that $\partial \mathbb{D} \setminus \overline{Q} \not\subset \mathcal{F}(g)$ which implies that $(\zeta, \theta) \cap \mathcal{J}(g) \neq \emptyset$. □

Proof of Lemma 3.9. Assume that there exists $x \in \partial D \setminus \overline{\phi(Q)}$. We may suppose that $x \in \mathbb{C}$. Since D is a completely invariant component of the Fatou set of f we have that g is non-Möbius, which implies that $\mathcal{J}(g)$ is a perfect set. Let $k := 2\text{card}(\mathcal{C}) + 1$. From [18, Theorem 1.7 and Corollary 6.14], we conclude that there exist $\zeta_1 := \exp(i\alpha_1), \ldots, \zeta_k := \exp(i\alpha_k)$ such that $0 \leq \alpha_1 < \cdots < \alpha_k < 2\pi$ and

(*) $\phi(\zeta_j) := \lim_{r \to 1} \phi(r\zeta_j)$ exists, for each $j \in \{1, \ldots, k\}$, and is not an exceptional point of f.

(**) $(\zeta_j, \zeta_{j+1}) \cap \mathcal{J}(g) \neq \emptyset$, for each $j \in \{1, \ldots, k\}$, where $\zeta_{k+1} := \zeta_1$.

We can find $r \in (0,1)$ such that $K := \bigcup_{j=1}^{k} \phi(\zeta_j[r,1])$ is a compact set which does not contain an exceptional point of f. Let U be a neighbourhood of x such that $U \subset \mathbb{C} \setminus \overline{\phi(Q)}$. Since $x \in \partial D \subset \mathcal{J}(f)$ we conclude that there exists $n \in \mathbb{N}$ such that $K \subset f^n(U)$. Hence we can find $s \in (r, 1)$ and curves $\delta_j : [s,1] \to U$, for each $j \in \{1, \ldots, k\}$, such that $f^n(\delta_j(t)) = \phi(t\zeta_j)$, for each $t \in [s,1]$. Let $j \in \{1, \ldots, k\}$. Since D is completely invariant we have that $\delta_j(t) \in D$, for each $t \in [s,1)$. Hence

$$\gamma_j : [s,1) \to \mathbb{D}, t \mapsto \phi^{-1}(\delta_j(t))$$

is a well defined curve in $\mathbb{D} \setminus \overline{Q}$. From [18, Proposition 2.14], we know that $\theta_j := \lim_{t \to 1} \gamma_j(t) \in \partial \mathbb{D}$ exists.

Assume that $\theta_j \in \overline{Q}$. Since $\gamma([s,1)) \cap \overline{Q} = \emptyset$ we conclude that there exists a crosscut $C \in \mathcal{C}$ such that θ_j is an endpoint of C. Since $\delta_j(1) = \lim_{t \to 1} \phi(\gamma_j(t))$ exists we conclude from the Lehto-Virtanen theorem that the angular limit of ϕ at θ_j exists and is equal to $\delta_j(1)$. Then the assumption yields that

$$\lim_{z \to \theta_j} \phi|_C = \delta_j(1),$$

which implies that $\delta_j(1) \in \overline{\phi(Q)}$. This is a contradiction to $\delta_j(1) \in U$. Thus $\theta_j \in \partial \mathbb{D} \setminus \overline{Q} \subset \mathcal{F}(g)$.

Furthermore, for each $t \in [s,1)$,

$$\phi(g^n(\gamma_j(t))) = f^n(\phi(\gamma_j(t))) = f^n(\delta_j(t)) = \phi(t\zeta_j),$$

and hence $g^n(\gamma_j(t)) = t\zeta_j$. Since $\theta_j \in \mathcal{F}(g)$ we have that g^n is analytic in θ_j and conclude that $g^n(\theta_j) = \zeta_j$.

It is easy to see that there are at most $2\text{card}(\mathcal{C})$ connected components of $\partial \mathbb{D} \setminus \overline{Q}$. Since $k > 2\text{card}(\mathcal{C})$ there exists a component I of $\partial \mathbb{D} \setminus \overline{Q}$ and $l, m \in \{1, \ldots, k\}$ such that $l \neq m$ and $\{\theta_l, \theta_m\} \subset I$. Since $I \subset \partial \mathbb{D} \setminus \overline{Q} \subset \mathcal{F}(g)$ and $\{\zeta_l, \zeta_m\} = \{g^n(\theta_l), g^n(\theta_m)\} \subset g^n(I)$, we conclude from Theorem 2.17 that $(\zeta_l, \zeta_m) \subset \mathcal{F}(g)$ or $(\zeta_m, \zeta_l) \subset \mathcal{F}(g)$. This is a contradiction to $(**)$. □

As a consequence of Lemma 3.9 and Theorem 3.8 we obtain the following result. Here, for each open subset Q of $\mathbb{D}$ and each $\zeta \in \partial \mathbb{D}$, $C_Q(\phi; \zeta)$ denotes the set of all $w \in \hat{\mathbb{C}}$ for which there is a sequence $(q_n)_{n \in \mathbb{N}}$ in Q such that $\lim_{n \to \infty} q_n = \zeta$ and $\lim_{n \to \infty} \phi(q_n) = w$.

Lemma 3.10. *Let f be a transcendental entire function. Suppose that there exists a completely invariant Baker domain D of f. Let $\phi : \mathbb{D} \to D$ be a Riemann map and let $g := \phi^{-1} \circ f \circ \phi$ be the corresponding inner function.*

Suppose that I is a component of $\mathcal{F}(g) \cap \partial \mathbb{D}$. Let $\zeta, \theta \in \partial \mathbb{D}$ be such that $I = (\zeta, \theta)$ and let

$$S := \{r\eta : r \in (0, 1), \eta \in I\} \quad \text{and} \quad T := \mathbb{D} \setminus \overline{S}.$$

Then

$$\bigcup_{\eta \in \overline{I}} C_S(\phi; \eta) \subset C_T(\phi; \zeta) \cup C_T(\phi; \theta),$$

and, for each $\xi \in \partial \mathbb{D} \setminus \overline{I}$,

$$\bigcup_{\eta \in \overline{I}} C_S(\phi; \eta) \cap C(\phi; \xi) \subset \{\infty\}.$$

Proof. Since $\mathcal{J}(g)$ is a perfect set we have that $\zeta \neq \theta$. According to Lemma 1.7 there exist sequences $(\zeta_n)_{n \in \mathbb{N}}$ and $(\theta_n)_{n \in \mathbb{N}}$ in $\partial \mathbb{D} \setminus \overline{I}$ such that $\lim_{r \to 1} \phi(r\zeta_n) = \lim_{r \to 1} \phi(r\theta_n) = \infty$, for each $n \in \mathbb{N}$, and $\lim_{n \to \infty} \zeta_n = \zeta$, and $\lim_{n \to \infty} \theta_n = \theta$. We may assume that both sequences are monotone, i.e. $\zeta_{n+1} \in (\zeta_n, \zeta)$ and $\theta_{n+1} \in (\theta, \theta_n)$, for each $n \in \mathbb{N}$. Define

$$S_n := \{r\eta : r \in (0, 1), \eta \in (\zeta_n, \theta_n)\} \quad \text{and} \quad T_n := \mathbb{D} \setminus \overline{S_n},$$

for each $n \in \mathbb{N}$. It is easy to see from the Jordan curve theorem that $\phi(S_n) \cap \phi(T_n) \subset \{\infty\}$, for each $n \in \mathbb{N}$. This implies the second statement of the conclusion. Furthermore, we conclude from Lemma 3.9 that

$$\bigcup_{\eta \in \overline{I}} C_S(\phi; \eta) \subset \bigcup_{\eta \in \partial \mathbb{D} \setminus I} C_T(\phi; \eta),$$

which together with the second statement of the conclusion implies the first statement of the conclusion. □

3.3. Further results about boundaries of components of the Fatou set. Using Lemma 3.9 we obtain the following theorem.

Theorem 3.11. *Let f be a transcendental entire function. Suppose that D is a completely invariant component of the Fatou set of f. Let $G \subset \mathbb{C}$ be a bounded Jordan domain such that $G \cap \mathcal{J}(f) \neq \emptyset$. Then $\partial G \cap D$ has infinitely many components.*

Proof. Let $\phi : \mathbb{D} \to D$ be a Riemann map and let $g := \phi^{-1} \circ f \circ \phi$ be the corresponding inner function. Since D is completely invariant we have that $\partial D = \mathcal{J}(f)$ and D is unbounded. This implies that $G \cap D \neq \emptyset$ and $(\mathbb{C} \setminus \overline{G}) \cap D \neq \emptyset$. Hence $\partial G \cap D \neq \emptyset$.

Assume that $\partial G \cap D$ consists of only finitely many components. Then $\partial G \cap D$ is the union of finitely many pairwise disjoint crosscuts of D. We conclude from [18, Proposition 2.14] that there exists a finite set $\mathcal{C}$ of pairwise disjoint crosscuts of $\mathbb{D}$ such that

$$\partial G \cap D = \bigcup_{C \in \mathcal{C}} \phi(C)$$

and condition (2) of Lemma 3.9 is satisfied. Let $Q := \phi^{-1}(D \setminus \overline{G})$. Then Q is an open subset of $\mathbb{D}$ such that condition (1) of Lemma 3.9 is satisfied. Since

$$\partial D \cap G = \mathcal{J}(f) \cap G \neq \emptyset$$

we conclude from Lemma 3.9 that $(\partial \mathbb{D} \setminus \overline{Q}) \cap \mathcal{J}(g) \neq \emptyset$. From Lemma 1.6 we conclude that there exists $\zeta \in \partial \mathbb{D} \setminus \overline{Q}$ such that $\infty \in C(\phi; \zeta)$. Thus $\phi(\mathbb{D} \setminus \overline{Q})$ is unbounded. This is a contradiction to $\phi(\mathbb{D} \setminus Q) \subset \overline{G}$. □

This leads to the following result.

Theorem 3.12. *Let f be a transcendental entire function. Suppose that there exists an unbounded component of $\mathcal{F}(f)$. Let $G \subset \mathbb{C}$ be a bounded Jordan domain such that $G \cap \mathcal{J}(f) \neq \emptyset$. Then $\partial G \cap \mathcal{F}(f)$ has infinitely many components.*

Proof. By Theorem 3.11 we can assume that there is no completely invariant component of any iterate of f. This implies that there are infinitely many unbounded components of the Fatou set of f. Let $\mathcal{E}$ be the set of components E of $\mathcal{F}(f)$ for which $E \cap \partial G \neq \emptyset$. It remains to show that $\mathrm{card}(\mathcal{E}) \geq n$, for each $n \in \mathbb{N}$.

Let $n \in \mathbb{N}$ and consider distinct unbounded components $D_1, \ldots, D_n$ of $\mathcal{F}(f)$. For each $j \in \{1, \ldots, n\}$ let $x_j \in D_j$ be a non-exceptional point of f. Since $G \cap \mathcal{J}(f) \neq \emptyset$ we can find $k \in \mathbb{N}$ and $y_1, \ldots, y_n \in G$ such that $f^k(y_j) = x_j$, for each $j \in \{1, \ldots, n\}$. Let G_j be the component of $\mathcal{F}(f)$ which contains y_j. Since $D_1, \ldots, D_n$ are unbounded and pairwise distinct and $f^k(G_j) \subset E_j$ for each $j \in \{1, \ldots, n\}$, we conclude that $G_1, \ldots, G_n$ are unbounded and pairwise distinct. Then $G_j \cap G \neq \emptyset \neq G_j \cap (\mathbb{C} \setminus \overline{G})$ which implies that $G_j \cap \partial G \neq \emptyset$, for each $j \in \{1, \ldots, n\}$. Thus we conclude that $n \leq \operatorname{card}(\mathcal{E})$. □

As a first consequence of Theorem 3.12, we obtain a result about the injectivity of the finite radial limits of a Riemann map of a component of the Fatou set. This result applies also to polynomials or those transcendental entire functions for which there is no unbounded component of the Fatou set. For these cases we make use of the following observation which is easily derived from the maximum principle and Montel's theorem.

Remark 3.13. *Suppose that f is an entire function and $G \subset \mathbb{C}$ is a bounded Jordan domain such that $\bigcup_{n \in \mathbb{N}} f^n(\partial G)$ is bounded. Then $G \subset \mathcal{F}(f)$.*

Theorem 3.14. *Let f be an entire function. Let D be a component of the Fatou set of f. Suppose that at least one of the following conditions is satisfied:*

(1) *f is transcendental and there exists an unbounded component of the Fatou set of f, or*
(2) *$\bigcup_{n \in \mathbb{N}} f^n(D)$ is bounded.*

Let $\phi : \mathbb{D} \to D$ be a Riemann map. Let $\operatorname{Acc}(D)$ be the set of finite accessible boundary points of D, and let Z be the set of all $\zeta \in \partial \mathbb{D}$ such that $\phi(\zeta) := \lim_{r \to 1} \phi(r\zeta)$ exists and is finite. Then the map

$$Z \to \operatorname{Acc}(D), \ \zeta \mapsto \phi(\zeta)$$

is a bijection.

Proof. It follows from [18, Proposition 2.14] and the Lehto-Virtanen theorem that the map is surjective. Let $w \in \operatorname{Acc}(D)$ and $\zeta_1, \zeta_2 \in Z$ such that $\phi(\zeta_j) = w$, for each $j \in \{1, 2\}$.

Assume that $\zeta_1 \neq \zeta_2$. Then $\gamma := \phi(\zeta_1[0,1] \cup \zeta_2[0,1])$ is a Jordan curve in $\mathbb{C}$. It follows from Theorem 3.12 and Remark 3.13 that the bounded component G of $\mathbb{C} \setminus \gamma$ is contained in $\mathcal{F}(f)$. We may assume that $\phi(Q) \subset G$, where $Q := \{r\eta : r \in (0,1), \eta \in (\zeta_1, \zeta_2)\}$. Since $\lim_{r \to 1} \phi(r\eta)$ exists and $\neq w$ almost everywhere on $\partial \mathbb{D}$ (see [18, Theorem 1.7 and Corollary 6.14]) we can find $\eta \in (\zeta_1, \zeta_2) \cap Z$ such that $\phi(\eta) \neq w$. Then

$$\phi(\eta) \in (\partial D \cap \overline{G}) \setminus \{w\} \subset \mathcal{J}(f) \cap G.$$

This is a contradiction to $G \subset \mathcal{F}(f)$. □

For the case of a Siegel disk of an entire function we can easily deduce the following result.

Corollary 3.15. *Let f be an entire function. Suppose that D is a Siegel disk for f. Then*

(1) *There is no periodic point of f in ∂D which is an accessible boundary point of D.*
(2) *f is univalent on the set of finite accessible boundary points of D.*

The first part of the corollary has been shown already for polynomials by W. Schmidt; see [19, Korollar 7.9].

Proof. There exists a Riemann map $\phi : \mathbb{D} \to D$ and $\lambda \in \exp(2\pi i(\mathbb{R} \setminus \mathbb{Q}))$ such that $f(\phi(z)) = \phi(\lambda z)$, for each $z \in \mathbb{D}$. Let x be a finite accessible boundary point of D and let $\zeta \in \partial \mathbb{D}$ such that $\lim_{r \to 1} \phi(r\zeta) = x$.

First, assume that there exists $n \in \mathbb{N}$ such that $f^n(x) = x$. Then we conclude that
$$\phi(r\lambda^n\zeta) = f^n(\phi(r\zeta)) \to x \quad \text{as } r \to 1.$$
Hence Theorem 3.14 implies that $\lambda^n\zeta = \zeta$. Thus we have that $\lambda^n = 1$, which is a contradiction to $\lambda \in \exp(2\pi i(\mathbb{R} \setminus \mathbb{Q}))$.

To prove (2), suppose that y is another finite accessible boundary point of D such that $f(x) = f(y)$. Then there exists $\theta \in \partial \mathbb{D}$ such that $\lim_{r \to 1} \phi(r\theta) = y$. For each $\eta \in \{\zeta, \theta\}$ we have that
$$\phi(r\lambda\eta) = f(\phi(r\eta)) \to f(x) = f(y) \quad \text{as } r \to 1.$$
By Theorem 3.14 we conclude that $\lambda\zeta = \lambda\theta$, which implies that $\zeta = \theta$ and hence $x = y$. $\square$

As a second consequence of Theorem 3.12, we obtain a result about the number of common accessible points of different components of the Fatou set.

Theorem 3.16. *Let f be an entire function. Let $\mathcal{E}$ be a finite set of components of the Fatou set of f. Suppose that at least one of the following conditions is satisfied:*

(1) *f is transcendental and there exists an unbounded component of the Fatou set of f, or*
(2) *$\bigcup_{n \in \mathbb{N}} f^n(D)$ is bounded, for each $D \in \mathcal{E}$.*

Then there are at most $\mathrm{card}(\mathcal{E}) - 1$ points in $\mathbb{C}$ which are common accessible boundary points of at least two components in $\mathcal{E}$.

Proof. We confine ourselves to the case when $n := \mathrm{card}(\mathcal{E}) = 2$. Let $D_1, D_2 \in \mathcal{E}$ such that $D_1 \neq D_2$. Assume that there are two distinct points $x, y \in \mathbb{C}$ which are accessible boundary points of D_1 and D_2. Since D_j is simply connected there exist a Riemann map $\phi_j : \mathbb{D} \to D_j$ and $\zeta_j, \theta_j \in \partial \mathbb{D}$ such that $\phi_j(\zeta_j) := \lim_{r \to 1} \phi_j(r\zeta_j)$ exist and are equal to x and $\phi_j(\theta_j) := \lim_{r \to 1} \phi_j(r\theta_j)$ exist and are equal to y, for each $j \in \{1, 2\}$. Then
$$\gamma := \bigcup_{j=1}^{2} \phi_j(\zeta_j[0,1] \cup \theta_j[0,1])$$
is a Jordan curve in $\mathbb{C}$ such that $\gamma \subset \{x, y\} \cup D_1 \cup D_2$. It follows from Theorem 3.12 and Remark 3.13 that the bounded component G of $\mathbb{C} \setminus \gamma$ is contained in $\mathcal{F}(f)$.

We may assume that $\phi_1(Q) \subset G$, where $Q := \{r\eta : r \in (0,1), \eta \in (\zeta_1, \theta_1)\}$. Since $\lim_{r \to 1} \phi_1(r\eta)$ exists and is neither equal to x nor to y almost everywhere on $\partial \mathbb{D}$ (see [18, Theorem 1.7 and Corollary 6.14]) we see that there exists $\eta \in (\zeta_1, \theta_1)$ such that $w := \lim_{r \to 1} \phi_1(r\eta)$ exists and $w \neq x, y$. Since $w \in \overline{G} \cap \partial D_1$ we conclude that $w \in G \cap \mathcal{J}(f)$. This is a contradiction to $G \subset \mathcal{F}(f)$. □

REFERENCES

[1] L.V. Ahlfors, *Complex analysis* McGraw-Hill, 1966.
[2] I.N. Baker, The domains of normality of an entire function, *Ann. Acad. Sci. Fenn. Ser. A. I. Math.* 1 (1975) 277–283.
[3] I.N. Baker and P. Domínguez, Boundaries of unbounded Fatou components of entire functions, *Ann. Acad. Sci. Fenn. Math.* 24 (1999) 437–464.
[4] I.N. Baker and J. Weinreich, Boundaries which arise in the daynamics of entire functions, *Rev. Roumaine Math. Pures Appl.* 36 (1991) 413-420.
[5] D. Bargmann, Simple proofs of some fundamental properties of the Julia set, *Ergodic Theory Dynam. Systems* 19 (1999) 553–558.
[6] A. F. Beardon, *Iteration of rational functions,* Springer, 1991.
[7] W. Bergweiler, Iteration of meromorphic functions, *Bull. Amer. Math. Soc.* 29 (1993) 151–188.
[8] W. Bergweiler, Invariant domains and singularities, *Math. Proc. Camb. Phil. Soc.* 117 (1995) 525–532.
[9] W. Bergweiler, On the Julia set of analytic self-maps of the punctured plane, *Analysis* 15 (1995) 251–256.
[10] F. Berteloot and J. Duval, Une démonstration directe de la densité des cycles répulsifs dans l'ensemble de Julia, *Complex analysis and geometry* (Paris 1997), Progr. Math. 188 (Birkhäuser, Basel, 2000) 221–222.
[11] P. Bonfert. On iteration in planar domains. *Michigan Math. J.* 44 (1997) 47–68.
[12] L. Carleson and T. W. Gamelin, *Complex dynamics,* Springer, 1993.
[13] E. F. Collingwood and A. J. Lohwater, *The theory of cluster sets,* Cambridge University Press, 1966.
[14] C. Cowen, Iteration and the solution of functional equations for functions analytic in the unit disk, *Trans. Amer. Math. Soc.* 265 (1981) 69–95.
[15] A. Denjoy, Sur l' itération des fonctions analytique, *C. R. Acad. Sci. Paris* 182 (1926) 255–257.
[16] M. Kisaka, On the boundaries of Baker domains, *Science Bulletin of Josai University* 4 (1998) 17–23.
[17] H. König, Conformal conjugacies in Baker domains, *J. London Math. Soc.* 59 (1999) 153–170.
[18] C. Pommerenke, *Boundary behaviour of conformal maps,* Springer, 1992.
[19] W. Schmidt, Fixpunkte im Rand stabiler Gebiete, *Dissertation an der Universität Dortmund* (1995).
[20] W. Schwick, Repelling periodic points in the Julia set, *Bull. London Math. Soc.* 29 (1997) 314–316.
[21] N. Steinmetz, *Rational iteration,* de Gruyter, 1993.
[22] J. Wolff, Sur l' itération des fonctions, *C. R. Acad. Sci. Paris* 182 (1926) 42–43, 200–201.

HSH N FINANCIAL MARKETS ADVISORY AG, KÜTERSTR. 8-12, D-24103 KIEL, GERMANY
   E-mail address: detlef.bargmannfma-ag.com

CONFORMAL AUTOMORPHISMS OF FINITELY CONNECTED REGIONS

A.F. BEARDON AND D.MINDA

In memory of Noel, a wonderful friend, teacher and colleague.

1. INTRODUCTION

This paper is about the group of automorphisms of a finitely connected region in the extended complex plane, and the conformal maps between two such regions. A *region* is a nonempty, open, connected subset of the extended complex plane $\mathbb{C}_\infty$. A region Ω has connectivity c if $\mathbb{C}_\infty \backslash \Omega$ has exactly c components, and we denote the connectivity of Ω by $c(\Omega)$. A surjective homeomorphism $f : \Omega \to \Omega$ is a *conformal automorphism* of Ω if it is analytic in Ω, and it is an *anti-conformal automorphism* of Ω if $\bar{f}$ is analytic in Ω, where $\bar{f}(z) = \overline{f(z)}$. The group of all automorphisms (conformal or anti-conformal) of Ω is denoted by $\mathcal{A}(\Omega)$, and the subgroup of conformal automorphisms is denoted by $\mathcal{A}^+(\Omega)$. Conformally and anti-conformally equivalent regions have isomorphic automorphism groups. More information on automorphism groups can be found in [3], [6], [15], [17] and [20] (see also [12]).

If Ω is simply connected then either $\mathcal{A}^+(\Omega)$ is the group $\mathcal{M}^+$ of conformal Möbius maps, or it is conjugate in $\mathcal{M}^+$ to the group of conformal Euclidean similarities, or isomorphic to the group of conformal hyperbolic motions acting on the open unit disc $\mathbb{D}$, according as Ω is $\mathbb{C}_\infty$, or $\mathbb{C}_\infty$ punctured at one point, or conformally equivalent to $\mathbb{D}$. In the latter case, it is possible for $\mathcal{A}^+(\Omega)$ to contain a non-linear polynomial p (for example, when Ω is a Siegel disc for p that is described in complex dynamics). In this case $\mathcal{A}^+(\Omega)$ contains all of the iterates of p, and also $p^{-1} : \Omega \to \Omega$ which is an algebraic function that does not have a single-valued extension to $\mathbb{C}$. If Ω is doubly connected, then $\mathcal{A}^+(\Omega)$ is isomorphic to the group of Möbius maps that leave an annulus invariant, and here $\mathcal{A}^+(\Omega)$ may contain a rational function of degree at least two (as in the case of a Herman ring). There is a simple characterization of simple and double connectivity in terms of $\mathcal{A}^+(\Omega)$: the region Ω is simply or doubly connected if and only if $\mathcal{A}^+(\Omega)$ contains a subgroup that is isomorphic as a topological group to the multiplicative group $\{z : |z| = 1\}$ (this is equivalent to the result in [4]). In a similar vein, there exists a region Ω of connectivity c with a transitive group of Möbius automorphisms if and only if c is 1 or 2 [3].

The second author was supported by a Taft Faculty Fellowship and wishes to thank the University of Cambridge for its hospitality during his visit November 2004 - April, 2005.

The situation is simpler if $3 \leq c(\Omega) < +\infty$ because then $\mathcal{A}(\Omega)$ is finite. In 1946 Heins obtained sharp upper bounds on the order of $\mathcal{A}^+(\Omega)$ as a function of $c(\Omega)$ when $3 \leq c(\Omega) < +\infty$, namely

$$|\mathcal{A}^+(\Omega)| \leq \begin{cases} 2c(\Omega) & \text{if } c(\Omega) \neq 4, 6, 8, 12, 20; \\ 12 & \text{if } c(\Omega) = 4; \\ 24 & \text{if } c(\Omega) = 6, 8; \\ 60 & \text{if } c(\Omega) = 12, 20 \end{cases}$$

(see [8], [11] and [19, p.427]). There is a relationship between the work of Heins and this paper; in both cases a finitely connected region is reduced to a generalized circular region (a region bounded by points and circles). However, Heins performs a further reduction to regions bounded only by points; this can increase the size of the conformal automorphism group. Our work is different in that we consider both conformal and anti-conformal automorphisms, and are interested in the automorphism group of a region, not just the maximal possible order of the conformal automorphism group as a function of the connectivity. It is straightforward to paraphrase our main results. We consider circular regions (all boundary components circles) and punctured spheres (all boundary components points) with a fixed ordering of the boundary components. A set of easily computed Möbius invariant numbers (inversive distances, cross-ratios or absolute cross-ratios) is associated with the region. Each automorphism of the region induces a permutation of the boundary components that leaves this set of numbers invariant. Our main result is: *conversely, each permutation of the boundary components that leaves the set of numbers invariant (an admissible permutation) corresponds to an automorphism of the region.* This result enables us to determine the isomorphism type of the automorphism group for these types of regions without knowing any automorphisms. At the same time our methods enable us to construct the automorphisms associated with admissible permutations.

We comment briefly on the analogous problem for regions of infinite connectivity and for Riemann surfaces. If Ω is of infinite connectivity then $\mathcal{A}^+(\Omega)$ can be trivial (for example, when $\Omega = \mathbb{C}\backslash\mathbb{Z}^+$), and it can be infinite (for example, when $\Omega = \mathbb{C}\backslash\mathbb{Z}$). Examples of compact Riemann surfaces with trivial automorphism groups are given in [7]. It is known that if G is any countable group then there exists a Riemann surface S such that $\mathcal{A}^+(S)$ is isomorphic to G (see, for example, [9], [13] and [14]). We do not know whether or not the same is true for plane regions.

This paper divides naturally into four parts: Sections 2–5, Sections 6–8, Sections 9–14 and Section 15–16. Section 2 contains a brief discussion of anti-conformal Möbius maps. Section 3 concerns some remarks about automorphisms of generalized circular regions with a clear statement of the connection between automorphisms and permutations of the boundary components. In Section 4 we state and prove one of our main results, namely that given any integer c, where $c \geq 3$, almost all regions of connectivity c have no conformal symmetry, and hence a trivial automorphism group. While this result is part of the folk-lore, we do not know of a published proof. As the concept of 'almost all regions' is not invariant under a change of parameters, we provide a 'natural' set of parameters with respect to which the claim is true. Our proof is elementary apart from an initial application of a result of Koebe on conformal mapping. Koebe's theorem permits us to restrict our attention to regions bounded by c circles, and the rest of the proof is based on the inversive distance between two disjoint circles in $\mathbb{C}_\infty$. The inversive distance is a powerful theoretical and computational tool which plays a similar role for circles as the cross-ratio plays for points. Roughly speaking, a region bounded by c mutually exterior circles has $3c$ real parameters, and a conformal Möbius map has six real parameters; thus, up to Möbius equivalence, a region bounded by c circles has $3c - 6$ real parameters. There are $\binom{c}{2}$ inversive distances between pairs of circles, and although $\binom{c}{2} \geq 3c - 6$, the inversive distances are more convenient parameters because they are invariant under Möbius maps. Although the inversive distance is a geometric quantity we have tried to avoid geometry in this part of this paper. In Section 5 we indicate why it is sufficient to consider regions which are bounded only by circles, or only by points, and compute the automorphism groups for several exceptional cases not covered by our general method.

The second part of the paper (Sections 6–8) deals with regions Ω which are bounded by circles $\Gamma_1, \ldots, \Gamma_c$, where $c \geq 3$. In this part we prove Theorem 4.1 which gives a necessary and sufficient condition, in terms of inversive distances, for two circular regions to be equivalent by a Möbius map. To each circular region bounded by c circles we associate a set of $\binom{c}{2}$ numbers d_{ij}, $1 \leq i < j \leq c$, that are the inversive distances between pairs of the boundary circles. The d_{ij} tell us whether or not the region possesses automorphisms, and they enable us to identify $\mathcal{A}(\Omega)$ with a subgroup $S(\Omega)$ of the group S_c of permutations of $\{1, \ldots, c\}$. This identification is actually *an algorithm to determine the isomorphism type of $\mathcal{A}(\Omega)$ of any circular region with $c(\Omega) \geq 3$ and, moreover, this can be done without knowing any of the automorphisms.* An effective method for computing the automorphisms after the group $S(\Omega)$ of permutations has been determined is indicated. We use this algorithm to determine, in Sections 7 and 8, the conformal automorphism groups of circular regions of connectivity three and four, respectively. To illustrate the nature of our results in this part of the paper we state two results now. First, this method yields the following result (and much more).

Theorem 1.1. *Suppose that the region Ω is bounded by $c \geq 3$ Euclidean circles of the same radii, with centres $z_1, \ldots, z_c$. If the numbers $|z_i - z_j|$, where $i \neq j$, are distinct, then $\mathcal{A}^+(\Omega)$ is the trivial group.*

Next, consider the region $\Omega(r)$ bounded by the four circles

(1.1)
$$\{x+iy : x = 0\}, \quad \{z : |z - (7+2i)| = 1\},$$
$$\{z : |z - (7-2i)| = 1\}, \quad \{z : |z - 12| = r\},$$

where r is chosen so that the circles are exterior to each other (see Figure 1).

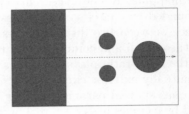

Figure 1

Obviously $z \mapsto \bar{z}$ is an anti-conformal automorphism of $\Omega(r)$, but for which values of r does $\Omega(r)$ possess conformal symmetry? Our results enable us to answer such questions and to find the automorphisms of $\Omega(r)$ for each value of r.

We shall see that (in a sense to be described later) the set of regions that have at least one boundary component that is a *puncture* (that is, a point) has measure zero. Nevertheless, there are interesting results for regions all of whose boundary components are punctures, and we call such regions *punctured spheres*. The third part of the paper (Sections 9–14) is concerned exclusively with the automorphism groups of punctured spheres. It is of interest to note that, within this restricted set of regions, the three and four punctured spheres *always* have a non-trivial automorphism group, unlike circular regions of connectivity three and four. The conformal automorphism group of a three-punctured sphere is (trivially) always the permutation group S_3. A sphere with four punctures always has non-trivial conformal symmetry since any set of four distinct points can be mapped by a Möbius transformation to $\{1, -1, w, -w\}$ for some w, and the new region is invariant under $z \mapsto -z$. Better still, we shall see later (by geometry and by algebra) that the conformal automorphism group of a four–punctured sphere always contains a copy of the Klein 4-group. The techniques used in this part of the paper are based on the cross-ratio and absolute cross-ratio of four points, and this plays an analogous role to that played by the inversive distance for regions with circular boundary components. The properties of the cross-ratio that we need are presented (carefully) in Section 10. Section 11 contains another of our main results, Theorem 11.1, which tells us, in terms of cross-ratios, when two punctured spheres are conformally equivalent by a Möbius map. This result is illustrated for four–punctured spheres in Section 12. For $c \geq 5$ we prove the generic sphere with c punctures has trivial conformal automorphism group in Section 13. In Section 14 the general case of Möbius maps and absolute cross-ratios is considered.

The problem of determining automorphism groups originated in complex analysis, but once we have used Koebe's theorem and passed to regions bounded by circles, the problem lies in inversive geometry. The fourth part of the paper (Sections 15–16) is devoted to a closer look at the role played by geometry. The numbers 12, 24 and 60 in Heins' result suggest that there is a close connection with the five Platonic solids. This is indeed so, and most of this paper (and Heins' result) is underpinned by the following well-known result (which we shall discuss in Section 16).

Theorem A. *Any finite group of conformal Möbius maps is isomorphic to one of the groups C_n, D_n, A_4, S_4 or A_5.*

Here, C_n is the cyclic group of order n, D_n is the dihedral group of order n (so that D_4 is the Klein 4-group, and $D_6 = S_3$), and S_n and A_n are the permutation and alternating groups, respectively, of $\{1, 2, \ldots, n\}$. It is well-known that A_4 (of order 12) is the conformal symmetry group of a regular tetrahedron, S_4 (of order 24) is the conformal symmetry group of a cube and a regular octahedron, and A_5 (of order 60) is the conformal symmetry group of a regular dodecahedron and a regular icosahedron. A study of the action of these groups on $\mathbb{C}_\infty$ leads to the next result.

Theorem 1.2. *Let Ω be any region for which $c(\Omega)$ is odd, at least three, and finite. Then $\mathcal{A}^+(\Omega)$ is either cyclic or dihedral.*

2. Möbius maps

We shall use the term *Möbius transformation* (or map) to denote any map of either of the forms $g(z)$ or $g(\bar{z})$, where $g(z) = (az+b)/(cz+d)$, $ad-bc \neq 0$. Often, this term is reserved for the map g, and here we shall refer to g as a *conformal Möbius map*. The maps of the form $g(\bar{z})$ are the *anti-conformal Möbius maps*. As usual, we regard a Euclidean straight line as a circle through ∞, and we use 'reflection' to denote both the inversion across a Euclidean circle, and the reflection across a straight line. Every Möbius map is the composition of at most four reflections. We let $\mathcal{M}$ be the Möbius group (the group generated by reflections), and $\mathcal{M}^+$ be the subgroup of conformal Möbius maps. Throughout, I denotes the identity map. For brevity we say two regions are *Möbius equivalent* if there is a Möbius map that carries one region onto the other.

Although the classification of conformal Möbius maps is well known, that of anti-conformal Möbius maps is not. As for conformal Möbius maps, we are content to classify the anti-conformal maps up to conjugacy. Let f be an anti-conformal Möbius map, and write $f(z) = g(\bar{z})$, where g is a conformal Möbius map. Let $N(f)$ be the number of fixed points of f. Suppose first that $N(f) \geq 3$. Select three fixed points of f, let Γ be the circle through these points, and let r be the reflection across Γ. Then fr is a conformal Möbius map with at least three fixed points, so that $fr = I$, or $f = r$. Thus *an anti-conformal Möbius map with at least three fixed points is a reflection across a circle.*

Now suppose that $N(f) = 2$. Then we may assume (by conjugation) that f fixes 0 and ∞. Thus g also fixes 0 and ∞, and so $f(z) = k\bar{z}$ for some nonzero k. Further conjugations (by a rotation about 0 and then, if necessary, by $z \mapsto 1/z$) allow us to assume that $k > 1$; thus *an anti-conformal Möbius map with exactly two fixed points is conjugate to $f(z) = k\bar{z}$, where $k > 1$*; this is called a *stretch reflection*.

Next, suppose that $N(f) = 1$; then we may assume that f fixes ∞, and only ∞. In this case $g(\infty) = \infty$, so that $f(z) = a\bar{z} + b$, say. By a further conjugation (by a rotation about 0) we may assume that $a > 0$. If the second iterate f^2 of f has a finite fixed point ζ, then f maps the segment $[\zeta, f(\zeta)]$ onto itself because f is affine, so that f has a fixed point in this segment, and this is false. Thus f^2 has no finite fixed point and, as $f^2(z) = a^2 z + a\bar{b} + b$, this implies that $a = 1$, and then (using $a = 1$) that b is not purely imaginary. Thus $f(z) = \bar{z} + u + iv$, where u and v are real and $u \neq 0$. Let $g(z) = uz + iv/2$. Then $g^{-1}fg(z) = \bar{z} + 1$; thus *an anti-conformal Möbius transformation with exactly one fixed point is conjugate to the glide reflection $z \mapsto \bar{z} + 1$.*

Finally, suppose that $N(f) = 0$. As f^2 is a conformal Möbius map it has a fixed point, say w, so that $f(w) \neq w$, and f interchanges w and $f(w)$. We may suppose (by passing to a conjugate) that f, and hence g, interchanges 0 and ∞; thus $f(z) = k/\bar{z}$, for some k. By a further conjugation (by a rotation about the origin) we may suppose that $|k| = 1$. Also, $k \neq 1$ (else $N(f) = \infty$). Thus f is conjugate to $z \mapsto k/\bar{z}$, where $|k| = 1$ and $k \neq 1$. Such an f is called a *rotary reflection*.

The anti-conformal Möbius maps that are involutions (of order two) are of interest. As the order of f is invariant under conjugation, we see that *an anti-conformal Möbius map has order two if and only if either f is a reflection across a circle, or f is conjugate to the map $z \mapsto -1/\bar{z}$* (which is inversion in $\{|z| = 1\}$ followed by a rotation of angle π about the origin). We remark that the map $z \mapsto -1/\bar{z}$ corresponds under stereographic projection to the antipodal map of the unit sphere.

3. Reduction to generalized circular regions

A plane region of finite connectivity c is a *generalized circular region* if it is obtained from $\mathbb{C}_\infty$ by removing c pairwise disjoint compact discs $K_1, \ldots, K_c$, some of which may be points. If each K_j has positive radius, we say that the region is a *circular region*; if all of the K_j are points, we say that the region is a *c-punctured sphere*. It is a fundamental fact that *a conformal map between two circular regions of the same finite connectivity is a Möbius* *transformation*; see [5, p.102], and [19, p.426]. This is proved for circular regions of finite connectivity, but it obviously extends to generalized circular regions because punctures are removable singularities of conformal maps. It is also true that an anti-conformal map f between two generalized circular regions is an anti-conformal Möbius map for $\bar{f}$ is a conformal map between generalized circular regions and so is Möbius. This property of generalized circular regions is fundamental to this paper, and we now discuss some of its important consequences.

First, it implies that *two generalized circular regions of finite connectivity are conformally equivalent if and only if they are conformally Möbius equivalent*. This implies that the automorphism group of a circular region of finite connectivity is a group of Möbius maps. Now a theorem of Koebe ([5, pp.100-106], [8, p.279-283], and [19, p.424]) states that a plane region of finite connectivity is conformally equivalent to a generalized circular region. As a consequence, it suffices to study the group of (Möbius) automorphisms of a generalized circular region of finite connectivity. Moreover, because any conformal map between two such regions is a Möbius map, it follows that if Ω and Ω' are two conformally equivalent generalized circular regions of finite connectivity, then $\mathcal{A}^+(\Omega)$ and $\mathcal{A}^+(\Omega')$ are conjugate subgroups of $\mathcal{M}^+$ (and similarly for the full automorphism groups). This geometric statement says more than that they are isomorphic as abstract groups.

The passage to generalized circular regions via Koebe's Theorem implies that every automorphism becomes a Möbius map. Conversely, if every conformal automorphism of Ω is Möbius, and if $c(\Omega) \leq 2$, then Ω is a generalized circular region ([3]). This result fails for $c \geq 3$. For instance, let $\omega = \exp(2\pi i/c)$ and let Ω be the extended plane with the c radial slits $[\frac{1}{2}\omega^j, 2\omega^j]$, $j = 1, \ldots, c$, removed. Ω is not a generalized circular region. Note that $f(z) = \omega z$ and $g(z) = 1/z$ are conformal automorphisms of order c and two, respectively, and together generate a subgroup $\langle f, g \rangle$ of $\mathcal{A}^+(\Omega)$ of order $2c$. For $c \neq 4, 6, 8, 12, 20$, Heins' bounds imply $\langle f, g \rangle = \mathcal{A}^+(\Omega)$. Thus, for $c \neq 4, 6, 8, 12, 20$, Ω has the maximal number of conformal Möbius automorphisms, yet it is not a generalized circular region. Recently, He and Schramm [10] have extended Koebe's result to the case of regions of countable connectivity. Maskit [15] has shown that every region is conformally equivalent to a region whose conformal automorphism group is a subgroup of $\mathcal{M}^+$.

When Ω is a generalized circular region of finite connectivity we can view the action of the group $\mathcal{A}(\Omega)$ in different ways. First, it acts on the region Ω in the obvious way. However, the observation that an automorphism of a generalized circular region of finite connectivity is Möbius shows that *each automorphism of Ω acts on $\mathbb{C}_\infty$, and not only on Ω*. Now suppose that

(3.1) $$\Omega = \mathbb{C}_\infty \backslash (K_1 \cup \cdots \cup K_c),$$

where the $K_1, \ldots, K_c$ are pairwise disjoint compact Euclidean discs or points. As any automorphism of Ω acts on $\mathbb{C}_\infty$, it also acts as a permutation of $\{K_1, \ldots, K_c\}$. Thus $\mathcal{A}(\Omega)$ provides a group action that acts on the finite set $\{K_1, \ldots, K_c\}$. Of course, an automorphism of a general finitely connected region does not necessarily extend to the boundary, but it still induces (from topological arguments) an action on the finite number of boundary components. Nevertheless, many of the arguments that follow will only apply to generalized circular regions.

Lemma 3.1. *The group $\mathcal{A}^+(\Omega)$ is of index 1 or 2 in $\mathcal{A}(\Omega)$, and so $\mathcal{A}(\Omega)$ is a finite group.*

Proof. The mapping of $\mathcal{A}(\Omega)$ into the multiplicative group $\{1, -1\} \cong C_2$ that assigns to $f \in \mathcal{A}(\Omega)$ the number $1, -1$ according as f is orientation preserving, or orientation reversing, respectively, is a group homomorphism with kernel $\mathcal{A}^+(\Omega)$. □

Example 8.2 gives a region for which $\mathcal{A}^+(\Omega) = \mathcal{A}(\Omega)$ while $\mathcal{A}^+(\Omega)$ is a proper normal subgroup of $\mathcal{A}(\Omega)$ for triply connected circular regions (see Section 7), so both cases in Lemma 3.1 can occur.

There is a simpler, but equivalent, way of expressing the action of $\mathcal{A}(\Omega)$ on $\{K_1, \ldots, K_c\}$ by referring to the action on the set $\mathbb{Z}_c = \{1, \ldots, c\}$ of suffices of the K_j as a subgroup of S_c, the permutation group of $\mathbb{Z}_c$. For each automorphism f we define a map $f_* : \mathbb{Z}_c \to \mathbb{Z}_c$ by $f(K_j) = K_{f_*(j)}$. Now f_* is injective since $f_*(s) = f_*(t)$ implies that $f(K_s) = f(K_t)$ and hence (as f^{-1} exists) that $s = t$. Moreover, it is trivial to check that $(fg)_* = f_* g_*$; thus the function $\Theta_* : \mathcal{A}(\Omega) \to S_c$ given by $\Theta_*(f) = f_*$ is a homomorphism of $\mathcal{A}(\Omega)$ into S_c. Let $S(\Omega), S^+(\Omega)$ denote the image of $\mathcal{A}(\Omega), \mathcal{A}^+(\Omega)$ under this homomorphism. These are subgroups of S_c with $S^+(\Omega) \subset S(\Omega)$.

The kernel $\mathcal{K}(\Omega) = \{f \in \mathcal{A}(\Omega) : f_* = \iota\} = \{f \in \mathcal{A}(\Omega) : f(K_j) = K_j, j = 1, \ldots, c\}$ of this homomorphism is particularly simple when $c \geq 3$. First, $\mathcal{K}(\Omega) \cap \mathcal{A}^+(\Omega)$, the kernel of Θ_* restricted to $\mathcal{A}^+(\Omega)$, contains only the identity I. To see this note that if $f \in \mathcal{K}(\Omega)$ satisfies $f(K_j) = K_j$ for each j, then f has a fixed point in each K_j. If, in addition, $f \in \mathcal{A}^+(\Omega)$, then f is a conformal Möbius map with $c \geq 3$ fixed points and so $f = I$. This is equivalent to the statement that the action of $\mathcal{A}^+(\Omega)$ on $\{K_1, \ldots, K_c\}$ is faithful (that is, each conformal automorphism of Ω is completely determined by its action on $\{K_1, \ldots, K_c\}$). (We note that the action of $\mathcal{A}^+(\Omega)$ on $\{K_1, \ldots, K_c\}$ is not faithful when $c(\Omega) \leq 2$.) This also shows that the restriction of Θ_* to $\mathcal{A}^+(\Omega)$ is always an isomorphism onto $S^+(\Omega)$. Now suppose that f and g are anticonformal automorphisms in $\mathcal{K}(\Omega)$. As $\mathcal{K}(\Omega)$ does not contain any non-trivial conformal automorphisms, we see that $fg = I$. If we take $g = f$ we deduce $f^2 = I$, so f is an anti-conformal involution. Then, returning to the case of a general g, we have $g = f^{-1} = f$. We conclude that $\mathcal{K}(\Omega)$ is either trivial, or is of the form $\{I, f\}$, where f is an anti-conformal involution. If $\mathcal{K}(\Omega) = \{I\}$, then Θ_* is an isomorphism of $\mathcal{A}(\Omega)$ onto $S(\Omega)$.

Next, we turn to the case in which $\mathcal{K}(\Omega) = \{I, f\}$, where f is an anti-conformal involution. An anti-conformal involution is either reflection across a circle or conjugate to $z \mapsto -1/\bar{z}$. It is easy to see that the only invariant circles under the map $z \mapsto -1/\bar{z}$ are the lines through the origin, and the unit circle, so $\mathcal{K}(\Omega)$ cannot contain an involution that is a rotary reflection. Thus, if $\mathcal{K}(\Omega)$ is not trivial, then $\mathcal{K}(\Omega) = \{I, r\}$, where r is a reflection across some circle Γ. This case has a simple geometric interpretation. Since $r \in \mathcal{K}(\Omega)$, $r(K_j) = K_j$ for each j. We conclude that each K_j must be symmetric about Γ. In this circumstance we say that Ω is *strongly symmetric*, that Γ is the *axis of symmetry* of Ω, and that r is the *canonical reflection* of Ω. Conversely, if Ω is strongly symmetric, then $r_* = \iota$ and so $\mathcal{K}(\Omega) = \{I, r\}$. For strongly symmetry region and for any $f \in \mathcal{A}^+(\Omega)$, $rfr \in \mathcal{A}^+(\Omega)$ and $(rfr)_* = f_*$, so $rfr = f$, or $rf = fr$ since Θ_* is injective on $\mathcal{A}^+(\Omega)$. Thus, the canonical reflection commutes with every conformal automorphism when a region is strongly symmetric. Also, $S^+(\Omega) = S(\Omega)$ when Ω is strongly symmetric. To see this, note that for any $\rho \in S(\Omega)$ there exists $f \in \mathcal{A}(\Omega)$ with $f_* = \rho$. Now, $(rf)_* = f_*$ and one of rf and f is conformal. The same type of argument shows Θ_* is a 2-to-1 map of $\mathcal{A}(\Omega)$ onto $S(\Omega)$. Our final comment is that

$$\mathcal{A}(\Omega) = \mathcal{A}^+(\Omega)\mathcal{K}(\Omega) = \{fg : f \in \mathcal{A}^+(\Omega) \text{ and } g \in \mathcal{K}(\Omega)\}$$

and the representation of every element of $\mathcal{A}(\Omega)$ in the form fg with $f \in \mathcal{A}^+(\Omega)$ and $g \in \mathcal{K}(\Omega)$ is unique since $\mathcal{A}^+(\Omega) \cap \mathcal{K}(\Omega) = \{I\}$. Likewise, $\mathcal{A}(\Omega) = \mathcal{K}(\Omega)\mathcal{A}^+(\Omega)$. In group theoretic terms $\mathcal{A}(\Omega)$ is the *internal direct product* of $\mathcal{A}^+(\Omega)$ and $\mathcal{K}(\Omega)$. The mapping $fg \mapsto (f, g)$ defines an isomorphism of $\mathcal{A}(\Omega)$ onto the direct product $\mathcal{A}^+(\Omega) \times \mathcal{K}(\Omega)$, so $\mathcal{A}(\Omega)$ is isomorphic to $\mathcal{A}^+(\Omega) \times C_2$. The general question of when $\mathcal{A}(\Omega)$ can be written as the internal direct product of $\mathcal{A}^+(\Omega)$ and a subgroup isomorphic to C_2 is discussed in the final section of the paper.

These facts are summarized in the following theorem.

Theorem 3.2. *Suppose Ω is a generalized circular region with connectivity $c(\Omega) \geq 3$, Θ_* is the homomorphism of $\mathcal{A}(\Omega)$ onto $S(\Omega)$ defined by $\Theta_*(f) = f_*$ and $\mathcal{K}(\Omega)$ is the kernel of Θ_*.*
(a) *The restriction of Θ_* to $\mathcal{A}^+(\Omega)$ is injective. Therefore, $\mathcal{A}^+(\Omega)$ is isomorphic to the subgroup $S^+(\Omega)$ of S_c, and so is one of the groups listed in Theorem A.*
(b) *The following are equivalent:*
   (i) *Θ_* is injective,*
   (ii) *$\mathcal{K}(\Omega) = \{I\}$,*
   (iii) *$\mathcal{A}(\Omega)$ is isomorphic to $S(\Omega)$,*
   (iv) *Ω is not strongly symmetric.*

(c) *The following are equivalent:*
  (i) $\mathcal{K}(\Omega) = \{I, r\}$, *where r is reflection across some circle,*
  (ii) Ω *is strongly symmetric.*
Moreover, when Ω is strongly symmetric, Θ_* is an isomorphism of $\mathcal{A}^+(\Omega)$ onto $S(\Omega) = S^+(\Omega)$ and $\mathcal{A}(\Omega) = \mathcal{A}^+(\Omega)\mathcal{K}(\Omega) = \mathcal{K}(\Omega)\mathcal{A}^+(\Omega)$.

This representation of the automorphisms of a generalized circular region of finite connectivity $c \geq 3$ by permutations in S_c is the basis of most of the results in this paper. At the moment, it is little more than notation, but later we shall show how to find the image groups $S(\Omega)$ and $S^+(\Omega)$ *without finding the Möbius automorphisms groups $\mathcal{A}(\Omega)$ and $\mathcal{A}^+(\Omega)$ explicitly*. Then we can use the groups $S(\Omega)$ and $S^+(\Omega)$ to construct the automorphisms of Ω.

Together Lemma 3.1 and Theorem 3.2 show that the index of $\mathcal{A}^+(\Omega)$ in $\mathcal{A}(\Omega)$ is either one or two, and $\mathcal{K}(\Omega)$ has order one or two. In theory this gives four possibilities. If the index is one, then Theorem 3.2(c) implies $\mathcal{K}(\Omega)$ must be trivial, so one of the four cases cannot occur. The case of index two and $\mathcal{K}(\Omega)$ trivial arises in Example 8.3. The combination of index two and $|\mathcal{K}(\Omega)| = 2$ occurs for any triply connected circular region; see Section 7.

4. CONFORMALLY TRIVIAL GENERALIZED CIRCULAR REGIONS

To each pair of disjoint circles Γ_1 and Γ_2 in $\mathbb{C}_\infty$ we can associate a Möbius invariant *inversive distance* (Γ_1, Γ_2), and this plays a fundamental role in our argument. For a systematic treatment of the inversive distance in n-dimensions, see [2, pp.28-30], but the definitions given here suffice for our purposes. If the Γ_j, $j = 1, 2$, are given by $|z - a_j| = r_j$, respectively, then

$$(4.1) \qquad (\Gamma_1, \Gamma_2) = \left| \frac{r_1^2 + r_2^2 - |a_1 - a_2|^2}{2r_1 r_2} \right|,$$

while if Γ_2 is a Euclidean line, then

$$(\Gamma_1, \Gamma_2) = \frac{\text{dist}(a_1, \Gamma_2)}{r_1},$$

where $\text{dist}(a_1, \Gamma_2)$ denotes the Euclidean distance from a_1 to Γ_2. Notice that if the Γ_j are disjoint circles that are exterior to each other, then

$$(\Gamma_1, \Gamma_2) = \frac{|a_1 - a_2|^2 - r_1^2 - r_2^2}{2r_1 r_2} > \frac{(r_1 + r_2)^2 - r_1^2 - r_2^2}{2r_1 r_2} = 1,$$

and, as the inversive distance is Möbius invariant, this inequality holds for all pairs of disjoint circles. For such pairs it is natural to write $(\Gamma_1, \Gamma_2) = \cosh \delta$, and we shall see later that δ has a simple interpretation as a hyperbolic distance. In view of Theorem 3.2 it is of interest to know whether a given circular region is strongly symmetric or not. Now the two circles Γ_1 and Γ_2 are orthogonal if and only if $|a_1 - a_2|^2 = r_1^2 + r_2^2$, and this is precisely the condition that $(\Gamma_1, \Gamma_2) = 0$. There is also an analytic interpretation of the inversive distance. If Γ_1 and Γ_2 are concentric circles, then $(\Gamma_1, \Gamma_2) = \cosh(2\pi m)$, where m is the modulus of the annulus bounded by Γ_1 and Γ_2. As both the inversive distance and the modulus are invariant under a Möbius map, this is true for all disjoint pairs of circles.

We now state a result which implies that the inversive distances between all pairs of boundary circles of a circular region determine the automorphism group of the region. We shall give a proof later, but we only need a special (and simple) case here. As we wish this first part of the paper to be self-contained, we give a (short) proof of the special case now. This special case contains Theorem 1.1 since in Theorem 1.1 the inversive distance between the circles with centres z_i and z_j is $(|z_i - z_j|^2/2r^2) - 1$, where r is the common value of the radii of the circles.

Theorem 4.1. *Suppose that Ω and Ω' are circular regions bounded by circles $\Gamma_1, \ldots, \Gamma_c$ and $\Gamma'_1, \ldots, \Gamma'_c$, respectively, where $c \geq 2$. Then there is a Möbius transformation f with $f(\Omega) = \Omega'$, and $f(\Gamma_j) = \Gamma'_j$, $1 \leq j \leq c$, if and only if $(\Gamma_j, \Gamma_k) = (\Gamma'_j, \Gamma'_k)$ for all j and k with $1 \leq j < k \leq c$. For $c \geq 3$ f is unique when Ω is not strongly symmetric and unique up to the pre-composition with the canonical reflection r of Ω when Ω is strongly symmetric.*

Corollary 4.2. *Let Ω be a circular region in $\mathbb{C}_\infty$ that is bounded by c mutually exterior circles $\Gamma_1, \ldots, \Gamma_c$, where $c \geq 3$. Suppose that the $\binom{c}{2}$ inversive distances (Γ_i, Γ_j), $1 \leq i < j \leq c$, are distinct.*
(a) If Ω is not strongly symmetric, then $\mathcal{A}(\Omega)$ is the trivial group.
(b) If Ω is strongly symmetric, then $\mathcal{A}^+(\Omega)$ is the trivial group, and $\mathcal{A}(\Omega) = \{I, r\}$, where r is the canonical reflection.

Proof of Corollary 4.2 Suppose that $f \in \mathcal{A}(\Omega)$. The Möbius invariance of the inversive distance implies that for each distinct i and j, $(\Gamma_i, \Gamma_j) = (f(\Gamma_i), f(\Gamma_j)) = (\Gamma_{f_*(i)}, \Gamma_{f_*(j)})$. Because the numbers (Γ_i, Γ_j) are distinct, we deduce that for all distinct i and j, $\{i, j\} = \{f_*(i), f_*(j)\}$. Given j, we can (as $c \geq 3$) choose k and ℓ such that j, k and ℓ are distinct; then

$$\{f_*(j)\} = \{f_*(j), f_*(k)\} \cap \{f_*(j), f_*(\ell)\} = \{j, k\} \cap \{j, \ell\} = \{j\}.$$

We deduce that f_* is the identity permutation, and the result now follows from Theorem 3.2. □

We shall now parametrize the space of generalized circular regions of fixed connectivity c. Given any such region Ω, we can stereographically project Ω onto the unit sphere $\mathbb{S}$ in $\mathbb{R}^3$. Each component of $\mathbb{C}_\infty \backslash \Omega$ is mapped to a (spherical) disc on $\mathbb{S}$, and this disc has a unique spherical centre and a unique chordal radius. It follows that if we order the components of $\mathbb{C}_\infty \backslash \Omega$ in some way, say $K_1, \ldots, K_c$, then we can parametrize the circular regions of connectivity c on $\mathbb{S}$ by a vector $(\gamma'_1, \ldots, \gamma'_c, r'_1, \ldots, r'_c)$ in $\mathbb{S}^c \times [0, \pi)^c$, where K_j is projected to the disc on $\mathbb{S}$ with spherical centre γ'_j and chordal radius r'_j. Of course, these parameters depend on our labelling of the components of $\mathbb{C}_\infty \backslash \Omega$

and, strictly speaking, we should pass to a quotient of $\mathbb{S}^c \times [0,\pi)^c$ by the appropriate permutation group which identifies two parameter vectors if and only if they represent different labellings of the same region. An equivalent (but inelegant) approach would be to take $c!$ copies of each Ω (one for each labelling) and work with the union of these copies of $\mathbb{S}^c \times [0,\pi)^c$. In any event, we shall ignore this step. It is clear that the set $\mathcal{P}$ of parameters that correspond to a circular region of connectivity c is an open subset of $\mathbb{S}^c \times [0,\pi)^c$. As this space has a natural finite product measure, we may regard $\mathcal{P}$ as a probability space. We denote the associated probability measure by μ_0, and we can now state and prove our version of the folk-lore theorem.

Theorem 4.3. *Suppose that $c \geq 3$. The set of parameters in $\mathcal{P}$ that correspond to generalized circular regions that have a non-trivial conformal automorphism group has μ_0-measure zero.*

Proof. Observe that Ω is bounded by a Euclidean line if and only if $\infty \in \partial\Omega$, and this is so if and only if for some j, $||\gamma'_j - (0,0,1)||^2 = (r'_j)^2$. As this is an algebraic constraint on the parameters in $\mathcal{P}$, the set of such Ω has μ_0-measure zero. Similarly, the j-th boundary component of Ω is a point if and only if $r'_j = 0$; the set of parameters that satisfy these constraints also has μ_0-measure zero. Now let $\mathcal{P}_0$ be the set of parameters that correspond to Euclidean circular regions (bounded by circles of positive radii). Then $\mu_0(\mathcal{P}\backslash\mathcal{P}_0) = 0$, so that from now on we can focus on the probability space $(\mathcal{P}_0, \mu_0)$.

Now let $\mathcal{P}_1$ be the set of parameters that correspond to the circular regions such that the $\binom{c}{2}$ numbers (Γ_i, Γ_j), $1 \leq i < j \leq c$ are distinct. By Corollary 4.2, the regions corresponding to parameters in $\mathcal{P}_1$ have a trivial conformal automorphism group and, as $\mathcal{P}_1 \subset \mathcal{P}_0$, it suffices to show that $\mu_0(\mathcal{P}_0\backslash\mathcal{P}_1) = 0$. Now suppose that a region Ω has its parameters in $\mathcal{P}_0\backslash\mathcal{P}_1$; then Ω is bounded by c Euclidean circles, each of positive radius, and some pair of inversive distances (Γ_i, Γ_j) must be equal. The idea now is to show that this equality of two inversive distances imposes an algebraic constraint on the parameters. As the set of parameters that satisfy a given algebraic constraint has μ_0-measure zero, and as there are only a finite number of such constraints to be satisfied, the proof will be complete. We now sketch the details of this argument.

Let Γ' be a circle on $\mathbb{S}$ with spherical centre $(\alpha_1, \alpha_2, \alpha_3)$ (so that $\alpha_1^2 + \alpha_2^2 + \alpha_3^2 = 1$), and chordal radius ρ (so that $0 \leq \rho \leq 2$). Then $\Gamma = \Pi \cap \mathbb{S}$, where Π is the plane in $\mathbb{R}^3$ given by

$$\alpha_1 X_1 + \alpha_2 X_2 + \alpha_3 X_3 = (2-\rho^2)/2.$$

Now let Γ be the circle in $\mathbb{C}_\infty$ that stereographically projects onto Γ'. Thus if $z \in \Gamma$ and $z = x + iy$, then

$$\alpha_1\left(\frac{2x}{|z|^2+1}\right) + \alpha_2\left(\frac{2y}{|z|^2+1}\right) + \alpha_3\left(\frac{|z|^2-1}{|z|^2+1}\right) = \frac{2-\rho^2}{2}.$$

A little algebra now shows that (in general) Γ has equation $|z-a| = r$, where

(4.2) $$a = \frac{-2(\alpha_1 + i\alpha_2)}{2\alpha_3 + \rho^2 - 2}, \quad r = \frac{\rho\sqrt{4-\rho^2}}{|2\alpha_3 + \rho^2 - 2|}.$$

In the exceptional case, $2\alpha_3 + \rho^2 - 2 = 0$, and this is so if and only if $(0,0,1) \in \Pi$.

We return to the proof of Theorem 4.2. The discussion above shows that if Ω is such that $(\Gamma_i, \Gamma_j) = (\Gamma_p, \Gamma_q)$, where $\{i,j\} \neq \{p,q\}$ then, from (4.2), the constraint $(\Gamma_i, \Gamma_j)^2 = (\Gamma_p, \Gamma_q)^2$ is a non-trivial algebraic constraint on the parameters. As there are at most $\binom{c}{2}$ such constraints, the proof is complete. □

5. THE REDUCTION TO CIRCULAR REGIONS AND PUNCTURED SPHERES

Up to this point we have considered generalized circular regions. We now indicate why in the remainder of the paper we can restrict our attention to circular regions and to punctured spheres. Suppose the generalized circular region Ω is bounded by k circles $\Gamma_1, \ldots, \Gamma_k$ and p points $\gamma_{k+1}, \ldots, \gamma_c$, where $1 \leq k < c$, $p = c - k$ and $c \geq 3$. Let Ω_k be the region bounded by the circles, and let Ω_p be the complement of the points. Since a conformal automorphism of Ω must permute the circles among themselves, and also the points among themselves, we see that $\mathcal{A}^+(\Omega) = \mathcal{A}^+(\Omega_k) \cap \mathcal{A}^+(\Omega_p)$. The groups $\mathcal{A}^+(\Omega_k)$ and $\mathcal{A}^+(\Omega_p)$ are finite when k and p are at least three. If $k \geq 3$, or $p \geq 3$, we can use the methods that we develop later to determine either $\mathcal{A}^+(\Omega_k)$ (when $k \geq 3$) and $\mathcal{A}^+(\Omega_p)$ (when $p \geq 3$). We can then check whether the automorphisms in either of these groups permute the remaining points or circles. The cases not covered by this procedure are $(k,p) = (1,2), (2,1), (2,2)$ and we now discuss these exceptional cases.

The case $(k,p) = (1,2)$ (one disc and two punctures). The complement of the disc is the hyperbolic plane, and the two punctures are points in this plane. Clearly, the only non-trivial automorphism is a hyperbolic isometry that is a rotation of order two.

The case $(k,p) = (2,1)$ (two discs and one puncture). We may take the puncture at ∞, and then look at the Euclidean similarities that interchange two discs. In this case it is clear that the only non-trivial conformal automorphism is a Euclidean isometry that is a rotation of order two.

The case $(k,p) = (2,2)$ (two discs and two punctures). We can assume that Γ_1 is $|z| = 1/R$, Γ_2 is $|z| = R$, and $\Omega = A_R \backslash \{\gamma_3, \gamma_4\}$, where $A_R = \{z : 1/R < |z| < R\}$ and $R > 1$. Each conformal automorphism of Ω is also a conformal automorphism of A_R, and these latter automorphisms are rotations about the origin, and maps of the form $z \mapsto e^{i\theta}/z$. Each conformal automorphism f of Ω induces a permutation f_* of $\{1,2,3,4\}$ that leaves both sets $\{1,2\}$, $\{3,4\}$ invariant; thus the image $S^+(\Omega)$ of $\mathcal{A}^+(\Omega)$ (see Theorem 3.2) is a subgroup

of the Klein 4-group $\{\iota, (1\,2), (3\,4), (1\,2)(3\,4)\}$, where (here and elsewhere) ι is the identity permutation. Now it is easy to see that
(a) there exists f in $\mathcal{A}^+(\Omega)$ with $f_* = (3\,4)$ if and only if $\gamma_3 = -\gamma_4$ (in which case, $f(z) = -z$):
(b) there exists f in $\mathcal{A}^+(\Omega)$ with $f_* = (1\,2)(3\,4)$ if and only if $|\gamma_3||\gamma_4| = 1$ (in which case $f(z) = \gamma_3\gamma_4/z$);
(c) there exists f in $\mathcal{A}^+(\Omega)$ with $f_* = (1\,2)$ if and only if $\gamma_3 = -\gamma_4$ and $|\gamma_3| = 1$ (in which case $f(z) = \gamma_3\gamma_4/z$).

Given these (we omit the proofs), we see that $\mathcal{A}^+(\Omega)$ is isomorphic to one of the four groups $\{\iota\}$, $\{\iota, (3\,4)\}$, $\{\iota, (1\,2)(3\,4)\}$ and $\{\iota, (1\,2), (3\,4), (1\,2)(3\,4)\}$. The corresponding groups $\mathcal{A}^+(\Omega)$ are of the form $\{I\}$, $\{I, -z\}$, $\{I, e^{i\theta}/z\}$ and $\{I, -z, e^{i\theta}/z, -e^{i\theta}/z\}$.

6. MÖBIUS EQUIVALENCE AND INVERSIVE DISTANCE

In this section we prove Theorem 4.1, a special case of which was established in Section 4. The case $c = 3$ of Theorem 4.1 illustrates an important difference between points and circles. Möbius maps act transitively on ordered triples of distinct points. The analog is false for triples of circles; the equality of inversive distances is the constraint that must be satisfied for Möbius maps to act transitively on ordered triples of circles.

We begin with an elementary lemma that is essential in the proof of Theorem 4.1.

Lemma 6.1. *Suppose Γ_1 is $|z| = 1$, Γ_2 is $|z| = R$, where $R > 1$, and Γ_3 is $\{z : |z - a| = r\}$. If Γ_3 lies in the annulus bounded by Γ_1 and Γ_2, then $|a|$ and r are uniquely determined by the numbers (Γ_1, Γ_3) and (Γ_2, Γ_3).*

Proof. For $j = 1, 2$ let $d_j = (\Gamma_j, \Gamma_3)$. Then

(6.1) $\qquad 2rd_1 = |a|^2 - r^2 - 1, \quad 2rRd_2 = |a|^2 - r^2 - R^2.$

By eliminating $|a|^2 - r^2$ from these equations we obtain

$$r = \frac{R^2 - 1}{2(d_1 - Rd_2)}.$$

Then we can use either of the equations in (6.1) to solve for $|a|$. □

The proof of Theorem 4.1 The first statement in Theorem 4.1 is about the existence of a certain map; the second statement is about its uniqueness. Given the existence of the map, the second statement follows directly from Theorem 3.2 when $c \geq 3$; thus we need only consider the question of existence. If the map f in Theorem 4.1 exists then, as inversive distances are Möbius invariant, the equality of the pairs of inversive distances follows. Suppose now that $(\Gamma_j, \Gamma_k) = (\Gamma'_j, \Gamma'_k)$ for all j and k with $1 \leq j < k \leq c$. We shall now establish the existence of the required map f. Of course, given any Möbius maps g and h of our choice, we have

$$(g(\Gamma_i), g(\Gamma_j)) = (\Gamma_i, \Gamma_j) = (\Gamma'_i, \Gamma'_j) = (h(\Gamma'_i), h(\Gamma'_j)),$$

so it suffices to show that $g(\Omega)$ and $h(\Omega')$ are Möbius equivalent. For $i \neq j$ let $d_{ij} = (\Gamma_i, \Gamma_j) = (\Gamma'_i, \Gamma'_j)$. We consider the cases $c = 2, 3, 4$ first.

Suppose that $c = 2$. By a suitable choice of g and h we may assume that Γ_1 is $|z| = 1$, Γ_2 is $|z| = r$, Γ'_1 is $|z| = 1$, and Γ'_2 is $|z| = R$, where $r > 1$ and $R > 1$. As $(1+r^2)/2r = (\Gamma_1, \Gamma_2) = (\Gamma'_1, \Gamma'_2) = (1+R^2)/2R$, we find that $r = R$. Thus Ω and Ω' are Möbius equivalent. Notice that $d_{12} = (1+R^2)/2R > 1$, so that R is uniquely determined by d_{12}.

Now suppose that $c = 3$. The case $c = 2$ implies that there is a Möbius map taking Γ_j to Γ'_j for $j = 1, 2$; thus we may assume that $\Gamma_j = \Gamma'_j$ for $j = 1, 2$. By a suitable choice of g (with $h = g$) we may assume that $\Gamma_1 = \Gamma'_1 = \{z : |z| = 1\}$ and $\Gamma_2 = \Gamma'_2 = \{z : |z| = R\}$, where $R > 1$. By a further rotation we may assume that Γ_3 is a circle $|z - a| = r$, where a is positive and $1 < a - r < a < a + r < R$ and, similarly, Γ'_3 is $|z - b| = t$ where b is positive and $1 < b - t < b + t < R$. Then by Lemma 6.1 $a = b$ and $r = s$, so that with this normalization $\Omega = \Omega'$.

Next, suppose that $c = 4$. The proof for the case $c = 3$ implies that we may assume that $\Gamma_j = \Gamma'_j$ for $j = 1, 2, 3$. Next, by applying a suitable Möbius map we may assume that

(6.2)
$$\Gamma_1 = \Gamma'_1 = \{z : |z| = 1\}, \quad \Gamma_2 = \Gamma'_2 = \{z : |z| = R\},$$
$$\Gamma_3 = \Gamma'_3 = \{z : |z - a_3| = r_3\},$$

where $1 < a_3 - r_3 < a_3 + r_3 < R$, and where $R, a_3 > 0$ and r_3 are uniquely determined by the d_{ij}, $1 \leq i < j \leq 3$. Then

$$\Gamma_4 = \{z : |z - a_4| = r_4\}, \quad \Gamma'_4 = \{z : |z - a'_4| = r'_4\}$$

both lie between $|z| = 1$ and $|z| = R$. Since $(\Gamma_1, \Gamma_4) = (\Gamma_1, \Gamma'_4)$ and $(\Gamma_2, \Gamma_4) = (\Gamma_2, \Gamma'_4)$, Lemma 6.1 implies that $|a_4| = |a'_4|$ and $r_4 = r'_4$. From $(\Gamma_3, \Gamma_4) = (\Gamma_3, \Gamma'_4)$ and the formula for inversive distance, we obtain $|a_3 - a_4| = |a_3 - a'_4|$. This identity in conjunction with $|a'_4| = |a_4|$ implies that either $a'_4 = a_4$ or $a'_4 = \bar{a}_4$, so that Γ'_4 is either Γ_4 or its reflection across $\mathbb{R}$. Thus Ω' is either Ω or its reflection across $\mathbb{R}$.

Now suppose that $c \geq 5$. There are two cases to consider, namely (i) when Ω is strongly symmetric and (ii) when Ω is not strongly symmetric. Suppose first that Ω is strongly symmetric. From the truth of the result for $c - 1$ we may assume $\Gamma_j = \Gamma'_j$, $j = 1, \ldots, c-1$ and that (6.2) holds. The axis of Ω must be $\mathbb{R}$ since this is the unique circle orthogonal to Γ_1, Γ_2 and Γ_3. The argument in the case $c = 4$ applied to the four circles Γ_j, $j = 1, 2, 3, c$, enables us to conclude that Γ'_c is Γ_c since Γ_c is symmetric about $\mathbb{R}$. Thus $\Omega' = \Omega$. Now, consider (ii). Let Γ be the circle orthogonal to Γ_1, Γ_2 and Γ_3. Since Ω is not strongly symmetric, there is some j, $4 \leq j \leq c$ such that Γ_j is not symmetric about Γ. By renumbering we may assume Γ_4 is not symmetric about Γ. From the validity of the result for $c-1$ we may assume $\Gamma_j = \Gamma'_j$, $j = 1, \ldots, c-1$, and that (6.2) holds. Now the argument in the case $c = 4$ applied to the circles Γ_j, $j = 1, 2, 3, c$, enables us to conclude that Γ'_c is either Γ_c or $\bar{\Gamma}_c$, the reflection of Γ_c over the real axis. We want to show that $\Gamma'_c = \Gamma_c$, and this is true if $\Gamma_c = \bar{\Gamma}_c$. Suppose, then, that $\Gamma_c \neq \bar{\Gamma}_c$. Since Γ_4 is not symmetric about $\mathbb{R}$, we see that $(\Gamma_4, \Gamma_c) \neq (\Gamma_4, \bar{\Gamma}_c)$. But $(\Gamma_4, \Gamma'_c) = (\Gamma_4, \Gamma_c)$; thus $\Gamma'_c \neq \bar{\Gamma}_c$, so $\Gamma'_c = \Gamma_c$. In both cases, $\Gamma'_c = \Gamma_c$ so that $\Omega' = \Omega$. $\square$

As an illustration of Theorem 4.1, consider the following example.

Example 6.2. Let Ω be the region bounded by the three circles

$$\Gamma_1 = \{z : |z - 6i| = 2\}, \Gamma_2 = \{z : |z| = 2\}, \text{ and } \Gamma_3 = \{z : |z - 6| = 2\}.$$

We know that there is a Möbius map f such that $f(\Gamma_1) = \{z : |z| = 1\}$, $f(\Gamma_2) = \{z : |z-a| = r\}$ and $f(\Gamma_3) = \{z : |z| = R\}$, where $1 < a-r < a+r < R$. It is tedious to find f explicitly, but Theorem 4.1 implies that it is easy to find a, r and R explicitly. For example, $(1 + R^2)/2R = \bigl(f(\Gamma_1), f(\Gamma_3)\bigr) = (\Gamma_1, \Gamma_3) = 8$, so that $R = 8 + 3\sqrt{7}$.

Example 6.3. Suppose that Ω and Ω' are circular regions as in Theorem 4.1 with $c \geq 3$ with $(\Gamma_i, \Gamma_j) = (\Gamma'_i, \Gamma'_j)$ for $i \neq j$. It is possible to construct a Möbius map f such that $f(\Gamma_j) = \Gamma'_j$, $j = 1, \ldots, c$.

We begin with $c = 3$ since, as we shall see, the general case reduces to this. We need to use the fact that there is a unique circle Γ, called the *radical circle*, that is orthogonal to Γ_1, Γ_2 and Γ_3. Therefore, the Möbius map f will not be unique. We indicate how to construct this mutually orthogonal circle. Suppose $\Gamma_j = \{z : |z - a_j| = r_j\}$. If the centres a_1, a_2 and a_3 are collinear, then Γ is the line through the centres. If the centres are not collinear, then we determine the orthogonal circle $\Gamma = \{z : |z - a| = r\}$ by solving the simultaneous equations $(\Gamma, \Gamma_j) = 0$, or

$$\begin{aligned}(6.3) \quad 0 &= |a - a_j|^2 - r^2 - r_j^2 \\ &= |a|^2 - r^2 - 2\operatorname{Re}(\bar{a}_j a) + |a_j|^2 - r_j^2.\end{aligned}$$

By eliminating $|a|^2 - r^2$ from a pair of these equations we obtain

$$(6.4) \qquad -2\operatorname{Re}(\bar{a}_i - \bar{a}_j)ai = |a_j|^2 - |a_i|^2 - r_j^2 + r_i^2.$$

The line defined by (6.4) is called the *radical axis* for Γ_i and Γ_j. The radical axes are parallel if and only if the centres of the three Γ_j are collinear. Otherwise, the three radical axes are concurrent at a unique point a, called the *radical centre*. Then the value of r can be determined from (6.3). If two circles have equal radii, then their radical axis is the perpendicular bisector of the segment joining the centres. For example, the radical centre in Example 6.2 is $3 + 3i$.

Now, suppose Γ, Γ' is the radical circle for Γ_j, Γ'_j, respectively. Let Γ meet Γ_j at b_j and c_j, where the points b_1, c_1, b_2, c_2, b_3, c_3 occur in cyclic order along Γ. Likewise, Γ' meets Γ'_j at b'_j and c'_j.

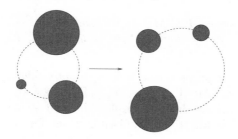

Figure 2

Let g be the unique conformal Möbius map with $g(b_1) = b'_1$, $g(c_1) = c'_1$ and $g(b_2) = b'_2$. We prove that g and gr, where r is the canonical reflection across Γ, both map Γ_j to Γ'_j, $j = 1, 2, 3$. Let f be a Möbius map whose existence is guaranteed by Theorem 4.1, so $f(\Gamma_j) = \Gamma'_j$, $j = 1, 2, 3$. Then $f(\Gamma) = \Gamma'$ because $f(\Gamma)$ is orthogonal to Γ'_j, $j = 1, 2, 3$. This implies that $f(b_j) = b'_j$ and $f(c_j) = c'_j$. Then the Möbius map $g^{-1}f$ fixes b_1, c_1 and b_2, so it is either the identity or r, reflection across Γ. Hence, either $f = g$ or $f = gr$. Both g and gr induce the correct correspondence of boundary circles.

Finally, suppose $c \geq 4$. Construct g as in the case $c = 3$ using just the first three circles. Again either $f = g$ or $f = gr$ since $g^{-1}f$ fixes b_1, c_1 and b_2. Now it is possible that only one of g and gr gives the correct correspondence of the remaining boundary circles.

The method of constructing g in the case $c = 3$ has an interesting consequence. There is a Möbius map f such that

$$f(\Gamma_1) = \{z : |z| = 1\}, \ f(\Gamma_2) = \{z : |z| = R\} \text{ and } f(\Gamma_3) = \{z : |z - a| = r\},$$

where $1 < a - r < a + r < R$. In this case the radical axis of the image region is the real axis and $b'_1 = -1$, $c'_1 = 1$, $b_2 = R$ and $c_2 = -R$. Therefore,

$$[b_1, c_1, b_2, c_2] = [-1, 1, R, -R]$$
$$= \frac{1}{2}\left(\frac{1 + R^2}{2R} + 1\right)$$
$$= \frac{1}{2}((\Gamma_1, \Gamma_2) + 1).$$

This expresses the inversive distance (Γ_1, Γ_2) as a cross-ratio.

Theorem 4.1 implies that if Ω is a circular region with connectivity c then the $\binom{c}{2}$ inversive distances (Γ_i, Γ_j), $1 \leq i < j \leq c$, determine Ω up to a Möbius transformation. Actually, when Ω is not strongly symmetric fewer inversive distances are required; one can show that $4c - 10$ (suitably chosen) inversive distances are sufficient to determine Ω up to Möbius equivalence. Also, Theorem 4.1 suggests a natural question: *does the conclusion of Theorem 4.1 remain valid when $\Gamma_1, \ldots, \Gamma_c$ are any set of c distinct circles in $\mathbb{C}_\infty$?*

Theorem 6.4. *Suppose that Ω is a circular region bounded by the circles $\Gamma_1, \ldots, \Gamma_c$, where $c \geq 3$. Then $S(\Omega) = \{\rho \in S_c : (\Gamma_i, \Gamma_j) = (\Gamma_{\rho(i)}, \Gamma_{\rho(j)})$ for all $i \neq j\}$.*

Proof. We say that a permutation ρ in S_c is *admissible (in the sense of inversive distance)* for Ω if $(\Gamma_i, \Gamma_j) = (\Gamma_{\rho(i)}, \Gamma_{\rho(j)})$ for all $i \neq j$. The Möbius invariance of the inversive distance implies that if $f \in \mathcal{A}(\Omega)$, then f_* is admissible. On the other hand, if ρ is admissible, then Theorem 4.1 with $\Omega' = \Omega$, and $\Gamma'_j = \Gamma_{\rho(j)}$ implies there exists $f \in \mathcal{A}(\Omega)$ with $f_* = \rho$. □

Theorem 6.4 provides an algorithm for determining (up to isomorphism) the automorphism group of any circular region. Given a circular region Ω in $\mathbb{C}_\infty$ bounded by the circles $\Gamma_1, \ldots, \Gamma_c$, calculate the $\binom{c}{2}$ inversive distances (Γ_i, Γ_j) for $i \neq j$. Then determine the admissible permutations ρ in S_c. Note that the method of Example 6.3 can be used to calculate the Möbius automorphism associated with an admissible permutation. We shall exploit these ideas in the next two sections.

7. Circular Regions with Connectivity Three

We determine the groups $\mathcal{A}^+(\Omega)$ and $\mathcal{A}(\Omega)$ for any triply connected circular region Ω.

Theorem 7.1. *Let Ω be the circular region bounded by the three mutually exterior circles Γ_1, Γ_2 and Γ_3, and let $\mathcal{C}$ be the cardinality of the set $\{(\Gamma_1, \Gamma_2), (\Gamma_1, \Gamma_3), (\Gamma_2, \Gamma_3)\}$ of inversive distances. Then*

$$\mathcal{A}^+(\Omega) \cong \begin{cases} S_3 & \text{if } \mathcal{C} = 1; \\ C_2 & \text{if } \mathcal{C} = 2; \\ \{I\} & \text{if } \mathcal{C} = 3. \end{cases}$$

In particular, the generic region of this type has no conformal symmetries. In all cases, $\mathcal{A}(\Omega)$ is isomorphic to $\mathcal{A}^+(\Omega) \times C_2$.

Proof. Any triply connected circular region is strongly symmetric because three mutually exterior circles have a common orthogonal circle. Thus $\mathcal{A}(\Omega)$ is isomorphic to $\mathcal{A}^+(\Omega) \times C_2$ by Theorem 3.2. If $\mathcal{C} = 1$ then $(\Gamma_1, \Gamma_2) = (\Gamma_1, \Gamma_3) = (\Gamma_2, \Gamma_3)$, and every permutation in S_3 is admissible. Thus $\mathcal{A}^+(\Omega)$ is S_3. If $\mathcal{C} = 3$ then, by Corollary 4.2, $\mathcal{A}^+(\Omega)$ is trivial. Finally, assume $\mathcal{C} = 2$. We may suppose that $(\Gamma_1, \Gamma_2) = (\Gamma_1, \Gamma_3) \neq (\Gamma_2, \Gamma_3)$. It is clear from this that any admissible permutation ρ in S_3 must leave the pair $\{2, 3\}$ invariant, and hence it must also fix 1. Theorem 6.4 implies that $S(\Omega) = \{\iota, (2\ 3)\}$, so $\mathcal{A}^+(\Omega)$ is cyclic of order two. □

Theorem 7.1 identifies the isomorphism type of $\mathcal{A}(\Omega)$. We shall now show how we can use geometry to determine the conformal automorphisms of Ω. Let Γ be the axis of symmetry for Ω and r denote reflection over Γ. By Theorem 3.2 $\mathcal{A}(\Omega) = \mathcal{A}^+(\Omega)\{I, r\}$, so it suffices to determine $\mathcal{A}^+(\Omega)$. First, suppose that $(\Gamma_1, \Gamma_2) = (\Gamma_1, \Gamma_3) \neq (\Gamma_2, \Gamma_3)$. Then there is a unique circle, say Γ_{23}, that is orthogonal to Γ_1, and such that the reflection r_{23} across Γ_{23} interchanges Γ_2 and Γ_3. It is now clear that $f = r_{23}r$ is an involutory conformal Möbius automorphism of Ω such that $f_* = (2\,3)$, so that $\mathcal{A}^+(\Omega) = \{I, f\}$.

Now assume $\mathcal{C} = 1$. Let Γ_{ij} denote the unique circle such that the reflection r_{ij} across Γ_{ij} interchanges Γ_i with Γ_j and fixes the remaining Γ_k. The anti-conformal Möbius automorphism r_{ij} corresponds to the admissible permutation $(i\,j)$, and f_{ij}, defined as $r_{ij}r$, is a conformal automorphism with $f_* = (i\,j)$. As transpositions generate S_3, the f_{ij} generate $\mathcal{A}^+(\Omega)$.

For a triply connected circular region we can specify explicit algebraic conditions that determine the conformal symmetry type of the region. We may assume that Ω is bounded by the three circles $|z| = 1$, $|z| = R$, where $R > 1$, and $|z - a| = r$, where $a \in (1, R)$ and $1 < a - r < a + r < R$. It is then clear from the earlier argument that $\mathcal{A}^+(\Omega)$ is trivial if and only if the three numbers
$$\frac{1+R^2}{2R}, \quad \frac{a^2-1-r^2}{2r}, \quad \frac{a^2-R^2-r^2}{2rR}$$
are distinct. If exactly two of these numbers are equal to each other then $\mathcal{A}^+(\Omega)$ is isomorphic to C_2; if all three numbers are equal to each other, then $\mathcal{A}^+(\Omega)$ is isomorphic to S_3. In fact, the parameters a, r and R lie in an open subset G of $\mathbb{R}^3$ such that every point of G corresponds to a triply-connected circular region. The set of parameters for which two of the inversive distances are equal defines an algebraic variety V in $\mathbb{R}^3$, and $V \cap G$ has measure zero.

8. Circular regions with connectivity four

Theorem 6.4 can be used to determine the automorphism group of circular regions with connectivity four, and this section is devoted to examples of this.

Example 8.1. Let Ω be the region bounded by the circles
$$\Gamma_1 = \{z : |z - 12| = 4\}, \ \Gamma_2 = \{z : |z - 3i| = 1\},$$
$$\Gamma_3 = \{z : |z + 12| = 4\} \text{ and } \Gamma_4 = \{z : |z + 3i| = 1\}.$$
It is easy to see that Ω is not strongly symmetric. Direct calculations give $(\Gamma_i, \Gamma_j) = 17$ for all $i \neq j$. In this situation every permutation in S_4 is admissible, so $\mathcal{A}(\Omega)$ is isomorphic to S_4.

Example 8.2. This example shows that it is possible to have $\mathcal{A}(\Omega) = \mathcal{A}^+(\Omega)$. Let Ω be the region bounded by the circles
$$\Gamma_1 = \{z : |z - 2| = 1\}, \ \Gamma_2 = \{z : |z - (2 + 4i)| = 1\},$$

$$\Gamma_3 = \{z : |z+2| = 1\} \text{ and } \Gamma_4 = \{z : |z + (2+4i)| = 1\}.$$

It is easy to check that Ω is not strongly symmetric. Obviously, $g(z) = -z$ is an involutory conformal automorphism of Ω and $g_* = (1\,3)(2\,4)$. We shall show that there are no other non-trivial automorphisms of Ω. A calculation of the inversive distances gives

$$(\Gamma_1, \Gamma_2) = (\Gamma_3, \Gamma_4) = 7, \quad (\Gamma_1, \Gamma_4) = (\Gamma_2, \Gamma_3) = 15,$$
$$(\Gamma_2, \Gamma_4) = 39, \quad (\Gamma_1, \Gamma_3) = 7.$$

Let f be an automorphism. Clearly f_* leaves $\{2,4\}$ and $\{1,3\}$ invariant. If $f_*(2) = 2$ then, by considering the inversive distances that are equal to 7, we see that $f_*(1) = 1$. If $f_*(2) = 4$ then, by considering the inversive distances that are equal to 15, we see that $f_*(3) = 1$. Thus f_* is either ι or $(1\,3)(2\,4)$, which is g_*. This implies that $\mathcal{A}(\Omega) = \{I, g\} = \mathcal{A}^+(\Omega)$.

Example 8.3. In this example we answer the question posed in the Introduction. Here, $\Omega(r)$ is the region defined by the circles in (1.1):

$$\Gamma_1 = \{x + iy : x = 0\}, \quad \Gamma_2 = \{z : |z - (7+2i)| = 1\},$$
$$\Gamma_3 = \{z : |z - (7-2i)| = 1\}, \quad \Gamma_4 = \{z : |z - 12| = r\}.$$

It is clear that $\Omega(r)$ is not strongly symmetric, and a calculation shows that

(8.1)
$$(\Gamma_1, \Gamma_2) = (\Gamma_1, \Gamma_3) = (\Gamma_2, \Gamma_3) = 7,$$
$$(\Gamma_2, \Gamma_4) = (\Gamma_3, \Gamma_4) = \frac{28 - r^2}{2r}, \quad (\Gamma_1, \Gamma_4) = \frac{12}{r}.$$

We determine the admissible permutations $\rho \in S_4$; that is, all ρ such that $(\Gamma_{\rho(i)}, \Gamma_{\rho(j)}) = (\Gamma_i, \Gamma_j)$ for $i \neq j$. We consider cases depending on the value of r.

(1) *First assume that r is such that the three values of the inversive distances in (8.1) are distinct.* Then an admissible ρ must leave $\{1,4\}$, and hence also $\{2,3\}$, invariant. By considering the pairs with inversive distance 7, it is now clear that $\rho(4) = 4$, and so $\rho(1) = 1$. Thus the admissible permutations are ι and $(2\,3)$, and so $\mathcal{A}(\Omega)$ is isomorphic to C_2. The permutation $(2\,3)$ corresponds to reflection across the real axis.

(2) *Next, assume that $r = 12/7$:* then $(\Gamma_1, \Gamma_4) = 7 \neq (\Gamma_2, \Gamma_4) = (\Gamma_3, \Gamma_4)$. In this case an admissible ρ must map each pair $\{2,4\}$ and $\{3,4\}$ to itself or the other pair, so that $\rho(1) = 1$. By considering all pairs of circles that are an inversive distance 7 apart, we see that ρ must permute the set $\{\{1,2\}, \{1,3\}, \{1,4\}, \{2,3\}\}$; thus the pair $\{2,3\}$ is invariant. Hence, $\rho(4) = 4$, and again the admissible permutations ρ are ι and $(2\,3)$, and $\mathcal{A}(\Omega)$ is isomorphic to C_2.

(3) *Assume now that $r = \sqrt{77} - 7$.* Then $(\Gamma_2, \Gamma_4) = (\Gamma_3, \Gamma_4) = 7 \neq (\Gamma_1, \Gamma_4)$ and ρ must map the pair $\{1,4\}$ onto itself, and there are no other constraints.

In this case the admissible permutations ρ are ι, $(1\,4)$, $(2\,3)$ and $(1\,4)(2\,3)$. Thus $\mathcal{A}(\Omega)$ is the Klein four-group.

(4) *Finally, assume that* $r = 2$. Then $(\Gamma_2, \Gamma_4) = (\Gamma_3, \Gamma_4) = (\Gamma_1, \Gamma_4) = 6$ and each admissible permutation ρ must permute each of the sets of indices, $\{\{1,2\}, \{1,3\}, \{2,3\}\}$ and $\{\{2,4\}, \{3,4\}, \{1,4\}\}$. This implies that $\rho(4) = 4$, so that the admissible permutations form the group of all permutations of $\{1, 2, 3\}$. In this case $\mathcal{A}(\Omega)$ is isomorphic to S_3.

We can say more and, as an illustration, we examine the case $r = 2$ in more detail. The reflection r_{23} across $\mathbb{R}_\infty$ leaves Γ_1 and Γ_4 invariant, and interchanges Γ_2 and Γ_3; thus it corresponds to the permutation $(2\,3)$. Next, let $\Gamma_{12} = \{z : |z - (8 + 2i)| = 4\}$, and let r_{12} be the reflection across Γ_{12}. The circle Γ_{12} is orthogonal to Γ_3 and Γ_4, so r_{12} leaves these circles invariant. Also, r_{12} interchanges Γ_1 and Γ_2 so that r_{12} corresponds to the permutation $(1\,2)$. Similarly, the reflection r_{13} across the circle $\{z : |z - (8 - 2i)| = 4\}$ corresponds to the permutation $(1\,3)$. The maps r_{12}, r_{13} and r_{23} generate $\mathcal{A}(\Omega)$; explicitly,

$$r_{23}(z) = \bar{z}, \quad r_{12}(z) = \frac{(8+2i)\bar{z} - 52}{\bar{z} - (8-2i)}, \quad r_{13}(z) = \frac{(8-2i)\bar{z} - 52}{\bar{z} - (8+2i)}.$$

Finally, the product

$$f_{123}(z) = r_{12}r_{23}(z) = \frac{(8+2i)z - 52}{z - (8-2i)}$$

is a conformal automorphism of Ω; it has order three, and it corresponds to the permutation (123). f_{123}^2 is associated with the permutation $(3\,2\,1)$.

The case $r = 2$ also illustrates another point; $\mathcal{A}(\Omega(2))$ is not isomorphic to $\mathcal{A}^+(\Omega(2)) \times C_2$ even though the quotient group $\mathcal{A}(\Omega(2))/\mathcal{A}^+(\Omega(2))$ is isomorphic to C_2. This is true for a simple group-theoretic reason. Since $\mathcal{A}^+(\Omega(2)) \cong C_3$, $\mathcal{A}^+(\Omega(2)) \times C_2 \cong C_3 \times C_2$. This latter group contains four elements of order 3 and one element of order 2 while $\mathcal{A}(\Omega(2))$ contains two elements of order 3 and three elements of order 2.

Example 8.4. Let $\Gamma_1, \Gamma_2, \Gamma_3, \Gamma_4$ be circles of radius $1, \frac{1}{2}, \frac{1}{3}, \frac{1}{4}$, respectively, located as follows: Γ_1 is exterior to the unit circle $\partial \mathbb{D}$, and tangent to it at 1; Γ_2 is exterior to the unit circle $\partial \mathbb{D}$, and tangent to it at i; Γ_3 is interior to the unit circle $\partial \mathbb{D}$, and tangent to it at -1; Γ_4 is interior to the unit circle $\partial \mathbb{D}$, and tangent to it at $-i$; and let Ω be the region bounded by the circles Γ_j, $j = 1, 2, 3, 4$: see Figure 3 (where the unit circle is drawn for the convenience of the reader).

First, we note that the region Ω is strongly symmetric; to see this one has only to check that the circle Γ with centre $\frac{1}{2} + \frac{i}{3}$ and radius $\frac{7}{6}$ is orthogonal to each Γ_j. Hence, the reflection r across Γ is an anti-conformal automorphism of Ω, $\mathcal{A}^+(\Omega) \cong S(\Omega)$ and $\mathcal{A}(\Omega)$ is the internal direct product of $\mathcal{A}^+(\Omega)$ and

$\{I, r\}$. Also, observe that $s(z) = 1/\bar{z}$, an involutory rotary reflection, is an anti-conformal Möbius automorphism of Ω.

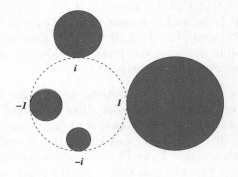

Figure 3

We determine all admissible permutations. A computation of inversive distances gives

$$(\Gamma_1, \Gamma_2) = (\Gamma_3, \Gamma_4) = 5, \quad (\Gamma_1, \Gamma_4) = (\Gamma_2, \Gamma_3) = 7,$$
$$(\Gamma_2, \Gamma_4) = 19, \quad (\Gamma_1, \Gamma_3) = 9.$$

Any admissible permutation ρ must
(a) map each of $\{1, 3\}$ and $\{2, 4\}$ to itself, and
(b) permute each of the sets $\{\{1, 2\}, \{3, 4\}\}$ and $\{\{1, 4\}, \{2, 3\}\}$.
If ρ leaves $\{1, 2\}$ invariant, then (a) implies that ρ fixes 1. Then ρ fixes each element of $\mathbb{Z}_4$ and so $\rho = \iota$. The other possibility is that ρ maps $\{1, 2\}$ to $\{3, 4\}$. From (a) this means ρ interchanges 1 and 3 and also switches 2 and 4; that is, $\rho = (1\,3)(2\,4)$. Thus, the only admissible permutations are ι and $(1\,3)(2\,4)$, so $\mathcal{A}^+(\Omega) \cong C_2$. The anti-conformal Möbius map s also corresponds to the permutation $(1\,3)(2\,4)$, so the product $rs = sr$ is the conformal Möbius map that corresponds to $(1\,3)(2\,4)$. Hence,

$$\mathcal{A}^+(\Omega) = \{\iota, rs = sr\} \text{ and } \mathcal{A}(\Omega) = \{\iota, r, s, rs = sr\}.$$

We remark that the determination of the common orthogonal circle in Example 8.4 follows from the approach suggested at the end of Section 3. It is of interest, however, to note that this is also a part of the theory of the inversive distance. First, it is clear that two circles Γ_1 and Γ_2 are orthogonal if and only if $(\Gamma_1, \Gamma_2) = 0$. More generally, if Γ_1 meets Γ_2 at an angle θ, then $(\Gamma_1, \Gamma_2) = \cos\theta$. In fact, if Γ_1 and Γ_2 do not meet, then they bound two hyperbolic planes in three-dimensional hyperbolic space, and if these planes are at a distance δ apart, then $(\Gamma_1, \Gamma_2) = \cosh\delta$. This strongly suggests that any discussion of automorphism groups of circular regions should lie in geometry rather than in complex analysis.

9. Thrice-punctured spheres

We now turn to the consideration of punctured spheres. For the next several sections we shall deal with conformal automorphism groups of punctured spheres and cross-ratios. In Section 14 we consider the automorphism groups of punctured spheres and absolute cross-ratios. We remind the reader that Lemma 3.1 and Theorem 3.2 are valid for punctured spheres. Also a punctured sphere is strongly symmetric if and only if the punctures are concyclic. The case of triply punctured spheres is exceptional in the sense that our general method only applies when there are at least four punctures. Therefore, we treat the case of three punctures first.

A region of the form $\mathbb{C}_\infty \setminus \{a, b, c\}$, where a, b and c are distinct, is a *thrice-punctured sphere*, and every such region is Möbius equivalent to the region $\Omega_0 = \mathbb{C}_\infty \setminus \{0, 1, \infty\}$ which is strongly symmetric with the real line as axis of symmetry, so $r(z) = \bar{z}$ is the canonical reflection. As $\mathcal{M}^+$ acts triply-transitively on $\mathbb{C}_\infty$, every permutation of $\{0, 1, \infty\}$ extends to a conformal Möbius automorphism of Ω_0; thus $\mathcal{A}_0^+ = \mathcal{A}^+(\Omega_0)$ is isomorphic to S_3. Explicitly, $\mathcal{A}_0^+ = \{f_1, f_2, f_3, f_4, f_5, f_6\}$, where

$$f_1(z) = z, \quad f_2(z) = 1 - z, \quad f_3(z) = \frac{z}{z-1},$$

$$f_4(z) = \frac{1}{z}, \quad f_5(z) = \frac{z-1}{z}, \quad f_6(z) = \frac{1}{1-z}.$$

The group $\mathcal{A}_0 = \mathcal{A}(\Omega_0)$ is obtained by adjoining r to $\mathcal{A}_0^+$, and $\mathcal{A}_0$ is isomorphic to $S_3 \times C_2$.

The fixed points in Ω_0 of the elements of $\mathcal{A}_0^+$ and the associated stabilizers in $\mathcal{A}_0^+$ are readily determined. The stabilizer of η and $\bar\eta$, where $\eta = \exp(\pi i/3)$, is $\{f_1, f_5, f_6\}$, while $\mathrm{stab}^+(\frac{1}{2}) = \{f_1, f_2\}$, $\mathrm{stab}^+(2) = \{f_1, f_3\}$, and $\mathrm{stab}^+(-1) = \{f_1, f_4\}$. A similar analysis is needed for the full group $\mathcal{A}_0$. $\bar{f}_1$ fixes $\Gamma_1 = \mathbb{R}$, $\bar{f}_2$ fixes $\Gamma_2 = \{z : \mathrm{Re}\, z = \frac{1}{2}\}$, $\bar{f}_3$ fixes $\Gamma_3 = \{z : |z-1| = 1\}$, and $\bar{f}_4$ fixes $\Gamma_4 = \{z : |z| = 1\}$ in the strong sense that each point of the circle is fixed. Thus $\bar{f}_j$ is reflection over Γ_j. The automorphisms $\bar{f}_5$ and $\bar{f}_6$ have no fixed points. For $\lambda \in \Gamma_j \setminus \{\frac{1}{2}, 2, -1, \eta, \bar\eta\}$, $\mathrm{stab}(\lambda) = \{f_1, \bar{f}_j\}$. Also, $\mathrm{stab}(\frac{1}{2}) = \{f_1, \bar{f}_1, f_2, \bar{f}_2\}$, $\mathrm{stab}(2) = \{f_1, \bar{f}_1, f_3, \bar{f}_3\}$, $\mathrm{stab}(-1) = \{f_1, \bar{f}_1, f_4, \bar{f}_4\}$ and $\mathrm{stab}(\eta) = \mathrm{stab}(\bar\eta) = \{f_1, \bar{f}_1, f_5, \bar{f}_5, f_6, \bar{f}_6\}$.

The rational function

$$F(z) = \frac{4(z^2 - z + 1)^3}{27 z^2 (z-1)^2}$$

maps Ω_0 onto $\mathbb{C}$, and it is an automorphic function for $\mathcal{A}_0^+$; that is, $F(z) = F(w)$ if and only if z and w are in the same $\mathcal{A}_0^+$-orbit ([1, p.282]). In particular,

(a) F has poles at and only at 0, 1 and ∞ (all of multiplicity two);
(b) F has zeros at and only at η and $\bar\eta$ (each of multiplicity three);
(c) $F(z) - 1$ has zeros at and only at -1, $\frac{1}{2}$ and 2 (each of multiplicity two).

We emphasize that *we shall use the notation Ω_0, $\mathcal{A}_0$, $\mathcal{A}_0^+$ and F throughout the remainder of the paper.*

10. THE CROSS-RATIO FUNCTION

As much of the remaining discussion depends on cross-ratios, it seems prudent to recall how cross-ratios change when we permute the entries in the cross-ratio. We need to treat the cross-ratio as a function rather carefully. Although this is well known, it is often stated in a rather casual way and there are advantages in taking a more formal approach than usual. An *injective four-tuple* is an injective map

$$s : \{1,2,3,4\} \to \mathbb{C}_\infty,$$

and as usual we denote this by the symbol (s_1, s_2, s_3, s_4). The set of injective four-tuples is denoted by $\mathcal{F}$. The *cross-ratio function* is the function $\chi : \mathcal{F} \to \Omega_0$ defined by

$$\chi(s) = \frac{(s_1 - s_3)(s_2 - s_4)}{(s_1 - s_2)(s_3 - s_4)},$$

and χ is surjective because the cross-ratio can take every value in $\mathbb{C}_\infty$ except $0, 1, \infty$. The following facts are fundamental and well known (where $\circ$ denotes the composition of functions).

Theorem 10.1. (a) *If $s, t \in \mathcal{F}$, then there is a conformal Möbius map f with $f \circ s = t$ (equivalently, $f(s_j) = t_j$ for $j = 1, 2, 3, 4$) if and only if $\chi(s) = \chi(t)$.*
(b) *A bijection f of $\mathbb{C}_\infty$ onto itself is a conformal Möbius map if and only if $\chi(f \circ s) = \chi(s)$ for all s in $\mathcal{F}$.*
(c) *For each ρ in S_4 there is a unique ρ^+ in $\mathcal{A}_0^+$ such that for every s in $\mathcal{F}$, $\chi(s \circ \rho) = \rho^+\big(\chi(s)\big)$.*
(d) *For all ρ and σ in S_4, $(\rho \circ \sigma)^+ = \sigma^+ \circ \rho^+$.*

It may be helpful to see the actions of these maps as follows:

$$\{1,2,3,4\} \xrightarrow{\rho} \{1,2,3,4\} \xrightarrow{s,t} \mathbb{C}_\infty \xrightarrow{f} \mathbb{C}_\infty,$$

$$\mathcal{F} \xrightarrow{\chi} \mathbb{C}_\infty \backslash \{0, 1, \infty\} \xrightarrow{\rho^+} \mathbb{C}_\infty \backslash \{0, 1, \infty\}.$$

We comment on Theorem 10.1. First, pars (a) and (b) are familiar properties of cross-ratios under conformal Möbius transformations, but note that we must use $f \circ s$ and *not* $f(s)$. Part (c) says that if we permute the entries in a cross-ratio whose value is λ, then the resulting cross-ratio is a conformal Möbius function ρ^+ of λ, regardless of the entries in the cross-ratio. Note, however, that the permutation takes place *on the set* $\{1, 2, 3, 4\}$ and *not* on the set $\{s_1, s_2, s_3, s_4\}$. The proof of part (d) is by direct computation from (c), and it says that $\rho \mapsto \rho^+$ is an 'anti-homomorphism'.

For reference we explicitly give the correspondence $\rho \mapsto \rho^+$.

(10.1)
$$\begin{aligned}
K &= \{\iota, (1\,2)(3\,4), (1\,3)(2\,4), (1\,4)(2\,3)\} \mapsto & f_1(z) &= z; \\
(1\,2)K &= \{(1\,2), (3\,4), (1\,3\,2\,4), (1\,4\,2\,3)\} \mapsto & f_2(z) &= 1 - z; \\
(1\,3)K &= \{(1\,3), (2\,4), (1\,2\,3\,4), (1\,4\,3\,2)\} \mapsto & f_3(z) &= z/(z - 1); \\
(1\,4)K &= \{(1\,4), (2\,3), (1\,2\,4\,3), (1\,3\,4\,2)\} \mapsto & f_4(z) &= 1/z; \\
(1\,2\,3)K &= \{(1\,2\,3), (1\,3\,4), (1\,4\,2), (2\,4\,3)\} \mapsto & f_5(z) &= (z-1)/z; \\
(1\,2\,4)K &= \{(1\,2\,4), (1\,3\,2), (1\,4\,3), (2\,3\,4)\} \mapsto & f_6(z) &= 1/(1-z).
\end{aligned}$$

Here K is the Klein four-group D_4. Set $G_j = K \cup (1\,j)K$ for $j = 2, 3, 4$. The G_j are conjugate subgroups of S_4, and the map $\rho \mapsto (i\,j)\rho(i\,j)$ is a conjugation of G_i onto G_j. Also, $K \cup (1\,2\,3)K \cup (1\,2\,4)K$ is A_4, the alternating subgroup of S_4.

There is a result analogous to Theorem 10.1(c) for conjugate cross-ratios. For each ρ in S_4 there is a unique ρ^- in $\mathcal{A}_0^- = \mathcal{A}_0 \setminus \mathcal{A}_0^+$ such that for every s in $\mathcal{F}$, $\overline{\chi(s \circ \rho)} = \rho^-(\chi(s))$. In fact, $\rho^- = \overline{\rho^+}$, so the correspondence $\rho \mapsto \rho^-$ is the same as the list (10.1) except that f_j must be changed to $\bar{f}_j$.

We shall now revert to the usual notation for cross-ratios, namely $[a, b, c, d]$ instead of $\chi(a, b, c, d)$.

11. Conformal Möbius equivalence and cross-ratios

This section is the analog of Section 6 for punctured spheres.

Theorem 11.1. *Let $\Omega = \mathbb{C}_\infty \setminus \{\gamma_1, \ldots, \gamma_c\}$ and $\Omega' = \mathbb{C}_\infty \setminus \{\gamma'_1, \ldots, \gamma'_c\}$, where $c \geq 4$. The following are equivalent:*
(a) *There is a unique conformal Möbius map f with $f(\gamma_j) = \gamma'_j$, $j = 1, \ldots, c$,*
(b) $[\gamma_i, \gamma_j, \gamma_k, \gamma_\ell] = [\gamma'_i, \gamma'_j, \gamma'_k, \gamma'_\ell]$ *for all distinct $i, j, k, \ell \in \mathbb{Z}_c$,*
(c) $[\gamma_1, \gamma_2, \gamma_3, \gamma_\ell] = [\gamma'_1, \gamma'_2, \gamma'_3, \gamma'_\ell]$, *for $\ell = 4, \ldots, c$.*

Proof. (a) $\Rightarrow$ (b) This is the invariance of the cross-ratio under conformal Möbius maps.
(b) $\Rightarrow$ (c) This is trivial.
(c) $\Rightarrow$ (a) Suppose (c) holds, and let f be the unique conformal Möbius transformation such that $f(\gamma_j) = \gamma'_j$, $j = 1, 2, 3$. Then, as f is Möbius and (c) holds,

$$\begin{aligned}
[\gamma'_1, \gamma'_2, \gamma'_3, \gamma'_\ell] &= [\gamma_1, \gamma_2, \gamma_3, \gamma_\ell] \\
&= [f(\gamma_1), f(\gamma_2), f(\gamma_3), f(\gamma_\ell)] \\
&= [\gamma'_1, \gamma'_2, \gamma'_3, f(\gamma_\ell)].
\end{aligned}$$

Hence, $f(\gamma_\ell) = \gamma'_\ell$ for $\ell = 4, \ldots, c$. $\square$

The force of Theorem 10.1 is that it shows that the equality of the $c-3$ cross-ratios in (c) implies the equality of all of the $\binom{c}{4}$ cross-ratios of four points chosen from $\{\gamma_1, \ldots, \gamma_c\}$ with the corresponding cross-ratios from $\{\gamma'_1, \ldots, \gamma'_c\}$.

As the proof of Theorem 11.1 shows, it is elementary to construct the conformal Möbius map f. If f is the unique conformal Möbius map with $f(\gamma_j) = \gamma'_j$ for $j = 1, 2, 3$, then f also maps γ_j to γ'_j for $j = 4, \ldots, c$.

Next, consider the two four-punctured spheres
$$\mathbb{C}_\infty \backslash \{\gamma_1, \gamma_2, \gamma_3, \gamma_4\} \quad \text{and} \quad \mathbb{C}_\infty \backslash \{\gamma'_1, \gamma'_2, \gamma'_3, \gamma'_4\}.$$
These are conformally equivalent if and only if there is a Möbius map of $\{\gamma_1, \gamma_2, \gamma_3, \gamma_4\}$ onto $\{\gamma'_1, \gamma'_2, \gamma'_3, \gamma'_4\}$, and this is so if and only if there is a Möbius map f, and a permutation ρ in S_4 such that $f(\gamma_j) = \gamma'_{\rho(j)}$ for $j = 1, 2, 3, 4$. Thus we have the following result which has the advantage of giving a criterion for conformal equivalence that is independent of the labelling of the points.

Corollary 11.2. *The two four-punctured spheres,*
$$\mathbb{C}_\infty \backslash \{\gamma_1, \gamma_2, \gamma_3, \gamma_4\} \quad \text{and} \quad \mathbb{C}_\infty \backslash \{\gamma'_1, \gamma'_2, \gamma'_3, \gamma'_4\},$$
are conformally equivalent if and only if $F([\gamma_1, \gamma_2, \gamma_3, \gamma_4]) = F([\gamma'_1, \gamma'_2, \gamma'_3, \gamma'_4])$.

Theorem 11.3. . *Let* $\Omega = \mathbb{C}_\infty \backslash \{\gamma_1, \ldots, \gamma_c\}$, *where* $c \geq 4$ *and the* γ_i *are distinct points in* $\mathbb{C}_\infty$. *Then*
$$S^+(\Omega) =$$
$$\{\rho \in S_c : [\gamma_i, \gamma_j, \gamma_k, \gamma_\ell] = [\gamma_{\rho(i)}, \gamma_{\rho(j)}, \gamma_{\rho(k)}, \gamma_{\rho(\ell)}] \text{ for all distinct } i, j, k, \ell \in \mathbb{Z}_c\}.$$

Proof. Call a permutation $\rho \in S_c$ admissible (in the cross-ratio sense) for Ω if
$$[\gamma_i, \gamma_j, \gamma_k, \gamma_\ell] = [\gamma_{\rho(i)}, \gamma_{\rho(j)}, \gamma_{\rho(k)}, \gamma_{\rho(\ell)}]$$
for all distinct $i, j, k, \ell \in \mathbb{Z}_c$. Theorem 11.1 implies that ρ is admissible for Ω if and only if there is a conformal Möbius automorphism of Ω with $f_* = \rho$. $\square$

In fact, $\rho \in S_c$ is admissible if and only if
$$[\gamma_{\rho(1)}, \gamma_{\rho(2)}, \gamma_{\rho(3)}, \gamma_{\rho(\ell)}] = [\gamma_1, \gamma_2, \gamma_3, \gamma_\ell], \quad \ell = 4, \ldots, c.$$

12. Four punctured spheres

We can use Theorem 11.3 to determine the conformal automorphism groups of four punctured spheres. We have seen (in Section 1) that the sphere with four punctures always has non-trivial conformal symmetry. There is, however, a better argument that is based on three-dimensional hyperbolic geometry. Given four distinct points a, b, c, d let γ_{ab} and γ_{cd} be the geodesics in $\mathbb{H}^3$ with end-points a, b and c, d respectively. Then γ_{ab} and γ_{cd} have a common orthogonal geodesic, say α, in $\mathbb{H}^3$ (any two geodesics with distinct end-points have a common orthogonal), and the rotation of order two about α is a Möbius map whose action on $\{a, b, c, d\}$ is the permutation $(ab)(cd)$. A similar argument holds for $(ac)(bd)$ and $(ad)(bc)$, thus we have the stronger result that *if Ω is a four-punctured sphere, then $\mathcal{A}^+(\Omega)$ contains a copy of the Klein four-group*.

Theorem 12.1. Suppose that $\Omega = \mathbb{C}_\infty \setminus \{\gamma_1, \gamma_2, \gamma_3, \gamma_4\}$ and $\lambda = [\gamma_1, \gamma_2, \gamma_3, \gamma_4]$.
(a) If λ is not one of the numbers $-1, \frac{1}{2}, 2, \eta, \bar{\eta}$, then $\mathcal{A}^+(\Omega)$ is isomorphic to the Klein four-group D_4.
(b) If $\lambda \in \{\frac{1}{2}, 2, -1\}$, then $\mathcal{A}^+(\Omega)$ is isomorphic to D_8.
(c) If $\lambda \in \{\eta, \bar{\eta}\}$, then $\mathcal{A}^+(\Omega)$ is isomorphic to A_4.

Proof. Taken together, Theorem 10.1(c), the results of Section 9 and the correspondences (10.1), make it easy to determine the admissible permutations. For any $\rho \in S_4$, Theorem 10.1(c) gives $[\gamma_{\rho(1)}, \gamma_{\rho(2)}, \gamma_{\rho(3)}, \gamma_{\rho(4)}] = \rho^+([\gamma_1, \gamma_2, \gamma_3, \gamma_4]) = \rho^+(\lambda)$. Hence, ρ is admissible if and only if λ is fixed by ρ^+, so that $S^+(\Omega) = \{\rho : \rho^+ \text{ fixes } \lambda\}$. From (10.1) $\rho^+ = I$ if and only if $\rho \in K$. Thus, $K \subset S^+(\Omega)$ independent of the value of λ, as was noted prior to the statement of the theorem.

If λ is not one of the numbers $-1, \frac{1}{2}, 2, \eta, \bar{\eta}$, then the only function in $\mathcal{A}_0^+$ that fixes λ is $f_1 = I$, and so $S^+(\Omega) = K$.

Next, assume λ is one of $\frac{1}{2}$, 2 and -1. Since $\text{stab}^+(\frac{1}{2}) = \{f_1 = I, f_2\}$, $S^+(\Omega) = G_2$ when $\lambda = \frac{1}{2}$. Likewise, $\text{stab}^+(2) = \{f_1 = I, f_3\}$ implies $S^+(\Omega) = G_3$ when $\lambda = 2$. Finally, $\text{stab}^+(-1) = \{f_1 = I, f_4\}$ gives $S^+(\Omega) = G_4$ when $\lambda = -1$. In all three cases the group $\mathcal{A}^+(\Omega)$ is isomorphic to D_8. This group can also be obtained more directly. By Corollary 11.2 two four punctured spheres with associated cross-ratio in $\{\frac{1}{2}, 2, -1\}$ are conformally equivalent, so it suffices to determine the conformal automorphism group for a particular region. Let $\Omega = \mathbb{C}_\infty \setminus \{1, i, -1, -i\}$. Now, $[-1, i, 1, -i] = \frac{1}{2}$, $[1, i, -1, -i] = 2$, and $[i, 1, -1, -i] = -1$. The conformal automorphism group $\mathcal{A}^+(\Omega)$ obviously contains

$$\{z, iz, -z, -iz, 1/z, i/z, -1/z, -i/z\}$$

which is isomorphic to D_8. Because the order of the automorphism group is eight, this is the group of conformal automorphisms.

The last possibility is that $\lambda \in \{\eta, \bar{\eta}\}$. Recall that $\text{stab}^+(\eta) = \text{stab}^+(\bar{\eta}) = \{f_1 = I, f_5, f_6\}$. Therefore, $S^+(\Omega) = A_4$. We can see this answer geometrically. Let $\Omega = \mathbb{C}_\infty \setminus \{0, , \sqrt{2}, \sqrt{2}\omega, \sqrt{2}\omega^2\}$. Note that $[0, \sqrt{2}, , \sqrt{2}\omega, \sqrt{2}\omega^2] = \eta$, where $\omega = \eta^2$. The chordal distance between each pair of points in the complement of Ω is $2\sqrt{\frac{2}{3}}$; the points are the vertices of a regular tetrahedron on the sphere. The conformal automorphism group for the sphere punctured at the vertices of a regular tetrahedron is A_4. $\square$

Example 12.2. Suppose $\Omega = \mathbb{C}_\infty \setminus \{0, 1, 2, 3\}$. Since $[0, 1, 2, 3] = 4$, $\mathcal{A}^+(\Omega)$ is the Klein four-group. Explicitly,

$$\mathcal{A}^+(\Omega) = \left\{z, \ 3-z, \ \frac{3(z-1)}{2z-3}, \ \frac{3(z-2)}{2z-3}\right\},$$

where these conformal automorphisms correspond in order to the permutations ι, $(03)(12)$, $(01)(23)$ and $(02)(13)$ of the punctures.

13. CONFORMALLY TRIVIAL PUNCTURED SPHERES

Unlike four punctured spheres, the generic sphere with $c \geq 5$ punctures has no conformal symmetry.

In order to work with the cross-ratio we need to discuss the issue of ordered versus unordered sets of four points. Four distinct points $\{\gamma_1, \gamma_2, \gamma_3, \gamma_4\}$ can be ordered in 4! ways. We can form the cross-ratio of each ordered quadruple; the set of distinct values of the cross-ratios is usually six, but can be two or three. We can avoid these matters by associating the complex number $F([\gamma_1, \gamma_2, \gamma_3, \gamma_4])$ with the unordered quadruple. The value obtained is independent of the ordering selected.

It is simple to formulate a condition for a lack of conformal symmetry in terms of the cross-ratio.

Theorem 13.1. *Suppose Ω is a sphere punctured at the distinct points $\gamma_1, \ldots, \gamma_c$, where $c \geq 5$. If*

(13.1) $$[\gamma_i, \gamma_j, \gamma_k, \gamma_\ell] \neq [\gamma_p, \gamma_q, \gamma_r, \gamma_s]$$

whenever $\{i, j, k, \ell\}$, $\{p, q, r, s\}$ are distinct subsets of four elements from $\mathbb{Z}_c$, then $\mathcal{A}^+(\Omega)$ is trivial.

Proof. Each $f \in \mathcal{A}^+(\Omega)$ induces a permutation $f_* \in S^+(\Omega)$. For each subset $\{i, j, k, \ell\}$ of four distinct elements from $\mathbb{Z}_c$,

$$[\gamma_i, \gamma_j, \gamma_k, \gamma_\ell] = [f(\gamma_i), f(\gamma_j), f(\gamma_k), f(\gamma_\ell)]$$
$$= [\gamma_{f_*(i)}, \gamma_{f_*(j)}, \gamma_{f_*(k)}, \gamma_{f_*(\ell)}],$$

because $f \in \mathcal{M}^+$. The hypothesis gives $\{i, j, k, \ell\} = \{f_*(i), f_*(j), f_*(k), f_*(\ell)\}$, so f_* maps $\{i, j, k, \ell\}$ onto itself. Hence, f_* maps each four element subset of $\mathbb{Z}_c$ into itself. This implies f_* fixes every element in $\mathbb{Z}_c$. To see this consider any $m \in \mathbb{Z}_c$. Because $c \geq 5$, there exist distinct $i, j, k, \ell \in \mathbb{Z}_c \setminus \{m\}$. Now, $\{m\} = \{i, j, k, m\} \setminus \{i, j, k, \ell\}$, and so

$$\{f_*(m)\} = f_*(\{i, j, k, m\} \setminus \{i, j, k, \ell\})$$
$$= f_*(\{i, j, k, m\}) \setminus f_*(\{i, j, k, \ell\})$$
$$= \{i, j, k, m\} \setminus \{i, j, k, \ell\}$$
$$= \{m\}.$$

Thus, $f_* = \iota$ and so $f = I$. Consequently, $\mathcal{A}^+(\Omega)$ is trivial. $\square$

Example 13.2. Theorem 13.1 makes it easy to exhibit explicit punctured spheres with no conformal symmetry. Let $\Omega = \mathbb{C}_\infty \setminus \{0, 1, 3, 6, 10\}$. Then

$$[0, 1, 3, 6] = 5, \quad [1, 3, 6, 10] = \frac{35}{8}, \quad [0, 1, 3, 10] = \frac{27}{7},$$
$$[0, 1, 6, 10] = \frac{27}{2}, \quad [0, 3, 6, 10] = \frac{7}{3}.$$

Theorem 13.1 requires that all of the cross ratios determined by permuting these points must also be distinct. This will be true if the images of these five numbers under the function F are distinct. Rather than calculating these values, we note the formula for F' shows it to be positive on the interval $(2, \infty)$. Because the five cross-ratios above are distinct and larger than 2, we know their images under F are distinct. Hence, Ω has no conformal symmetry.

Theorem 13.3. *The generic sphere with at least five punctures has a trivial conformal automorphism group.*

Proof. Fix $c \geq 5$. The parameter space is the set $(\gamma_1, \ldots, \gamma_c)$ of ordered c-tuples of distinct points in $\mathbb{C}_\infty$. The parameter space is an open subset $\mathcal{P}$ of $\mathbb{C}_\infty^c$. Let ν denote the pull-back to $\mathbb{C}_\infty$ via stereographic projection of the surface measure on $\mathbb{S}$. Then ν is a finite measure on $\mathbb{C}_\infty$ and so induces a finite measure μ on the product space $\mathbb{C}_\infty^c$. Let μ_0 denote the restriction of μ to $\mathcal{P}$. The set of c-tuples in $\mathcal{P}$ that have at least one entry equal to ∞ has μ_0 measure zero. Let $\mathcal{P}_0$ denote the intersection of $\mathcal{P}$ with $\mathbb{C}^c$. This is again an open set and has the same μ_0-measure as $\mathcal{P}$. For a pair, $\{i,j,k,\ell\}$ and $\{p,q,r,s\}$, of distinct subsets of $\mathbb{Z}_c$ the equation

$$[z_i, z_j, z_k, z_\ell] = [z_p, z_q, z_r, z_s]$$

defines an algebraic subvariety of complex co-dimension at least one in $\mathbb{C}^c$. If $\mathcal{P}_1$ is the union of these subvarieties over all pairs, $\{i,j,k,\ell\}$ and $\{p,q,r,s\}$, of distinct subsets of $\mathbb{Z}_c$, then $\mathcal{P}_1$ has measure zero. Finally, we show that if $(\gamma_1, \ldots, \gamma_c) \in \mathcal{P}_0 \backslash \mathcal{P}_1$, then $\mathcal{A}^+(\Omega)$ is trivial. This is simple to see. The assumption that $(\gamma_1, \ldots, \gamma_c) \in \mathcal{P}_0 \backslash \mathcal{P}_1$ means that inequality (13.1) holds for all pairs, $\{i,j,k,\ell\}$ and $\{p,q,r,s\}$, of distinct subsets of $\mathbb{Z}_c$. Therefore, f is the identity and $\mathcal{A}^+(\Omega)$ is trivial. $\square$

Example 13.4. In this example we investigate the conformal symmetries of the region $\Omega(u) = \mathbb{C}_\infty \backslash \{0, 1, 2, 3, u\}$, where $u \in \mathbb{R} \cup \{\infty\} \backslash [0, 3]$. Let g be a conformal automorphism of $\Omega(u)$. As isolated points are removable singularites of g, we see that g is a Möbius map that preserves $\mathbb{R}_\infty$ (which is $\mathbb{R} \cup \infty$). Now in general, if f is an elliptic Möbius map, and if Γ is a circle that is invariant under f, then either (a) the fixed points of f are on Γ, or (b) the fixed points of f are inverse with respect to Γ. Moreover, in case (a), f is of order two and it interchanges the two intervals of Γ obtained by removing the fixed points from Γ. This is true when f fixes 0 and ∞, and as the result is invariant under conjugation, it is true in general.

If g satisfies (a) then it must interchange two pairs of points and fix the fifth point. There are five possibilities for the fixed point, and each possibility gives rise to a unique choice of g as g must interchange the two punctures that are 'adjacent' to the fixed puncture, and also the two punctures that are 'farthest' from the fixed puncture (again, consider the conjugate situation in which $g(z) = -z$). It follows that there are exactly five possibilities for g and u and these are as follows:

(1) $u = -6$, $g(0) = 0$ and $g(z) = 6z/(5z-6)$;
(2) $u = -1$, $g(1) = 1$ and $g(z) = 2-z$;
(3) $u = 4$, $g(2) = 2$ and $g(z) = 4-z$;
(4) $u = 9$, $g(3) = 3$ and $g(z) = (9z-9)/(5z-9)$;
(5) $u = \infty$, $g(u) = u$ and $g(z) = 3-z$.

Now consider any g that arises in case (b). The cyclic group $\langle g \rangle$ generated by g acts as a permutation of $\{0, 1, 2, 3, u\}$ without fixed points, so that each point lies in an orbit whose length is the order of g. Thus the order of g divides five, whence g is of order five. It follows that for some integer m, we have $g^m(0) = 1$. As the relative order of the five punctures on $\mathbb{R}_\infty$ is preserved, we must also have $g^m(1) = 2$ and $g^m(2) = 3$. However, these conditions imply that $g^m(z) = z+1$ which is not elliptic; thus there are no conformal automorphisms that are in case (b). It follows that $\Omega(u)$ has a non-trivial conformal symmetry if and only if $u \in \{-6, -1, 4, 9, \infty\}$. In each case $\mathcal{A}^+(\Omega)$ is cyclic of order two, and $\mathcal{A}(\Omega)$ is the Klein four-group.

14. MÖBIUS EQUIVALENCE AND ABSOLUTE CROSS-RATIOS

This section is the analog of Section 11 for Möbius maps and absolute cross-ratios. In Section 6 we identified $S(\Omega)$ for circular regions of finite connectivity at least three, but provided no desription of $S^+(\Omega)$. In Section 11 the opposite was accomplished for punctured spheres. In this section we determine $S(\Omega)$ for punctured spheres.

The *absolute cross-ratio* of four points γ_j, $j = 1, 2, 3, 4$, in $\mathbb{C}_\infty$ is

$$|\gamma_1, \gamma_2, \gamma_3, \gamma_4| = \big|[\gamma_1, \gamma_2, \gamma_3, \gamma_4]\big|.$$

The absolute cross-ratio is invariant under the group $\mathcal{M}$ of all Möbius maps of $\mathbb{C}_\infty$ [2, p.32].

Theorem 14.1. *Suppose* $\mathbb{C}_\infty \setminus \{\gamma_1, \ldots, \gamma_c\}$ *and* $\mathbb{C}_\infty \setminus \{\gamma'_1, \ldots, \gamma'_c\}$ *are punctured spheres with* $c \geq 4$. *There is a Möbius map* f *with* $f(\gamma_j) = \gamma'_j$ *for* $j = 1, \ldots, c$ *if and only if*

(14.1) $\qquad |\gamma_i, \gamma_j, \gamma_k, \gamma_\ell| = |\gamma'_i, \gamma'_j, \gamma'_k, \gamma'_\ell|$ *for all distinct* $i, j, k, \ell \in \mathbb{Z}_c$.

If the points $\{\gamma_1, \ldots, \gamma_c\}$ *are not concyclic, then* f *is unique. If* $\{\gamma_1, \ldots, \gamma_c\}$ *lie on a circle* Γ, *then* f *is unique up to precomposition with reflection across* Γ.

Proof. If f exists, then the Möbius invariance of cross-ratios implies that (14.1) holds. Now, suppose (14.1) holds; we establish the existence of f by induction.

First, assume $c = 4$. Without loss of generality we may suppose $\gamma_1 = \gamma'_1 = 0$, $\gamma_2 = \gamma'_2 = 1$ and $\gamma_3 = \gamma'_3 = \infty$. Then

$$|\gamma_4| = |0, 1, \gamma_4, \infty| = |0, 1, \gamma'_4, \infty| = |\gamma'_4|$$

and
$$|1-\gamma_4| = |1,0,\gamma_4,\infty| = |1,0,\gamma_4',\infty| = |1-\gamma_4'|.$$

If $\gamma_4 = \gamma_4'$, we choose $f = I$. Otherwise, the identities above show that 0 and 1 are equidistant from γ_4 and γ_4', so $\mathbb{R}$ is the perpendicular bisector of the segment joining γ_4 and γ_4'. This implies $\gamma_4' = \bar{\gamma}_4$. In this situation we take $f(z) = \bar{z}$.

Next, suppose $c \geq 5$. The proof falls into cases according to whether or not the points $\gamma_1, \ldots, \gamma_c$ are concyclic. We begin with the case in which these points are concyclic. By using the induction hypothesis, we can assume that $\gamma_j = \gamma_j'$ for $j = 1, \ldots, c-1$. By making use of a preliminary Möbius transformation we can assume $\gamma_1 = \gamma_1' = 0$, $\gamma_2 = \gamma_2' = 1$, and $\gamma_3 = \gamma_3' = \infty$. Because the points γ_j are concyclic, all are real. By using the same argument as in the case $c = 4$ we now deduce γ_c' is γ_c since γ_c is real. Thus, $\Omega' = \Omega$.

Now, suppose $\gamma_1, \ldots, \gamma_c$ are not concyclic. Let Γ be the circle determined by γ_1, γ_2 and γ_3. Because the points γ_j are not concyclic, there exists j, $4 \leq j \leq c$, such that γ_j does not lie on Γ. By renumbering if necessary we may assume $j = 4$. By the induction hypothesis we may assume $\gamma_j = \gamma_j'$ for $j = 1, \ldots, c-1$, and we may further assume $\gamma_1 = \gamma_1' = 0$, $\gamma_2 = \gamma_2' = 1$, $\gamma_3 = \gamma_3' = \infty$. Now, γ_4 is not real. As before, we can show γ_c' is either γ_c or $\bar{\gamma}_c$. If γ_c is real then $\gamma_c' = \gamma_c$. Now suppose that γ_c is not real. Since
$$|\gamma_4, 0, \gamma_c, \infty| = \frac{|\gamma_4 - \gamma_c|}{|\gamma_4|}, \quad |\bar{\gamma}_4, 0, \gamma_c, \infty| = \frac{|\bar{\gamma}_4 - \gamma_c|}{|\bar{\gamma}_4|},$$
and γ_4 is not real, $|\gamma_4, 0, \gamma_c, \infty| \neq |\bar{\gamma}_4, 0, \gamma_c, \infty|$. Hence, $\gamma_c' = \bar{\gamma}_c$ is not possible, so $\gamma_c' = \gamma_c$ and $\Omega' = \Omega$. □

Corollary 14.2. *Suppose $\Omega = \mathbb{C}_\infty \setminus \{\gamma_1, \ldots, \gamma_c\}$, where $c \geq 4$. Then*
$$S(\Omega) =$$
$$\left\{\rho \in S_c : |\gamma_i, \gamma_j, \gamma_k, \gamma_\ell| = |\gamma_{\rho(i)}, \gamma_{\rho(j)}, \gamma_{\rho(k)}, \gamma_{\rho(\ell)}| \text{ for all distinct } i, j, k, \ell \in \mathbb{Z}_c\right\}$$
and $S(\Omega) = S^+(\Omega) \cup S^-(\Omega)$, where
$$S^-(\Omega) =$$
$$\left\{\rho \in S_c : [\gamma_i, \gamma_j, \gamma_k, \gamma_\ell] = \overline{[\gamma_{\rho(i)}, \gamma_{\rho(j)}, \gamma_{\rho(k)}, \gamma_{\rho(\ell)}]} \text{ for all distinct } i, j, k, \ell \in \mathbb{Z}_c\right\}.$$

Proof. Call a permutation $\rho \in S_c$ *admissible (in the absolute cross-ratio sense)* for Ω if
$$|\gamma_i, \gamma_j, \gamma_k, \gamma_\ell| = |\gamma_{\rho(i)}, \gamma_{\rho(j)}, \gamma_{\rho(k)}, \gamma_{\rho(\ell)}|$$
for all distinct $i, j, k, \ell \in \mathbb{Z}_c$. Theorem 14.1 implies that ρ is admissible if and only if there exists $f \in \mathcal{A}(\Omega)$ with $f_* = \rho$.

To show $S(\Omega) = S^+(\Omega) \cup S^-(\Omega)$, consider $\rho \in S(\Omega)$. Then there exists $f \in \mathcal{A}(\Omega)$ with $f_* = \rho$, so $\rho \in S^+(\Omega)$ or $\rho \in S^-(\Omega)$, according as f is conformal or anti-conformal. □

As an application of Corollary 14.2 we determine the possible automorphism groups for four punctured spheres.

Theorem 14.3. *Suppose* $\Omega = \mathbb{C}_\infty \setminus \{\gamma_1, \gamma_2, \gamma_3, \gamma_4\}$ *and* $\lambda = [\gamma_1, \gamma_2, \gamma_3, \gamma_4]$. *Set* $E = \mathbb{R} \cup_{j=2}^4 \Gamma_j$, *where* $\Gamma_2 = \{z : \operatorname{Re} z = \frac{1}{2}\}$, $\Gamma_3 = \{z : |z-1| = 1\}$ *and* $\Gamma_4 = \{z : |z| = 1\}$.
(a) *If* $\lambda \in \mathbb{C} \setminus E$, *then* $\mathcal{A}(\Omega) \cong D_4$, *the Klein four-group.*
(b) *If* $\lambda \in \mathbb{R} \setminus \{\frac{1}{2}, 2, -1\}$, *then* $\mathcal{A}(\Omega) \cong D_4 \times C_2$.
(c) *If* $\lambda \in \{\frac{1}{2}, 2, -1\}$, *then* $\mathcal{A}(\Omega) \cong D_8 \times C_2$.
(d) *If* $\lambda \in (\Gamma_1 \cup \Gamma_2 \cup \Gamma_3) \setminus \{\frac{1}{2}, 2, -1, \eta, \bar{\eta}\}$, *then* $\mathcal{A}(\Omega) \cong D_8$.
(e) *If* $\lambda \in \{\eta, \bar{\eta}\}$, *then* $\mathcal{A}(\Omega) \cong S_4$.

Proof. This result can be established in the same manner as the proof of Theorem 12.1 by using the results of Sections 9 and 10. We omit the details. We do point out that $D_4 \times C_2$ is not isomorphic to D_8 because D_8 has four elements of order 2 while all of the nontrivial elements in $D_4 \times C_2$ have order 2. □

Example 14.4. Let $\Omega = \mathbb{C}_\infty \setminus \{2, -2, i, -i\}$. Then

$$[2, -2, i, -i] = \frac{1}{2} + \frac{3i}{8} \in \Gamma_2, \quad [i, -2, 2, -i] = \frac{32}{25} - \frac{24i}{25} \in \Gamma_3,$$

$$[-i, -2, i, 2] = -\frac{7}{25} - \frac{24i}{25} \in \Gamma_4,$$

and so $\mathcal{A}(\Omega) \cong D_8$. In fact, $\mathcal{A}^+(\Omega) = \{z, -z, 2i/z, -2i/z\}$ and $\mathcal{A}(\Omega)$ is obtained by adjoining the reflection $\bar{z}$. The rotary reflections $2i/\bar{z}$ and $-2i/\bar{z}$ are the only elements of order 4.

There is a simple connection between inversive distance and absolute cross-ratio that should be mentioned. For four points a_j in $\mathbb{C}$ and $r_j > 0$ such that the four circles $\Gamma_j = \{z : |z - a_j| = r_j\}$ are mutually exterior, one finds

$$\frac{(\Gamma_1, \Gamma_3)(\Gamma_2, \Gamma_4)}{(\Gamma_1, \Gamma_2)(\Gamma_3, \Gamma_4)} = \frac{(|a_1 - a_3|^2 - r_1^2 - r_3^2)(|a_2 - a_4|^2 - r_2^2 - r_4^2)}{(|a_1 - a_2|^2 - r_1^2 - r_2^2)(|a_3 - a_4|^2 - r_3^2 - r_4^2)}.$$

Then as all $r_j \to 0$

$$\frac{(\Gamma_1, \Gamma_3)(\Gamma_2, \Gamma_4)}{(\Gamma_1, \Gamma_2)(\Gamma_3, \Gamma_4)} \to |a_1, a_2, a_3, a_4|^2,$$

so the absolute cross-ratio is a limiting case of ratios of inversive distances. By allowing some, but not all of the r_j, tend to zero one obtains a Möbius invariant expression that expresses a "distance" involving k circles and $4 - k$ points. This quantity could be useful for investigating generalized circular regions.

15. The internal direct product of automorphisms

A group G with identity e is the *internal direct product* of two of its subgroups H and K if (i) every element g of G can be expressed in the form $g = hk$, where $h \in H$ and $k \in K$, (ii) $H \cap K = \{e\}$, and (iii) every element of H commutes with every element of K. In this case the expression hk for g is unique, and $(h, k) \mapsto hk$ is an isomorphism of the external direct product $H \times K$ onto G. We show that in certain circumstances $\mathcal{A}(\Omega)$ is the internal direct product of $\mathcal{A}^+(\Omega)$ and a subgroup of order two. In view of (iii) we must investigate when a conformal Möbius map commutes with an anti-conformal involution. Now we have seen that an anti-conformal Möbius involution is is either a reflection across a circle, or a rotary reflection that is conjugate to $z \mapsto -1/\bar{z}$. Each Möbius map has a natural extension to the upper half $\mathbb{H}^3$ of $\mathbb{R}^3$, where it acts as a hyperbolic isometry, and the next lemma shows the importance of viewing Möbius maps in this way.

Lemma 15.1. *Suppose that h is an anti-conformal Möbius map of order two, and let Σ_h be the set of fixed points of h in $\mathbb{H}^3$. Then h commutes with a conformal Möbius map f if and only if $f(\Sigma_h) = \Sigma_h$.*

Proof. Suppose first that $hf = fh$, and take any point x in Σ_h. Then $h(f(x)) = f(h(x)) = f(x)$, so that h fixes $f(x)$. Thus $f(\Sigma_h) \subset \Sigma_h$. As f^{-1} also commutes with h we see that $f^{-1}(\Sigma_h) \subset \Sigma_h$, so that $f(\Sigma_h) = \Sigma_h$.

Now suppose that $f(\Sigma_h) = \Sigma_h$, and consider the two cases: (i) when h is a reflection across a circle Γ, and (ii) when h is an involutary rotary reflection with unique fixed point ζ in $\mathbb{H}^3$. In case (i) $\Gamma \subset \partial \Sigma_h$. Now take any z on Γ. Then $f(z) \in \Gamma$ so that $hf(z) = f(z) = fh(z)$. Thus $h^{-1}f^{-1}hf$ is a conformal Möbius map that fixes each point of Γ, and so is I. Thus in case (i), $hf = fh$. In case (ii) we have $f(\zeta) = \zeta$. By conjugation, we may assume that $h(z) = -1/\bar{z}$; then $\zeta = (0,0,1)$. Now f fixes ζ if and only if $f(z) = (az + b)/(-\bar{b}z + \bar{a})$, for some a and b with $|a|^2 + |b|^2 = 1$ [2, pp. 17 and 63]. It is now easy to check that any such f commutes with h. □

We can now apply this lemma to internal direct products of automorphisms.

Theorem 15.2. *Suppose that h is an anti-conformal automorphism of a circular region Ω. Then $\mathcal{A}(\Omega)$ is the internal direct product $\mathcal{A}^+(\Omega) \times \{I, h\}$ if and only if either*

(a) *h is a reflection across some circle Γ, and $f(\Gamma) = \Gamma$ for every f in $\mathcal{A}^+(\Omega)$,*

or

(b) *h is a rotary reflection of order two with (unique) fixed point ζ in $\mathbb{H}^3$, and $f(\zeta) = \zeta$ for every f in $\mathcal{A}^+(\Omega)$.*

Proof. Suppose that $\mathcal{A}(\Omega)$ is the internal direct product $\mathcal{A}^+(\Omega) \times \{I, h\}$. Since $\{I, h\}$ is a group, $h^2 = I$, so that h is an anti-conformal automorphism of Ω. Thus h is either a reflection across some circle Γ, or an involutary rotary reflection. As h commutes with every f in $\mathcal{A}^+(\Omega)$, Lemma 15.1 implies that either (a) or (b) holds.

Conversely, suppose that either (a) or (b) holds. In both cases h is an anti-conformal Möbius involution and, by assumption, h is an automorphism of Ω. By Lemma 15.1, h commutes with every element of $\mathcal{A}^+(\Omega)$. The coset decomposition $\mathcal{A}(\Omega) = \mathcal{A}^+(\Omega) \cup \mathcal{A}^+(\Omega)h$, and the fact that $\mathcal{A}^+(\Omega) \cap \{I, h\} = I$ ensure that $\mathcal{A}(\Omega)$ is the given internal direct product. □

It is possible for $\mathcal{A}(\Omega)$ to be an internal direct product $\mathcal{A}^+(\Omega) \times \{I, h\}$ even when Ω is not strongly symmetric. For example, consider a region Ω that is not strongly symmetric, but that is symmetric about one, and only one, circle Γ. Let h be the reflection across Γ. If $f \in \mathcal{A}^+(\Omega)$ then, by the uniqueness of Γ, $f(\Gamma) = \Gamma$, so that $\mathcal{A}(\Omega)$ is the internal direct product $\mathcal{A}^+(\Omega) \times \{I, h\}$. Note that if Ω is not strongly symmetric then $\{I, h\}$ is not the kernel of the homomorphism Θ_*. Finally, as there are essentially only five possibilities for $\mathcal{A}^+(\Omega)$, Theorem 15.2 would enable us to give a complete characterization of those situations in which $\mathcal{A}(\Omega)$ is the internal direct product $\mathcal{A}^+(\Omega) \times \{I, h\}$. We leave this to the interested reader.

16. A GEOMETRIC VIEW

In this last section of the paper we consider the geometry that lies behind our earlier work (for more details see [2], Theorem 3.4.1 and pp. 84-86, [8], [18] and [19]). We begin by describing the ideas behind Theorem A, and Heins' theorem, and many of our arguments. The Möbius group $\mathcal{M}$ acting on $\mathbb{C}_\infty$ is generated by reflections across circles. More generally, the group $\mathcal{M}_3$ acting on $\mathbb{R}^3 \cup \{\infty\}$ is generated by reflections in spheres (and planes) in $\mathbb{R}^3$, and the subgroup consisting of the even numbers of composition of reflections is denoted by $\mathcal{M}_3^+$. A circle in $\mathbb{C}_\infty$ can be written as $\Pi \cap \mathbb{C}_\infty$, where Π is a sphere (or plane) in $\mathbb{R}^3$ that is orthogonal to $\mathbb{C}_\infty$, and the reflection across C is the restriction of the reflection across Π to $\mathbb{C}_\infty$. Thus each Möbius map in $\mathcal{M}$ acts on $\mathbb{C}_\infty$, but is also the restriction of some element of $\mathcal{M}_3^+$ that leaves the upper half-space $\mathbb{H}^3 = \{(x_1, x_2, x_3) : x_3 > 0\}$ invariant. The half-space $\mathbb{H}^3$ is a model of hyperbolic space, and each Möbius map in $\mathcal{M}$ is an isometry with respect to its hyperbolic metric $ds = |dx|/x_3$.

The proof that *any finite subgroup of* $\mathcal{M}^+$ *is conjugate in* $\mathcal{M}_3^+$ *to a finite subgroup of* SO(3), and hence is isomorphic to one of the groups in Theorem A, is classical. Let G be any finite subgroup of $\mathcal{M}^+$. Choose any point in $\mathbb{H}^3$ and let its G-orbit be $\mathcal{O}$. As G is finite, $\mathcal{O}$ is a finite subset of $\mathbb{H}^3$. A simple compactness argument shows that there is a unique smallest closed hyperbolic sphere $\mathcal{S}$ in $\mathbb{H}^3$ that contains $\mathcal{O}$, and as $\mathcal{O}$ is invariant under G, so too is $\mathcal{S}$. It follows that the hyperbolic centre, say ζ, of $\mathcal{S}$ is also invariant under G; thus ζ is fixed by every element of G in its action on $\mathbb{H}^3$.

Exactly as in the two-dimensional case, the unit ball $\mathbb{B}^3$ is also a model of three-dimensional hyperbolic space with metric $2|dx|/(1 - ||x||^2)$, and there is an element g, say, of $\mathcal{M}_3^+$ that maps $\mathbb{H}^3$ onto $\mathbb{B}^3$ and preserves hyperbolic distances. We can construct g so that $g(\zeta) = 0$, and then the conjugate group gGg^{-1} is a group of hyperbolic isometries of $\mathbb{B}^3$, every element of which fixes the origin. Any hyperbolic isometry of $\mathbb{B}^3$ that fixes 0 is an orthogonal map of $\mathbb{R}^3$ onto itself, so that G is conjugate in $\mathcal{M}_3^+$ to to a finite subgroup of SO(3). It is well-known (and classical) that the finite subgroups of SO(3) are the symmetry groups of the five Platonic solids and the plane lamina, and these are the groups listed in Theorem A.

The relevant facts about the geometric actions of the finite subgroups of SO(3) are as follows. Let G be a finite subgroup of SO(3) acting on the unit sphere $\mathbb{S}$, and suppose that $\mathcal{O}_1, \ldots, \mathcal{O}_s$ is a complete list of the (finite number of) orbits with a non-trivial stabilizer. Obviously, $\mathcal{O}_1 \cup \ldots \cup \mathcal{O}_s$ is G-invariant. Let n_j be the order of the stabilizer of each point in $\mathcal{O}_j$; thus for each j, $|G| = n_j |\mathcal{O}_j|$ (where $|E|$ denotes the cardinality of a set E). Then there are only five possibilities for G, and these are as follows.

(1) $G = C_m$, of order m: there are two orbits of length 1, and all other orbits have length m.

(2) $G = D_{2m}$, of order $2m$: there is one orbit of length 2, there are two orbits of length m, and all other orbits have length $2m$.

(3) $G = A_4$, of order 12: there is one orbit of length 6, there are two orbits of length 4, and all other orbits have length 12.

(4) $G = S_4$, of order 24: there is one orbit of length 6, one of length 8, one of length 12, and all other orbits have length 24.

(5) $G = A_5$, of order 60: there is one orbit of length twelve, one of length twenty, one of length thirty, and all other orbits have length sixty.

We can now obtain information on the group $\mathcal{A}^+(\Omega)$. In the conjugation described above, any closed disc in $\mathbb{C}_\infty$ is transferred to a Eulidean disc on the unit sphere $\mathbb{S}$. If Ω, of connectivity c, is a circular region in $\mathbb{C}_\infty$ then $\mathcal{A}^+(\Omega)$ is conjugate to a finite subgroup, say $\mathcal{A}$, of SO(3) that permutes c given closed discs on $\mathbb{S}$. Let p_i denote the point components, and D_j be the circles (of positive radius) that form the complement of the image of Ω on $\mathbb{S}$. As the elements of $\mathcal{A}$ are Euclidean isometries, we can replace each disc D_j by its centre, say p'_j, and so $\mathcal{A}$ now acts on $\mathbb{S}$ as a finite group of rotations of $\mathbb{R}^3$ that permutes each of two (disjoint) sets $\{p_i\}$ and $\{p'_j\}$ on $\mathbb{S}$. Of course, one of these sets may be empty. This discussion leads immediately to our next theorem (which includes Theorem 1.2).

Theorem 16.1. *Let Ω be the circular region given by (3.1). If some K_j is invariant, then $\mathcal{A}^+(\Omega)$ is cyclic. If the orbit of some K_j contains q components, where q is odd and $q \geq 3$ then $\mathcal{A}^+(\Omega)$ is either cyclic or dihedral. In particular, if $c(\Omega)$ is odd, then $\mathcal{A}^+(\Omega)$ is either cyclic or dihedral.*

Proof. We need only consider the group $\mathcal{A}$ which acts on $\mathbb{S}$, and which permutes the disjoint sets $\{p_i\}$ and $\{p'_j\}$ on $\mathbb{S}$ (see the discussion above). If some K_j is invariant under $\mathcal{A}^+(\Omega)$ then one of the points p_i or p'_j is invariant under

$\mathcal{A}$ so that $\mathcal{A}$, and hence $\mathcal{A}^+(\Omega)$, is cyclic. If the orbit of some K_j has cardinality q, where q is odd and $q \geq 3$, then the action of $\mathcal{A}$ on $\mathbb{S}$ has an orbit of cardinality q. An examination of the possible groups and their geometric actions shows that $\mathcal{A}$ cannot be any of A_4, S_4 or A_5. The only alternatives are C_n or D_n for some n. If $c(\Omega)$ is odd then some orbit of some K_j has odd cardinality and the last result follows immediately. $\square$

Now consider a circular region Ω bounded by $\Gamma_1, \ldots, \Gamma_c$. Each circle Γ_j determines a unique sphere Σ_j in $\mathbb{R}^3$ that is orthogonal to $\mathbb{R}^2$. Let $\Sigma_j^+ = \Sigma_j \cap \mathbb{H}^3$. Then each Σ_j^+ is a hyperbolic plane in $\mathbb{H}^3$, and it is known that $(\Gamma_i, \Gamma_j) = \cosh d_h(\Sigma_i^+, \Sigma_j^+)$, where $d_h(\Sigma, \Sigma')$ denotes the hyperbolic distance between the hyperbolic planes Σ and Σ'. The region in $\mathbb{H}^3$ that is bounded by the Σ_j^+ is, in fact, an unbounded hyperbolic polyhedron Π, and the inversive distances (Γ_i, Γ_j) are, in some sense, the geometric parameters of this polyhedron. As each conformal automorphism of Ω extends to a hyperbolic isometry that preserves the set of Σ_j^+, we may regard $\mathcal{A}^+(\Omega)$ as the group of hyperbolic rigid motions of the hyperbolic polyhedron Π. Thus, just as we have already invoked the symmetry groups of certain Euclidean polyhedra, we now do the same for hyperbolic polyhedra.

These ideas lead us to an alternative view of strong symmetry. Suppose that Ω is a strongly symmetric circular region, and let Γ be the axis of strong symmetry. Let Δ be one of the discs bounded by Γ. Then $\Omega \cap \Delta$ is a hyperbolic polygon bounded by the geodesics $\Gamma_j \cap \Delta$, and $\mathcal{A}^+(\Omega)$ is just the group of conformal hyperbolic symmetries of the polygon $\Gamma_j \cap \Delta$. Thus, in this sense, we may regard the situation when Ω is strongly symmetric as a lower dimensional version of the problem of the symmetries of an unbounded hyperbolic polyhedron in $\mathbb{H}^3$.

In conclusion, we mention four problems for further investigations.

Question 1 *Is the conclusion of Theorem 4.1 valid when $\Gamma_1, \ldots, \Gamma_c$ are any set of c distinct circles in $\mathbb{C}_\infty$?* Here the Γ_j are allowed to be intersecting, or tangent, to each other.

Question 2 A Möbius map acting on $\mathbb{R}^n \cup \{\infty\}$ is a composition of reflections in $(n-1)$-dimensional spheres, and the inversive distance between two such spheres is a Möbius invariant. *Which of our results remain true in all dimensions?* When $n \geq 3$ the only conformal maps in $\mathbb{R}^n$ are the Möbius maps, so the original question about the conformal automorphisms of an arbitrary region is considerably less tractable in higher dimensions.

Question 3 Suppose that h is a conformal mapping of a region Δ onto Ω, where Ω is a circular region of finite connectivity c. Let the boundary component of Δ and Ω be X_i and Γ_i, respectively, where h maps X_i onto Γ_i in sense of boundary correspondence. Now define the inversive distance (X_i, X_j) to be (Γ_i, Γ_j); then (X_i, X_j) does not depend on the mapping h, and it is conformally invariant. Note that here the X_i need not be circles. *How can we give an intrinsic description of the conformal invariant (X_i, X_j)?*

Question 4 Suppose that Ω and Ω' are two circular regions of the same connectivity, and suppose that φ is a K-quasiconformal map of Ω onto Ω'. Is it possible to obtain bounds on K in terms of the inversive distances between pairs of circles in $\partial\Omega$, and in $\partial\Omega'$?

References

[1] Ahlfors, L.V., *Complex Analysis*, Third Edn., McGraw-Hill, 1979.
[2] Beardon, A.F., *The geometry of discrete groups*, Springer-Verlag, 1983.
[3] Beardon, A.F., Conformal automorphisms of plane domains, *J. London Math. Soc.* (2), **32** (1985), 245-253.
[4] Burckel, R.B. and Minda, D., A characterization of simple and double connectivity, *Exposition. Math.*, **16** (1998), 133-144.
[5] Conway, J.B., *Functions of One Complex Variable II*, Springer-Verlag, New York 1995.
[6] Erkama, T., Möbius automorphisms of plane domains, *Ann. Acad. Sci. Fenn. Ser. A I Math.*, **10** (1985), 155-162.
[7] Everitt, B., A family of conformally asymmetric Riemann surfaces, *Glasgow Math. J.*, **39** (1997), 221-225.
[8] Ford, L.R., *Automorphic functions*, Chelsea 1951.
[9] Greenberg, L., Conformal transformations of Riemann surfaces, *Amer. J. Math.*, **82** (1960), 749-760.
[10] He, Zheng-Xu and Schramm, O., Fixed points, Koebe uniformization and circle packings, *Ann. of Math.* (2), **137** (1993), 369-406.
[11] Heins, M., On the number of 1-1 directly conformal maps, which a multiply plane connected region of finite connectivity $p > 2$ admits onto itself, *Bull. Amer. Math. Soc.*, **52** (1946), 454-457.
[12] Heins, M., A note on a theorem of Radó concerning the $(1, m)$ conformal maps of a multiply-connected region into itself, *Bull. Amer. Math. Soc.*, **47** (1941), 128-130.
[13] Macbeath, A.M., On a curve of genus 7, *Proc. London Math. Soc.* (3), **15** (1965), 527-542.
[14] MacLachlan, C., Abelian groups of automorphisms of compact Riemann surfaces, *Proc. London Math. Soc.* (3), **15** (1965), 699-712.
[15] Maskit, B., The conformal group of a plane domain, *Amer. J. Math.*, **90** (1968), 718-722.
[16] Moore, E.H., The cross-ratio group of $n!$ Cremona transformations of order $n - 3$ in Flat Space of $n - 3$ dimensions, *Amer. J. Math.*, **22** (1900), 279-291.
[17] Mueller, C. and Rudin, W., Proper holomorphic self-maps of plane regions, *Complex Variables*, **17** (1991), 113-121.
[18] Ratcliffe, J.G., *Foundations of Hyperbolic Manifolds*, Graduate Texts 149, Springer-Verlag, 1994.
[19] Tsuji, M., *Potential theory in modern function theory*, Maruzen, 1959.
[20] Volynec, I.A., Groups of conformal automorphisms of plane domains in the uniform metric, *Complex Variables*, **19** (1992), 195-203.

Centre for Mathematical Sciences, University of Cambridge, Wilberforce Road, Cambridge CB3 0WB, England
E-mail address: A.F.Beardon@dpmms.cam.ac.uk

Department of Mathematical Sciences, University of Cincinnati, Cincinnati, Ohio 45221-0025, USA
E-mail address: David.Minda@math.uc.edu

MEROMORPHIC FUNCTIONS WITH TWO COMPLETELY INVARIANT DOMAINS

WALTER BERGWEILER AND ALEXANDRE EREMENKO

Dedicated to the memory of Professor I. N. Baker

ABSTRACT. We show that if a meromorphic function has two completely invariant Fatou components and only finitely many critical and asymptotic values, then its Julia set is a Jordan curve. However, even if both domains are attracting basins, the Julia set need not be a quasicircle. We also show that all critical and asymptotic values are contained in the two completely invariant components. This need not be the case for functions with infinitely many critical and asymptotic values.

1. INTRODUCTION AND MAIN RESULT

Let f be a meromorphic function in the complex plane $\mathbb{C}$. We always assume that f is not fractional linear or constant. For the definitions and main facts of the theory of iteration of meromorphic functions we refer to a series of papers by Baker, Kotus and Lü [2, 3, 4, 5], who started the subject, and to the survey article [8]. For the dynamics of rational functions we refer to the books [7, 11, 21, 25].

A *completely invariant domain* is a component D of the set of normality such that $f^{-1}(D) = D$. There is an unproved conjecture (see [4, p. 608], [8, Question 6]) that a meromorphic function can have at most two completely invariant domains. For rational functions this fact easily follows from Fatou's investigations [15], and it was first explicitly stated by Brolin [10, §8]. Moreover, if a rational function has two completely invariant domains, then their common boundary is a Jordan curve on the Riemann sphere, and each of the domains coincides with the basin of attraction of an attracting or superattracting fixed point, or of an attracting petal of a neutral fixed point with multiplier 1; see [15, p. 300-303] and [10]. All critical values of f are contained in the completely invariant domains.

In this paper we extend these results to a class of transcendental meromorphic functions in $\mathbb{C}$. This class S consists of meromorphic functions with finitely many critical and asymptotic values. Let $A = A(f)$ be the set of

---

Supported by the German-Israeli Foundation for Scientific Research and Development (G.I.F.), grant no. G -643-117.6/1999.

Supported by NSF grants DMS-0100512 and DMS-0244421, and by the Humboldt Foundation.

critical and asymptotic values. We also call the elements of A *singular values* of f. For $f \in S$ the map
$$f : \mathbb{C}\backslash f^{-1}(A) \to \overline{\mathbb{C}}\backslash A$$
is a covering. By $J = J(f) \subset \mathbb{C}$ we denote the Julia set of f.

Baker, Kotus and Lü [4, Theorem 4.5] proved that functions of the class S have at most two completely invariant domains. Cao and Wang [12, Theorem 1] have shown that if a function in the class S has two completely invariant domains, then its Fatou set is the union of these domains. We complement these results with the following

Theorem. *Let f be a function of the class S, having two completely invariant domains D_j, $j = 1, 2$. Then*

(i) *each D_j is the basin of attraction of an attracting or superattracting fixed point, or of a petal of a neutral fixed point with multiplier 1,*
(ii) $A(f) \subset D_1 \cup D_2$,
(iii) *each D_j contains at most one asymptotic value, and if a is an asymptotic value and $0 < \epsilon < \text{dist}\,(a, A\backslash\{a\})$, then the set $\{z : |f(z)-a| < \epsilon\}$ has only one unbounded component,*
(iv) $J \cup \{\infty\}$ *is a Jordan curve in $\overline{\mathbb{C}}$.*

A simple example of a meromorphic function of class S with two completely invariant domains is $f(z) = \tan z$, for which the upper and lower half-planes are completely invariant, and each of these half-planes is attracted to one of the two petals of the fixed point $z = 0$.

More examples will be given later in §3.

In the case that f is rational and both D_1 and D_2 are attracting or superattracting basins, Sullivan [27, Theorem 7] and Yakobson [28] proved that J is a quasicircle. Steinmetz [26] extended this result to the case that both completely invariant domains are basins of two petals attached to the same neutral fixed point. We will construct examples of transcendental functions in S for which D_1 and D_2 are attracting basins, or basins of petals attached to the same neutral fixed point, but where J is not a quasicircle; see Examples 1 and 2 in §3.

On the other hand, Keen and Kotus [16, Corollary 8.2] have shown that for the family $f_\lambda(z) = \lambda \tan z$ there exists a domain Ω containing $(1, \infty)$ such that f_λ has two completely invariant attracting basins and $J(f_\lambda)$ is a quasicircle for $\lambda \in \Omega$. Meromorphic functions for which the Julia set is contained in a quasicircle were also considered by Baker, Kotus and Lü [2, §5].

Baker [1] proved that an entire function f can have at most one completely invariant component of the set of normality, and that such a domain contains all critical values. Eremenko and Lyubich [14, §6] proved that a completely invariant domain of an entire function also contains all asymptotic values of a certain type, namely those associated with direct singularities of f^{-1}. On the other hand, Bergweiler [9] constructed an entire function with a completely invariant domain D, and such that some asymptotic value belongs to the Julia set $J = \partial D$. Example 3 in §3 shows that meromorphic functions with two completely invariant components of the set of normality can have asymptotic values on their Julia sets. So (ii) does not hold for general meromorphic functions, without the assumption that $f \in S$.

2. Proof of the Theorem

We shall need the following result of Baker, Kotus and Lü [4, Lemmas 4.2 and 4.3] which does not require that $f \in S$. Here and in the following all topological notions are related to $\mathbb{C}$ unless $\overline{\mathbb{C}}$ is explicitly mentioned.

Lemma 1. *Let f be meromorphic with two completely invariant components D_1 and D_2 of the set of normality. Then D_1 and D_2 are simply-connected and $J = \partial D_1 = \partial D_2$. In particular, J is a connected subset of $\mathbb{C}$.*

Proof of the Theorem. As the result is known for rational f, we assume that our f is transcendental.

Statement (i) follows from the classification of dynamics on the Fatou set for meromorphic functions of the class S, given in [5, Theorems 2.2 and 2.3], [8, Theorem 6], and [24, p. 3252].

To prove (ii), we consider for $j = 1, 2$ the finite sets $A_j = A \cap D_j$. Let Γ_j be a Jordan curve in D_j which separates A_j from ∂D_j. Let G_j be the Jordan regions bounded by the Γ_j. Let $G = \overline{\mathbb{C}} \backslash (\overline{G_1} \cup \overline{G_2})$ be the doubly connected region in $\overline{\mathbb{C}}$ bounded by Γ_1 and Γ_2. Notice that G contains the Julia set J.

We define $\gamma_j = f^{-1}(\Gamma_j)$. Then

(1) $$f : \gamma_j \to \Gamma_j, \; j = 1, 2,$$

are covering maps.

We claim that each $\gamma_j \subset \mathbb{C}$ is a single simple curve tending to infinity in both directions, which means that $\gamma_j \cup \{\infty\}$ is a Jordan curve in $\overline{\mathbb{C}}$.

To prove the claim, we fix j and consider the full preimages $H_j = f^{-1}(G_j)$ and $F_j = f^{-1}(D_j \backslash \overline{G_j})$. Then $D_j = F_j \cup H_j \cup \gamma_j$. The boundary of each component of F_j contains a component of γ_j, and this gives a bijective correspondence between components of F_j and components of γ_j.

We notice that H_j is connected. Indeed, by complete invariance of D_j, we have $H_j \subset D_j$, so every two points z_1 and z_2 in H_j can be connected by a curve β in D_j, so that β does not pass through the critical points of f. The image $f(\beta)$ of this curve begins and ends in G_j, and does not pass through the critical values of f. By a small perturbation of β we achieve that $f(\beta)$ does not pass through asymptotic values. Using the fact that

$$f : D_j \backslash f^{-1}(A_j) \to D_j \backslash A_j,$$

is a covering and that $A_j \subset G_j$, we can deform β into a curve in H_j which still connects z_1 and z_2. This proves that the H_j are connected.

It follows that H_j is unbounded, as it contains infinitely many preimages of a generic point in G_j.

It is easy to see that the boundary of each component F_j' of F_j intersects the Julia set.

For each component F_j' of F_j, the intersection $\partial F_j' \cap \overline{H_j}$ is a component γ_j' of γ_j. This component γ_j' divides the plane into two parts, one containing H_j and the other containing F_j'. We conclude that every component of γ_j is unbounded, because H_j is unbounded, and ∂F_j intersects the Julia set which is unbounded and connected by Lemma 1. (A similar argument for unboundedness of each component of γ_j was given in [4]).

For every component γ_j' of γ_j, the component of $\mathbb{C}\backslash\gamma_j'$ that contains F_j' intersects the Julia set. Since the Julia set is connected by Lemma 1, we conclude that F_j and γ_j are connected. So the map (1) is a universal covering by a connected set γ_j, for each $j = 1, 2$. Thus $\gamma_1 \cup \{\infty\}$ and $\gamma_2 \cup \{\infty\}$ are Jordan curves in $\overline{\mathbb{C}}$ whose intersection consists of the single point ∞. This proves our claim.

As a corollary we obtain that the point ∞ is accessible from each D_j, and so all poles of all iterates f^n are accessible from each D_j. (This fact was established in [4].)

Next we note that the set $\gamma_1 \cup \gamma_2 \cup \{\infty\}$ separates the sphere into three simply connected regions. We denote by W that region whose boundary in $\overline{\mathbb{C}}$ is $\gamma_1 \cup \gamma_2 \cup \{\infty\}$. Then

$$(2) \qquad f^{-1}(G) = W,$$

in particular, W contains the Julia set J.

To complete the proof of (ii) we choose an arbitrary point $w \in J$ and show that w is neither a critical value nor an asymptotic value.

Fix an arbitrary point $w_1 \in \Gamma_1$. The preimage $f^{-1}(w_1)$ consists of infinitely many points $a_k \in \gamma_1$, which we enumerate by all integers in a natural order on γ_1. Let ϕ_k be the branches of f^{-1} such that $\phi_k(w_1) = a_k$. We find a simple curve Δ from w_1 to some point $w_2 \in \Gamma_2$ such that $\Delta\backslash\{w_1, w_2\}$ is contained in $G\backslash\{w\}$, and such that all branches ϕ_k have analytic continuation along Δ to the point w_2. We define

$$G' = G\backslash\Delta \subset \overline{\mathbb{C}}.$$

The full preimage $f^{-1}(\Delta)$ consists of infinitely many disjoint simple curves δ_k starting at the points a_k and ending at some points $b_k \in \gamma_2$. The open curves $\delta_k\backslash\{a_k, b_k\}$ are disjoint from $\gamma_1 \cup \gamma_2$.

For every integer k, let Q_k be the Jordan region bounded by δ_k, δ_{k+1}, the arc (a_k, a_{k+1}) of γ_1 and the arc (b_k, b_{k+1}) of γ_2. Then f maps Q_k into G', and $f(\partial Q_k) \subset \partial G'$. So

$$(3) \qquad f : Q_k \to G'$$

is a ramified covering, continuous up to the boundary. Furthermore, the boundary map is a local homeomorphism. As each point of $\Gamma_1\backslash\{w_1\}$ has only one preimage on ∂Q_k, we conclude that (3) is a homeomorphism. Now it follows that the restriction $f : (b_k, b_{k+1}) \to \Gamma_2\backslash\{w_2\}$ is a homeomorphism and

$$W = \bigcup_{k=-\infty}^{\infty} Q_k \cup \delta_k \backslash \{a_k, b_k\}.$$

It follows that there are no critical points over w, so w is not a critical value.

If w were an asymptotic value, there would be a curve α in W which tends to infinity, and such that $f(z) \to w$ as $z \to \infty$, $z \in \alpha$. But this curve α would intersect infinitely many of the curves δ_k, so its image $f(\alpha)$ would intersect Δ infinitely many times, which contradicts the assumption that $f(\alpha)$ tends to w.

This completes the proof of (ii). The proof actually shows that $f : W \to G$ is a universal covering, a fact which we will use later.

To prove (iii), let us assume that D_1 contains two asymptotic values, or that $\{z \in D_1 : |f(z) - a| < \epsilon\}$ has two unbounded components for some asymptotic value $a \in D_1$. Then there exists a curve $\alpha \subset D_1$, tending to infinity in both directions, such that $f(z)$ has limits as $z \to \infty$, $z \in \alpha$, in both directions, where these limits are the two asymptotic values in the first case, and where both limits are equal to a in the second case, but the two tails of the curve α are in different components of $\{z \in D_1 : |f(z) - a| < \epsilon\}$. Now one of the regions, say R, into which α partitions the plane does not intersect the Julia set J (because J is connected by Lemma 1), and thus $R \subset D_1$. We want to conclude that f has a limit as $z \to \infty$ in R.

To do this, we choose an arbitrary point $b \in D_2$ and consider the function $g(z) = (f - b)^{-1}$ which is holomorphic and bounded in D_1. This function has limits when $z \to \infty$, $z \in \alpha$, so by a theorem of Lindelöf [22], these limits coincide and g has a limit as $z \to \infty$ in R. This proves (iii).

To prove (iv), we distinguish several cases, according to the dynamics of f in each D_j.

1. Suppose first that both D_1 and D_2 are basins of attraction of attracting or superattracting points. Then we choose the curves Γ_j as above, but with the additional property that $f(\Gamma_j) \subset G_j$, that is $f(\Gamma_j) \cap \overline{G} = \emptyset$. To achieve this, we denote by z_j the attracting or superattracting fixed point in D_j, choose G_j to be the open hyperbolic disc centered at z_j, of large enough hyperbolic radius, so that $A_j \subset G_j$, and put $\Gamma_j = \partial G_j$. Then the G_j are f-invariant, and moreover $f(\overline{G_j}) \subset G_j$ for $j = 1, 2$, because f is strictly contracting with respect to the hyperbolic metric in D_j. It follows that the closure of $W = f^{-1}(G)$ is contained in G. Let h be the hyperbolic metric in G, and $|f'(z)|_h$ the infinitesimal length distortion by f at the point $z \in W$ with respect to h. By the Theorem of Pick [21, Theorem 2.11] there exists $K > 1$

such that

(4) $$|f'(z)|_h \geq K, \quad z \in W.$$

Now we consider successive preimages $W_n = f^{-n}(W)$. Note that $\infty \notin A(f)$ by (ii) which implies that the components of $f^{-1}(\gamma_j)$ are bounded for $j = 1, 2$. We deduce that every component of W_n is a Jordan domain whose boundary consists of two cross-cuts, one of D_1 and the other of D_2. These crosscuts meet at two poles of f^n. It follows from (4) that the diameter (with respect to the metric h) of every component of W_n is at most CK^{-n}, where $C > 0$ is a constant. Now we notice that

$$J = \bigcap_{n=1}^{\infty} \overline{W_n}$$

and prove that every point $z \in J$ is accessible both from D_1 and D_2.

The accessibility of poles of the iterates f^n was already noticed before. Now we assume that z is not a pole of any iterate. Let V_n be the component of W_n that contains z. Then $V_1 \supset V_2 \supset \ldots$. The intersection $V_k \cap D_j$ is connected (its relative boundary with respect to D_j is a cross-cut in D_j), so one can choose a sequence $z_{k,j} \in V_k \cap D_j$ and connect $z_{k,j}$ with $z_{k+1,j}$ by a curve $\ell_{k,j}$ in $V_k \cap D_j$. The union of these curves gives a curve in D_j which tends to z.

The proof of (iii) in the attracting case is completed by an application of Schoenflies' theorem that if each point of a common boundary of two domains on the sphere is accessible from both domains then this common boundary is a Jordan curve [23].

2. To prove (iv) in the remaining cases, suppose, for example, that D_1 is the domain of attraction of a petal associated with a neutral fixed point a. We need several lemmas.

Lemma 2. *There exists a Jordan domain G_1 such that $\overline{G_1} \subset D_1 \cup \{a\}$, $f(\overline{G_1}) \subset G_1 \cup \{a\}$, $A_1 \subset G_1$, and G_1 is absorbing, that is, for every compact $K \subset D_1$ there exists a positive integer n such that $f^n(K) \subset G_1$.*

Proof. It is well known (see, e. g. [21, §10]) that there exists a domain G_1 having all properties mentioned except possibly $A_1 \subset G_1$. Such a domain is called an *attracting petal*.

Let P be an attracting petal. Choose a point $z_0 \in D_1$ and let $r > 0$ be so large that the open hyperbolic disc $B(z_0, r)$ of radius r centered at z_0 contains A_1, and $z_1 = f(z_0) \in B(z_0, r)$. Then put

$$G_1' = P \cup \left(\bigcup_{k=0}^{\infty} B(z_k, r) \right), \quad z_k = f^k(z_0).$$

Then G_1' is absorbing because the petal P is absorbing. Notice that for every neighborhood V of a, all but finitely many discs $B(z_k, r)$ are contained in V. This easily follows from the comparison of the Euclidean and hyperbolic metrics near a, or, alternatively, from the local description of dynamics near a neutral fixed point with multiplier 1. It is easy to see that $f(\overline{G'_1}) \subset G'_1 \cup \{a\}$. Now we fill the holes in G'_1: let X be the unbounded component of $\overline{G'_1}$ and $G_1 = \mathbb{C} \backslash \overline{X}$. It is easy to see that G_1 is a Jordan domain (its boundary is a union of arcs of hyperbolic circles which is locally finite, except at the point a, plus some boundary arcs of the petal). □

Now we fix the following notation till the end of the proof of the Theorem. If D_j is a basin of an attracting or superattracting fixed point, let G_j be the Jordan region constructed in the first part of the proof of (iv). If D_j is a basin of a petal, let G_j be the region from Lemma 2. We define $\Gamma_j = \partial G_j$. This is a Jordan curve in D_j or in $D_j \cup \{a\}$ which encloses all singular values in D_j.

Next we define $G = \overline{\mathbb{C}} \backslash (\overline{G_1} \cup \overline{G_2})$. This region is simply connected in the case that both D_1 and D_2 are basins of two petals associated with the *same* fixed point, and doubly connected in all other cases. If G is doubly connected, we make a simple cut δ disjoint from the set A of singular points, as in the proof of (ii), to obtain a simply connected region $G' = G \backslash \delta$. If G is simply connected we set $G' = G$. All branches of f^{-n} are holomorphic in G'. Let $\gamma_j = f^{-1}(\Gamma_j)$.

Lemma 3. *There exists a repelling fixed point $b \in J$ which is accessible from both D_1 and D_2 by simple curves β_j which begin at some points of Γ_j and do not intersect G_j, and which satisfy $f(\beta_j) \cap \overline{G'} = \beta_j$, for $j = 1, 2$.*

Proof. We use the notation introduced before the statement of the Lemma. Fix one of the components Q, of $f^{-1}(G')$, such that $\overline{Q} \subset G'$. Let ϕ be the branch of f^{-1} which maps G' onto Q. Then ϕ has an attracting fixed point $b \in Q$. Let $z_0 \in \Gamma_1$ and $z_1 = \phi(z_0) \in \partial Q$. We connect z_0 and z_1 by a simple curve β in $(G' \backslash Q) \cap D_1$. Such a curve exists because $z_1 \in \gamma_1$, and the component of $\mathbb{C} \backslash \gamma_1$ that contains G_1 is completely contained in D_1.

Now

$$\beta_1 = \bigcup_{k=1}^{\infty} \phi^k(\beta)$$

is a curve in D_1 tending to b which satisfies $f(\beta_1) \cap \overline{G'} = \beta_1$. Similarly a curve β_2 in D_2 is constructed. □

Now, if G is doubly connected, we set $G''' = G \backslash (\beta_1 \cup \beta_2 \cup \{b\})$. If G is simply connected then $G''' = G$. Then G''' is a simply connected region which contains no singular values of f. Let $\{\phi_k\}_{k \in \mathbb{N}}$ be the set of all branches of f^{-1} in G'''. These branches map G''' onto Jordan regions $T_k \subset G'''$. These regions T_k are of two types: the regions of the first type are contained in G''' with their closures, while the regions of the second type have common boundary points with G'''.

We claim that there are only finitely many regions of the second type. To study these regions T_k, we first observe that the full preimage of Γ_j is a curve γ_j which can have at most one point in common with Γ_j, namely the neutral fixed point on Γ_j. Thus the region $W = f^{-1}(G)$ bounded by γ_1 and γ_2 is a simply connected region contained in G, and the boundary ∂W has at most two common points with ∂G, namely the neutral fixed points. The full preimage of the cross-cut

$$\alpha = \beta_1 \cup \beta_2 \cup \{b\}$$

constructed in Lemma 2 consists of countably many disjoint curves $\alpha_k \subset \overline{W}$. Each α_k connects a point on γ_1 to a point on γ_2. One of the α_k, say α_1, is contained in α while all others are disjoint from α. Thus our regions T_k are curvilinear rectangles, similar to the Q_k used in the proof of (ii). In particular, they cluster only at ∞ so that only finitely many of them are of the first type.

It is easy to see that every region of the second type has on its boundary exactly one of the following points: a neutral fixed point or the repelling fixed point b. Indeed, let T be a region of the second type, and $\phi : G''' \to T$ the corresponding branch of the inverse. Then the iterates $\phi^n(z)$ converge to a unique point $c \in \overline{T}$ by the Denjoy–Wolff Theorem. (This theorem is usually stated for the unit disk, but it follows for Jordan domains like T by the Riemann mapping theorem, using that the Riemann map extends homeomorphically to the boundary.) On the other hand, it follows from the local dynamics near the repelling fixed point b and a neutral fixed point a that there exists $\epsilon > 0$ such that $\phi^n(z) \to b$ if $|z - b| < \epsilon$ and $\phi^n(z) \to a$ if $|z - a| < \epsilon$, $z \in T$.

If ϕ_j and ϕ_k are two different branches of f^{-1} in G''', whose images are of the second type, then the images $(\phi_k \circ \phi_j)(G''')$ are compactly contained in G'''. There exists a compact subset set $K \subset G'''$ which contains all regions T of the first type as well as all images $(\phi_k \circ \phi_j)(G''')$ where $j \neq k$.

Now consider the hyperbolic metric in G''' and let $|\phi'|_h$ stand for the infinitesimal length distortion of a branch ϕ with respect to this hyperbolic metric. Then we have for all $z \in G'''$ and some $\lambda \in (0, 1)$:

(5) $$|\phi'_k(z)|_h < \lambda \quad \text{for all } k \text{ of the first type}$$

and

(6) $$|(\phi_j \circ \phi_k)'(z)|_h < \lambda \quad \text{for all } j \neq k.$$

Let $W_n = f^{-n}(W)$. Then the Julia set can be represented as the intersection of a decreasing sequence of closed sets $J = \bigcap_{n=1}^{\infty} \overline{W_n}$. The points of the Julia set are divided into the following categories:

a) poles of f and their preimages,
b) neutral fixed points and their preimages,
c) the repelling point b and its preimages
d) those points of J which are interior to all $f^{-n}(G''')$.

We have already seen that all points of the categories a)-c) are accessible from each of the domains D_1 and D_2.

The proof that the points of the type d) are accessible is similar to the argument in the case that both D_1 and D_2 are attracting basins: we will show that each such point z can be surrounded by a nested sequence of Jordan curves whose diameter tends to zero.

Indeed, each point z of the class d) can be obtained as a limit
$$z = \lim_{n \to \infty} (\phi_{k_1} \circ \phi_{k_2} \circ \ldots \circ \phi_{k_n})(w),$$
where $w \in G'''$. For a point z of the category d), the sequence $k_1, k_2, \ldots$ is uniquely defined. We will call this sequence the *itinerary* of z. Let us consider the domains
$$T_n(z) = (\phi_{k_1} \circ \phi_{k_2} \circ \ldots \circ \phi_{k_n})(G'''),$$
in other words, $T_n(z)$ is that component of $f^{-n}(G''')$ which contains z. The boundary of $T_n(z)$ is a Jordan curve which intersects the Julia set at a finite set of points of categories a)-c). The complementary arcs of these points are cross-cuts of D_1 and D_2. Thus, to show that z is accessible from D_1 and D_2, it is enough to show that the diameter of $T_n(z)$ tends to zero as $n \to \infty$. Let $z \in J$ be a point of category d), and $k_1, k_2, \ldots$ its itinerary. Then the sequence $k_1, k_2, \ldots$ cannot have an infinite tail consisting of the branch numbers of the second type. Indeed, the iterates of any branch of the second type converge to a boundary point x of G''' (a neutral fixed point or the point b). In this case, z will be a preimage of x.

Since the itinerary does not stabilize on a branch number of the second type, we can use (5) and (6) to conclude that $\operatorname{diam} T_n(z) \to 0$.

This completes the proof. □

3. Examples

Example 1. Let
$$g(z) = \frac{1}{1 + a \cos \sqrt{z}}$$
where $0 < a < \frac{1}{5}$. Then there exists $b < 0$ such that
$$f(z) = \frac{g(z+b) - g(b)}{g'(b)}$$
has a parabolic fixed point at zero, with two completely invariant parabolic basins attached to it. Moreover, $f \in S$ and the Julia set of f is a Jordan curve, but not a quasicircle.

Verification. Note that g has no poles on the real axis. We have
$$g'(z) = \frac{a \sin \sqrt{z}}{\sqrt{z}(1 + a \cos \sqrt{z})^2}$$

and
$$g''(z) = -\frac{a^2(\cos\sqrt{z})^2 - a\cos\sqrt{z} - 2a^2}{z(1 + a\cos\sqrt{z})^3} - \frac{a\sin\sqrt{z}}{z\sqrt{z}(1 + a\cos\sqrt{z})^2}.$$

It follows that
$$\lim_{x \to -\infty} g''(x)x\cos\sqrt{x} = -\frac{1}{4a} < 0$$

so that $g''(x) > 0$ if x is negative and of sufficiently large modulus. On the other hand,
$$g''(0) = \frac{a(5a - 1)}{2(a + 1)^3} < 0.$$

Thus there exists $b \in (-\infty, 0)$ with $g''(b)=0$ and $g''(x) > 0$ for $x < b$.

The critical points of g are given by $(k\pi)^2$ where $k \in \mathbb{N}$, and g has a maximum there for odd k and a minimum for even k. It follows that $g'(x) > 0$ for $x < \pi^2$ and thus in particular for $x \le b$. Thus f has the critical points $(k\pi)^2 - b$, with corresponding critical values
$$d_\pm = \frac{(1 \pm a)^{-1} - g(b)}{g'(b)}.$$

Moreover, f has the asymptotic value $c = -g(b)/g'(b)$, which is also a Picard exceptional value of f, and it has no other asymptotic values. Therefore $A(f) = \{c, d_+, d_-\}$.

Next we note that $f(0) = 0$, $f'(0) = 1$ and $f''(0) = 0$. Also, because $f'(x) = g'(x + b)/g'(b)$ we have $0 < f'(x) < 1$ for $x < 0$. It follows from the mean value theorem that if $x < 0$, then $x < f(x) < 0$. Thus $(-\infty, 0)$ lies in a parabolic basin U attached to the parabolic point b. In particular, U contains the value $c = -g(b)/g'(b)$ which is a Picard exceptional value of f. We note that $f^{-1}(D(c,r))$ is connected for sufficiently small $r > 0$, and thus U is completely invariant.

Since $f''(0) = 0$ there is at least one parabolic basin V different from U attached to the parabolic point 0. As f has a completely invariant domain, every component of the set of normality is simply connected. Thus V is simply connected. Now V must contain a singularity of f^{-1}. Thus V contains one of the critical values d_+ and d_-, and in fact a corresponding critical point $\xi = (k\pi)^2 - b$. Since $f^n(\xi) \in \mathbb{R} \cap V$ and $f^n(\xi) \to 0$ as $n \to \infty$, and since V is simply connected and symmetric with respect to the real axis, we conclude that $(0, \xi] \subset V$. Since $f((0, \infty)) \subset (0, d_-] = f(\pi^2 - b)$ we conclude that the positive real axis is contained in V.

We now show that V is completely invariant. Suppose that W is a component of $f^{-1}(V)$ with $W \ne V$. Since W contains no critical points of f, and V contains no asymptotic values, there exists a branch φ of f^{-1} which maps V to W. This functions φ can be continued analytically to any point in $\overline{\mathbb{C}} \setminus \{c\}$. By the monodromy theorem, φ extends to a a meromorphic function from $\overline{\mathbb{C}} \setminus \{c\}$ to $\mathbb{C}$. But this implies that f is univalent, a contradiction.

It follows from part (iv) of our Theorem that the Julia set of f is a Jordan curve. On the other hand we note that if $w = u + iv$ with $|v| < T$, then $(\mathrm{Im}(w^2))^2 = (2uv)^2 \leq 4T^2u^2 < 4T^2(u^2 - v^2) + 4T^4 = 4T^2\,\mathrm{Re}(w^2) + 4T^4$. It follows that if $4T^2\,\mathrm{Re}\,z \leq (\mathrm{Im}\,z)^2 - 4T^4$, then $|\cos\sqrt{z}| \geq \sinh T$ and thus $z \in U$, if T is large enough. Thus the Julia set of f is contained in the domain $\{z \in \mathbb{C} : 4T^2\,\mathrm{Re}\,z > (\mathrm{Im}\,z)^2 - 4T^4\}$ if T is large enough. This implies that it is not a quasicircle.

Remark. It seems that g'' has only one negative zero. But since we have not proved this, we have just defined b to be the smallest zero of g''. The values a and b are related by

$$a = \frac{\sqrt{b}\cos\sqrt{b} - \sin\sqrt{b}}{\sqrt{b} + \sqrt{b}\sin^2\sqrt{b} - \sin\sqrt{b}\cos\sqrt{b}}$$

For example, if $b = -1$, then $a = 0.16763487\ldots$, $g(b) = 0.764166\ldots$ and $1/g'(b) = 16.083479\ldots$ so that

$$f(z) = 16.083479\left(\frac{1}{1 + 0.16763487\cos\sqrt{z-1}} - 0.764166\right).$$

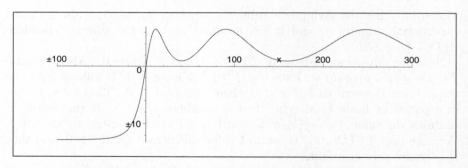

FIGURE 1. The graph of the function f from Example 1 with $b = -1$.

Example 2. Let g and f be as in Example 1 and let $\alpha > 1$. Then there exists $\alpha_0 = \alpha_0(a) > 1$ such that if $1 < \alpha < \alpha_0$, then $f_\alpha(z) = \alpha f(z)$ has two completely invariant attracting basins.

Verification. It is not difficult to see that if α is sufficiently close to 1, then f_α does indeed have two attracting fixed points $\xi_+ > 0$ and and $\xi_- < 0$, with $\xi_\pm \to 0$ as $\alpha \to 1$. The verification that their immediate attracting basins are completely invariant is analogous to that in Example 1.

Remark. We consider again the case $b = -1$. Then f_α has the form

$$f_\alpha(z) = \beta\left(\frac{1}{1 + 0.16763487\cos\sqrt{z-1}} - 0.764166\right)$$

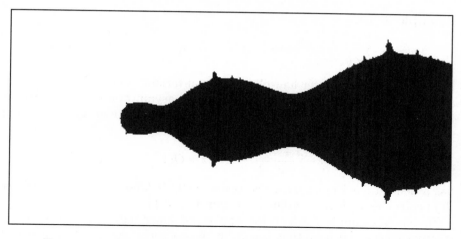

FIGURE 2. The parabolic basin of the function f from Example 1 with $b = -1$ which contains the positive real axis is shown in black. The range shown is $-100 < \operatorname{Re} z < 300, |\operatorname{Im} z| < 100$.

with $\beta > 16.083479\ldots$. For $\beta = 26.712615\ldots$ the positive attracting fixed point coincides with the critical point $1 + \pi^2$ and thus is superattracting.

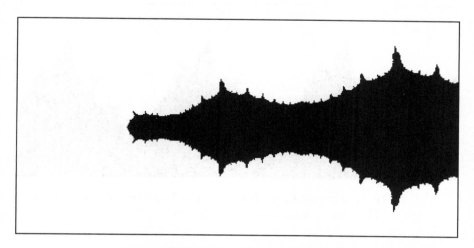

FIGURE 3. The black region is the superattracting basin of the function f from Example 2, with $b = -1$ and α chosen such that $1 + \pi^2$ is a superattracting fixed point. The range shown is $-100 < \operatorname{Re} z < 300, |\operatorname{Im} z| < 100$.

Example 3. Let
$$g(z) = \sum_{k=0}^{\infty} \frac{1}{a_k - z}, \quad \text{where} \quad 0 < a_0 < a_1 < \ldots, \quad \sum_{k=0}^{\infty} \frac{1}{a_k} < \infty.$$
Both the upper and lower half-planes are g-invariant, and we have $g(x) \to 0$ as $x \to -\infty$ along the negative ray, so 0 is an asymptotic value. Evidently, the second derivative g'' changes sign on (a_0, a_1), so there exists $c \in (a_0, a_1)$ such that $g''(c) = 0$. Then the function
$$f(z) = \frac{g(z+c) - g(c)}{g'(c)} = z + O(z^3), \quad z \to 0$$
has a neutral fixed point with two petals at 0. It follows that the Julia set $J(f)$ coincides with the real line, and thus $0 \in J(f)$.

To get an example f where the upper and lower half-plane are superattracting basins, we note that g can be chosen such that g' has a non-real zero τ, and with $a = \operatorname{Im} \tau / \operatorname{Im} g(\tau)$ and $b = \operatorname{Re} \tau - a \operatorname{Re} g(\tau)$ the function $f(z) = ag(z) + b$ satisfies $f(\tau) = \tau$ and $f'(\tau) = 0$, as well as $f(\overline{\tau}) = \overline{\tau}$ and $f'(\overline{\tau}) = 0$.

Example 4. For $a = -3.7488381 - 1.3843391i$ the function
$$f(z) = a \tan z / \tan a$$
has fixed points $\pm a$ of multiplier 1. The Julia set is a Jordan curve by our theorem, but clearly not a quasicircle.

FIGURE 4. The parabolic basins of the function from Example 4. The range shown is $|\operatorname{Re} z| < 5, |\operatorname{Im} z| < 2$.

Example 5. For $a = 1/(1 - \tanh^2 1) = 2.3810978$ the function
$$f(z) = a \tan z - a \tan i + i$$
has the fixed point i of multiplier 1 and the attracting fixed point $-3.1864112i$. Again the Julia set is a Jordan curve, but not a quasicircle.

FIGURE 5. The parabolic basin of the function from Example 5 is shown in white, the attracting one in black. The range shown is $|\operatorname{Re} z| < 5, |\operatorname{Im} z| < 2$.

Example 6. Our final example has two completely invariant half-planes, but unlike $\tan z$, it has no asymptotic values. Another feature of this example is that it has minimal possible growth among the functions of class S, namely

(7) $$T(r, f) = O((\log r)^2), \quad r \to \infty,$$

where T is the Nevanlinna characteristic. Langley [18, 19] proved that meromorphic functions with the property $T(r, f) = o((\log r)^2)$ have infinitely many singular values.

Let h be the branch of the arccosine which maps the 4-th quadrant

$$Q_4 = \{z : \operatorname{Re} z > 0, \operatorname{Im} z < 0\}$$

onto the half-strip

$$H = \{z : \operatorname{Re} z \in (0, \pi/2), \operatorname{Im} z > 0\}.$$

Let g be the conformal map of the rectangle

$$R = \{z : \operatorname{Re} z \in (0, \pi/2), \operatorname{Im} z \in (0, a)\},$$

with $a > 0$, onto Q_4, such that $g(\pi/2) = 0$ and $g(\pi/2 + ia) = \infty$, and $g(ia) > g(0) > 0$. By the Reflection Principle, g has an analytic continuation to the half-strip H and maps this half-strip into the left half-plane. It is easy to see that g is an elliptic function.

The composite function $f = g \circ h$ maps the positive ray into itself, and applying the reflection again we conclude that it maps the right half-plane into itself. The boundary values on the imaginary axis belong to the imaginary axis, so by another reflection f extends to a meromorphic function in the plane. We see that the right and left half-planes are both completely invariant.

The function f has 4 critical values, $\pm g(ia)$ and $\pm g(0)$, two in the right half-plane and two in the left half-plane.

To estimate the growth of f it is enough to notice that
$$\arccos z = i\log z + O(1) \quad \text{as } z \to \infty,$$
in the lower half-plane and in the upper half-plane. Taking into account that g is an elliptic function we obtain (7).

Our function f satisfies the differential equation
$$(1-z^2)(f')^2 = c(f^2 - p^2)(f^2 - q^2),$$
where $p = g(ia)$, $q = g(0)$ and c is a real constant.

A similar differential equation was considered by Bank and Kaufman [6]; see also [17, 20].

Acknowledgment. We thank the referee for useful comments.

REFERENCES

[1] I. N. Baker, Completely invariant domains of entire functions, in *Mathematical essays dedicated to A. J. Macintyre*, edited by H. Shankar, Ohio University Press, Athens, Ohio, 1970, 33–35.

[2] I. N. Baker, J. Kotus, and Y. Lü, Iterates of meromorphic functions I, *Ergodic Theory Dynam. Systems* 11 (1991), 241–248.

[3] I. N. Baker, J. Kotus, and Y. Lü, Iterates of meromorphic functions II: Examples of wandering domains, *J. London Math. Soc.* (2) 42 (1990), 267–278.

[4] I. N. Baker, J. Kotus, and Y. Lü, Iterates of meromorphic functions III: Preperiodic domains, *Ergodic Theory Dynam. Systems* 11 (1991) 603–618.

[5] I. N. Baker, J. Kotus, and Y. Lü, Iterates of meromorphic functions IV: Critically finite functions, *Results Math.* 22 (1992), 651–656.

[6] S. B. Bank and R. P. Kaufman, On meromorphic solutions of first-order differential equations, *Comment. Math. Helv.* 51 (1976), 289–299.

[7] A. F. Beardon, *Iteration of rational functions*. Springer, New York, 1991.

[8] W. Bergweiler, Iteration of meromorphic functions, *Bull. Amer. Math. Soc. (N. S.)* 29 (1993), 151–188.

[9] W. Bergweiler, A question of Eremenko and Lyubich concerning completely invariant domains and indirect singularities, *Proc. Amer. Math. Soc.* 130 (2002), 3231–3236.

[10] H. Brolin, Invariant sets under iteration of rational functions, *Ark. Mat.* 6 (1967), 103–141.

[11] L. Carleson and T. W. Gamelin, *Complex dynamics*, Springer, New York, Berlin, Heidelberg 1993.

[12] C.-L. Cao and Y.-F. Wang, On completely invariant Fatou components, *Ark. Mat.* 41 (2003), 253–265.

[13] A. Douady and J. H. Hubbard, Étude dynamique des polynômes complexes I & II, *Publ. Math. Orsay* 84–02 (1984) & 85–04 (1985).

[14] A. E. Eremenko and M. Yu. Lyubich, Dynamical properties of some classes of entire functions, *Ann. Inst. Fourier* 42 (1992), 989–1020.

[15] P. Fatou, Sur les équations fonctionelles, *Bull. Soc. Math. France* 47 (1919), 161–271; 48 (1920), 33–94, 208–314.

[16] L. Keen and J. Kotus, Dynamics of the family $\lambda \tan z$. *Conform. Geom. Dyn.* 1 (1997), 28–57.

[17] I. Laine, *Nevanlinna theory and complex differential equations*, Walter de Gruyter, Berlin 1993.

[18] J. K. Langley, On the multiple points of certain meromorphic functions, *Proc. Amer. Math. Soc.* 123 (1995), 1787–1795.
[19] J. K. Langley, On differential polynomials, fixpoints and critical values of meromorphic functions, *Result. Math.* 35 (1999) 284–309.
[20] J. K. Langley, Critical values of slowly growing meromorphic functions, *Comput. Methods Funct. Theory* 2 (2002), 537–547.
[21] J. Milnor, *Dynamics in One Complex Variable.* Vieweg, Braunschweig, Wiesbaden, 1999.
[22] R. Nevanlinna, *Analytic functions*, Springer, Berlin, Heidelberg, New York 1970.
[23] M. H. A. Newman, *Elements of the topology of plane sets of points.* 2nd ed., Cambridge, At the University Press, 1951.
[24] P. J. Rippon and G. M. Stallard, Iteration of a class of hyperbolic meromorphic functions. *Proc. Amer. Math. Soc.* 127 (1999), 3251–3258.
[25] N. Steinmetz, *Rational iteration*, Walter de Gruyter, Berlin 1993.
[26] N. Steinmetz, Jordan and Julia, *Math. Ann.* 307 (1997), 531–541.
[27] D. Sullivan, Conformal dynamical systems, in *Geometric dynamics (Rio de Janeiro, 1981)*, 725–752, Lecture Notes in Math., 1007, Springer, Berlin, 1983.
[28] M. V. Yakobson, Boundaries of certain domains of normality for rational mappings (Russian), *Uspekhi Mat. Nauk* 39 (1984), no. 6 (240), 211–212.

MATHEMATISCHES SEMINAR, CHRISTIAN–ALBRECHTS–UNIVERSITÄT ZU KIEL, LUDEWIG–MEYN–STR. 4, D–24098 KIEL, GERMANY
E-mail address: bergweiler@math.uni-kiel.de

DEPARTMENT OF MATHEMATICS, PURDUE UNIVERSITY, WEST LAFAYETTE, IN 47907, USA
E-mail address: eremenko@math.purdue.edu

A FAMILY OF MATINGS BETWEEN TRANSCENDENTAL ENTIRE FUNCTIONS AND A FUCHSIAN GROUP

SHAUN BULLETT AND MARIANNE FREIBERGER

For Noel Baker

ABSTRACT. Let Γ denote the subgroup of the modular group generated by $z \to -1/z$ and $z \to z+2$. We present a family of holomorphic correspondences $\mathcal{F}$ on the Riemann sphere which realise "matings" between maps of the form $z \to \mu \sin^2(\pi\sqrt{z})$ and Γ, in the sense that the Riemann sphere is partitioned into dynamically invariant regions Ω and $\Lambda = \Lambda_+ \cup \Lambda_-$ with the following properties:

(i) Ω is open, simply-connected and the action of $\mathcal{F}$ on Ω is conjugate to that of a generating set of elements of Γ;

(ii) Λ_- and Λ_+ are each homeomorphic to a set obtained from the filled Julia set of $\mu \sin^2(\pi\sqrt{z})$ by "splitting infinity and shrinking dynamical rays to points". $\mathcal{F}$ restricted to Λ_- as domain and range is single-valued and conjugate to $\mu \sin^2(\pi\sqrt{z})$ on this modified filled Julia set. The action of $\mathcal{F}^{-1}$ restricted to Λ_+ is conjugate to that of $\mathcal{F}$ on Λ_-.

While we have investigated matings between polynomial maps and Kleinian groups in earlier work, the examples presented here are the first to involve transcendental entire functions. We think of our construction as "resolving" the essential singularity of $z \to \mu \sin^2(\pi\sqrt{z})$ by replacing the point at infinity with a region on which the dynamics is group-like.

1. THE GROUP Γ

Let Γ be the subgroup of $PSL(2, \mathbb{Z})$ generated by the matrices:

$$\begin{pmatrix} 0 & 1 \\ -1 & 0 \end{pmatrix} \quad \text{and} \quad \begin{pmatrix} 1 & 2 \\ 0 & 1 \end{pmatrix}.$$

The action of Γ on the upper half-plane by Möbius transformations is proper and discontinuous, with limit set $\hat{\mathbb{R}} = \mathbb{R} \cup \{\infty\}$. As an abstract group Γ is the free product of two cyclic groups, one of infinite order, generated by

$$\rho = \begin{pmatrix} 0 & 1 \\ -1 & -2 \end{pmatrix},$$

and the other of order two, generated by

$$\sigma = \begin{pmatrix} 0 & 1 \\ -1 & 0 \end{pmatrix}.$$

Both authors thank the UK EPSRC for their support through grant no. GR/R73232/01.

We view Γ as the limit as $n \to \infty$ of the *Hecke groups* H_n generated by the matrices

$$\sigma = \begin{pmatrix} 0 & 1 \\ -1 & 0 \end{pmatrix} \quad \text{and} \quad \rho_n = \begin{pmatrix} 0 & 1 \\ -1 & -2\cos(\pi/n) \end{pmatrix}.$$

Let $\mathbb{R}_+$ denote the set of positive real numbers and let $\hat{\mathbb{R}}_+ = \mathbb{R}_+ \cup \{0, \infty\}$. For each $j \in \mathbb{N}$ we have

$$\sigma\rho^j(\hat{\mathbb{R}}_+) = \left[\frac{j+1}{j}, \frac{j}{j-1}\right] \quad \text{and} \quad \sigma\rho^{-j}(\hat{\mathbb{R}}_+) = \left[\frac{j-1}{j}, \frac{j}{j+1}\right].$$

To each $x \in \hat{\mathbb{R}}$, which is not in the orbit of 1, we associate an infinite sequence $s(x)$ of non-zero integers, which has initial segment $s_1 s_2 \ldots s_j$ if $x \in \sigma\rho^{s_1}\sigma\rho^{s_2}\ldots\sigma\rho^{s_j}(\hat{\mathbb{R}}_+)$. Since each $\sigma\rho^j$ is a contraction in the spherical metric, every sequence specifies a unique point. However, points in the orbit of 0 and ∞ have more than one associated sequence, while points in the orbit of 1 have no sequence at all.

We overcome this difficulty as follows. Let $\mathcal{S}$ denote the set of infinite sequences $(s_1, s_2, \ldots)$ with each $s_j \in \mathbb{Z} - \{0\}$, and let $\sim$ be the relation defined on $\mathcal{S}$ by

$$(s_1, \ldots, s_n, -1, -1, -1, \ldots) \sim (s_1, s_2, \ldots, s_n + 1, 1, 1, 1, \ldots)$$

for $s_n \neq -1$. Next, let $\mathcal{S}_{fin}$ denote the set of finite sequences $(s_1, \ldots, s_n)$. By convention, we shall write each such sequence as $(s_1, \ldots, s_n, \infty)$. Equip the union $(\mathcal{S}/\sim) \cup \mathcal{S}_{fin}$ with the following topology: if s is a sequence in $\mathcal{S}/\sim$, then a sequence t is close to s if the entries of t and s agree to a large number of places, regardless of whether t is in $\mathcal{S}/\sim$ or in $\mathcal{S}_{fin}$. If $s \in \mathcal{S}_{fin}$, with $s = (s_1, \ldots, s_n, \infty)$, then t is close to s if t is of the form $(s_1, \ldots, s_n, s_{n+1}, \ldots)$ with $|s_{n+1}|$ large, again regardless of whether t is in $\mathcal{S}/\sim$ or in $\mathcal{S}_{fin}$.

Proposition 1. *With the topology defined above, $(\mathcal{S}/\sim) \cup \mathcal{S}_{fin}$ is homeomorphic to $\hat{\mathbb{R}}_+$.*

Proof. This is a straightforward consequence of the action of Γ described above. Sequences in $\mathcal{S}$ correspond to points not in the orbit of 1, and the relation $\sim$ deals with the difficulty that each point in the orbit of 0 or ∞ has two associated sequences. Sequences in $\mathcal{S}_{fin}$ correspond to points in the orbit of 1: thus $(s_1, \ldots, s_n, \infty)$ represents the point $\sigma\rho^{s_1} \ldots \sigma\rho^{s_n}(1)$. □

With this representation, the action of $\sigma\rho^j$ on $\hat{\mathbb{R}}_+$ becomes "shift to the right and insert the non-zero integer j". The sequence (or sequences) $s(x)$ associated to a point $x \in \hat{\mathbb{R}}_+$ will be called the *itinerary* (or *itineraries*) of x.

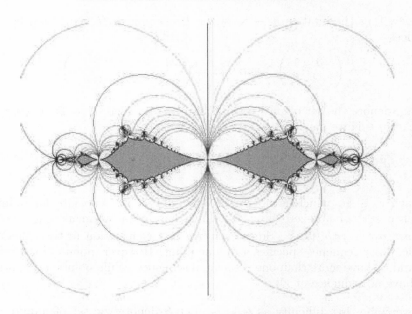

FIGURE 1. A mating between Γ and a holomorphic map. The grey region is the set Λ. Its complement Ω is tiled by fundamental domains for Γ. See Remark 5 at the end of Section 4 for details of the correspondences illustrated in Figures 1 and 2.

2. Correspondences and matings

A *correspondence* on the Riemann sphere $\hat{\mathbb{C}}$ is a multivalued map $\mathcal{F} : \hat{\mathbb{C}} \to \hat{\mathbb{C}}$. The *graph* of $\mathcal{F}$ is the subset $\{(z,w) : w \in \mathcal{F}(z)\} \subset \hat{\mathbb{C}} \times \hat{\mathbb{C}}$. We allow the sets $\mathcal{F}(z)$ and $\mathcal{F}^{-1}(z)$ to have infinite cardinality. If this is the case for generic z, we describe the correspondence as having *bidegree* $(\infty : \infty)$.

Definition 1. *We say that an $(\infty : \infty)$ correspondence $\mathcal{F}$ is a mating between Γ and a holomorphic map if:*
- *there exists a conformal involution J on $\hat{\mathbb{C}}$ conjugating $\mathcal{F}$ to $\mathcal{F}^{-1}$;*
- *$\hat{\mathbb{C}}$ is partitioned into sets Ω and $\Lambda = \Lambda_- \cup \Lambda_+$, where*
  (i) Ω is open, simply-connected and invariant under both $\mathcal{F}$ and J; moreover there is a conformal homeomorphism ϕ from the open complex upper half-plane to Ω, conjugating the action of σ to that of $J|_\Omega$, and the action of $\bigcup_{j \in \mathbb{Z}-\{0\}} \sigma \rho^j$ to that of $\mathcal{F}|_\Omega$, and extending to a continuous map at 0 and ∞, sending both to a single fixed point p of J, which lies in $\partial \Omega$;
  (ii) Λ_- is closed and backward invariant under $\mathcal{F}$, and $\mathcal{F}$ restricted to Λ_- as domain and codomain is a (single-valued) holomorphic map;
  (iii) $\Lambda_+ = J(\Lambda_-)$ and $\Lambda_- \cap \Lambda_+$ consists of the fixed point p of J.

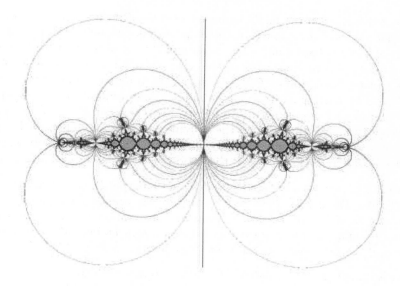

FIGURE 2. A mating between Γ and another holomorphic map.

Examples are plotted in Figures 1 and 2 (the figures are plotted in coordinates where J is the involution $z \leftrightarrow -z$). These examples lie in the complex one-parameter family of correspondences $\mathcal{F}_a$ defined below.

We say that a correspondence f is a *sub-correspondence* of the correspondence $\mathcal{F}$ if $f(z) \subset \mathcal{F}(z)$ for all $z \in \hat{\mathbb{C}}$. We say that f is *holomorphic* if it is defined by an equation

$$w \in f(z) \quad \Leftrightarrow \quad p(z, w) = 0$$

where $p(z, w)$ is a polynomial in z and w. The graph of f is a Riemann surface, the algebraic curve defined by the equation $p(z, w) = 0$.

As our candidates for matings we shall identify a complex one-parameter family of correspondences $\mathcal{F} = \mathcal{F}_a$, each of which splits into a countable union of holomorphic (2 : 2) sub-correspondences $\{f_j\}_{j \geq 1}$ having $f_j|_\Omega$ conjugate to $\{\sigma\rho^j, \sigma\rho^{-j}\}$. We shall show that when a lies in a certain subset of the parameter space, $\mathcal{F}$ is a mating between Γ and a holomorphic map $\mathcal{F}|_{\Lambda_-}$, and that the holomorphic map $\mathcal{F}|_{\Lambda_-}$ has the combinatorics of a scalar multiple μR of the transcendental entire function

$$R(z) = \sin^2(\pi\sqrt{z}).$$

For particular classes of such examples we prove (and for wider classes we conjecture) that for an appropriate value of μ there is a conformal conjugacy between the maps $\mathcal{F}|_{\Lambda_-}$ and μR on their respective Fatou sets. This conjugacy extends to a continuous bijection associating certain points in the limit set of Γ to "dynamical rays" in the Julia set of the entire function μR. For the particular values of μ we consider, the map μR has a repelling fixed point and a super-attracting fixed point. The closure of the set of points with bounded orbits under μR, which we will call the *filled Julia set* of μR, is the whole of $\hat{\mathbb{C}}$. We picture the action of $\mathcal{F}$ on $\hat{\mathbb{C}}$ as obtained from two copies of the action of μR on $\mathbb{C}$ by cutting along the dynamical ray from infinity to the repelling fixed point in each copy to give two half-planes which can be glued to form a single copy of $\mathbb{C}$. We then retract the two filled Julia sets to a bounded region of $\mathbb{C}$, at the same time contracting dynamical rays to points. Finally we glue into the newly created space in $\hat{\mathbb{C}}$ the action of Γ on the upper half-plane.

Remark 1. *Since $z \to \lambda^2 \sin^2(\pi\sqrt{z})$ is double-covered by $\zeta \to \lambda\sin(\pi\zeta)$, via $z = \zeta^2$, the Fatou and Julia sets of the former are double-covered by the Fatou and Julia sets of the latter. Thus our descriptions below of the Fatou and Julia sets of the functions $\mu R(z)$ are easily translatable into descriptions of those of the functions $\lambda\sin(\pi\zeta)$, and vice versa, where $\mu = \lambda^2$.*

Remark 2. *In [2], we constructed matings between Hecke groups H_n and scalar multiples of the degree $n-1$ polynomial R_{n-1} defined by*

$$R_{n-1}(z) = \frac{1}{2}(1 - T_{n-1}(1 - 2z)),$$

where T_{n-1} is the $(n-1)$st Chebyshev polynomial (the polynomial which expresses $\cos((n-1)\theta)$ in terms of $\cos\theta$). The polynomial R_n is both conjugate to T_n and double-covered by it, in the sense that $R_n(z) = T_n^2(\sqrt{z})$. When we make the analogous construction in the infinite degree case (see Section 3), the polynomial T_n is replaced by the function $T(z) = \sin(\pi z)$, and the polynomial R_n is replaced by the function $R(z) = \sin^2(\pi\sqrt{z})$, which is double-covered by $T(z)$. The group Γ is the algebraic limit of the Hecke groups H_n. The current paper is independent of [2] in the sense that we shall present a direct construction of the $(\infty : \infty)$ correspondences. However we shall omit details of some proofs and refer the reader to the proofs in [2] of analogous results for the finite case.

3. A FAMILY OF CORRESPONDENCES

The correspondences constructed in [2] had the form $J \circ Cov_0^Q$ (this notation will be explained later), where the rational map Q is the factored quotient map

$$\hat{\mathbb{C}} = \hat{\mathbb{C}}/S \to \hat{\mathbb{C}}/D_{2n} = \hat{\mathbb{C}}$$

obtained from the standard action of the dihedral group D_{2n} on $\hat{\mathbb{C}}$, and S is one of the pair of involutions generating this group. In suitable coordinates, Q is the Chebyshev polynomial T_n. Below we make the analogous construction in the case $n = \infty$.

Consider the action of the infinite dihedral group D_∞ on $\hat{\mathbb{C}}$ generated by the involution $S : \zeta \to -\zeta$ and the translation $T : \zeta \to \zeta + 1$. The limit set of this action consists of the single point ∞.

For the quotient map $\mathbb{C} \to \mathbb{C}/D_\infty$ we may take the map $\zeta \to \cos(2\pi\zeta)$, or, more conveniently, the map

$$\zeta \to \frac{1}{2}(1 - \cos(2\pi\zeta)) = \sin^2(\pi\zeta)$$

since, writing $z = \zeta^2$, this gives us the explicit form

$$R(z) = \sin^2(\pi\sqrt{z})$$

for the factored quotient map

$$\mathbb{C}/S \to \mathbb{C}/D_\infty.$$

The *covering correspondence* Cov^R, defined by

$$z \to w \quad \Leftrightarrow \quad R(z) = R(w),$$

splits into $(2 : 2)$ correspondences on $\mathbb{C}/S$, each coming from a pair of maps $\zeta \to \zeta \pm j$ on $\mathbb{C}$ (where j is a positive integer), together with a $(1 : 1)$ correspondence coming from $\zeta \to \zeta$. It is an elementary calculation (substitute z for ζ^2) that these $(2 : 2)$ correspondences have the form

$$g_j : z \to w \quad \Leftrightarrow \quad w^2 - 2(z + j^2)w + (z - j^2)^2 = 0.$$

Thus

$$w \in Cov^R(z) \quad \Leftrightarrow \quad z = w \text{ or } g_j(z, w) = 0 \text{ for some } j.$$

Despite the fact that $R(z)$ is undefined at $z = \infty$, each of the correspondences g_j has $g_j(\infty) = \{\infty\}$. Thus we may adopt the convention that $Cov^R(\infty) = \{\infty\}$ and think of Cov^R as a correspondence defined on the whole of $\hat{\mathbb{C}}$, with graph the union of the graphs of the $(2 : 2)$ holomorphic sub-correspondences g_j and that of the identity. We denote by Cov_0^R (the *deleted covering correspondence of R*) the correspondence obtained from Cov^R by deleting the graph of the identity.

We note that g_1 is a *symmetric* $(2 : 2)$ correspondence ($w \in g_1(z) \Leftrightarrow z \in g_1(w)$). It follows that for g_1 one can define "iteration without backtracking", under which if z maps to w we do not allow w to map directly back to z unless z is the *unique* element of $g_1(w)$. Denote the jth iterate of g_1 without backtracking by $g_1^{[j]}$.

Lemma 1. $g_j = g_1^{[j]}$.

Proof. Direct calculation (or the observation that g_j is $\{T^j, T^{-j}\}$ "pushed down from $\hat{\mathbb{C}}$ to $\hat{\mathbb{C}}/S$"). □

Given an involution J we use the notation $J \circ Cov_0^R$ to denote the correspondence obtained by composing the involution J with Cov_0^R, namely

$$w \in J \circ Cov_0^R(z) \quad \Leftrightarrow \quad (Jw)^2 - 2(z+j^2)(Jw) + (z-j^2)^2 = 0 \text{ for some } j \geq 1.$$

Thus the graph of $J \circ Cov_0^R$ is the union of the graphs of the holomorphic $(2:2)$ sub-correspondences f_j defined by

$$w \in f_j(z) \quad \Leftrightarrow \quad (Jw)^2 - 2(z+j^2)(Jw) + (z-j^2)^2 = 0.$$

Writing $\mathcal{F}$ for $J \circ Cov_0^R$ we see that $\{f_1, J\}$ generates $\{\mathcal{F}, J\}$ in the following sense.

Lemma 2. *The grand orbits obtained by mixed iteration of f_1, f_1^{-1} and J are identical to those obtained by mixed iteration of $\mathcal{F}$, $\mathcal{F}^{-1}$ and J.*

Proof. This follows at once from the previous lemma and the fact that J is an involution. □

Remark 3. *Our candidates for matings will be among correspondences $\mathcal{F}$ of the form $J \circ Cov_0^R$. We note that if such an $\mathcal{F}$ is a mating then restricted to Ω each $g_j|_\Omega$ is conjugate to $\{\rho^j, \rho^{-j}\}$ on the upper half-plane (by Lemma 1) and hence each $f_j|_\Omega$ is conjugate to $\{\sigma\rho^j, \sigma\rho^{-j}\}$.*

3.1. Fundamental domains and singular points.
The behaviour of the points $z = 0$ and $z = 1/4$ plays an important part in the dynamics of correspondences $\mathcal{F}$. The point 0 is a *singular* point for each f_j in the sense that $f_j(0)$ is the single point $J(j^2)$. The point $1/4$ has the property that for each j the intersection $f_j(1/4) \cap f_{j+1}(1/4)$ is non-empty. In fact, this intersection is the single point $J((j+1/2)^2)$. We also note here that the point $1/4$ is fixed by g_1, and sent to $J(1/4)$ by f_1.

Definition 2. *A transversal D_R for the function R is a maximal domain in $\mathbb{C}$ on which R is injective.*

A transversal plays the same role for the correspondence Cov_0^R as does a fundamental domain for a Kleinian group. Any transversal for our function R is an unbounded subset of $\mathbb{C}$, or equivalently it contains ∞ on its boundary. See Figure 3.

Proposition 2. *If $\mathcal{F}$, of the form $J \circ Cov_0^R$ with R as above, is a mating between Γ and a map, then one of the fixed points of J coincides with the fixed point $1/4$ of $g_1 = J \circ f_1$, and the singular point 0 lies in Λ_+. Moreover, there exists a closed disc D such that*
- *D is a fundamental domain of the involution J;*
- *D is contained in the closure of a transversal D_R of R with $\partial D \cap \partial D_R = \{1/4\}$;*
- *$f_1(D) \subset D$ with $\partial f_1(D) \cap \partial D = \{1/4\}$;*
- *$f_j(D)$ is properly contained in D for all $j > 1$, and $f_j(D) \cap f_{j+1}(D) = \{f_{j+1}(1/4)\}$;*

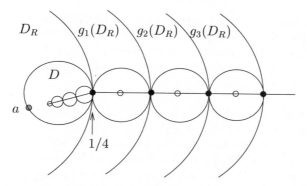

FIGURE 3. D_R is a maximal domain on which R is injective; when it is symmetric about the real axis (as pictured) it is bounded by the line $z = \{(1/2 + it)^2 : t \in \mathbb{R}\}$, which crosses the real axis at $1/4$, and the images of D_R under Cov_0^R are the "strips" $g_j(D_R)$ bounded by the lines $z = \{(j \pm 1/2 + it)^2 : t \in \mathbb{R}\}$. The disc D is a fundamental domain for J (which has fixed points $1/4$ and a). The discs shown within D are the images of D under $\mathcal{F}$. The discs shown outside D are the images of D under $Cov_0^R = J \circ \mathcal{F}$. The black points are $1/4$ and its images under Cov_0^R and the white points are the images of 0 under Cov_0^R.

Proof. By Definition 1 and Remark 3, there exists a homeomorphism ϕ from the upper half-plane to the set Ω, conjugating $\{\sigma\rho, \sigma\rho^{-1}\}$ to f_1 and σ to J. Moreover, ϕ extends continuously to 0 and ∞, sending both to a fixed point p of J, which lies in $\partial\Omega$. The parabolic map ρ sends ∞ to 0. Since ϕ extends to these points and since ρ is conjugate to one of the branches of g_1, we deduce that p is the unique fixed point $1/4$ of g_1.

We next show that the singular point 0 lies in Λ_+. The point 0 has only one image v_1 under g_1, but v_1 has two distinct inverse images under g_1. As there is no point in the upper half-plane which has this property under $\{\rho, \rho^{-1}\}$, it follows that 0 does not lie in Ω. But neither can 0 lie in Λ_-, for then at least one of the $f_j(0) = J(v_j)$ would also lie in Λ_- (since $\mathcal{F}$ restricted to Λ_- as domain and range is a well-defined map), but as points sufficiently close to 0 have two distinct but nearby images under f_j this would contradict the single-valuedness of $\mathcal{F}|\Lambda_-$. The only remaining possibility is that $0 \in \Lambda_+$.

Now consider the imaginary axis $\mathcal{I}$ in the upper half-plane. Since $\phi(0) = \phi(\infty)$, the image $\phi(\mathcal{I})$ of $\mathcal{I}$ is a closed curve in Ω separating Λ_+ and Λ_- and dividing the sphere into a fundamental domain for J and its image. Let D denote the component of $\hat{\mathbb{C}} - \phi(\mathcal{I})$ containing $\phi(\mathcal{R})$, where $\mathcal{R} = \{z = x + iy : x, y > 0\}$, so $D - (\Omega \cap D)$ is either Λ_+ or Λ_-. In fact $D - (\Omega \cap D) = \Lambda_+$, as we shall now see. Any point $x \in \hat{\mathbb{R}}$ which is of the form $\sigma\rho^j(0)$ or $\sigma\rho^j(\infty)$, for $j \in \mathbb{Z} - \{0\}$, lies on the positive real line, which is part of the boundary of $\mathcal{R}$. Since the conjugacy ϕ extends to these points, it follows that $f_j(\phi(x))$ lies in D. Hence f_j is not single-valued as a map from $D - (\Omega \cap D)$ to itself, and so $D - (\Omega \cap D)$ cannot be Λ_-.

For a fundamental domain for the action of ρ on the upper half-plane we may take the region D_ρ bounded by the geodesic L_1 from -1 to ∞, together with the geodesic L_2 from -1 to 0 and the positive real axis. Since ϕ identifies 0 with infinity, $\phi(L_1) \cup \phi(L_2)$ forms a closed Jordan curve. Now D_ρ contains $\mathcal{R}$, so $D_R = \Lambda_+ \cup \phi(D_\rho)$ is a topological disc containing D and Λ_+. Note that $\partial D \cap \partial D_R = \{1/4\}$. Since Cov_0^R acts on $\phi(D_\rho)$ as $\{\phi\rho^j\phi^{-1}\}_{j \in \mathbb{Z}-\{0\}}$ and also sends Λ_+ to Λ_-, we deduce that Cov_0^R maps D_R outside itself, and hence that D_R is contained in a transversal for R. But L_1 is sent to L_2 by ρ, so D_R is a *maximal* set on which R is injective. Moreover, since $D \subset D_R$ and D is a fundamental domain for J, we deduce that f_j (which is the composition $J \circ g_j$) sends D into itself for each integer $j \geq 1$.

It only remains to show that $\partial f_1(D) \cap \partial D = \{1/4\}$, that $\partial f_j(D) \cap \partial D = \emptyset$ for each $j > 1$, and that $f_j(D)$ meets $f_{j+1}(D)$ in the single point $f_{j+1}(1/4)$ for each $j \geq 1$. First observe that from the conjugacy between g_j on Ω and $\{\rho^j, \rho^{-j}\}$ on the upper half-plane and the fact that ϕ identifies 0 with ∞, we may deduce that ϕ identifies $\rho^j(0)$ with $\rho^{-j}(\infty)$ for each j. Next, from the same conjugacy, we may deduce that for each $j \geq 1$ the set $g_j(D_R)$ is a pinched annulus, bounded by the circles $\phi(\rho^{1-j}L_1) \cup \phi(\rho^{j-1}L_2)$ and $\phi(\rho^{-j}L_1) \cup \phi(\rho^j(L_2)$ (which touch at $\phi(-1)$, which is the point at infinity on the Riemann sphere $\hat{\mathbb{C}}$), and that these pinched annuli ("strips" in Figure 3), together with the disc D_R, partition $\hat{\mathbb{C}}$. The set $g_j(D) \subset g_j(D_R)$ is a disc, bounded by a circle $\rho^{-j}(\mathcal{I}) \cup \rho^j(\mathcal{I})$. Since this circle touches ∂D_R only in the case $j = 1$, and then at the point $1/4$, and since it touches $\partial g_j(D_R)$ only at the points $g_j(1/4) = (j-1/2)^2$ and $g_{j+1}(1/4) = (j+1/2)^2$, we deduce that $\partial g_1(D) \cap \partial D = \{1/4\}$, that $\partial g_j(D) \cap \partial D = \emptyset$ for each $j > 1$, and that $g_j(D)$ meets $g_{j+1}(D)$ in the single point $g_{j+1}(1/4)$ for each $j \geq 1$. Composing with the involution J we obtain the corresponding results for $\{f_j\}_{j \geq 1}$. □

As we shall see below, the converse of Proposition 2 also holds.

Definition 3. *We define J_a to be the involution of $\hat{\mathbb{C}}$ which has fixed points $1/4$ and a, and $\mathcal{M}$ to be the set of parameters $a \in \mathbb{C}$ such that the correspondence $\mathcal{F}_a = J_a \circ Cov_0^R$ satisfies the conclusions of Proposition 2.*

The set $\mathcal{M}$ is non-empty. For example, if a is such that $J_a(j^2) = 0$ for some integer $j \geq 1$, then the point 0 is fixed by f_j, and one can show that J_a satisfies the conclusions of Proposition 2.

Now, let $\mathcal{F} = \mathcal{F}_a$ where $a \in \mathcal{M}$ and define

$$\Lambda_+ = \overline{\bigcap_{n=0}^{\infty} \mathcal{F}^n(D)}, \quad \Lambda_- = J(\Lambda_+), \quad \Omega = \hat{\mathbb{C}} - (\Lambda_- \cup \Lambda_+).$$

Lemma 3. Λ_+ *is simply-connected.*

Proof. Since $0 \in \Lambda_+ \subset D$, we have that $f_j(D)$ is a topological disc for each j, and each $f_j(D) \cap f_{j+1}(D)$ consists of exactly one point in the orbit of $1/4$. In particular, $\overline{\mathcal{F}(D)}$ is simply-connected (see Figure 3). Using an induction one can show that since $0 \in \mathcal{F}^n(D)$ for all n, $\overline{\mathcal{F}^n(D)}$ is simply-connected for all n. It follows that Λ_+ is simply-connected. □

We can now prove:

Theorem 1. $\mathcal{F}_a = J_a \circ Cov_0^R$ *is a mating between* Γ *and a map if and only if* $a \in \mathcal{M}$.

Proof. One implication is given by Proposition 2. The proof of the other implication is very similar to the proof of Proposition 7.1 in [2], and we shall only give an outline here.

Suppose that $a \in \mathcal{M}$. First we note that since Λ_+ is simply-connected, so are $\Lambda_+ \cup \Lambda_-$ and Ω. It follows that there exists a Riemann map ϕ from the upper half-plane to Ω. Clearly, $J_a|_\Omega$ is conjugate via ϕ to an involution, which we may take to be $\sigma : z \to -1/z$, on the upper half-plane.

By Lemma 1 we know that $J_a \circ \mathcal{F}_a|_\Omega$ is conjugate via ϕ to a cyclic group generated by an element ρ_a of $PSL(2, \mathbb{R})$ which has infinite order, and that $\phi^{-1}(D_R \cap \Omega)$ is a fundamental domain for ρ_a. Moreover, ∂D_R lies in Ω, by the same argument as used in the proof of Proposition 2, and contains ∞. Thus, ∞ is accessible via a path in Ω, and so there exists a unique point q in the real line such that ϕ extends continuously to q and sends it to ∞. Now $q = \phi^{-1}(\infty)$ is a fixed point of ρ_a. Examining the behaviour of Cov_0^R, we see that the images of $D_R - \Lambda_+$ under branches of Cov_0^R accumulate at ∞. Thus, the images of the fundamental domain of ρ_a under all the powers of ρ and ρ^{-1} accumulate at the fixed point $\phi^{-1}(\infty)$, and hence ρ_a is parabolic.

To show that $\rho_a = \rho$, we first notice that the point $1/4$ is accessible from Ω, via the curve ∂D_R. A small neighbourhood of $1/4$ intersects Ω in two connected components. Thus, there exist two points a and b in the real line, such that ϕ continuously extends to both and sends them both to $1/4$. Moreover, a and b are end-points of the boundary of a fundamental domain of ρ_a. Similarly, there exist two points c and d in the real line, such that ϕ continuously extends to both and sends them both to $1/4$, and which are end-points of the boundary of a fundamental domain of σ. Using an argument explained in detail in [2], Proposition 7.1, one can show that $a = b$ and $b = c$. This is sufficient to prove that $\rho = \rho_a$. □

4. Transcendental matings

We claim that correspondences of the form $\mathcal{F}_a = J_a \circ Cov_0^R$ with $a \in \mathcal{M}$ can be regarded as matings between Γ and transcendental entire functions of the form $\mu R(z)$ where $R(z) = \sin^2(\pi\sqrt{z})$ and μ is some scalar. In order to justify this claim we shall show that the infinite-to-one map $\mathcal{F}^{-1} : \Lambda_+ \to \Lambda_+$ resembles μR (for an appropriate value of μ) in a very strong sense. We cannot hope for a conjugacy between $\mathcal{F}^{-1}$ on the whole of Λ_+, and μR on the whole of its filled Julia set, since the latter is the whole complex plane. However, we define below the notion of the *nth truncation* of $\mathcal{F}$, with associated limit set $\Lambda_+^n \subset \Lambda_+ = \overline{\bigcup_n \Lambda_+^n}$ (pictorially, Λ_+^n is obtained from Λ_+, as shown in Figure 1, by "cutting off the ends of branches"). We then show in the remaining sections of the paper that there is a conjugacy from $\mathcal{F}^{-1}$ on $\bigcup_n \Lambda_+^n$ (*not* its closure) to μR on the analogous subset of its filled Julia set, and that the remaining points in Λ_+ correspond either to "dynamical rays" of μR or to the point at infinity.

For simplicity, we shall restrict our attention to parameters a for which the singular point 0 is fixed by f_m for some $m \geq 1$, ie $0 = J_a(m^2)$ for some $m \geq 1$. We denote the correspondence satisfying this condition by $\mathcal{F}_m$.

We write Λ'_+ for the set $\bigcap_k \mathcal{F}^k(D)$. Let P denote the point on the real line where the images $f_j(D)$ accumulate. So P is ∞ in the parametrization $\mathcal{F}_m = J \circ Cov_0^R$ and a finite real point in the coordinates adopted in Figures 1 and 2.

Lemma 4. $\Lambda_+ - \Lambda'_+$ consists of points in the orbit of P.

Proof. Clearly, $\Lambda_+ - \Lambda'_+$ contains all the points in the orbit of P. Now let $\mathcal{P}_k = \bigcup_{i=0}^k \mathcal{F}^i(P)$, for $k \geq 0$, so the orbit of P is equal to the union $\bigcup_k \mathcal{P}_k$. Then

$$\Lambda'_+ \cup \bigcup_k \mathcal{P} = \bigcap_k (\mathcal{F}^k(D) \cup \mathcal{P}_{k-1}) = \overline{\bigcap_k \mathcal{F}^k(D)} = \Lambda_+. \quad \square$$

The structure of Λ'_+ is easily understood: each point which is not in the orbit of $1/4$ lies in the intersection of a sequence of nested discs D_k, each D_k being a connected component of the interior of $\mathcal{F}_m^k(D)$. The point $1/4$ lies on the boundary of infinitely many such discs, and it is the only point at which the boundaries of these discs touch. The same is true for any point x in the orbit of $1/4$, and since $f_j(1/4) \cap f_{j+1}(1/4)$ is non-empty, we have two sequences of discs associated to such an x.

We see that if $m > 1$, the sub-correspondence $f_m^{-1} : f_m(D) \to D$ is quadratic-like and hybrid equivalent to $z \to z^2$. If $m = 1$, then this map is a *simply-pinched* quadratic-like map [2]. Thus, by the Straightening Theorem [6] (or an ad hoc argument when $m = 1$), the set $\mathcal{K} = \bigcap f_m^n(D)$ is homeomorphic to the filled Julia set of $z \to z^2$. The homeomorphism conjugates f_m to $z \to \pm\sqrt{z}$ and is conformal on the interior of $\mathcal{K}$.

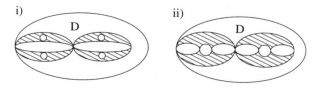

FIGURE 4. Figure 4(i) shows D and its images under the truncated map $(\mathcal{F}_1|_2)^i(D)$ for $i = 1, 2$. Figure 4(ii) shows D and its images under the truncated map $(\mathcal{F}_2|_2)^i(D)$ for $i = 1, 2$.

Theorem 2. Λ_+ *is the closure of a countable union of copies of $\mathcal{K}$.*

Proof. By Lemma 4 it is sufficient to show that each point in the interior of Λ'_+ lies in one of the images of $\mathcal{K}$ under forward iterates of $\mathcal{F}_m$. Suppose $x \in \Lambda'_+$ and suppose that x does not lie in one of the images of $\mathcal{K}$. Then x lies in the intersection $\mathcal{I}$ of a sequence of nested discs as described above. Let $\{D_n\}$ denote that sequence, and let $A_n = D_n - \overline{D_{n+1}}$, a sequence of nested annuli. Using techniques explained in detail in the proof of Lemma 7.1 in [2], one can show that there exists a subsequence $\{A_{n_k}\}$ of annuli with the property that some iterate of $\mathcal{F}^{-1}$ maps each A_{n_k} onto an annulus B_k, which is one of the sets $D - \overline{f_j(D)}$, where $j > 1$. Moreover, this iterate is either a degree one or a degree two map when restricted to A_{n_k}. Thus, the modulus of each A_{n_k} is either equal to that of B_k, or it is half the modulus of B_k. Now, the moduli of the annuli $D - \overline{f_j(D)}$, $j > 1$, are clearly bounded below, since the discs $f_j(D)$ accumulate at $J(\infty)$. Hence, the moduli of the A_{n_k} are bounded below, so their sum diverges. Now the annulus $D_1 - \mathcal{I}$ contains all the annuli A_{n_k}. By the Groetzsch inequality, the modulus of $D_1 - \mathcal{I}$ is infinite, and hence $\mathcal{I}$ consists of a single point, namely our point x.

Each of the discs D_n is an image of D and therefore contains an image of $\mathcal{K}$, so x is an accumulation point of such images. $\square$

It will be useful to regard Λ_+ as a limit of sets Λ_+^n associated to *truncated correspondences*: for each positive integer n, we denote by $\mathcal{F}_m|_n$ the correspondence $f_1 \cup f_2 \cup \ldots \cup f_n$, and by Λ_+^n the set $\bigcap_k (\mathcal{F}_m|_n)^k(D)$. The configuration of the discs $(\mathcal{F}_m|_n)^k(D)$ depends on the position of 0 within $(\mathcal{F}_m|_n)^{k-1}(D)$. See Figure 4. Clearly, $\Lambda_+ = \overline{\bigcup_n \Lambda_+^n}$.

Remark 4. *There are two sets of points that lie in Λ_+ but not in $\bigcup_n \Lambda_+^n$. The first set consists of points in the orbit of P. The second consists of points x of Λ'_+ with the property that the set $\{k : \mathcal{F}_m^{-l}(x) \in f_k(D) \text{ for some } l\}$ is not bounded above. A point of this type is said to have an "unbounded itinerary".*

Remark 5. *Figure 1 shows $\mathcal{F}_1$; that is, the parameter a is chosen such that $J_a(1) = 0$. Figure 2 shows a correspondence with the parameter a chosen to make 0 an attracting point of period 2 under f_1^{-1} - a case which could be analysed by generalising the methods given here.*

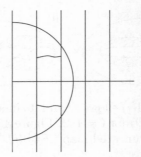

FIGURE 5. The discs D_+^t and U_2^t for some $t \geq t(2)$.

5. DYNAMICS OF THE MAP $\mu_m R(z)$

In this section we describe the dynamics of $z \to \mu_m R(z) = \mu_m \sin^2(\pi\sqrt{z})$, where $\mu_m = (m - 1/2)^2$ and m is a positive integer. This map has critical points j^2, $j \geq 1$, all mapping to the fixed point zero, and $(j - 1/2)^2$, $j \geq 1$, all mapping to the critical value μ_m. Thus, the point $(m - 1/2)^2$ is a superattracting fixed point. Let $\lambda_m = m - 1/2$ and define $r_{\lambda_m}(z) = \lambda_m \sin(\pi z)$. Then r_{λ_m} has critical points $j - 1/2$, $j \geq 1$, mapping to $\pm \lambda_m$.

Proposition 3. *Let m be an odd positive integer, and let D_+^t denote the set $\{z = x + iy : |z| < t, x > 0\}$. Then*

(i) *For each odd positive integer $n > 1$, there exists a real number $t(n)$ such that for all $t \geq t(n)$ there is a disc U_n^t such that*
   - *The critical point $n - 1/2$ is contained in U_n^t,*
   - $\bar{U}_n^t \subset D_+^t$,
   - $r_{\lambda_m}(U_n^t) = D_+^t$. *Thus $r_{\lambda_m}|_{U_n^t}$ is quadratic-like.*

(ii) *For each even positive integer the same is true with r_{λ_m} replaced by $-r_{\lambda_m}$.*

(iii) *For $n = 1$, there exists a real number $t(1)$ such that for all $t \geq t(1)$ there is a disc U_1^t such that*
   - *The critical point $1/2$ is contained in U_1^t,*
   - $\bar{U}_1^t \subset \bar{D}_+^t$, *and the $\partial U_1^t \cap \partial D_+^t$ consists of a subinterval of the imaginary axis,*
   - $r_{\lambda_m}(U_1^t) = D_+^t$.

Proof. Assume that n is odd (for n even just replace r_{λ_m} by $-r_{\lambda_m}$). For any $t \in \mathbb{R}$, the inverse image of D_+^t which contains $n - 1/2$ consists of a quadrilateral region which has vertical segments at $n - 1$ and n as two of its boundary components. A calculation shows that for large enough t, these vertical segments and the other two boundary components are properly contained in D_+^t. See Figure 5. □

Remark 6. *If m is even, then the same result holds, but with r_{λ_m} replaced by $-r_{\lambda_m}$.*

Proposition 4. *Let m be a positive integer. Then for each positive integer n there exists a topological disc D_n such that the interior of $(\mu_m R)^{-1}(D_n)$ consists of infinitely many connected components V_j^n, such that:*

- *each V_j^n contains the critical point $(j - 1/2)^2$;*
- *∂V_j^n and ∂V_{j+1}^n meet in the point j^2 which is an inverse image of 0;*
- *$\mu_m R$ restricted to each V_j^n is a two-to-one map. For $1 < j \leq n$ the disc V_j^n is properly contained in D_n and hence this restriction is quadratic-like. It is simply-pinched quadratic-like if $j = 1$.*

Proof. Given m and n, let $t \geq t_n$ and consider the disc D_+^t. We note that $t(n) > t(1)$, so by Proposition 3 the inverse image of D_+^t contains the quadrilateral region U_1^t, which contains $1/2$ and has a subinterval I of the imaginary axis as a boundary component. Let W denote the image of $U_1^t \cup I$ under $z \to z^2$. Then the map $\mu_m R$ restricted to W is a quadratic-like map, 0 is a repelling fixed point and the landing point of the external ray of argument zero of this map. By "cutting out" the external ray of argument zero from W_n we can create a topological disc S such that $\mu_m R$ restricted to S is simply-pinched quadratic-like with contact point 0. Define $D_n = \mu_m R(S)$. Then D_n has the required properties. □

6. The Fatou set of $\mu_m R$

We fix a positive integer m and continue to study the dynamics of the map $\mu_m R$. For any positive integer n, we define the *truncated map* $\mu_m R|_n$ to be the restriction of $\mu_m R$ to the discs V_i^n, $1 \leq i \leq n$, defined in the previous Proposition. Then each $\overline{V_i^n}$ is contained in $\overline{D_n}$, and the boundaries touch only if $i = 1$, in which case the contact point is 0. We define

$$K_n = \bigcap_k (\mu_m R|_n)^{-k}(D_n).$$

For n large enough, one of the V_i^n contains the critical point $(2m - 1)^2/4$, which is a fixed point. Thus, by the same reasoning as for the truncated correspondences, the set K_n contains a copy $\mathcal{H}$ of the filled Julia set of $z \to z^2$ and the map restricted to this copy is conjugate to $z \to z^2$, the conjugacy being conformal on the interior. We note that $\mathcal{H}$ does not depend on n.

Theorem 3. *The union of the inverse images of $\mathcal{H}$ under $\mu_m R$ is the Fatou set of $\mu_m R$.*

Proof. Each Fatou component of $\mu_m R$ is pre-periodic, and eventually lands on either an attracting or parabolic basin, or a Siegel disc (Herman rings, Baker domains and wandering domains do not occur since $(\mu_m R)^{-1}$ has finitely many singular points). Each of these Fatou components "uses up" a critical point of $\mu_m R$ (see [9], Theorems 2.4.1, 2.4.2 and 2.4.6.). However, all critical points map to the same critical value μ_m, so $\mathcal{H}$ is the only forward invariant Fatou component. □

The similarities between the dynamics of the truncated map $\mu_m R|_n$ and the truncated correspondence $\mathcal{F}_m|_n$ are clear: for each k the sets $(\mu_m R|_n)^k(D_n)$ and $(\mathcal{F}_m|_n)^k(D)$ are topologically the same (see Figure 4). The points zero and μ_m in the dynamical plane of $\mu_m R$ play the same roles as the points $1/4$ and 0 (respectively) in the correspondence plane. We will see that there is a conformal conjugacy between the two maps restricted to their truncated filled Julia sets. The difference between the two situations is that this conjugacy does not extend to the sets $K = \overline{\bigcup_n K_n}$ and $\Lambda_+ = \overline{\bigcup_n \Lambda_+^n}$. In the $\mu_m R$ situation, the discs D_n vary with n. In fact these discs increase in size as n increases, and K is the whole of $\mathbb{C}$. $K - \bigcup_n K_n$ consists of a set of curves, so-called dynamical rays, while $\Lambda_+ - \bigcup_n \Lambda_+^n$ consists of a set of points (see Remark 4). In other words, while a sequence of points in $\bigcup_n \Lambda_+^n$ might accumulate at a point, the corresponding sequence in $\bigcup_n K_n$ might accumulate along the whole of a curve. We shall discuss this possibility further in Section 8. But first we shall establish a conjugacy between actions on truncated filled Julia sets.

7. Conjugacies on "truncated filled Julia sets"

Consider the correspondence $\mathcal{F}_m$ and the map $\mu_m R$. We construct a bijection

$$\Psi : \bigcup_n \Lambda_+^n \to \bigcup_n K_n$$

which is a homeomorphism on each Λ_+^n, sending it onto K_n. We have a conjugacy $\Psi : \mathcal{K} \to \mathcal{H}$, which is conformal on the interior of $\mathcal{K}$, coming from the conjugacy to $z \to z^2$ of the relevant restrictions of the map and correspondence. We can extend Ψ to the inverse images of the closed Fatou component $\mathcal{K}$ to give a conjugacy on these. Now consider any point $x \in \bigcup_n \Lambda_+^n$ which does not lie in an image of $\mathcal{K}$ under iterates of $\mathcal{F}_m$. Then x lies in Λ_+^n for some n, and it is in the intersection of a nested sequence $\{D_k\}$ of discs, as described in Section 4. To each disc D_k we can associate a unique disc D_k' in $(\mu_m R|_n)^k(D_n)$, as the dynamics of the truncated map and correspondence are topologically the same. Hence we obtain a sequence $\{D_k'\}$ of nested discs in the dynamical plane of $\mu_m R|_n$. By Theorem 2 we have $\bigcap_k D_k = \{x\}$. It is clear that $\bigcap_k D_k'$ is not an inverse image of the Fatou component $\mathcal{H}$. We can use the methods in the proof of Theorem 2 to prove that $\bigcap_k D_k'$ consists of a single point y. We define $\Psi(x) = y$.

It is easy to check that Ψ is a homeomorphism restricted to each Λ_+^n, that it is conformal on interiors, has image K_n and conjugates $\mathcal{F}_m$ to $\mu_m R$. Ψ restricted to the union of the Fatou components of Λ_+ is a conformal conjugacy between $\mathcal{F}_m$ and $\mu_m R$ restricted to its Fatou set.

8. Dynamical rays

8.1. $\partial \Lambda_+$ and the group. Recall that the correspondence $\mathcal{F}_m$ represents a mating between a map and the group Γ. Thus, there exists a conformal homeomorphism ϕ from the upper half-plane to $\Omega = \hat{\mathbb{C}} - \Lambda$ which conjugates the action of the group to that of $\mathcal{F}_m|_\Omega$.

Theorem 4. *For each $m > 1$ the homeomorphism ϕ from the upper half-plane to Ω can be extended to a continuous surjection $\phi : \hat{\mathbb{R}} \to \partial \Lambda$, with $\phi(\hat{\mathbb{R}}_+) = \partial \Lambda_+$. Hence $\partial \Lambda_+$ is locally connected.*

Proof. We shall only give a very brief indication of the method here. The analogous result with the Hecke group H_4 in place of Γ is stated as Theorem 9.1 in [2] and proved in detail there.

Each point $x \in \hat{\mathbb{R}}_+$ is the unique point in the intersection of a nested sequence of closed discs in $\hat{\mathbb{C}}$ (these nested sequences correspond to the nested sequences of closed intervals on $\hat{\mathbb{R}}_+$ discussed in Section 1). The parts of the boundaries of these discs which lie in the upper half-plane are carried by ϕ to curves in Ω. By a combination of arguments involving nested discs in $\Omega \cup \Lambda$ and the hyperbolicity of the quadratic maps involved in the restrictions of $\mathcal{F}_m^{-1}$, it is possible to show that the end-points of these curves converge to a unique point $y \in \partial \Lambda_+$ which we then define to be $\phi(x) = y$. $\square$

Remark 7. *The restriction of this result to $m > 1$ is for technical reasons, involving the fact that for $m = 1$ the relevant quadratic map is only pinched-equivalent to z^2 rather than hybrid-equivalent to it. However we conjecture that Theorem 4 is equally valid in the case $m = 1$, and indeed we have an outline argument for a proof of this using pinching techniques.*

The homeomorphism ϕ gives rise to an equivalence relation $\sim$ on $\hat{\mathbb{R}}$ by $x \sim y$ if and only if $\phi(x) = \phi(y)$; thus $\hat{\mathbb{R}}/\sim$ is homeomorphic to $\partial \Lambda_+$. We give an explicit definition of ϕ.

Definition 4. *Define the critical interval $\mathcal{C}$ to be the interval with endpoints given by the itineraries $(-m, -m, -m, \ldots)$ and $(m, m, m, \ldots)$ (see Section 1).*

Definition 5. *Let $\sim$ be the closure of the following equivalence relation defined inductively on the orbit of 0 and ∞: let $0 \sim \infty$ and suppose $x \sim y$.*
- *If $y = 1/x$ and $x, y \notin \mathcal{C}$ define $\sigma \rho^i(x) \sim \sigma \rho^{-i}(y) = \frac{1}{\sigma \rho^i(x)}$ for any $i \neq 0$.*
- *Otherwise define $\sigma \rho^i(x) \sim \sigma \rho^i(y)$.*
- *For $x, y \in \mathbb{R}_-$ define $x \sim y$ if and only if $\sigma(x) \sim \sigma(y)$.*

Theorem 5. *For any two points x and y in $\hat{\mathbb{R}}$ we have $\phi(x) = \phi(y)$ if and only if $x \sim y$.*

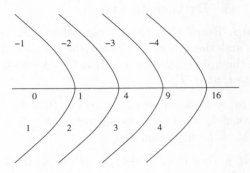

FIGURE 6. The division of the plane into labelled strips.

Proof. Once again we give only the briefest of indications of the proof: see [2] Lemmas 9.1 and 9.2 and Proposition 9.1 for details in the analogous case where H_4 plays the role of Γ here. Let $\approx$ be the equivalence relation $x \approx y$ if and only if $\phi(x) = \phi(y)$. The two branches of f_j correspond to the pair $\{\sigma\rho^j, \sigma\rho^{-j}\}$. From this one can show that points in $\partial \mathcal{K}$ map via ϕ^{-1} to the set T consisting of points whose itineraries are infinite sequences in m and $-m$ only. These are identified under $\approx$ to form a circle, since $\mathcal{K}$ is homeomorphic to the filled Julia set of $z \to z^2$. One can derive the definition of $\approx$ inductively on the orbit of 0 and ∞: if $x \approx y$ then either $\sigma\rho^i(x) \approx \sigma\rho^i(y)$ or $\sigma\rho^i(x) \approx \sigma\rho^{-i}(y)$. The fact that points in T get identified by $\approx$ to form a circle imposes combinatorial restrictions on which of the two possibilities can occur in each case. Closer inspection shows that the rule prescribing which possibility occurs is exactly that given in the definition of $\sim$. □

8.2. **The structure of $J(\mu_m R)$.** By adapting [1] and [4] we assign an *itinerary* to each point in the Julia set $J(\mu_m R)$, and we show that $J(\mu_m R)$ is a disjoint union of curves, each labelled by a sequence. First divide the plane into labelled strips as follows (Figure 6):

- Let $S(-n)$ denote the open region bounded by the real interval $[(n-1)^2, n^2]$ and by the lines $\{(x+iy)^2 \ : \ x = (n-1), y \geq 0\}$ and $\{(x+iy)^2 \ : \ x = n, y \geq 0\}$;
- Let $S(n)$ denote the open region bounded by the real interval $[(n-1)^2, n^2]$ and by the lines $\{(x+iy)^2 \ : \ x = (n-1), y \leq 0\}$ and $\{(x+iy)^2 \ : \ x = n, y \leq 0\}$.

Each strip is an inverse image of $\mathbb{C} - (\mathbb{R}_- \cup [0, \mu_m])$, where $\mathbb{R}_-$ denotes the negative real line. The non-real boundary components of each strip map onto $\mathbb{R}_-$, and the real boundary component maps two-to-one onto the interval $[0, \mu_m]$.

If z is a point in the Julia set whose orbit never lands on $\mathbb{R}_- \cup [0, \mu_m]$ then each $z_k = (\mu_m R)^k(z)$ lies inside one of the open strips. We give z an *itinerary* $s(z) = \{s_n\}_{n \geq 0}$ by defining $s_i = n$ if and only if $(\mu_m R)^i(z)$ lies in the strip $S(n)$, for some $n \in \mathbb{Z} - \{0\}$.

Next consider $z \in \mathbb{R}_+$ whose orbit never lands on 0. Then all the images of z are real and positive and hence each lies on the common boundary of two strips. We give z two itineraries $s(z)$ and $s'(z)$ inductively as follows: if $z = z_0$ lies on the boundary of strips $S(n)$ and $S(-n)$ define $s_0 = n$ and $s'_0 = -n$. Now suppose that s_i and s'_i have been defined for all $i \leq k$ and that z_k and z_{k+1} lie on the common boundary of the strips $S(l)$ and $S(-l)$, and $S(m)$ and $S(-m)$ respectively. Then if z_k lies in the open interval $((l-1)^2, (l-1/2)^2)$ we define $s_{k+1} = -m$ and $s'_{k+1} = m$. Otherwise we define $s_{k+1} = m$ and $s'_{k+1} = -m$.

If z is a point whose orbit eventually lies on $\mathbb{R}_+$ but does not include 0, we define the itinerary accordingly. Now we have itineraries for all points in the Julia set which do not eventually land on $\mathbb{R}_- \cup \{0\}$.

Note that points in $\partial(\bigcup_n K_n)$ have bounded orbits, and therefore the itinerary of any such point has the property that the set of integers j such that j or $-j$ occurs in the itinerary of z is bounded above. We call such itineraries *bounded itineraries*.

Theorem 6. *For any sequence s of non-zero integers, the set h_s of points in the Julia set which have itinerary s is either empty or homeomorphic to a ray $[y_s, \infty]$. If s is bounded, then the end-point (the point corresponding to y_s) lies on $\partial(\bigcup_n K_n)$ and it is the only point in $h_s \cap \partial(\bigcup_n K_n)$.*

Proof. We will only give an outline of the proof and refer the reader to [1] and [4] for details. Firstly we notice that the partition of the plane that gives rise to the sets h_s is double-covered by a partition of the plane which respects the dynamics of the map $r_{\lambda_m}(z) = \lambda_m \sin(\pi z)$, $\lambda_m = \sqrt{\mu_m}$, and that it is sufficient to prove an analogous result for $r_{\lambda_m}(z)$. Our choice of parameter makes it possible to show that the set $A = \{z : |(r_{\lambda_m})'(z)| \leq 1\}$ lies within the union of the Fatou components of r_{λ_m} which meet the real line: it consists of topological disks A_j, $j \in \mathbb{Z}$, each containing exactly one critical point $j - 1/2$. A calculation shows that each A_j maps under r_{λ_m} either into the disc A_m containing the fixed critical point λ_m or into the disc A_{-m} containing the fixed critical point $-\lambda_m$. It follows that each A_j lies in the basin of attraction of either λ_m or $-\lambda_m$. This result enables us to use exactly the same methods as in [1] and [4] to prove that for each itinerary the set of points with that itinerary is either empty or homeomorphic to a ray. We transfer this result to the $\mu_m R$-plane to get the required result.

Now suppose that $z \in \partial(\bigcup_n K_n)$ has itinerary s, so the set h_s is non-empty. Suppose that z does not lie on the boundary of the Fatou component $\mathcal{H}$ or any of its pre-images under $\mu_m R$. Then it lies in the intersection of a sequence of nested discs as described earlier. Any other point in $\partial(\bigcup_n K_n)$ with the same itinerary s lies in the same intersection of nested discs. But this consists of a single point, so z is the unique point in $\partial(\bigcup_n K_n)$ with itinerary s.

Suppose that $z \in \partial \mathcal{H}$. Recall that $\mathcal{H}$ is the filled Julia set of a quadratic-like restriction of $\mu_m R$ to a disc D (this restriction is pinched if $m = 1$). Thus z lies in the intersection of infinitely many inverse images of the cut disc $D - [0, \mu_m]$. Again, one can check that any other point in $\partial(\bigcup_n K_n)$ with itinerary s lies in the same intersection. For $m > 1$, the quadratic-like restriction is hybrid-equivalent to $z \to z^2$ which is expanding on a neighbourhood of its Julia set. Therefore z is the only point in $\partial(\bigcup_n K_n)$ with itinerary s. If $m = 1$ we still have a topological conjugacy between $z \to z^2$ and $\mu_m R$ on $\partial \mathcal{H}$, so the result holds in this case as well. This implies the result for points lying in an inverse image of $\partial \mathcal{H}$. □

So far, we have ignored points whose orbits land on $\mathbb{R}_- \cup \{0\}$. Clearly these lie on a set of curves, the inverse images of $\mathbb{R}_-$, each with end-point an inverse image of 0. The half-line $\mathbb{R}_-$ and its pre-images form the boundary components of adjacent strips, so we can give such points several itineraries, leading to an identification of sequences in non-zero integers. We note that we have encountered the same identification before, namely in the parametrisation of points in $\hat{\mathbb{R}}_+$ by infinite sequences, given in Section 1.

We call each curve h_s a *dynamical ray* (or *hair*) of $\mu_m R$ and note that the Julia set is the union of the dynamical rays. Note that two dynamical rays might share a common end-point. Moreover, it is not true that there exists a dynamical ray h_s for each itinerary s: each dynamical ray is the intersection of infinitely many inverse images of $\mathbb{C} - [0, \mu_m]$ and this intersection might consist only of the point at infinity. Let us call a sequence *realisable* if it has a non-empty dynamical ray associated to it. A precise condition for realisability is given in [10], namely that a sequence is realisable if and only if it is *exponentially bounded* (see [10] for a definition of this term).

Thus we have a picture of $J(\mu_m R)$ as consisting of a Cantor set of dynamical rays corresponding to realisable sequences, each ray leading from a point on the boundary of the Fatou set to the point at infinity. Restricting attention at first to *bounded* sequences, and recalling the bijection

$$\Psi : \bigcup_n \Lambda_+^n \to \bigcup_n K_n$$

of Fatou sets defined in Section 7, a close look at the action of $\mathcal{F}_m^{-1}$ on $\partial \Lambda_+$ and that of $\mu_m R(z)$ on $\bigcup_n \partial K_n$ shows that for each $z \in \bigcup_n \partial \Lambda_+^n$, and each of the inverse images x of z under ϕ, the point $\Psi(z) = w$ is the end-point of a dynamical ray with itinerary $s(x)$.

It might be tempting to extend Ψ to a map $\partial\Lambda_+ \to \partial(\bigcup_n K) \cup \{\infty\}$ by assigning to each $z \in \partial\Lambda_+$ the end-point of a dynamical ray with itinerary $s(x)$, where $x \in \phi^{-1}(z)$, if such a ray exists, and assigning to z the value infinity if there is no such ray. But such a Ψ could not be continuous as it would send a dense set in $\partial\Lambda_+$ to ∞. Points $z_n \in \partial\Lambda_+$ which accumulate at some point z can have images $\Psi(z_n)$ accumulating not at $\Psi(z)$, but along a ray that lands at $\Psi(z)$. The best we can say formally is that if we give the space of dynamical rays the topology associated to the sequences labelling them, then there is a homeomorphism between points in $\partial\Lambda_+$ with realisable sequences, and the quotient of the space of realisable rays obtained by identifying rays which land at the same point.

Informally, we think of the action of $\mathcal{F}_m^{-1}$ on Λ_+ as a "compactified" version of the action of $\mu_m R$ on $\mathbb{C}$. Each dynamical ray gets shrunk to a point and the point at infinity is split into infinitely many points, one for each empty dynamical ray, and one for each approach to ∞ through inverse images of the real line (corresponding to points in the orbit of P). The essential singularity at ∞ has been "blown up" and in its place we have glued an open disc carrying the action of the group Γ.

9. Remarks and generalisations

Firstly we remark that each truncated map $\mu R|_n$ has the combinatorics of a restriction of a scalar multiple of the Chebyshev polynomial of odd degree $2n + 1$. For results on how the filled Julia sets K_n of these polynomials converge as $n \to \infty$ see [8]. For further properties of the Fatou and Julia sets of maps $z \to \lambda \sin(z)$ see [5] and [10].

Secondly, we remark that in this paper we have only considered the simplest matings involving the group Γ, namely those for which the point $(2m-1)^2/4$ is fixed by $\mathcal{F}^{-1}$. This means that the restriction f_m^{-1} of $\mathcal{F}^{-1}$ is hybrid-equivalent to $z \to z^2$. However, using the methods of [7], one can prove that for each real c such that $z \to z^2 + c$ has a connected filled Julia set, there exists a correspondence $\mathcal{F}_{a(c)}$ depending on a real parameter $a(c)$, such that the restriction f_m^{-1} of $\mathcal{F}_{a(c)}^{-1}$ is hybrid-equivalent to $z \to z^2 + c$. Similarly, it can be shown that there exists a parameter μ_c such that the relevant restriction of $\mu_c R$ is hybrid-equivalent to $z \to z^2 + c$. Provided $z \to z^2 + c$ is hyperbolic, the methods used in this paper can be adapted to show that $\mathcal{F}_c$ represents a mating between Γ and $\mu_c R$. Note that the parameters μ we consider here lie in the limb $\mathcal{L}$ of the connectedness locus of the family μR in which 0 is a non-attracting fixed point at which a single dynamical ray lands.

In general, we conjecture that the moduli space $\mathcal{M}$ of correspondences in our family is the closure of a countable union of copies of the quadratic Mandelbrot set: any composition $f_{\epsilon_1} \circ \ldots \circ f_{\epsilon_n}$ maps D onto a set of discs and its inverse restricted to each disc is quadratic-like. We conjecture that for each such restriction and for each $z \to z^2 + c$ with connected Julia set there exists $a \in \mathcal{M}$ such that the relevant restriction of $\mathcal{F}_a$ is hybrid-equivalent to $z \to z^2 + c$. Thus we obtain a copy of the Mandelbrot set for each such restriction: in the language of [2] we have a copy of the Mandelbrot set for each *critical address*.

Using exactly the same reasoning, we expect that the limb $\mathcal{L}$ of the connectedness locus of the family $\{\mu R\}$ is the closure of a countable union of copies of the Mandelbrot set, and that $\mathcal{M}$ and $\mathcal{L}$ are homeomorphic (it is known that $\mathcal{L}$ does contain infinitely many copies of the Mandelbrot set lying along the real axis). We conjecture that for each μ in the interior of $\mathcal{L}$ the map μR can be mated with the group Γ by the corresponding $\mathcal{F}_a$ with a in the interior of $\mathcal{M}$.

References

[1] JM Aarts and LG Oversteegen, *The geometry of Julia sets*, Trans. Amer. Math. Soc., 338 (1993), no. 2, 897-918.

[2] S Bullett and M Freiberger, *Holomorphic correspondences mating Chebyshev-like maps with Hecke groups*, Ergodic Theory Dynam. Systems, 25 (2005), no. 4, 1057-1090.

[3] S Bullett and M Freiberger, *Hecke groups, polynomials and matings*, Int. J. Modern Physics B, 17 (2003), nos. 22-24, 3922-3931.

[4] R Devaney, Se^x: *Dynamics, topology and bifurcations of complex exponentials*, Topology and its Applications, 110 (2001), 133-161.

[5] P Domínguez and N Fagella, *Residual Julia sets of rational and transcendental functions*, Transcendental dynamics and complex analysis, Cambridge University Press, 2008.

[6] A Douady and J H Hubbard, *On the dynamics of polynomial-like mappings*, Ann. Sci. Éc. Norm. Sup., 18 (1985), 287-343.

[7] M Freiberger, *PhD Thesis 2001*, Queen Mary, University of London.

[8] B Krauskopf and H Kriete, *Hausdorff convergence and the limit shape of unicorns*, Experimental Math., 6 (1997), 117-135.

[9] S Morosawa, Y Nishimura, M Taniguchi, T Ueda, *Holomorphic Dynamics*, Cambridge Studies in Advanced Mathematics No 66, Cambridge University Press, 2000.

[10] G Rottenfusser and D Schleicher, *Escaping points of the cosine family*, Transcendental dynamics and complex analysis, Cambridge University Press, 2008.

School of Mathematical Sciences, Queen Mary, University of London, Mile End Road, London E1 4NS, UK
E-mail address: S.R.Bullett@qmul.ac.uk

MMP, Centre for Mathematical Sciences, Wilberforce Road, Cambridge CB3 0WA, UK
E-mail address: M.Freiberger@damtp.cam.ac.uk

SINGULAR PERTURBATIONS OF z^n

ROBERT L. DEVANEY
MATT HOLZER
DANIEL M. LOOK
MÓNICA MORENO ROCHA
DAVID UMINSKY

1. Introduction

Our goal in this paper is to describe the topology of and dynamics on certain Julia sets of functions drawn from the family of rational maps of the complex plane given by

$$F_\lambda(z) = z^n + \frac{\lambda}{z^d}$$

where $n, d \in \mathbb{Z}^+$. When $\lambda = 0$, these maps reduce to $z \mapsto z^n$ and the dynamical behavior in this case is well understood: the Julia set of F_λ is just the unit circle and all other orbits tend either to ∞ or to the superattracting fixed point at 0.

When $\lambda \neq 0$, several things happen. First of all, the map F_λ now has degree $n + d$ rather than n. Secondly, the origin is a pole rather than a fixed point. And, finally, there are $n + d$ new critical points in addition to the original critical points at 0 and ∞. As we discuss below, the orbits of all of these new critical points behave symmetrically, so we essentially have only one additional "free" critical orbit for each of these maps. As is well known in complex dynamics, the behavior of this critical orbit determines much of the structure of the Julia sets of these maps.

One of our main goals in this paper is to describe what happens to the Julia set when the parameter λ is nonzero but small. In this case, the map F_λ is called a *singular perturbation* of z^n. The reason for the interest in such a perturbation arises from Newton's method. Suppose we are applying Newton's method to find the roots of a family of polynomials P_λ which has a multiple root at, say, the parameter $\lambda = 0$. For example, consider the especially simple case of $P_\lambda(z) = z^2 + \lambda$. When $\lambda = 0$ this polynomial has a multiple root at 0 and the Newton iteration function is simply $N_\lambda(z) = z/2$. However, when $\lambda \neq 0$, the Newton iteration function becomes

$$N_\lambda(z) = \frac{z^2 - \lambda}{2z}$$

and we see that, as in the family F_λ, the degree jumps as we move away from $\lambda = 0$. In addition, instead of a fixed point at the origin, after the perturbation, there is a pole at the origin.

For the families F_λ, there are a number of different cases to consider depending on the values of n and d. When $n \geq 2$, the point at ∞ is a superattracting fixed point whereas when $n = 1$ this point is a parabolic fixed point. Since much of the interesting dynamical behavior occurs when the free critical points tend to or land at ∞, the singular perturbations therefore behave very differently in these two cases.

One of the main results that we describe below is the following. When $n \geq 2$, we have an immediate basin of attraction B_λ of the superattracting fixed point at ∞. Note that F_λ is n to 1 on a neighborhood of ∞ in B_λ. Since 0 is a pole of order d, the only preimages of points in this neighborhood lie in a neighborhood of the origin. We let T_λ be the preimage of B_λ surrounding the origin. (The sets B_λ and T_λ may or may not be disjoint.) As we shall show, for λ small, it is possible that the critical orbits eventually land in B_λ and hence tend to ∞. In this case, we have the following result described in Section 3.

Theorem (The Escape Trichotomy). *Suppose $n \geq 2$ and that the orbits of the free critical points of F_λ tend to ∞. Then*

(1) *If one of the critical values lies in B_λ, then $J(F_\lambda)$ is a Cantor set and $F_\lambda \mid J(F_\lambda)$ is a one-sided shift on $n + d$ symbols. Otherwise, the preimage T_λ is disjoint from B_λ.*
(2) *If one of the critical values lies in $T_\lambda \neq B_\lambda$, then $J(F_\lambda)$ is a Cantor set of simple closed curves (quasicircles).*
(3) *If one of the critical values lies in a preimage of B_λ different from T_λ, then $J(F_\lambda)$ is a Sierpinski curve.*

Several Julia sets illustrating this trichotomy and drawn from the family where $n = d = 3$ are included in Figure 1.

A *Sierpinski curve* is a very interesting topological space. By definition, a Sierpinski curve is a planar set that is homeomorphic to the well-known Sierpinski carpet fractal. But a Sierpinski curve has an alternative topological characterization: any planar set that is compact, connected, locally connected, nowhere dense, and has the property that any two complementary domains are bounded by disjoint simple closed curves is known to be homeomorphic to the Sierpinski carpet [24]. Moreover, such a set is a universal planar set in the sense that it contains a homeomorphic copy of any compact, connected, one-dimensional subset of the plane.

When $n \geq 2$, there are certain cases of this Theorem that may or may not hold, depending on the value of d. For example, if n and d satisfy

$$\frac{1}{n} + \frac{1}{d} < 1,$$

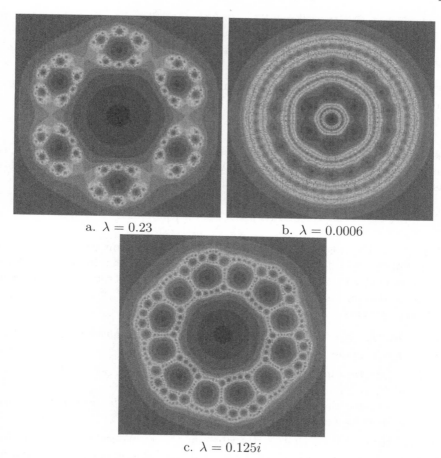

a. $\lambda = 0.23$ b. $\lambda = 0.0006$

c. $\lambda = 0.125i$

FIGURE 1. Some Julia sets for $z^3 + \lambda/z^3$: if $\lambda = 0.23$, $J(F_\lambda)$ is a Cantor set; if $\lambda = 0.0006$, $J(F_\lambda)$ is a Cantor set of circles; and if $\lambda = 0.125i$, $J(F_\lambda)$ is a Sierpinski curve.

then there is a neighborhood of $\lambda = 0$ for which the critical values all lie in $T_\lambda \neq B_\lambda$ and so the Julia set is a Cantor set of simple closed curves. This phenomenon was first observed by McMullen for small λ (see [14]) and so we call the regime in the λ-plane where this occurs the *McMullen domain*. There is no McMullen domain if this inequality does not hold, i.e., if n or d is equal to 1 or if $n = d = 2$. Instead, in the special cases where $n = d = 2$ or $n > 1, d = 1$, we have the following result which is described in Section 4:

Theorem. *Suppose $n = d = 2$ or $n > 1$, $d = 1$. Then, in every neighborhood of the origin in the parameter plane, there are infinitely many disjoint open sets $\mathcal{O}_j$ with $j = 1, 2, 3, \ldots$ of parameters having the following properties:*

(1) If $\lambda \in \mathcal{O}_j$, then the Julia set of F_λ is a Sierpinski curve, so that if $\lambda \in \mathcal{O}_j$ and $\mu \in \mathcal{O}_k$, the Julia sets of F_λ and F_μ are homeomorphic;
(2) But if $k \neq j$, the maps F_λ and F_μ are not topologically conjugate on their respective Julia sets.

The case where $n = 1$ is fundamentally different from the other cases since the function

$$F_\lambda(z) = z + \frac{\lambda}{z^d}$$

has a parabolic fixed point at ∞. Furthermore, any map of this form is linearly conjugate to the case where $\lambda = 1$. So, instead of considering this case, we adjust the family slightly to deal instead with the family

$$F_\lambda(z) = \lambda \left(z + \frac{1}{z^d} \right)$$

when $n = 1$. For this family ∞ is an attracting fixed point when $|\lambda| > 1$ and is repelling when $|\lambda| < 1$. So when $|\lambda| < 1$, ∞ is in the Julia set, and we may have that the critical orbits map onto ∞. In this case, the Julia set is the entire Riemann sphere. Much else occurs near $\lambda = 0$, for, as we show in Section 5, we have:

Theorem. Let $F_\lambda(z) = \lambda(z + 1/z)$. Then, in any neighborhood of $\lambda = 0$ in the parameter plane:

(1) There are infinitely many parameter values λ for which the Julia set of F_λ is the entire Riemann sphere;
(2) There are also infinitely many parameter values for which the critical orbit is superattracting.

Unlike the situation that is described in the previous two theorems for λ near 0, in the case where we have a McMullen domain, the dynamical behavior of F_λ is the same for any λ sufficiently close to 0. However, away from this region, F_λ exhibits a rich array of different dynamical behavior. For example, in Section 6 we show that there are many different ways that the Julia sets may be Sierpinski curves. In the previous Sierpinski curve examples, the complement of $J(F_\lambda)$ was simply B_λ together with all of its preimages. However, there are parameter values in these families for which the Julia set is a Sierpinski curve whose complementary domains consist of a variety of different attracting basins (not just B_λ) together with their preimages. Again, while these Julia sets are all homeomorphic to one another, the dynamics on different pairs of these sets is often quite different.

There is another famous Sierpinski "object" in fractal geometry, namely, the Sierpinski gasket or triangle. Objects similar in construction to this shape also occur in these families. In Section 7 we construct infinitely many "Sierpinski gasket-like" Julia sets for F_λ. Unlike the Sierpinski curves, each pair of these Julia sets are topologically as well as dynamically distinct.

This paper is respectfully dedicated to the memory of Professor Noel Baker. Professor Baker's numerous contributions to the field of complex dynamics have been an inspiration to all of us.

2. Preliminaries

We consider the maps

$$F_\lambda(z) = z^n + \frac{\lambda}{z^d}$$

where $n, d \in \mathbb{Z}^+$. The *Julia set* of F_λ, $J(F_\lambda)$, is defined to be the set of points at which the family of iterates of F_λ fails to be a normal family in the sense of Montel. Equivalently, the Julia set is the closure of the set of repelling periodic points for F_λ or, alternatively, the set of points on which F_λ behaves chaotically. The complement of the Julia set is called the *Fatou set*.

There are $n + d$ finite and nonzero critical points for F_λ and all are of the form $\omega^k c_\lambda$ where c_λ is one of the critical points and $\omega^{n+d} = 1$. Similarly, the critical values are arranged symmetrically with respect to $z \mapsto \omega z$, though there need not be $n + d$ of them. For example, if $n = d$, the $n + d$ critical points are given by $\lambda^{1/2n}$, while there are only two critical values given by $\pm 2\sqrt{\lambda}$. There are $n + d$ prepoles at the points $(-\lambda)^{1/(n+d)}$.

Note that $F_\lambda(\omega z) = \omega^n F_\lambda(z)$. Hence the orbits of points of the form $\omega^j z$ all behave "symmetrically" under iteration of F_λ. For example, if $F_\lambda^i(z) \to \infty$, then $F_\lambda^i(\omega^k z)$ also tends to ∞ for each k. If $F_\lambda^i(z)$ tends to an attracting cycle, then so does $F_\lambda^i(\omega^k z)$. Note, however, that the cycles involved may be different depending on k and, indeed, they may even have different periods. Nonetheless, all points lying on this set of attracting cycles are of the form $\omega^j z_0$ for some $z_0 \in \mathbb{C}$. In particular, all $n + d$ critical points have orbits that behave symmetrically, so this is why there is only one free critical orbit for F_λ.

We now restrict attention to the case $n \geq 2$; the case $n = 1$ will be dealt with in Section 5. The point at ∞ is a superattracting fixed point for F_λ and it is well known that F_λ is conjugate to $z \mapsto z^n$ in a neighborhood of ∞, so we have an immediate basin of attraction B_λ at ∞. Since F_λ has a pole of order d at 0, there is an open neighborhood of 0 that is mapped d to 1 onto a neighborhood of ∞ in B_λ. If B_λ does not contain this neighborhood, then there is a disjoint open set about 0 that is mapped d to 1 onto B. This set is called the *trap door* and we denote it by T_λ. Since the degree of F_λ is $n + d$, all points in the preimage of B_λ lie either in B_λ or in T_λ.

Using the symmetry $F_\lambda(\omega z) = \omega^n F_\lambda(z)$, it is straightforward to check that all of B_λ, T_λ, and $J(F_\lambda)$ are symmetric under $z \mapsto \omega z$. We say that these sets possess $n + d$-fold symmetry. In particular, since the critical points are arranged symmetrically about the origin, it follows that if one of the critical points lies in B_λ (resp.,T_λ), then all of the critical points lie in B_λ (resp., T_λ).

For other components of the Fatou set, the symmetry situation is somewhat different: either a component contains $\omega^j z_0$ for a given z_0 in the Fatou set and all $j \in \mathbb{Z}$, or else such a component contains none of the $\omega^j z_0$ with $j \neq 0$ mod $n + d$:

Symmetry Lemma. *Suppose U is a connected component of the Fatou set of F_λ. Suppose also that both z_0 and $\omega^j z_0$ belong to U, where $\omega^j \neq 1$. Then in fact, $\omega^i z_0$ belongs to U for all i and, as a consequence, U has $n + d$-fold symmetry and surrounds the origin.*

See [8] for a proof of this fact.

3. THE ESCAPE TRICHOTOMY

For the well-studied family of quadratic maps $Q_c(z) = z^2 + c$ with c a complex parameter there is the well known Fundamental Dichotomy:

(1) If the orbit of the one free critical point at 0 tends to ∞, then the Julia set of Q_c is a Cantor set;
(2) If the orbit of 0 does not tend to ∞, then the Julia set is a connected set.

In this section we discuss a similar result for F_λ that we call the Escape Trichotomy. Unlike the family of quadratic maps Q_c, there exist three different "ways" that the critical orbit for F_λ can tend to infinity. If the critical orbit tends to infinity, then all of the critical values must lie in B_λ or one of its preimages. These three different scenarios lead to three distinct classes of Julia sets for F_λ that comprise the Escape Trichotomy.

3.1. Critical Values in B_λ.
We first assume that one of the critical values of F_λ lies in B_λ. In this case, $J(F_\lambda)$ is a Cantor set. We sketch a proof of this fact here (for more details, see [8]).

By symmetry, if one of the critical values lies in B_λ, then all of the critical values do so as well. Let v be a critical value of F_λ and let c be a critical point such that $F_\lambda(c) = v$. Let U be an open disk in B_λ containing both v and ∞ with $F_\lambda(U) \subset U$. We may assume that U has $(n+d)$-fold symmetry. Let V be the preimage of $F_\lambda(U)$ containing the origin. We may also assume that U and V are disjoint.

Let γ be an arc in U connecting v to ∞. The preimage of γ is an arc γ' that contains c and is mapped two-to-one onto γ. One portion of γ' connects c to ∞. The curve γ' must therefore also lie in B_λ, and so we see that c and hence all of the critical points must lie in B_λ.

Since c is a critical point, it follows that γ' contains a second preimage of ∞. One checks easily that this second preimage of ∞ is 0, not ∞, and so γ' extends all the way from 0 to ∞. In particular, γ' meets both U and V, and so both of these sets lie in B_λ. Therefore B_λ and T_λ are not disjoint sets. Let W be the preimage of U. It follows that W contains U, V, and a neighborhood of γ'.

Since v was an arbitrary critical value of F_λ we can repeat this process and obtain $n + d$ arcs connecting 0 and ∞ such that each arc contains a distinct critical point. Furthermore, these arcs may be chosen so that they do not intersect and are symmetric under $z \to \omega z$ where $\omega^{n+d} = 1$. Each of these arcs also lies in W and so W consists of the Riemann sphere with $n + d$ disjoint and symmetric disks A_j for $j = 1, \ldots, n + d$ removed. Finally, it is easy to check that each A_j in the complement of W is mapped univalently over the complement of U and hence over all of the other A_i. Therefore, each of the $n + d$ sets A_j contain preimages of all of the other A_i, and the Julia set is contained in the union of these $(n + d)^2$ sets. See Figure 2. Standard arguments then show that the Julia set is a Cantor set and F_λ is a one-sided shift on $n + d$ symbols on this set. Figure 1a displays an example of a Julia set for which the critical values lie in B_λ.

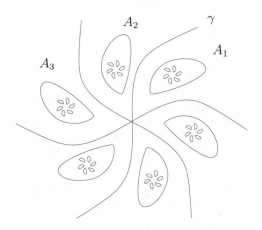

FIGURE 2. The sets A_j and their preimages.

3.2. Critical Values in T_λ.

Assume now that B_λ and T_λ are disjoint and that one, and hence all, of the critical values of F_λ now lie in T_λ. In this case, $J(F_\lambda)$ is a Cantor set of simple closed curves. To see this, note first that, since B_λ and T_λ are both open disks, the Riemann-Hurwitz formula shows that the preimage of T_λ is an open annulus surrounding the origin and located between $\overline{T}_\lambda$ and $\overline{B}_\lambda$. We denote this preimage of T_λ by T_λ^{-1} and the nth preimage of T_λ by T_λ^{-n}. The annulus T_λ^{-1} contains all of the critical points and its closure divides the region between $\overline{T}_\lambda$ and $\overline{B}_\lambda$ into two open subannuli that are mapped onto $\mathbb{C} - (\overline{B}_\lambda \cup \overline{T}_\lambda)$. We call these subannuli A_{in} and A_{out}, with A_{in} the subannulus bordering T_λ and A_{out} the subannulus bordering B_λ. Note that since the boundary of T_λ^{-1}, ∂T_λ^{-1}, is mapped onto ∂T_λ, whereas both ∂T_λ and ∂B_λ are both mapped onto ∂B_λ, it must be the case that ∂T_λ^{-1}

is disjoint from ∂T_λ and ∂B_λ. See Figure 3. Let A denote the union of the three annuli A_{in}, A_{out}, and T_λ^{-1}.

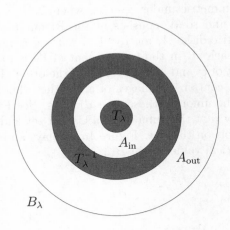

FIGURE 3. The sets A_{in}, A_{out}, T_λ, B_λ and T_λ^{-1}.

Since all of the critical points lie in T_λ^{-1}, the annuli A_{in} and A_{out} are mapped as coverings onto $\mathbb{C} - (\overline{B}_\lambda \cup \overline{T}_\lambda)$. Hence there exist preimages of T_λ^{-1} in each of these subannuli. Note that there will be two annular components of T_λ^{-2}, one in A_{in} and one in A_{out}. See Figure 4. Continuing in this fashion, we see that T_λ^{-n} consists of 2^{n-1} subannuli. In [8], quasiconformal surgery was used to show that the boundaries of B_λ, T_λ, and all of the preimages of T_λ are simple closed curves surrounding the origin. Hence the Julia set is given by a nested intersection of closed annuli and the result follows exactly as in the case described by McMullen in [14].

We remark that, by the covering properties of F_λ on A_{in} and A_{out}, we must have
$$\text{mod } A > \text{mod } A_{\text{in}} + \text{mod } A_{\text{out}} = \left(\frac{1}{d} + \frac{1}{n}\right) \text{mod } A$$
where mod A denotes the modulus of A. Hence, as in the McMullen result, we must have $1/d + 1/n < 1$ in order for v to lie in the trap door. Therefore, if $1/d + 1/n > 1$, then v cannot lie in the trap door, so part 2 of the Escape Trichotomy Theorem cannot occur if $d = n = 2$ or if either n or d is equal to 1. In Figure 1b we display a Julia set for which the critical values all lie in T_λ.

3.3. **Critical Values in a Preimage of T_λ.** We now describe the final case where the critical values have orbits that eventually escape through the trap door, but the critical values do not themselves lie in the trap door. In this case the Julia set is a Sierpinski curve. We first observe that the Julia set

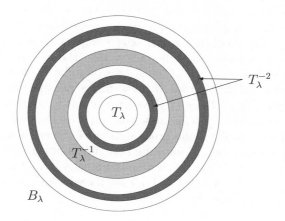

FIGURE 4. Inverse images of T_λ.

of F_λ is compact, connected, locally connected, and nowhere dense. Indeed, since we are assuming that the critical orbit eventually enters the basin of ∞, we have that the Julia set is given by $\mathbb{C} - \cup F_\lambda^{-j}(B_\lambda)$. That is, $J(F_\lambda)$ is $\mathbb{C}$ with countably many disjoint, simply connected, open sets removed. Hence $J(F_\lambda)$ is compact and connected. Since $J(F_\lambda) \neq \mathbb{C}$, $J(F_\lambda)$ cannot contain any open sets, so $J(F_\lambda)$ is also nowhere dense. Finally, since the critical orbits all tend to ∞ and hence do not lie in or accumulate on $J(F_\lambda)$, it follows that F_λ is hyperbolic on $J(F_\lambda)$ and standard arguments show that $J(F_\lambda)$ is locally connected (see [16]). In particular, since B_λ is a simply connected component of the Fatou set, it follows that the boundary of B_λ is locally connected. Hence $J(F_\lambda)$ fulfills the first four of the conditions to be a Sierpinski curve.

To finish showing that $J(F_\lambda)$ is a Sierpinski curve we need to show that the boundaries of B_λ as well as all of the preimages of B_λ are simple closed curves and that these boundary curves are pairwise disjoint. To see this, we first claim that $\mathbb{C} - \overline{B}_\lambda$ is a connected open set. This should be contrasted with the situation for quadratic polynomial Julia sets where $\mathbb{C} - \overline{B}_\lambda$ often consists of infinitely many disjoint open sets (consider the Julia sets known as the basilica or Douady's rabbit, for example). Assume that $\mathbb{C} - \overline{B}_\lambda$ has more than one component. Let W_0 be the component of $\mathbb{C} - \overline{B}_\lambda$ that contains the origin. Note that $T_\lambda \subset W_0$. Since $F_\lambda(\partial T_\lambda) = \partial B_\lambda \supset \partial W_0$, it follows that there are points in W_0 whose images also lie in W_0 and consequently $F_\lambda(W_0) \supset W_0$. Now if one of the prepoles lies in a component of $\mathbb{C} - \overline{B}_\lambda$ that is disjoint from W_0, then by symmetry all of the prepoles have this property. But this then gives us too many preimages of points in W_0, and so all of the prepoles must in fact lie in W_0. It then follows that all of the preimages of any point in W_0 lie in W_0.

If there were another component of $\mathbb{C} - \overline{B}_\lambda$, then the boundary of this set must eventually be mapped over the boundary of W_0 since $\partial W_0 \subset J(F_\lambda)$, and so there must be additional preimages of points in W_0. But again, this is impossible. Therefore W_0 is the only component of $\mathbb{C} - \overline{B}_\lambda$. Standard arguments using external rays then show that the boundary of W_0 must in fact be a simple closed curve. So too are the boundaries of all of the preimages of B_λ. One then checks that all of these curves are disjoint, for a point that lies in the intersection of one of these curves must either be a critical point or one of its preimages, but we know that all critical points have orbits that tend to ∞. This completes the proof that the Julia set is a Sierpinski curve.

In Figure 5 we show B_λ, T_λ and the first two preimages of T_λ in the special case where $n = d = 2$ and under the assumption that there are no critical points in T_λ^{-1} or T_λ^{-2}. An actual Julia set for which the critical points lie in T_λ^{-2} is depicted in Figure 1c.

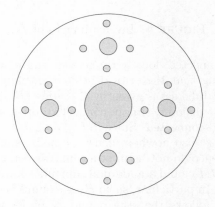

FIGURE 5. B_λ, T_λ, T_λ^{-1} and T_λ^{-2}.

In Figure 6, we show the λ plane in the case $n = d = 4$. The outside grey region in this image consists of λ-values for which $J(F_\lambda)$ is a Cantor set. The central grey region is the *McMullen domain* in which $J(F_\lambda)$ is a Cantor set of simple closed curves. The region between these two sets is called the connectedness locus as the Julia sets are always connected when λ lies in this region. The other grey regions in this figure correspond to *Sierpinski holes* in which the corresponding Julia sets are Sierpinski curves.

4. THE CASE $n = d = 2$

As mentioned earlier, the cases where $n = d = 2$ or $n > 1, d = 1$ are significantly different from the other cases where $n \geq 2$ because there is no McMullen domain in parameter space. In these cases, we instead have infinitely many open sets of parameters in any neighborhood $\lambda = 0$ in parameter space in which the critical orbits eventually enter B_λ and hence the Julia set

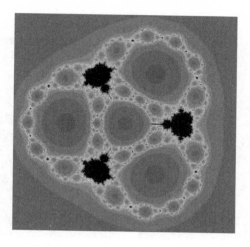

FIGURE 6. The parameter plane when $n = d = 4$.

is a Sierpinski curve. In each of these open sets the number of iterations that it takes for the critical orbit to enter B_λ is different, and so two maps drawn from different open sets are dynamically distinct in the sense that these maps are not topologically conjugate.

We sketch the proof of this when $n = d = 2$. We show that there are infinitely many open intervals in $\mathbb{R}^-$ in any neighborhood of the origin in parameter space in which the critical orbit eventually escapes. Similar results hold when $n > 1$, $d = 1$, though the real axis need not be the home of these open sets.

When $n = d = 2$, the four critical points and four prepoles of F_λ all lie on the circle of radius $|\lambda|^{1/4}$ centered at the origin. We call this circle the *critical circle*. The case $n = d = 2$ is especially simple since the second image of the critical points is given by

$$F_\lambda^2(c_\lambda) = 4\lambda + \frac{1}{4}$$

and so $\lambda \mapsto F_\lambda^2(c_\lambda)$ is an analytic function of λ that is a homeomorphism. If $-1/16 < \lambda < 0$, then one checks easily that the critical circle is mapped strictly inside itself. Therefore, as in the previous section, $J(F_\lambda)$ is a connected set and B_λ and T_λ are disjoint. In particular, the second image of the critical point lands on the real axis and lies in the complement of B_λ in $\mathbb{R}$.

Proposition. *There is an increasing sequence $\lambda_2, \lambda_3, \ldots$ in $\mathbb{R}^-$ with $\lambda_j \to 0$ and $F_{\lambda_j}^j(c_{\lambda_j}) = 0$.*

Proof: Since $F_\lambda^2(c_\lambda) = 4\lambda + 1/4$, this quantity increases monotonically toward $1/4$ as $\lambda \to 0$. Now the orbit of $1/4$ remains in $\mathbb{R}^+$ for all iterations of F_0 and

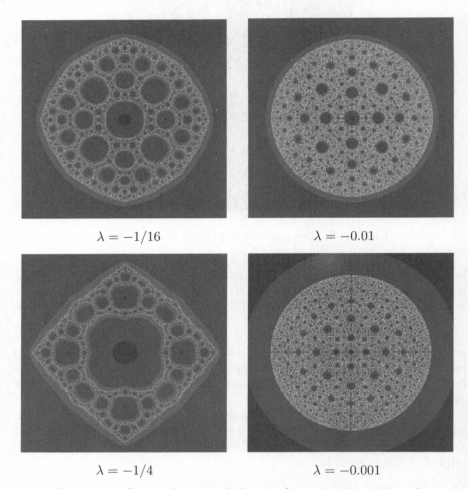

$\lambda = -1/16$ $\lambda = -0.01$

$\lambda = -1/4$ $\lambda = -0.001$

FIGURE 7. Sierpinski curve Julia sets for various negative values of λ when $n = d = 2$. All of these sets are homeomorphic, but the dynamics on each is different.

decreases monotonically to 0. Hence, given N, for λ sufficiently small, $F_\lambda^j(c_\lambda)$ also lies in $\mathbb{R}^+$ for $2 \leq j \leq N$ and moreover this finite sequence is decreasing.

Now suppose $\beta < \alpha < 0$. We have $F_\beta(x) < F_\alpha(x)$ for all $x \in \mathbb{R}^+$. Also, $F_\beta^2(c_\beta) < F_\alpha^2(c_\alpha) < 1/4$. Hence $F_\beta^j(c_\beta) < F_\alpha^j(c_\alpha)$ for all j for which $F_\beta^j(c_\beta) \in \mathbb{R}^+$. The result then follows by continuity of F_λ with respect to λ. □

Note that $\lambda_2 = -1/16$. Using the previous Proposition, we may find open intervals I_j about λ_j for $j = 2, 3, \ldots$ having the property that, if $\lambda \in I_j$, then $F_\lambda^j(c_\lambda) \in T_\lambda$, and so $F_\lambda^{j+1}(c_\lambda) \in B_\lambda$. Therefore, $F_\lambda^n(c_\lambda) \to \infty$ as $n \to \infty$, and the Escape Trichotomy then shows that $J(F_\lambda)$ is a Sierpinski curve.

Now let $C(c_\lambda)$ denote the component of the Fatou set of F_λ containing c_λ. The map F_λ is two-to-one on each of the four sets $C(c_\lambda)$ containing these critical points, and we have $F_\lambda^j(C(c_\lambda)) = T_\lambda$ for some j. Now suppose that $F_\lambda | J(F_\lambda)$ is conjugate to $F_\alpha | J(F_\alpha)$ for some $\alpha \in \cup I_k$ for some $k > 1$. This conjugacy must take the boundaries of B_λ and T_λ to the corresponding boundaries of B_α and T_α. Similarly the boundaries of the four regions $C(c_\lambda)$ must be mapped to one of the corresponding regions by the conjugacy, since these are the only complementary domains (besides B_λ and T_λ) on which F_λ is two-to-one. If, however, $\lambda \in I_j$ and $\alpha \in I_k$ with $j \neq k$, then these maps cannot be conjugate, since a conjugacy maps each of the jth preimages of T_λ to one of the jth preimages of T_α. Such a conjugacy would also have to map boundaries of domains on which F_λ and F_α were two-to-one to each other. Since $j \neq k$, this is impossible. We therefore have:

Theorem. *Let $\lambda \in I_j$ and $\alpha \in I_k$ with $j \neq k$. Then F_λ is not conjugate to F_α on their corresponding Julia sets.*

In Figure 7 we display several dynamically distinct Sierpinski curve Julia sets for λ close to 0.

In Figure 8 we display the parameter plane for the case $n = d = 2$ as well as a magnification around $\lambda = 0$. In contrast to the image in Figure 6, all of the internal grey regions in this image are Sierpinski holes. There is no McMullen domain when $n = d = 2$.

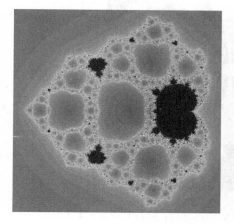 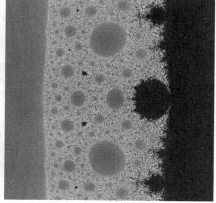

FIGURE 8. The parameter plane and a magnification when $n = d = 2$.

5. The case $n = 1$

In this section we restrict attention to the family of functions
$$F_\lambda(z) = z + \frac{\lambda}{z}$$
so that $n = 1$. The dynamics of these maps are quite different from those for which $n > 1$. First, one checks easily that, for each λ, the map F_λ is conjugate to the function
$$F_1(z) = z + \frac{1}{z}.$$
Hence this family does not really depend on a parameter. Therefore we change the family slightly so that we consider instead
$$F_\lambda(z) = \lambda\left(z + \frac{1}{z}\right).$$
This family is conjugate to the family
$$G_\lambda(z) = \lambda z + \frac{1}{z},$$
and so can be regarded as a linear perturbation of the involution $z \mapsto 1/z$.

The main difference between this family and our original family is that these functions have a repelling fixed point at infinity whenever $|\lambda| < 1$. Consequently, 0 lies in the Julia set and thus there is no trap door as in the case where $n > 1$.

FIGURE 9. The λ plane for the function $F_\lambda(z) = \lambda(z + \frac{1}{z})$.

As in the previous cases, we are mainly concerned with the case of λ small, so that we are perturbing away from the identically zero function. It has been shown by Yongcheng [25] that for $0 < |\lambda| \leq 1$ the Julia set is connected while if $|\lambda| > 1$, it is a Cantor set. The parameter space is plotted in Figure 9. Similar figures have been produced by Hawkins [12] and Milnor [17]. Note that most of the interesting behavior seems to occur as we approach the parameter 0 along the imaginary axis. In fact, it is easy to check that, for $0 < \lambda \leq 1$, $J(F_\lambda)$ is the imaginary axis, and all other points have orbits that are attracted to one of two attracting fixed points. For $-1 \leq \lambda < 0$, $J(F_\lambda)$ is the real axis, and this set separates the basins of an attracting two-cycle. In both of the large black circular regions in parameter space flanking the origin, the Julia sets are similar curves passing through the origin and ∞. In contrast, the dynamical behavior along the imaginary axis is much more complicated.

Given nonzero λ, the function F_λ is a degree-two rational map with two critical points at ± 1. The orbits of these critical points behave symmetrically under F_λ. For purely imaginary parameter values, this function has the desirable property that, in the dynamical plane, the real axis is mapped to the imaginary axis and vice versa. Therefore, for such parameter values we will consider the second iterate map restricted to the real axis, that is, we restrict attention to the behavior of $F_{i\lambda}^2$ on $\mathbb{R}$, where λ is now a real parameter.

We compute
$$F_{i\lambda}^2(x) = -\lambda^2 \left(x + \frac{1}{x}\right) + \frac{1}{x + \frac{1}{x}}.$$

Note that, for small λ, this second iterate map can be viewed as a perturbation of the λ-independent function
$$x \mapsto \frac{1}{x + \frac{1}{x}}.$$

When one and hence both of the critical points land on the repelling fixed point at ∞, the Julia set is known to be the entire Riemann sphere [16]. We will refer to such parameter values as *blowup* points, with the convention that a blowup point of order n is one such that $F_{i\lambda}^{2n}(1) = 0$. Parameter values for which this occurs are also known as m-ergodic rational maps (although m-ergodicity describes a larger set of maps than just those for which a critical point lands on a repelling cycle). Rees [21] has proved that m-ergodic maps comprise a set of positive Lebesgue measure in the parameter space of most rational maps. Hawkins [12] developed a computer algorithm for finding and plotting these parameter values. In that paper, it was shown numerically that the m-ergodic maps accumulate on the origin along the imaginary axis in parameter space. We formalize this observation via the following theorem.

Theorem. *For the family of functions $F_{i\lambda}(z) = i\lambda(z + 1/z)$, in any neighborhood of $\lambda = 0$, there exists:*

(1) *A countably infinite set of λ-values lying in $(-1, 1)$ for which the Julia set is the entire Riemann sphere;*

(2) A countably infinite set of λ-values lying in $(-1, 1)$ for which the critical point is part of a superattracting cycle.

Proof: To prove the first assertion, we will define a function $G_n : \mathbb{R} \to \mathbb{R}$ via $G_n(\lambda) = F_{i\lambda}^{2n}(1)$ where $\lambda \in \mathbb{R}$. For $\lambda_1 = .5$, $G_1(\lambda_1) = 0$. Also, $G_1(0) = 1/2$. Further, note that $G_m(\lambda)$ is continuous except at blowup points of order less than m. We now see that G_2 maps $(0, \lambda_1)$ to $(-\infty, 1/2)$. Thus, by continuity of G_2 in this interval, there exists a $\lambda_2 \in (0, \lambda_1)$ such that $G_2(\lambda_2) = 0$. If more than one λ value exists, we will chose the smallest to be λ_2. This ensures that G_3 will be continuous on $(0, \lambda_2)$. Iterating this process we obtain the desired sequence.

Now suppose that this sequence does not accumulate on the origin. In other words, there exists some interval $(0, \hat{\lambda})$ such that $G_n(\lambda) > 0$ for all n and $\lambda \in (0, \hat{\lambda})$. Since the graph of $F_{i\lambda}^2$ lies strictly below the diagonal on $(0, 1)$ and $F_{i\lambda}^2$ is monotonically increasing there, the interval $(0, 1)$ is mapped inside itself. Thus, by the contraction mapping principle there exists a fixed point in $(0, 1)$, which is a contradiction.

To prove the second part of the assertion, let λ_n and λ_m be blowup points of order n and m. Assume $n < m$. For fixed m there are a finite number of discontinuities of G_m in the interval (λ_m, λ_n). Furthermore, these discontinuities represent blowup points of order less than m. Therefore, we will restrict ourselves to a subinterval on which G_m is continuous and note that the result holding here is sufficient to establish the result in the general setting. Thus, without loss of generality, assume that G_m is continuous on (λ_m, λ_n). Therefore, $G_m(\lambda_n) = \infty$ and $G_m(\lambda_m) = 0$. By continuity of G_m there exists $\lambda_p \in (\lambda_m, \lambda_n)$ such that $G_m(\lambda_p) = 1$.

□

We will now briefly turn our attention to the case where $d > 1$. In this case the critical points are $c = d^{\frac{1}{d+1}}$. As in the $d = 1$ case the critical points do not depend on the parameter value. Also there exist lines, analogous to the imaginary axis for the case $d = 1$ and passing through the origin in parameter space for which F_λ^{d+1} is invariant over $\mathbb{R}$. The parameter planes for several of these functions are plotted in Figure 10. For this class of rational functions, the results of Rees [21] guarantee a set of positive Lebesgue measure in parameter space for which the Julia set is the whole Riemann sphere. However, it is unknown whether this behavior accumulates on the origin and hence whether a corollary to the Theorem is true for $d > 1$.

6. BURIED SIERPINSKI CURVES

In this section, we discuss an infinite collection of dynamically distinct Sierpinski curve Julia sets for the family F_λ where the Fatou components are quite different than those described in previous sections. Instead of being preimages of a single superattracting basin at ∞, we give examples where the

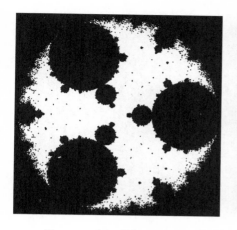 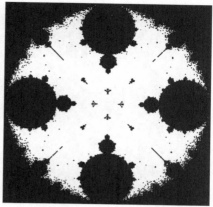

FIGURE 10. The λ plane for the functions $F_\lambda(z) = \lambda(z+1/z^2)$ and $F_\lambda(z) = \lambda(z+1/z^3)$

complementary domains consist of a collection of different attracting basins together with the basin at ∞ and all of the preimages of these basins. As before, we sketch a proof that the dynamics on these Julia sets are all distinct from one another as well as from those mentioned above, but again, all of these Julia sets are homeomorphic.

For simplicity, we restrict attention in this section to the special family $F_\lambda(z) = z^2 + \lambda/z$ with $\lambda \in \mathbb{R}^-$. In Figure 11, we display the Julia set of F_λ when $\lambda = -0.327$. For this map, there are attracting basins of period 3 and period 6 together with the basin at ∞. We also display the case where $\lambda = -0.5066$ for which there are three different attracting basins of period 4 together with the basin at ∞. The basins of the finite cycles are displayed in black.

There is a positive real fixed point for F_λ which we denote by $p(\lambda)$. Also, $c(\lambda) = (\lambda/2)^{1/3}$ is a critical point and

$$v(\lambda) = \frac{3}{2^{2/3}}\lambda^{2/3}$$

is a critical value. Note that, for $\lambda \in \mathbb{R}^-$, both $c(\lambda)$ and $v(\lambda)$ are real.

Let $\lambda^* = -16/27$. Straightforward calculations show that $p(\lambda^*) = 4/3$ and $p(\lambda^*)$ is repelling. Further, the real critical point $c(\lambda^*) = -2/3$ is pre-fixed, i.e., $F_{\lambda^*}(c(\lambda^*)) = 4/3 = p(\lambda^*)$. For λ-values slightly larger than λ^*, the real critical value lies to the left of $p(\lambda)$ and hence subsequent points on the orbit of the critical value begin to decrease. Graphical iteration shows that there is a sequence of λ-values tending to λ^* for which the critical orbit decreases along the positive axis and then, at the next iteration, lands back at $c(\lambda)$. See Figure 12. Thus, for these λ-values, we have a superattracting cycle. More precisely, we have:

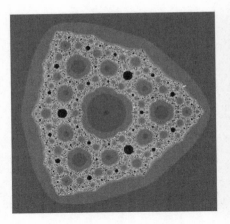
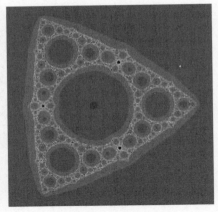

FIGURE 11. The Julia sets for $F_\lambda(z) = z^2 + \lambda/z$ where $\lambda = -0.327$ and $\lambda = -0.5066$.

Theorem. *There is a decreasing sequence $\lambda_n \in \mathbb{R}^-$ for $n \geq 3$, such that $\lambda_n \to \lambda^* = -16/27$, and having the property that F_{λ_n} has a superattracting cycle of period n given by $x_j(\lambda_n) = F_{\lambda_n}(x_{j-1}(\lambda_n))$, where*
(1) $x_0(\lambda_n) = x_n(\lambda_n) = c(\lambda_n)$, *and*
(2) $x_0 < 0 < x_{n-1} < x_{n-2} < \cdots < x_1 = v(\lambda_n) < p(\lambda_n)$.

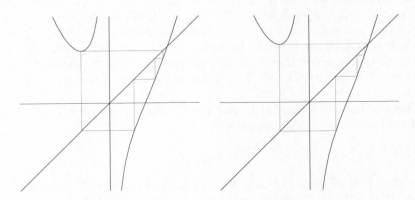

FIGURE 12. The graphs of $F_\lambda(x) = x^2 + \lambda/x$ where $\lambda = \lambda_4$ and $\lambda = \lambda_7$.

For a proof see [6]. Now fix a particular parameter value $\lambda = \lambda_n$ for which F_λ has a superattracting periodic point x_0 lying in $\mathbb{R}^-$ as described in the previous Theorem. We say that a basin of attraction of F_λ is *buried* if the boundary of this basin is disjoint from the boundaries of all other basins of attraction (including B_λ). Note that, if the basin of one point on an attracting cycle is buried, then so too are all forward and backward images of this basin, so the entire basin of the cycle is buried. In [6] the following was shown:

Theorem. *All of the basins of F_λ are buried and $J(F_\lambda)$ is a Sierpinski curve.*

As discussed earlier, any two Sierpinski curves are homeomorphic. Hence $J(F_{\lambda_n})$ is topologically equivalent to $J(F_{\lambda_m})$ for any n and m. However, each of these Julia sets is dynamically distinct from the others since the periods of the superattracting cycles are different.

In Figure 13 we display the parameter plane for the degree three family

$$F_\lambda(z) = z^2 + \frac{\lambda}{z}$$

together with a magnification of a certain region along the negative real axis.

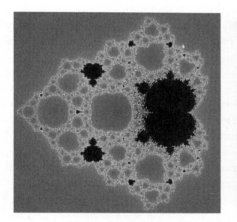
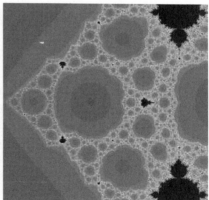

FIGURE 13. The parameter plane for the degree three family of rational maps and a magnification.

The grey holes in this parameter plane correspond to parameter values for which the critical orbit eventually escapes to ∞ through the trap door, so the Julia set is a Sierpinski curve as discussed in Section 3. These are the Sierpinski holes. Note the existence of a small copy of a Mandelbrot set along the negative real axis in this image. In fact, there are infinitely many such Mandelbrot sets converging to the left tip of the parameter space, which is the parameter λ^*. The parameters for which we have the superattracting cycles constructed above form the centers of the main cardioids of certain of these Mandelbrot sets.

We remark that there appear to be two very different types of baby Mandelbrot sets in this picture, some of which touch the outer boundary of the connectedness locus, and some that do not. It is known [3] that those Mandelbrot sets that touch the outer boundary actually touch infinitely many of the Sierpinski holes as well. We conjecture that the Mandelbrot sets corresponding to the λ_n in this section are also buried, this time in the sense that these sets do not touch any of the Sierpinski holes, nor the outer boundary.

7. SIERPINSKI GASKET-LIKE JULIA SETS

One of the outstanding theorems in the study of the families of polynomials $z \mapsto z^d + c$ is the Landing Theorem, due to A. Douady and J. H. Hubbard [11], which states that every external ray in the parameter plane whose external angle is rational lands at a unique point in the boundary of the connectedness locus. Recently, C. Petersen and G. Ryd [20] have shown that this result may be extended to many other one-parameter families of maps with a single free critical orbit, including the family F_λ when $n \geq 2$. In this section we will concentrate on λ-values that correspond to external rays whose external angles are of the special form p/n^j with $p, j \in \mathbb{Z}$. The Landing Theorem implies that such a λ-value is a parameter for which the critical orbits eventually land on a fixed point in the boundary of B_λ. We call the corresponding maps *Misiurewicz-Sierpinski maps*, or MS maps, for short.

In Figures 14 and 15 we display several examples of Julia sets corresponding to Misiurewicz parameters for $z \mapsto z^2 + \lambda/z$ and $z \mapsto z^2 + \lambda/z^2$ respectively. Clearly, these sets are no longer homeomorphic to the Sierpinski curve, as infinitely many complementary boundaries meet other complementary boundaries at one or more points. In particular, the Julia set in the left-hand side of Figure 14 is homeomorphic to the well-known *Sierpinski gasket* (or triangle). Although the second Julia set in Figure 14 looks similar to the Sierpinski gasket, these two Julia sets are not homeomorphic, as we explain below.

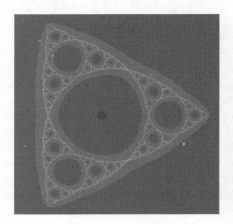 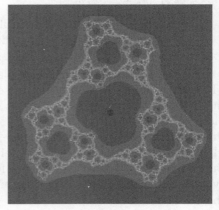

FIGURE 14. Julia sets from the family $z \mapsto z^2 + \lambda/z$ with $\lambda \approx -0.59257$ and $-0.03804 + i0.42622$.

 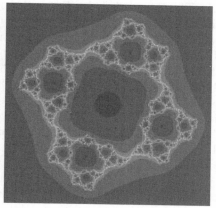

FIGURE 15. Julia sets from the family $z \mapsto z^2 + \lambda/z^2$ with $\lambda \approx -0.36428$ and $\lambda \approx -0.01965 + i0.2754$.

The Julia sets in Figure 15 can be thought of as generalizations of a Sierpinski gasket set with four distinguished vertices. We will see that these Julia sets are again not homeomorphic to each other. A *generalized Sierpinski gasket* set with four distinguished vertices is constructed as follows. Consider the closed unit disk in the plane from which we remove an open rectangular region whose vertices lie in the boundary of the disk. We assume that the removed rectangle is symmetric under rotation of the disk by angle $\pi/2$. We are left with four symmetric closed sets which we denote by I_0, I_1, I_2 and I_3. From each of the I_j we next remove an open "generalized" rectangle whose vertices lie on the boundary of I_j. We stipulate that exactly two of these vertices lie on the boundary of the previously removed rectangle and that the newly removed sets are all symmetrically arranged. This leaves sixteen sets whose only intersection points are vertices of the removed rectangles. We continue in this fashion by removing at each stage open generalized rectangles with exactly two vertices lying in the boundary of the previously removed rectangle. In the limit this produces a set which we call a generalized Sierpinski gasket or a *Sierpinski gasket-like* set.

For simplicity, in this section we consider only the special case where $n = d = 2$, although all of the results go over with minor modifications to the more general family of maps with $n \geq 2$, $d \geq 1$. See [9].

Theorem. *Let $F_\lambda(z) = z^2 + \lambda/z^2$ be an MS map. Then the Julia set $J(F_\lambda)$ is a generalized Sierpinski gasket-like set with four distinguished vertices. Moreover, if we assume that λ and μ are chosen so that F_λ and F_μ are MS maps from the same family, then their Julia sets are homeomorphic if and only if $\lambda = \overline{\mu}$.*

7.1. **Topology of Julia Sets.** Suppose F_λ is an MS map with $n = d = 2$. Since the post-critical orbit is finite, the map is sub-hyperbolic and thus the boundary of each Fatou component is locally connected [16]. Moreover, as shown in [8], there is only one component to the set $\mathbb{C} - \overline{B}_\lambda$ and the boundary of this set is a simple closed curve which is also the boundary of B_λ. Denote the boundary of B_λ by β_λ and the boundary of the trap door by τ_λ. Our assumption implies that the four finite and non-zero critical points $c_\lambda = \lambda^{1/4}$ lie in both β_λ and τ_λ. A straightforward argument given in [7] shows that if the set $\beta_\lambda \cap \tau_\lambda$ is non-empty, then the critical points are the only points in this intersection. We call these points the *corners* of the trap door. The four corners separate τ_λ into four *edges*.

Using the fact that F_λ is conjugate to $z \mapsto z^2$ in B_λ, and that this conjugacy extends to β_λ, there exist four disjoint smooth curves, γ_j for $j = 0, 1, 2, 3$, connecting each of the critical points c_j to ∞ in B_λ. The γ_j are the external rays landing at c_j. Let $H_\lambda(z) = \sqrt{\lambda}/z$. One checks easily that the two involutions H_λ interchange B_λ and T_λ and satisfy $F_\lambda((H_\lambda(z)) = F_\lambda(z)$. Let ν_j denote the image of γ_j under the involution H_λ that fixes c_j. Then the curve $\eta_j = \gamma_j \cup \nu_j$ connects 0 to ∞ and meets $J(F_\lambda)$ only at c_j. Moreover, the η_j are pairwise disjoint (except at 0 and ∞). Hence these four curves divide the Julia set into four symmetric pieces $I_0, \ldots, I_3$ where we assume that $c_j \in I_j$ but c_j does not lie in the other three regions. Let I_0 be the component that contains the repelling fixed point $p(\lambda)$ that lies in β_λ. Note that the I_j are neither open nor closed subsets of $J(F_\lambda)$.

Since there are no critical points in any of the preimages of the trap door, it follows that each of its preimages is mapped in one-to-one fashion onto the trap door by F_λ. Hence each component of $F_\lambda^{-k}(\tau_\lambda)$ also has four corners and edges, and each of these corners is mapped by F_λ^k onto a distinct critical point in τ_λ.

To see that $J(F_\lambda)$ is a Sierpinski gasket-like set, we require the following lemma.

Lemma. *Let τ_λ^k be the union of all of the components of $F_\lambda^{-k}(\tau_\lambda)$ and let A be a particular component in τ_λ^k with $k \geq 1$. Then exactly two of the corner points of A lie in a particular edge of a single component of τ_λ^{k-1}.*

Proof: The case $k = 1$ is seen as follows. We have that F_λ maps each I_j for $j = 0, \ldots, 3$ in one-to-one fashion onto all of $J(F_\lambda)$, with $F_\lambda(I_j \cap \beta_\lambda)$ mapped onto one of the two halves of β_λ lying between two critical values (which, by assumption, are not equal to any of the critical points). Hence $F_\lambda(I_j \cap \beta_\lambda)$ contains exactly two critical points. Similarly, $F_\lambda(I_j \cap \tau_\lambda)$ maps onto the other half of β_λ and so also meets two critical points. The preimages of these latter two critical points in τ_λ are precisely the corners of the component of τ_λ^1 that lies in I_j. Thus we see that each component in τ_λ^1 meets the boundary of one of the I_j's in two points lying in β_λ and two points lying in τ_λ. In particular, two of the corners lie in the edge of τ_λ that meets I_j.

Now consider a component in τ_λ^k with $k > 1$. F_λ^k maps each component in τ_λ^k onto τ_λ and therefore F_λ^{k-1} maps the components in τ_λ^k onto one of the four components of τ_λ^1. Since each of these four components meets a particular edge of τ_λ in exactly two corner points, it follows that each component of τ_λ^k meets an edge of one of the components of τ_λ^{k-1} in exactly two corner points as claimed.

□

We may now show the Julia set of an MS map is a gasket-like set as follows. Let $K_0 = \overline{\mathbb{C}} - B_\lambda$ and $K_1 = K_0 - T_\lambda$. Then K_1 consists of the union of the four sectors I_j which are mapped in a one-to-one fashion onto K_0. Define recursively the sets $K_{n+1} = K_n - F_\lambda^{-n}(T_\lambda)$. Each K_n is a nested collection of closed and connected subsets of the Riemann sphere with exactly 4^n generalized rectangles removed at each nth step. Moreover, the above lemma shows that for each n, the removed rectangles satisfy the two corner restriction given in the definition of gasket-like sets. Is not hard to see that $\cap_{n=0}^\infty K_n$ coincides with $J(F_\lambda)$ and hence is a Sierpinski gasket-like set.

7.2. Homeomorphisms Between Julia Sets. Before proceeding with the discussion of homeomorphisms between Julia sets of MS maps, we provide a topological characterization of the critical points and the corners of every τ_λ^k. Proofs of the following propositions may be found in [9].

Proposition. (The Disconnection Property.) *The four corners of the trap door are the only set of four points in the Julia set whose removal disconnects $J(F_\lambda)$ into exactly four components. Any other set of four points removed from $J(F_\lambda)$ will yield at most three components.*

Clearly the corners of each component A in τ_λ^k inherit the disconnection property when restricted to the largest connected component of τ_λ^{k-1} that contains A. A homeomorphism between Julia sets of MS maps must then preserve this topological invariant as described in the following result.

Proposition. *Suppose F_λ and F_μ are MS maps. If there exists a homeomorphism $h : J(F_\lambda) \to J(F_\mu)$, then*

(1) *The map h takes the corners of $F_\lambda^{-k}(\tau_\lambda)$ to the corners of $F_\mu^{-k}(\tau_\mu)$ when $k \geq 0$.*
(2) *For $k \geq 1$, each component of $F_\lambda^{-k}(\tau_\lambda)$ is mapped to a unique component of $F_\mu^{-k}(\tau_\mu)$.*

Suppose λ and μ are given parameters that correspond to MS maps of the degree four family. Unless these parameters are complex conjugate, the main theorem in this section states their Julia sets are not homeomorphic. To prove this assertion, we have developed a recursive algorithm based solely on the configuration of the corners of a finite number of preimages of τ_λ along β_λ. The configuration is completely determined by the itinerary associated to the finite critical orbit. If the itineraries for λ and μ disagree at the $(n+1)$st entry, then the algorithm shows that the corner configurations of τ_λ^n and τ_μ^n differ along the respective boundaries of the basin at infinity. Hence there is no homeomorphism between these Julia sets. We illustrate this algorithm with the two examples given in Figure 15.

Using the partition given by the sectors I_j we define the itinerary of a point $z \in J(F_\lambda)$ as the infinite sequence $S(z) = (s_0 s_1 s_2 \ldots) \in \{0, 1, 2, 3\}^N$ defined in the natural way by its orbit in the regions I_j. Hence the itinerary of the accessible fixed point p_λ is $\overline{0} = (000\ldots)$, while the itinerary of $-p_\lambda$ is $2\overline{0} = (2000\ldots)$, and so forth.

By assumption the itinerary of any critical point of a MS map ends with an infinite string of 0's. Due to the four-fold symmetry and the existence of a unique free critical orbit, we will only concentrate on the itinerary of the critical point c_λ that lies in the first quadrant.

The two examples displayed in Figure 15, in which $\lambda \approx -0.36428$ and $\mu \approx -0.01965 + 0.2754\,i$, correspond to the landing points of external rays with arguments $1/2$ and $1/4$ respectively. The extension of the Landing Theorem for the rational families implies that the external rays of the same argument must land in the dynamical plane at the second iterate of the critical point. Thus, the itinerary of $F_\lambda^2(c_\lambda)$ is $(2\overline{0})$ and the itinerary of $F_\mu^2(c_\mu)$ is $(12\overline{0})$. It follows that the itinerary of c_λ is $(112\overline{0})$ while the itinerary of c_μ is $(1112\overline{0})$.

Since these itineraries differ at the third entry, we only need to look at the configuration of the corners of the second preimage of the trap door.

We start with the case $\lambda \approx -0.36428$. The ray $1/8$ lands at the critical point $c_1 = c_\lambda$. By symmetry, the ray $7/8$ lands at c_0. Thus, the preimages of c_1 and c_0 in I_0 are landing points of the rays $1/16$ and $15/16$ respectively. Note that these points are two corners of the component of τ_λ^1 that lies in I_0. The remaining two corners of this component lie in the arc of τ_λ contained in I_0 and are mapped onto the critical points c_2 and c_3.

By four-fold symmetry, we can compute the external rays landing on the corners of each component of τ_λ^1 in each remaining sector I_j by adding a proper multiple of $\pi/2$. In particular, two corners of the component of τ_λ^1 in I_1 correspond to landing points of the rational rays $5/16$ and $3/16$.

Now we compute the configurations of the components of τ_λ^2. For our purposes it suffices to find the configuration of the corners of $B \subset \tau_\lambda^2$ lying along the arc $\gamma \subset \beta_\lambda$ bounded by the rays $1/16$ and $1/8$. Under F_λ, γ is mapped onto an arc bounded by rays $1/8$ and $1/4$. Since the ray $3/16$ lands at a corner of the component of τ_λ^1 in I_1, this implies that the ray $3/32$ lands at a corner of the component B in τ_λ^2 along γ. A similar analysis can be done to compute the locations of the remaining three corners of B. See Figure 16.

For the case $\mu \approx -0.01965 + i0.2754$, let $c_1 = c_\mu$ be the critical point lying in the first quadrant which is the landing point of the ray $1/16$. By symmetry, the ray 13/16 lands at c_0. Hence the first preimages of c_1 and c_0 in I_0 are landing points of the rays 1/32 and 29/32 respectively. We may compute the external rays of the remaining corners in τ_μ^1 by addition of a multiple of $\pi/2$ as before. In particular the external rays landing at corner points of τ_μ^1 in I_1 are 9/32 and 5/32.

Let γ denote the arc of β_μ bounded by the landing points of the rays 1/16 and 1/32. Then γ is mapped onto the arc bounded by the rays 1/8 and 1/16. In this case, the image of γ fails to contain a corner point of τ_μ^1 in I_1 as $1/16 < 5/32$. This implies that there is no corners of the component B in τ_μ^2 along γ. See Figure 16.

The previous proposition implies that a homeomorphism between $J(F_\lambda)$ and $J(F_\mu)$ must preserve the configurations shown in Figure 16, which is impossible. Therefore these Julia sets cannot be homeomorphic.

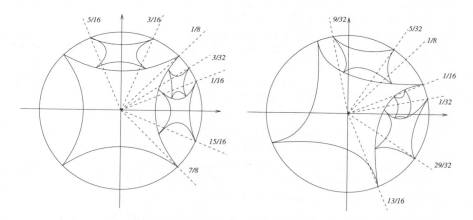

FIGURE 16. A schematic representation of the Julia set $J(F_\lambda)$ and $J(F_\mu)$, respectively, up to second preimage of the trap door. For clarity, only the relevant rational rays and certain preimages of the trap door in sectors I_0 and I_1 have been displayed.

References

[1] Blanchard, P., Devaney, R. L., Look, D. M., Seal, P., and Shapiro, Y. Sierpinski Curve Julia Sets and Singular Perturbations of Complex Polynomials. *Ergodic Theory and Dynamical Systems.* **25** (2005), 1047-1055.

[2] Devaney, R. L. Cantor and Sierpinski, Julia and Fatou: Complex Topology Meets Complex Dynamics. *Notices Amer. Math. Soc.* **51** (2004), 9-15.

[3] Devaney, R. L. Baby Mandelbrot Sets Adorned with Halos in Families of Rational Maps. In *Complex Dynamics: Twenty-Five Years After the Appearance of the Mandelbrot Set*. Contemporary Mathematics. Amer. Math. Soc. **396** (2006), 37-50.

[4] Devaney, R. L. The McMullen Domain: Satellite Mandelbrot Sets and Sierpinski Holes. To appear in *Conformal Geometry and Dynamics*.

[5] Devaney, R. L., Josic, K. and Shapiro, Y. Singular Perturbations of Quadratic Maps. *Intl. J. Bifurcation and Chaos*. **14** (2004), 161-169.

[6] Devaney, R. L. and Look, D. M. Buried Sierpinski Curve Julia Sets. *Discrete and Continuous Dynamical Systems*. **13** (2005), 1035-1046.

[7] Devaney, R. L. and Look, D. M. A Criterion for Sierpinski Curve Julia Sets for Rational Maps. *Topology Proceedings* **30** (2006), 163-179.

[8] Devaney, R. L., Look, D. M. and Uminsky, D. The Escape Trichotomy for Singularly Perturbed Rational Maps. *Indiana University Mathematics Journal* **54** (2005), 1621-1634.

[9] Devaney, R. L., Moreno Rocha, M. and Siegmund, S. Rational Maps with Generalized Sierpinski Gasket Julia Sets. To appear in *Topology and its Applications*.

[10] Douady, A. and Hubbard, J. Itération des Polynômes Quadratiques Complexes, *C.R. Acad. Sci. Paris*, t. 29, (1982), 123-126.

[11] Douady, A. and Hubbard, J. Etude Dynamique des Polynômes Complexes. Partie I, *Publ. Math. D'Orsay* **84-02** (1984) and Partie II, *Publ. Math. D'Orsay* **85-04** (1985).

[12] Hawkins, J. Lebesgue Ergodic Rational Maps in Parameter Space. *Intl. J. Bifurcation and Chaos*. **13** (2003), 1423-1447.

[13] Kuratowski, K. *Topology*. Vol. 2. Academic Press, New York (1968).

[14] McMullen, C. Automorphisms of Rational Maps. *Holomorphic Functions and Moduli*. Vol. 1. Math. Sci. Res. Inst. Publ. **10**. Springer, New York (1988).

[15] McMullen, C. The Classification of Conformal Dynamical Systems. *Current Developments in Mathematics*. Int'l. Press, Cambridge, MA (1995), 323-360.

[16] Milnor, J. *Dynamics in One Complex Variable*. Vieweg (1999).

[17] Milnor, J. Geometry and Dynamics of Quadratic Rational Maps. *Exper. Math.* **2** (1993), 37-83.

[18] Milnor, J. and Tan Lei. A "Sierpinski Carpet" as Julia Set. Appendix F in Geometry and Dynamics of Quadratic Rational Maps. *Experiment. Math.* **2** (1993), 37-83.

[19] Morosawa, S., Nishimura, Y., Taniguchi, M., and Ueda, T. *Holomorphic Dynamics*. Cambridge University Press (2000).

[20] Petersen, C. and Ryd, G. Convergence of Rational Rays in Parameter Spaces. In *The Mandelbrot set: Theme and Variations*, London Mathematical Society, Lecture Note Series 274, Cambridge University Press (2000), 161-172.

[21] Rees, M. Positive Measure Sets of Ergodic Rational Maps. *Ann. Sci. Ec. Norm. Sup.* **19** (1986), 383-407.

[22] Steinmetz, N. *Rational Iteration*. W. de Gruyter, Berlin (1993).

[23] Sullivan, D., Quasiconformal Maps and Dynamical Systems I, Solutions of the Fatou-Julia Problem on Wandering Domains. *Ann. Math.* **122** (1985), 401-418.

[24] Whyburn, G. T. Topological Characterization of the Sierpinski Curve. *Fund. Math.* **45** (1958), 320-324.

[25] Yongcheng, Y. On the Julia Sets of Quadratic Rational Maps. *Complex Variables.* **18** (1992), 141-147.

ROBERT L. DEVANEY, DEPARTMENT OF MATHEMATICS, BOSTON UNIVERSITY, 111 CUMMINGTON STREET, BOSTON MA 02215
E-mail address: bob@bu.edu.

RESIDUAL JULIA SETS OF RATIONAL AND TRANSCENDENTAL FUNCTIONS

PATRICIA DOMÍNGUEZ AND NÚRIA FAGELLA

ABSTRACT. The *residual Julia set*, denoted by $J_r(f)$, is defined to be the subset of those points of the Julia set which do not belong to the boundary of any component of the Fatou set. The points of $J_r(f)$ are called *buried points* of $J(f)$ and a component of $J(f)$ which is contained in $J_r(f)$ is called a *buried component*. In this paper we survey the most important results related to the residual Julia set for several classes of functions. We also give a new criterion to deduce the existence of buried points and, in some cases, of unbounded curves in the residual Julia set (the so-called *Devaney hairs*). Some examples are the sine family, certain meromorphic maps constructed by surgery and the exponential family.

Dedicated to Professor Noel Baker

1. INTRODUCTION

Given a map $f : X \to X$, where X is a topological space, the sequence formed by its iterates will be denoted by $f^0 := \text{Id}$, $f^n := f \circ f^{n-1}$, $n \in \mathbb{N}$. When f is a holomorphic map and X is a Riemann surface the study makes sense and is non-trivial when X is either the Riemann sphere $\widehat{\mathbb{C}}$, the complex plane $\mathbb{C}$ or the complex plane minus one point $\mathbb{C} \setminus \{0\}$. All other interesting cases can be reduced to one of these three.

In this paper we deal with the following classes of maps (partially following [12]).

- $\mathcal{R} = \{f : \widehat{\mathbb{C}} \to \widehat{\mathbb{C}} \mid f \text{ is rational of degree at least two}\}$.
- $\mathcal{E} = \{f : \mathbb{C} \to \mathbb{C} \mid f \text{ is transcendental entire}\}$.
- $\mathcal{M} = \{f : \mathbb{C} \to \widehat{\mathbb{C}} \mid f \text{ is transcendental meromorphic with at least one not omitted pole}\}$.

Note that functions in $\mathcal{M}$ have one single essential singularity. This class is usually called the general class of meromorphic functions (see [6]).

The first author was supported by CONACYT and the second author was partially supported by MEC's grants BFM2003-09504MTM2005-02139 and MTM2006-05849/Consolider (including a FEDER contribution) and CIRIT's grant 2005/SGR-01028.

2000 *Mathematics Subject Classification*: Primary 37F10, Secondary 30D05.

Key Words: iteration, Fatou set, Julia set, residual Julia set, buried points, buried components.

If f is a map in any of the classes above and we denote by X its domain of definition, the *Fatou set* $F(f)$ (or stable set) consists of all points $z \in X$ such that the sequence of iterates of f is well defined and forms a normal family in a neighborhood of z. The *Julia set* (or chaotic set) is its complement and it is denoted by $J(f) = \widehat{\mathbb{C}} \setminus F(f)$.

Classes $\mathcal{R}$ and $\mathcal{E}$ are classical and were initially studied by P. Fatou and G. Julia, and later by many other authors. Introductions to rational functions can be found in the books by Beardon [10], Carleson and Gamelin [16], Milnor [32] and Steinmetz [40]. Functions in $\mathcal{E}$ and $\mathcal{M}$ have been studied more recently. For a general survey including all the above classes we refer to Bergweiler [12] or [29].

Many properties of $J(f)$ and $F(f)$ are much the same for all classes above but different proofs are needed and some discrepancies arise. We recall some well-known facts, which hold for all these classes: by definition, the Fatou set $F(f)$ is open and the Julia set $J(f)$ is closed; the Julia set is perfect and non-empty; the sets $J(f)$ and $F(f)$ are completely invariant under f; for z_0 any non exceptional point, $J(f)$ coincides with the closure of the backward orbit of z_0; and finally the repelling periodic points are dense in $J(f)$.

The possible dynamics of a periodic connected component U of the Fatou set of f (i.e. $f^p(U) \subset U$, for some $p \geq 1$), is classified in one of the following possibilities: attracting domain, parabolic domain, rotation domain (Siegel disc or Herman ring) or Baker domain also called essentially parabolic domain. Herman rings do not exist for $f \in \mathcal{E}$. A Fatou component that is neither periodic nor pre-periodic is called a wandering domain. Neither Baker domains nor wandering domains exist for f in $\mathcal{R}$ or in $\mathcal{E}$ of *finite type* (i.e. such that the inverse function has only finitely many singularities).

We define the *residual Julia set* of f denoted by $J_r(f)$ as the set of those points of $J(f)$ which do not belong to the boundary of any component of the Fatou set $F(f)$. The points of $J_r(f)$ are called *buried points* of $J(f)$ and a component of $J(f)$ that belongs to $J_r(f)$ is called a *buried component*. This concept was first introduced in the context of Kleinian groups. Abikoff in [1, 2] defined the residual set $\Lambda_r(\Gamma)$ of a Kleinian group Γ to be the subset of those points of the limit set $\Lambda(\Gamma)$ which do not lie on the boundary of any component of the complement of $\Lambda(\Gamma)$. Abikoff gave examples where $\Lambda_r(\Gamma) \neq \emptyset$. In his well-known paper [41] Sullivan draws attention to the dictionary of correspondences between complex dynamics and Kleinian groups (see [34, Chapter 5] for a first version of what is called Sullivan's dictionary). Following this idea, in 1988 McMullen [31] defined a buried component of a rational function to be a component of the Julia set which does not meet the boundary of any component of the Fatou set. Similarly, for a buried point of the Julia set. McMullen gave an example of a rational function with buried components. Beardon studied this example in his book [10] and he also gave conditions under which the existence of buried components was assured (see [11]). After these results several mathematicians have studied buried components for rational functions (see [7], [33], [34] and [36]). The first discussion about residual Julia sets for functions in class $\mathcal{E}$ of finite type was given by Qiao in [35]. He also gave some conditions for $f \in \mathcal{E}$, of finite type, under which the Julia set contained buried points [36]. Different examples of $f \in \mathcal{E}$ with $J_r(f) \neq \emptyset$ and related results were given independently by Domínguez in [21]. In [7] Baker and Domínguez discussed some results of Morosawa and Qiao on conditions for $f \in \mathcal{R}$ to have buried points or buried components. They showed that these results can be extended to functions in class $\mathcal{M}$.

This paper describes some of the results mentioned above on the residual Julia set for different classes of functions such as rational, transcendental entire and transcendental meromorphic (see Sections 2, 3, 4 and 5).

In Section 6 we prove some new results about the residual Julia set of some classes of entire or meromorphic functions and apply them to the sine family and to some meromorphic functions constructed by surgery. We also give certain conditions under which all points on the so-called Devaney hairs (or rays) are buried points. In the particular case of exponential functions having an attracting periodic orbit of period greater than one, we show that all hairs except possibly a countable number of them are buried components. Moreover we characterize these exceptions in terms of the kneading sequence.

2. Basic properties of the residual Julia set

We give here some basic results about the residual Julia set which hold for functions which belong to any of the classes defined in the introduction. In this section, f will denote a function in $\mathcal{R}, \mathcal{E}$ or $\mathcal{M}$.

The first proposition deals with completely invariant components of the Fatou set.

Proposition 2.1. *If the Fatou set of f has a completely invariant component, then the residual Julia set is empty.*

Proof. If the Fatou set has a completely invariant component U, then $\overline{U}$ is also completely invariant. Hence, by the minimality of $J(f)$, we have $\partial U = J(f)$ so that the residual Julia set $J_r(f) = \emptyset$ by definition. □

In particular for a nonlinear polynomial P the unbounded Fatou component is completely invariant and the residual Julia set is empty. Perhaps this contributed to the relatively late recognition of the possible existence of residual Julia sets.

The following is a trivial observation.

Proposition 2.2. *If there exists a buried component of $J(f)$, then $J(f)$ is disconnected.*

Proof. Observe that buried components are defined as connected components of the Julia set which are all buried. For such components to exist, $J(f)$ must be disconnected. □

Notice that this result is only for buried components, not for buried points. There are examples (see the next section) of maps with a connected Julia set and a non-empty residual Julia set.

Finally, what follows is a result that was proven by Morosawa [33] for rational functions and by Baker and Domínguez [7] (in a slightly stronger form) for functions in the remaining classes.

Proposition 2.3. *If the residual Julia set of f is non-empty, then $J_r(f)$ is completely invariant, dense in $J(f)$ and uncountably infinite.*

Proof. The complete invariance of $J_r(f)$ follows easily from the complete invariance of $F(f)$. Then $\overline{J_r(f)}$ must be completely invariant and has more than three points. By the minimality of $J(f)$ we have that $J(f) = \overline{J_r(f)}$. To prove the second part, Morosawa constructs a Cantor set contained in $J_r(f)$, following the method of Abikoff in [2]. □

In their paper [7], Baker and Domínguez prove something slightly stronger, namely that $J_r(f)$ is residual in the sense of category theory. We recall that a residual set is the complement of a countable union of nowhere dense sets (which are in this case the boundaries of the Fatou components). Any residual set in this sense, must contain an intersection of open dense subsets. Since $J(f)$ is a complete metric space, it is a Baire space and therefore any residual subset is dense.

3. The residual Julia set for rational functions

The first example of a rational function with $J_r(f) \neq \emptyset$ was given by Mc-Mullen [31]. The idea behind this example is to have a Julia set that consists of a Cantor set of nested Jordan curves (see Figure 1). Such an object necessarily has plenty of buried components, as in the middle thirds Cantor set C, where the extreme points of each interval belong to the set but many other points in C are not extreme points and therefore do not belong to the boundary of any component of C^c.

More precisely let

$$R(z) = z^2 + \frac{\lambda}{z^3}, \lambda > 0.$$

One can see that ∞ is a super-attracting fixed point of R. It can be shown that zero and ∞ lie in different components of the Fatou set, say F_0 and F_∞ respectively. When λ is sufficiently small the following facts can be proved.

(a) F_0 and F_∞ are simply-connected, while other components of the Fatou set are doubly connected;

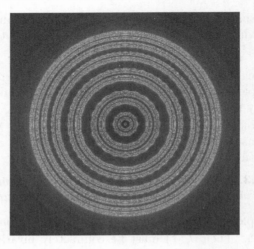

FIGURE 1. Dynamical plane of $R(z) = z^2 + \frac{\lambda}{z^3}$ where $\lambda = 10^{-8}$. The Julia set is a Cantor set of quasicircles.

(b) $R^n \to \infty$ on $F(R)$;
(c) ∞ attracts all critical points of $R(f)$;
(d) the Julia set is a Cantor set of nested Jordan curves;
(e) there are components of the Julia set which do not meet the boundary of any component of the Fatou set. Such components are quasicircles.

This example can also be found in [10, Chapter 5, p. 266]. In [11] Beardon proves the first general theorem about buried components for $f \in \mathcal{R}$.

Theorem 3.1 ([11]). *Let $f \in \mathcal{R}$. Suppose that $J(f)$ is disconnected, and that every component of the Fatou set has finite connectivity. Then $J(f)$ has a buried component, so $J_r(f) \neq \emptyset$.*

The proof of Theorem 3.1 is based on the following result.

Theorem 3.2. *Let $f \in \mathcal{R}$. If $J(f)$ is disconnected, then it has uncountably many components, and each point of $J(f)$ is an accumulation point of distinct components of $J(f)$.*

Indeed, if each component of $F(f)$ has finite connectivity, and if $J(f)$ is disconnected, then there are only countably many components of $J(f)$ which lie on the boundary of some component of the Fatou set. The theorem follows immediately. Qiao in [36] improves Beardon's result by proving the following statement.

Theorem 3.3 ([36]). *Let $f \in \mathcal{R}$ and $J(f) \neq \widehat{\mathbb{C}}$. The Julia set $J(f)$ contains buried components if and only if (i) $J(f)$ is disconnected and (ii) $F(f)$ has no completely invariant component.*

Proof. If all Fatou components are finitely connected, the result follows from Beardon's theorem. If not, Qiao proves the existence of a periodic connected component of the Julia set which can be surrounded by a closed curve γ in the Fatou set, such that the bounded component of $\mathbb{C}\setminus\gamma$ contains no periodic Fatou component. Such a component of the Julia set is necessarily buried. $\square$

Thus by Qiao's Theorem 3.3, it is not difficult to find examples of rational functions with degree greater than one for which the Julia set contains buried components, so the residual Julia set $J_r(f)$ is not empty. See [36] for the complete proof of Theorem 3.3 and examples.

In [7] the authors gave a proof of Theorem 3.3, differing from that in [36], which can be used also for transcendental entire functions with some changes.

Similar results were proved (independently) by Morosawa (see [33] and [34]).

One can ask if it is possible to have a rational function with connected Julia set and non-empty residual Julia set. Naturally, such a function will have buried points but not buried components. In what follows we describe this example due to Morosawa [33], [34].

A function $f \in \mathcal{R}$ is hyperbolic if each critical point of f has a forward orbit that accumulates at a (super) attracting cycle of f. The following theorem gives a characterization of those hyperbolic functions with non-empty residual Julia set.

Theorem 3.4 ([33]). *Let f be a hyperbolic rational function with degree at least two. Then the residual Julia set is empty if and only if either (i) $F(f)$ has a completely invariant component or (ii) consists of only two components.*

In his proof, Morosawa relies on the fact that the boundary of a Fatou component of a hyperbolic function is locally connected. This allows a great control on the union of all these sets.

The example of Morosawa consists of the following hyperbolic rational function:

$$R(z) = \frac{-2z+1}{(z-1)^2}.$$

Observe that the set $\{0, 1, \infty\}$ is a super-attracting cycle of $R(z)$, since the critical points of $R(z)$ are 0 and 1. Thus $F(R)$ has countably many components. Moreover, every component of $F(R)$ is eventually absorbed into this super-attracting cycle and R is hyperbolic. Hence the residual Julia set is not empty by Theorem 3.4. It can be shown that each component of the immediate basin of this super-attracting cycle is simply-connected. An arbitrary component of $F(R)$ except that of the immediate basin contains no critical point. Thus, every component of $F(R)$ is simply connected. Therefore, the Julia set of R is connected.

This class of examples with locally connected Fatou boundaries was generalized in the following theorem.

Theorem 3.5 ([37]). *Let $f \in \mathcal{R}$ with degree $d \geq 2$ and $J(f) \neq \mathbb{C}$. Suppose that $F(f)$ has no completely invariant components. Then either the residual Julia set is non-empty or the Julia set is not locally connected.*

In fact, this result proves, in the case of locally connected Julia sets, the stronger conjecture of Makienko which reads as follows.

Conjecture 3.6 (Makienko, [27]). *Let $f \in \mathcal{R}$. Then $J(f)$ has buried points if and only if $F(f)$ has no completely invariant components.*

4. Residual Julia sets for transcendental entire functions

Most of the results in Section 2 for rational functions can be extended to transcendental meromorphic functions, as we will see in Section 5. However, when we deal with functions in class $\mathcal{E}$, the statements are considerably different. We recall that a component U of the Fatou set that is neither periodic nor preperiodic is called a wandering domain. Sullivan in [41] proved the non existence of wandering domains for a rational function. This result was later extended to functions in class $\mathcal{E}$ of finite type in [26] and [28]. However functions in class $\mathcal{E}$ with infinitely many singularities of f^{-1} may have wandering domains. Examples in class $\mathcal{E}$ can be found in [5] such that $F(f)$ contains some wandering domains which roughly speaking form an unbounded sequence of concentric rings. For such functions $J(f)$ is not connected. Thus for functions in class $\mathcal{E}$ it is possible to have both cases of wandering and no wandering domains. This turns out to be an important difference when dealing with the residual Julia sets.

4.1. **Functions with no wandering domains.** An important class of transcendental entire functions is the class S of functions f of finite type. As mentioned above, functions in class S do not possess wandering domains. The main results related to the residual Julia set for functions in class S were given by Qiao [36] and Baker and Domínguez [7]. In [36], Qiao stated the following theorem.

Theorem 4.1. *Let $f \in S$ and $J(f) \neq \mathbb{C}$. The Julia set of f contains buried points if and only if $F(f)$ is disconnected.*

The "only if" part of Theorem 4.1 is immediate since $F(f)$ being connected would imply the existence of a unique connected component of the Fatou set which has to be completely invariant. In the "if" part Qiao separates the proof into two cases. If the components of the Fatou set are all bounded, then he shows that there exist continua of buried points, all tending to infinity under iteration (in fact these are the "Devaney hairs", see Section 6). The construction he uses is the same as in [20]. When some component of the Fatou set is unbounded, a new construction is needed. The author constructs a set of unbounded regions and shows that they all contain points of the Julia set, but at the same time one of them does not contain any periodic component of the Fatou set. Hence all repelling periodic points in this region must be buried.

The authors in [7] gave an alternative construction to show the following result. We remark that functions in class $\mathcal{E}$ can have at most one completely invariant component of $F(f)$ (see [4]) but, *a priori*, there could also be other Fatou components. If $f \in S$, any completely invariant component must form the whole Fatou set [26].

Theorem 4.2. *Let $f \in \mathcal{E}$ such that $F(f)$ is not connected. Suppose that there are no wandering domains, no completely invariant Fatou component and no Baker domains in which f is univalent.*

(i) If all periodic components of $F(f)$ are bounded, then $J_r(f) \neq \emptyset$.
(ii) If there are unbounded periodic Fatou components and ∞ is an accessible boundary point in one of these components, then there are buried components of $J(f)$ which are unbounded, so $J_r(f) \neq \emptyset$.

Before proving Theorem 4.2, we state the following useful lemma (for a proof see [7]).

Lemma 4.3. *If $f \in \mathcal{E}$ has no wandering domains and $J(f)$ has no buried components, then there is some periodic cycle of Fatou components $G_1, G_2, \ldots G_p$, such that $J(f) = \overline{K(G_j)}$ for all $1 \leq j \leq p$, where $K(G_j)$ denotes the union of all components of $J(f)$ which meet ∂G_j.*

Sketch of the proof of Theorem 4.2 (ii). Suppose that the Fatou set has an unbounded periodic component H such that $f^p(H) \subset H$ and ∞ is accessible in H. Since H is unbounded it is simply-connected and is the immediate basin of attraction of an attracting or parabolic fixed point of f^p, a Siegel disc or a Baker domain such that f^p is not univalent in H. Further, since H is not completely invariant there is a non-periodic component K such that $f^p(K) \subset H$.

The set $E = \{e^{i\theta} : \text{radial limit } \lim_{r \to 1} \Psi(re^{i\theta}) = \infty, \Psi : D(0,1) \to H \text{ is the Riemann map}\}$ contains infinitely many points by a result in [8]. Let us take values θ_j, $1 \leq j \leq 4$, such that $0 < \theta_1 < \theta_2 < \theta_3 < \theta_4 < 2\pi$ and radii $\lambda_j : z = re^{i\theta_j}$, $0 < r < 1$. Assume that $e^{i\theta_1}$ and $e^{i\theta_3}$ are in E and that the limits (as $r \to 1$) on the radii λ_2, λ_4 exist and are finite points α_2, α_4 in ∂H. It can be shown that there is a simple path $\Gamma \subset H$ which runs to ∞ at both ends. If g is an appropriate branch of f^{-p}, then consider the continuation of

g along Γ and obtain a simple path Γ' which runs to ∞ at both ends and divides K.

The points α_2, α_4 in ∂H are separated by Γ and the points $p_2 = g(\alpha_2)$, $p_4 = g(\alpha_4)$ in ∂K are separated by Γ'. If we assume that there are no buried components, then there is a periodic component L of the Fatou set such that $J(f) = \overline{T}$, where T is the union of all components of the Julia set which meet ∂L (see Lemma 4.3). Thus there are points q_2, q_4 of T so close to p_2, p_4 that they are separated by Γ'. The points q_2, q_4 belong to different components of the Julia set which are also separated by Γ'. These components contain points r_2, r_4 in ∂L. Then there are points in L which are also close to r_2, r_4 but they are separated by Γ'. This is impossible since the Fatou components L and K are distinct. □

Remark 4.4. Observe that when $f \in S$ and $F(f)$ is not connected there are no wandering domains, no completely invariant domains and no Baker domains (see [26] and [28]). Thus Theorem 4.2 gives the following corollary.

Corollary 4.5. *Let $f \in S$ such that $F(f)$ is not connected. If either all periodic components of $F(f)$ are bounded or there exists an unbounded periodic component in which ∞ is an accessible point, then $J_r(f) \neq \emptyset$. In the latter case there are buried components of $J(f)$.*

Examples of functions satisfying the hypotheses of the above results include the family $\lambda \sin z$ for some values of λ and the family λe^z for any λ such that there is an attracting p-cycle (see also Section 6).

Remark 4.6. We do not know any example of an unbounded Fatou component G such that ∞ is not accessible along any path in G. If there are no such examples then Corollary 4.5 would imply Theorem 4.1.

4.2. Functions with wandering domains.
It is well known for $f \in \mathcal{E}$ that a multiply-connected component of $F(f)$ must be wandering [5]. The Julia set for $f \in \mathcal{E}$ cannot be totally disconnected, as may happen for polynomials. However singleton components of the Julia set can occur. If the Julia set has a singleton component $\{\psi\} \neq \infty$ then $F(f)$ has a multiply-connected component. The converse of this statement is also true and we will see that this yields the existence of buried points in the Julia set. The main result in [21] is as follows.

Theorem 4.7. *If $f \in \mathcal{E}$ and $F(f)$ has a multiply-connected component, then the Julia set has singleton components and such components are buried and dense in $J(f)$.*

Sketch of the proof. The result of Baker [5] mentioned above states that if $F(f)$ has a multiply-connected component U, then U and all its iterates are bounded wandering components and $f^n \to \infty$ in U (as $n \to \infty$). Also there is a simple curve γ in U such that $f^n(\gamma)$ is a curve in $f^n(U) = U_n$ on which $|z|$ is large and winds round zero. In [21] the following results were shown: (i) $f^{n+1}(\gamma)$ is in $U_{n+1} \neq U_n$, winds round zero and must be outside γ_n, (ii) there is a component N_n of the Fatou set between U_n and U_{n+1} and an integer $m \geq 2$ such that $f^m(N_n) \subset U_n$ and (iii) the component N_n is a multiply-connected component which does not wind round zero i.e. zero is in the unbounded component of $\mathbb{C} \setminus \overline{N_n}$. Thus picking a repelling periodic point η with period p, N_n (as above) for five different values of n and choosing $m \in \mathbb{N}$ so that the spherical derivative of f^{mp} at η satisfies the hypothesis of Ahlfors' five island theorem (see [3]), it can be shown that the disc $D(\eta, R)$, $R > 0$ contains a sequence of different multiply-connected components N_k of the Fatou set with diameter tending to zero and N_{k+1} is inside one of the inner boundary components of N_k. A sequence $\psi_k \in \partial J(f)$ is a Cauchy sequence which converges to a point $\psi \in J(f)$. By construction, ψ is a buried component of the Julia set. Thus there is a dense subset G of $J(f)$ such that each $\alpha \in G$ is a buried singleton component of $J(f)$. □

5. Residual Julia sets for transcendental meromorphic functions

In Section 3 we discussed some results of Morosawa [33] and Qiao [36] on conditions for a rational function to have buried points or buried components. In this section we will focus on those results that can be extended to transcendental meromorphic functions.

For a meromorphic function in $\mathbb{C}$, Baker and Domínguez [7] proved the following result for functions with no wandering domains, which is a step towards a generalization of Theorem 3.3 to meromorphic maps.

Theorem 5.1. *Let f be meromorphic with no wandering domains. Assume that $J(f)$ is not connected and that $F(f)$ has no completely invariant component. Then the residual Julia set $J_r(f)$ is non-empty.*

In order to give the proof of Theorem 5.1 we start with the following lemma.

Lemma 5.2. *If f is meromorphic and U is a multiply-connected periodic Fatou component such that $\partial U = J(f)$, then U is completely invariant.*

Sketch of the proof. Assume that U is a multiply-connected Fatou component such that for some $p \in \mathbb{N}$ we have $f^p(U) \subset U$ and that $\partial U = J(f)$ but that U is not completely invariant. Observe that any other Fatou component is simply-connected because, otherwise, there would be part of the Julia set bounded away from ∂U. Since there is one component H such that $H \neq U$, $f(H) \subset U$ we obtain a contradiction at once in the case of rational functions, since then $f(H) = U$, $f(\partial H) = \partial U$ and $f(\partial H)$ is connected while ∂U is not. Thus we may assume that f is transcendental so that $J(f)$ and U are unbounded. We may suppose that there is a point $s \in H$ such that

$f(s) = q \in \gamma$, $f'(s) \neq 0$, and γ is a simple closed path in U which encloses some points of ∂U.

Let g denote the branch of f^{-1} such that $g(q) = s$. It can be proved that the continuation of g maps γ to a simple curve Γ in H which goes to ∞ at both ends. Then the Gross Star Theorem is used to construct paths from q to boundary components of U inside and outside γ, which show that $\partial U = J(f)$, and hence U meets both components of $\mathbb{C} \setminus \overline{\Gamma}$ which is impossible since $\Gamma \subset H \neq U$. □

Now we are able to prove Theorem 5.1.

Proof. Suppose that f is meromorphic without wandering domains, $J(f)$ is not connected and $J_r(f)$ is empty. Then there is a periodic Fatou component U such that $\partial U = J(f)$ (see [7] for a proof of this fact). Since $J(f)$ is not connected, then there exists a component of the Fatou set which is multiply-connected. But $\partial U = J(f)$, and hence U is multiply-connected. By Lemma 5.2, U is completely invariant which contradicts the hypotheses. □

Note that functions in $\mathcal{E}$ cannot satisfy the conditions of Theorem 5.1 since the existence of a multiply-connected component of $F(f)$ implies the existence of wandering domains [5].

For a function $f \in \mathcal{M}$ the following theorem covers almost all the cases when $J(f)$ is not connected (see [7] for the proof).

Theorem 5.3. *Let $f \in \mathcal{M}$ have no wandering domains and no completely invariant Fatou components, and suppose that either (a) $F(f)$ has a component of connectivity at least three, or (b) $F(f)$ has three doubly-connected components U_i, $1 \leq i \leq 3$, such that one of the following conditions holds.*

(i) Each U_i lies in the unbounded component of the complement of the other two.

(ii) Two components U_1, U_2 lie in the bounded component of U_3^c but U_1 lies in the unbounded component of U_2^c and U_2 lies in the unbounded component of U_1^c.

Then $J(f)$ contains singleton components which are dense in $J(f)$ and buried.

Examples of functions satisfying the hypotheses of Theorem 5.3 can be found in [7]. Such examples were obtained by using Runge's theorem. In Section 6.2 we provide an example of a meromorphic function obtained by surgery as in [23] satisfying the hypotheses of Theorem 5.3 and containing unbounded continua of buried points. It would be interesting to know under which conditions a function with connected Julia set must have non empty residual Julia set.

While this paper was under correction, the full generalization of Theorem 3.3 for meromorphic functions was proven by Ng, Zheng and Choi in [43]. Their result reads as follows.

Theorem 5.4. *Let $f \in \mathcal{M}$. The Julia set $J(f)$ has buried components if $F(f)$ has no completely invariant components and $J(f)$ is disconnected. Moreover, if $F(f)$ has an infinitely connected component, then the singleton buried components are dense in $J(f)$.*

Sketch of the proof. The proof is divided into two cases. In the first case the authors consider that each Fatou component of f has finite connectivity. As the Julia set is disconnected, then it has uncountably many components but there are only countably many components of $J(f)$ which lie on the boundary of some component of the Fatou set. Then the result follows.

In the second case the authors assume that f has at least one infinitely connected component of the Fatou set. By Theorem A in [22] the singleton components of $J(f)$ are dense in the Julia set. Taking an arbitrary open disk $D(a,r)$, where $a \in J(f)$ and r is a positive number, it is sufficient to prove that $D(a,r)$ contains at least one buried component of the Julia set. To prove this fact the authors make an inductive construction to obtain:
(a) a sequence of infinitely connected pairwise disjoint Fatou components $\{U_n\}$;
(b) a sequence of non-contractible closed curves $\{\gamma_n\} \subset U_n \subset F(f)$; and
(c) a sequence of singleton components $\{b_n\} \in \partial U_n$ (and therefore $\{b_n\} \in int(\gamma)$).

It follows that the sequence $\{b_n\}$ must have a limit, say $b \in J(f)$. Observe that b is a singleton component since it is in the interior of all the curves $\gamma_n \subset F(f)$ whose diameter tends to zero. Moreover b must be buried since each curve γ_n belongs to different component of the Fatou set (indeed if b belongs to the boundary of some Fatou component W, then W would have zero diameter). Thus $\{b\}$ is a singleton buried component in $D(a,r)$. As a and r were taken arbitrarily it follows that singleton buried components are dense in the Julia set. □

Kisaka and Shishikura [30] have shown that there are transcendental meromorphic functions for which the Julia set is disconnected, while the Fatou set has no completely invariant components and no infinitely connected components. These functions are not covered by Theorem 5.4 but they are covered by the following result which was proved in [22].

Theorem 5.5. *Let $f(z) \in \mathcal{M}$ and suppose that $F(f)$ has multiply-connected components $A_i, i \in \mathbb{N}$ all different, such that each A_i separates $0, \infty$ and $f(A_i) \subset A_{i+1}$ for $i \in \mathbb{N}$. Then $J(f)$ has a dense set of buried singleton components.*

The authors of Theorem 5.4 also prove an attempt to adapt Makienko's Conjecture (3.6) to meromorphic functions of finite type as a generalization of Theorem 3.5 (see [43, Section 5] for the proof).

Theorem 5.6. *Let* $f \in \mathcal{M}$ *be of finite type. Suppose that* $J(f) \setminus \{\infty\}$ *is locally connected. Then* $J_r(f)$ *is empty if and only if* $F(f)$ *has a completely invariant component or consists of only two components.*

6. Hairs in the residual Julia Set

In this section we deal with some classes of entire or meromorphic transcendental maps that contain unbounded continua of buried points in their Julia sets. These continua are the well-known *Devaney hairs*, also called *dynamic rays*.

If $f : \mathbb{C} \to \mathbb{C}$ is an entire transcendental function, a *hair of* f is defined as a curve $\gamma : (0, \infty) \to \mathbb{C}$ in the Julia set of f, such that $\gamma(t) \xrightarrow[t \to \infty]{} \infty$ and $f^n(\gamma(t)) \xrightarrow[n \to \infty]{} \infty$ for any $t \in (0, \infty)$. If the limit $\lim_{t \to 0} \gamma(t)$ exists (say, it equals z_0) and is finite we say that the hair *lands* at z_0. This point is called the *endpoint* of the hair. In other words, all points on the curve γ have orbits that tend (exponentially fast) towards the essential singularity at infinity, while endpoints might escape or not.

Hairs were initially described for the exponential function by Devaney and Krych in [19] and then extended to the exponential family, $z \mapsto \lambda \exp(z)$ (for several classes of parameter values) in [18] and later in [15]. More recently, Schleicher and Zimmer [39] have extended this description to most parameter values. For more general classes of entire transcendental functions the main reference is by Devaney and Tangerman [20], where they show that *Cantor bouquets*, i.e. Cantor sets of hairs, appear in the Julia set of any entire transcendental function of finite type that has at least one *hyperbolic asymptotic tract*. A hyperbolic asymptotic tract is an unbounded connected open set where orbits that remain in it behave in an "exponential fashion", that is, f increases the modulus and the derivative exponentially (see [20] for precise definitions). For example, the right half plane is a hyperbolic asymptotic tract for the exponential family, while the upper and the lower half plane are hyperbolic asymptotic tracts for any map in the sine family $z \mapsto \lambda \sin(z)$. It seems reasonable to think that the proof of the existence of Cantor bouquets can be generalized for entire transcendental maps of *bounded type* (the set of singularities of f^{-1} is contained in a bounded set) with at least one hyperbolic asymptotic tract. In fact, such an extension has been announced by Rottenfüsser, Rückert, Rempe and Schleicher [38] for an even larger class of functions of bounded type. Similar and other extensions can also be found in [9].

Transcendental meromorphic functions may also contain Devaney hairs in their Julia sets. Although no general theory is known, many examples can be constructed using, for example, surgery techniques or otherwise adding a pole to a known entire map (see Section 6.2). We will refer to the class of transcendental functions that possess such Devaney hairs in their Julia sets as $\mathcal{H}$.

In this section we want to give a new criterion to assure the existence of buried points in the Julia set, and also some conditions for a transcendental map under which we can assure that the Devaney hairs are in the residual Julia set, i.e., they are unbounded curves of buried points. If the endpoints are also buried then the hair and its endpoint form a buried component. The precise statement is as follows.

Proposition 6.1. *Let $f \in \mathcal{R} \cup \mathcal{E} \cup \mathcal{M}$, and $A \subsetneq \mathbb{C}$ be a closed set with non-empty interior. Suppose the following conditions are satisfied:*

a) $(\mathbb{C} \setminus A) \cap J(f) \neq \emptyset$, and

b) all the Fatou components of f eventually iterate inside A and never leave again. That is, if Ω is a Fatou component, $f^n(\Omega) \subset A$ for all $n > N$, where N depends on Ω.

Then the residual Julia set is non-empty. More precisely, the residual Julia set contains the set

$$\{z \in J(f) \mid f^n(z) \text{ is defined for all } n \text{ and } f^n(z) \notin A \text{ for infinitely many } n\}.$$

In particular if A is bounded, any point of the Julia set with an infinite unbounded forward orbit belongs to the residual Julia set.

Proof. Since any point on the boundary of a Fatou component maps to another point with the same property, we have that all these points (those with an infinite orbit) must eventually fall into A and never leave again. Thus no point of the Julia set which leaves A infinitely often can be in the boundary of a Fatou component.

Now, the complement of A is an open set which contains points of $J(f)$. Let $z \in J(f) \cap (\mathbb{C} \setminus A)$, and let U be a neighborhood of z entirely contained in $\mathbb{C} \setminus A$. Since periodic points are dense in $J(f)$, it follows that U must contain a periodic point of the Julia set. This point has to come back to itself infinitely often. Since it lies in the complement of A, this must be a point in the residual Julia set. □

Remark 6.2. (1) The hypotheses inmediately rule out the case of a completely invariant component. Indeed if U is a completely invariant component, then $J(f) = \partial U$. But U must be in A because it is invariant, which implies that the whole Julia set is in A, contradicting the fact that some point in the Julia set must be in the complement.

(2) The hypotheses allow for wandering domains (as long as they eventually wander inside A) or Baker domains of any kind (as long as they are contained in A)
(3) If A is bounded and we assume $f \in \mathcal{H}$, then it follows that all points in the Devaney hairs (except maybe the endpoint) are buried points.
(4) If A is bounded one might be tempted to deduce that all poles and prepoles are buried points. However, this is not necessarily the case (as pointed out by the referee) since one can not *a priori* exclude the possibility of a Fatou component U with a pole in the boundary such that $f(U)$ is unbounded and $f^2(U)$ is bounded and included in A. Even in the absence of finite asymptotic values we are not able at this point to rule out this situation.

In what follows we will analyze three examples which satisfy the hypotheses of Proposition 6.1. The first two examples – the sine family and an example of a meromorphic function constructed by surgery – have bounded Fatou components while the third example is the exponential family that has unbounded Fatou components, and we treat it separately.

6.1. The sine family. An important example is given by the sine family $S_\lambda(z) = \lambda \sin(z)$. This family has two critical values (with symmetric orbits) and no finite asymptotic values. Hence $S_\lambda \in S$ for all λ. Since both the far upper and the far lower half planes are hyperbolic asymptotic tracts, it follows from the standard arguments that $S_\lambda \in \mathcal{H}$ and all functions in this family have a pair of Cantor bouquets [20], one in the upper half plane and another one in the lower half plane. See Figure 2. How the landing of these hairs occurs depends on the parameter value.

For $|\lambda| < 1$, the fixed point $z = 0$ is attracting and its basin is completely invariant and therefore unbounded [24]. For $\lambda = 1$, the Fatou set consists of the parabolic basin of 0, all of whose connected components are bounded (see [14]). For $\lambda = e^{2\pi i \theta}$ with θ an irrational number of bounded type, the map S_λ has an invariant bounded Siegel disk around $z = 0$ and all other Fatou components are its preimages (see [42]). For other λ parameter values, it is not known to our knowledge whether the Fatou components of S_λ are bounded or not.

At the end of this section we prove the following.

Proposition 6.3. *Let $\lambda \in \mathbb{C}$ such that $|\text{Re}(\lambda)| \geq \frac{\pi}{2}$. Then all the Fatou components of S_λ are bounded.*

Remark 6.4. We conjecture that Proposition 6.3 is true for all $|\lambda| \geq 1$, but our construction only works for this smaller set (see Subsection 6.1.1).

Combining Proposition 6.1 and Proposition 6.3 we have the following corollary.

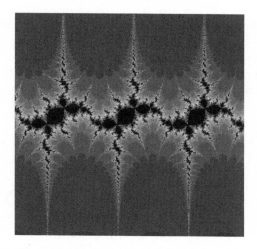

FIGURE 2. Dynamical plane of $S_\lambda(z) = \lambda \sin(z)$ where $\lambda = 1.88853 + i\, 0.673125$. There are two attracting orbits of period three. Their immediate basins of attraction are bounded and all other Fatou components are preimages of those. All Devaney hairs consist of buried points.

Corollary 6.5. *Let $\lambda \in \mathbb{C}$ such that $|\mathrm{Re}(\lambda)| \geq \frac{\pi}{2}$. Then all the Devaney hairs of S_λ consist of buried points.*

Proof. Members of the sine family have two critical values $\{\lambda, -\lambda\}$ and no asymptotic values. Hence there are no wandering domains, no Baker domains and at most two periodic cycles of Fatou components. By Proposition 6.3, they must be bounded. Let A be any bounded set containing all the periodic cycles of Fatou components. Then we are under the hypotheses of Proposition 6.1 and the corollary follows. □

Notice that if the Julia set is the whole plane then, trivially, all its points are buried.

6.1.1. *Boundedness of Fatou components (Proof of Proposition 6.3).* The main part of the proof is the following Proposition.

Proposition 6.6. *Let $\lambda \in \mathbb{C}$ be such that $|\mathrm{Re}(\lambda)| \geq \frac{\pi}{2}$. Then there is a fixed hair in the upper half plane whose endpoint is the repelling fixed point $z = 0$. More precisely, there exists an invariant curve $\{\gamma(t)\}_{0 \leq t < \infty}$ such that*

(1) $|\mathrm{Re}\,\gamma(t)| \leq \pi/2$ for all $t \geq 0$.
(2) $\lim_{t \to \infty} \gamma(t) = \infty$ and $\lim_{t \to 0} \gamma(t) = 0$.
(3) For all $t > 0$, $\lim_{n \to \infty} S_\lambda^n(\gamma(t)) = \infty$.

Proof. Let B consist of the vertical strip
$$B = \{z \in \mathbb{C} \mid -\frac{\pi}{2} < \operatorname{Re}(z) < \frac{\pi}{2}\}.$$
See Figure 3. The right (resp. left) vertical boundary of B is mapped to the ray segment starting at λ (resp. $-\lambda$) and going to infinity, given that
$$\lambda \sin(\pm\frac{\pi}{2} + yi) = \pm\lambda \cosh(y).$$
Any horizontal segment of the form $\{x + y_0 i\}_{|x| < \frac{\pi}{2}}$ joining the two boundaries

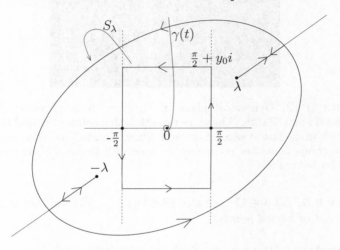

FIGURE 3. Setup of the proof of Proposition 6.6 for $|\operatorname{Re}(\lambda)| \geq \frac{\pi}{2}$.

is mapped under S_λ to half an ellipse of radii $|\lambda| \cosh(y_0)$ and $|\lambda| \sinh(|y_0|)$, rotated by λ. The symmetric segment is mapped to the other half.

It follows from these observations that B is mapped one to one to the whole plane except the 2 ray segments. Hence the standard constructions apply (see e.g. [20]) to prove, in particular, that the "tail" of the hair $\gamma(t)$ (i.e. for t large enough) exists in B.

In order to see that it lands at $z = 0$, observe that a box like the one in Figure 3 (in fact, independently of its height y_0) is mapped one to one to a set that completely covers it. Hence a well defined branch of the inverse exists inside the box and satisfies the hypothesis of the Schwarz lemma. It follows easily by iterating the inverse that $z = 0$ is the unique fixed point in the box and that $\gamma(t)$ must converge to zero as $t \to 0$. □

We proceed to see how the boundedness of the Fatou components follows from Proposition 6.6.

Having a fixed hair $\gamma(t)$ landing at $z = 0$ gives, by symmetry, another invariant hair $\widetilde{\gamma}(t) = -\gamma(t)$ coming from below landing at the same point.

These two hairs together with $\{0\}$ form an invariant curve Γ in the Julia set, contained in the strip

$$B = \{z \in \mathbb{C} \mid -\frac{\pi}{2} < \mathrm{Re}(z) < \frac{\pi}{2}\}.$$

Since the function S_λ is 2π-periodic, all the 2π-translations of Γ are vertical curves in the Julia set all mapped to Γ under one iteration.

Hence the Julia set divides the plane into infinitely many "vertical strips" and, as a consequence, no Fatou component can have unbounded real part, or else it would have to intersect the Julia set. Let us name these strips $\{R_k\}_{k \in \mathbb{Z}}$, where R_k intersects $\mathbb{R}$ in the interval $[2k\pi, 2(k+1)\pi]$.

Now, let us suppose that a periodic Fatou component U, (S_λ has no wandering domains) has unbounded imaginary part. Since U is periodic, all its images must be contained in a finite set of strips, say $R_{-N}, \ldots, R_0, \ldots, R_N$, for some $N \in \mathbb{N}$. We will show that no open set can remain forever in these strips under iteration, unless its imaginary part is bounded.

To that end, choose a point $z_0 = x_0 + iy_0 \in U$ with high enough imaginary part so that

(1) $$\sinh(y_0 - 2\pi) > \frac{4(N+1)\pi}{|\lambda|}.$$

Since U is open, there exists $\delta_0 = \delta_0(y_0)$ such that the round disk $D_0 = D(z_0, \delta_0)$ is contained in U, see Figure 4.

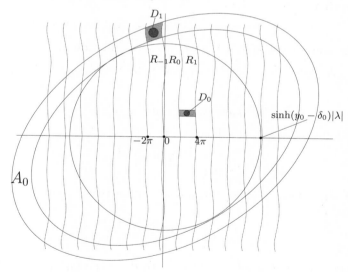

FIGURE 4. Sketch of the disks D_0 and D_1 in the case $N = 1$.

Assuming U is in R_{k_0}, with $|k_0| \leq N$, the "rectangle" $R_{k_0} \cap \{y_0 - \delta_0 < \mathrm{Im}(z) < y_0 + \delta_0\}$ is mapped to a huge annulus A_0 formed by two concentric

ellipses, both transversal to R_k for all $|k| \leq N$, since they contain a disk of radius R where
$$R = |\lambda|\sinh(y_0 - \delta) > |\lambda|\sinh(y_0 - 2\pi) > 4(N+1)\pi.$$
See Figure 4. Since $S_\lambda(D_0)$ belongs to the Fatou component $S_\lambda(U)$ which is constrained to lie in a strip R_{k_1} with $|k_1| \leq N$, it follows that D_0 must be mapped under S_λ to one of the two connected components of $A_0 \cap R_{k_1}$. Without loss of generality we suppose that $S_\lambda(D)$ is in the component that lies in the upper half plane.

Let $z_1 = x_1 + iy_1 = S_\lambda(z_0)$. Since S_λ is univalent in D_0, it follows from the Koebe distortion theorem that
$$\operatorname{dist}(z_1, \partial S_\lambda(D_0)) \geq \frac{1}{4}|S_\lambda'(z_0)|\operatorname{dist}(z_0, \partial D_0).$$
Hence, $S_\lambda(D_0)$ contains a round disk $D_1 = D(z_1, \delta_1)$ where
$$\delta_1 \geq \frac{1}{4}|\lambda||\cos(z_0)|\delta_0 \geq \frac{1}{4}|\lambda|\sinh(y_0)\delta_0,$$
given that $|\cos(x+iy)| \geq \sinh(y)$ for $y \geq 0$.

Now we can apply the same argument to D_1 and conclude that $S_\lambda^2(U)$ must contain a round disk around $z_2 = S_\lambda(z_1)$ of radius δ_2 with
$$\delta_2 \geq \frac{1}{4}|\lambda||\cos(z_1)|\delta_1 \geq \frac{1}{4^2}|\lambda|^2 \sinh(y_1)\sinh(y_0)\delta_0.$$
We claim that $y_1 \gg y_0$. Indeed
$$|x_1 + iy_1| = |z_1| = |\lambda||\sin(x_0 + iy_0)| \geq \sinh y_0 |\lambda|.$$
Using also that $|x_1| < 4(N+1)\pi$ we have by Equation (1)
$$|\lambda|^2 \sinh^2(y_0) \leq |z_1|^2 = x_1^2 + y_1^2 \leq |\lambda|^2 \sinh^2(y_0 - 2\pi) + y_1^2.$$
Hence
$$y_1^2 \geq |\lambda|^2(\sinh^2(y_0) - \sinh^2(y_0 - 2\pi)) \gg y_0^2$$
and the claim is proved. It follows that
$$\delta_2 \geq \left(\frac{|\lambda|\sinh(y_0)}{4}\right)^2 \delta_0.$$
Applying the same argument n times we obtain that $S_\lambda^n(U)$ contains a round disk of radius
$$\delta_n \geq \left(\frac{|\lambda|\sinh(y_0)}{4}\right)^n \delta_0,$$
which tends to ∞ as $n \to \infty$. This contradicts the fact that $S_\lambda^n(U)$ is constrained to lie for all n inside one of the strips R_k with $|k| \leq N$.

Once it has been established that all the periodic components of the Fatou set are bounded, it follows easily that the other components are too. Indeed, any preimage of a bounded component must lie within a vertical strip R_k

for some $k \in \mathbb{N}$. If such a component were unbounded then its image would also be unbounded which is a contradiction. Moreover, S_λ has no wandering domains. This concludes the proof of Proposition 6.3.

6.2. A meromorphic function constructed by surgery.
Using surgery methods exactly as in [23] one can construct meromorphic functions with buried Devaney hairs.

Let us consider as a first map for example the function $f(z) = \lambda \sin(z)$ where λ is chosen so that f has a bounded invariant Siegel disk. This can be accomplished for instance by taking an appropriate λ with $\text{Re}(\lambda) \geq \pi/2$ on the boundary of the main hyperbolic component attached to the unit disk at $\lambda = 1$ (which has period one and center $\lambda = \pi/2$, see [24]). As λ runs along the boundary of the component, the multiplier of one of the fixed points of f runs along the unit circle, so there will be invariant Siegel disks for infinitely many values of λ. Moreover, it follows from Proposition 6.3 that all Fatou components of f must be bounded.

We assume that the Siegel disk of f is centered at $z = 0$ (making an affine change of coordinates if necessary) and its rotation number is a certain number θ which we choose to be of bounded type.

As a second map consider the quadratic polynomial $\widetilde{f}(z) = \rho z(1-z)$ where $\rho = e^{-2\pi i \theta}$. It is well known that $\widetilde{f}$ has an invariant Siegel disk at $z = 0$ of rotation number $-\theta$.

To fix ideas, the surgery construction consists of "gluing" the two dynamical planes (or spheres) where f and $\widetilde{f}$ act, along one of the invariant curves in each of the Siegel disks, γ and $\widetilde{\gamma}$ (see Figure 5). The result, after performing the details of the surgery, is a new holomorphic function F which is conjugate to f on the unbounded component of the complement of γ, while it is conjugate to $\widetilde{f}$ on the bounded component of the complement of γ.

To make the above construction somewhat more precise, let γ (resp. $\widetilde{\gamma}$) be an invariant curve in the Siegel disk of f (resp. $\widetilde{f}$). We call B (resp. $\widetilde{B}$) the bounded component of the complement of γ (resp. $\widetilde{\gamma}$). Then, a new map G is set to be equal to f in $\mathbb{C} \setminus B$ and equal to $\Phi^{-1} \circ \widetilde{f} \circ \Phi$ on B, where $\Phi : \widehat{\mathbb{C}} \to \widehat{\mathbb{C}}$ is a quasiconformal homeomorphism satisfying:

1. $\Phi(\gamma) = \widetilde{\gamma}$,
2. $f = \Phi^{-1} \circ \widetilde{f} \circ \Phi$ on γ,
3. $\Phi : B \to \mathbb{C} \setminus \widetilde{B}$, $\Phi(0) = \infty$, $\Phi(\infty) = 0$, and
4. Φ is conformal everywhere except on a neighborhood of γ (inside the Siegel disk).

By the measurable mapping theorem, there exists a holomorphic map F conjugate to G (see Figure 5). We refer to [23] for the complete details.

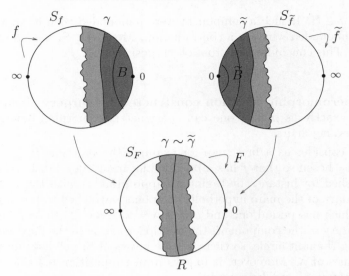

FIGURE 5. The surgery construction. We have drawn the three Riemann spheres $S_f, S_{\tilde{f}}$ and S_F as the dynamical planes of $f, \tilde{f}$ and F respectively.

The new map F (or an affine conjugate) still has a transcendental singularity at ∞ and no other. It has a bounded Herman ring around γ (since both Siegel disks were bounded), and the poles of F correspond to the zeros of $\tilde{f}$ (different from $z = 0$), so F has one single pole. The point $z = 0$ is now a superattracting point whose immediate basin is in the bounded component of the complement of the Herman ring. The function F also has two Cantor bouquets since the dynamics around ∞ have not changed. Note that no periodic Fatou component other than the Herman ring can exist, since F has only two critical orbits (f and $\tilde{f}$ have one each) and no asymptotic values (f and $\tilde{f}$ have none), and both orbits must accumulate on the respective boundaries of the Herman ring. We refer the reader to [23] for a detailed explanation of the properties of F.

Hence all the Fatou components of F eventually iterate on the Herman ring. Since there are no wandering domains, it follows that we are under the hypotheses of Proposition 6.1. By the observations above, the set A can be taken to be bounded (any bounded set containing the ring) and we may conclude that all points which tend to infinity under iteration are buried points. In particular all Devaney hairs are unbounded continua of buried points.

6.3. **The exponential family.** The discussions above deal with cases where all the periodic Fatou components are bounded. Notice that this excludes from the discussion an important family like the exponential family, since all hyperbolic members of the exponential family have unbounded immediate basins of attraction. See Figure 6.

FIGURE 6. Dynamical plane of $E_\lambda(z) = \lambda \exp(z)$, where λ is chosen so that there is an attracting periodic orbit of period 3.

With some work, we could see that hyperbolic exponential functions also satisfy the hypotheses of Proposition 6.1, with a set A which is unbounded and which contains the immediate basin of attraction. Hence we could conclude that the residual Julia set is non-empty. However, it is simpler and more profitable to study the functions closely to conclude something stronger, namely that most of the hairs in the Julia set are buried. More precisely we prove the following.

Proposition 6.7. *Let $\lambda \in \mathbb{C}$ be such that $E_\lambda(z) = \lambda e^z$ has an attracting periodic orbit. Then all hairs, except possibly a countable number of them, are buried components.*

To prove this proposition we will associate to the parameter λ an infinite integer sequence called the *kneading sequence* $K(\lambda) = 0k_1 \ldots k_{n-1}$ (observe that $k_0 = 0$ for all λ) and likewise to each point z in the Julia set another infinite sequence called the *itinerary of z* and denoted by $K(z)$. After seeing the precise construction, it will be clear that all points on a hair must share the same itinerary.

With this notation, Proposition 6.7 is a consequence of the following.

Proposition 6.8. *Let $\lambda \in \mathbb{C}$ such that $E_\lambda(z) = \lambda e^z$ has an attracting periodic orbit of period n. Suppose $K(\lambda) =: K = 0k_1 \ldots k_{n-1}$ is the kneading sequence. Then, all the hairs in the Julia set are buried components except* those with itinerary $T K u_1 K u_2 K u_3 K \ldots$, where T is any finite sequence, and $u_1, u_2, \ldots$ are arbitrary entries. Hence only countably many hairs can have points which are not buried.

The rest of this section is dedicated to proving Proposition 6.8. We start by sketching the construction that allows us to define the kneading sequence of a map. We refer the reader to [17] and [13] for all missing details.

6.3.1. *The fingers and the glove.* Let $z_0, z_1 = E_\lambda(z_0), \ldots, z_{n-1} = E_\lambda(z_{n-2})$ be the points of the attracting periodic orbit.

Let A^* denote the immediate basin of attraction of the periodic orbit and, for $0 \leq i \leq n-1$, define $A^*(z_i)$ to be the connected component of A^* which contains z_i. We name the points in the orbit so that the asymptotic value 0 belongs to $A^*(z_0)$.

We now construct geometrically and define what we call *fingers*. For $\nu \in \mathbb{R}$, let $H_\nu = \{z \mid \text{Re}(z) > \nu\}$.

Definition. An unbounded simply-connected $F \in \mathbb{C}$ is called a *finger* of *width c* if

a) F is bounded by a single simple curve $\gamma \subset \mathbb{C}$.
b) There exists ν such that $F \cap H_\nu$ is simply connected, extends to infinity, and satisfies

$$F \cap H_\nu \subset \{z \mid \text{Im}(z) \in [a - d/2, a + d/2]\} \text{ for some } a \in \mathbb{R},$$

and c is the infimum value for d.

Observe that the preimage of any finger which does not contain 0 consists of infinitely many fingers of width smaller than 2π which are $2\pi i$–translates of each other.

We begin the construction by choosing B to be a topological disk in $A^*(z_0)$ that contains both 0 and z_0, and having the property that B is mapped strictly inside itself under E_λ^n. This set can be defined precisely using linearizing coordinates.

We now take successive preimages of the disk B (see Figure 7). More precisely, let B_{n-1} be the open set in $\mathbb{C}$ which is mapped to B. Note that, since $0 \in B$, it follows that B_{n-1} has a single connected component which contains a left half plane, and whose image under E_λ wraps infinitely many times over $B \setminus \{0\}$. Note that the point z_{n-1} belongs to the set B_{n-1}, which lies inside $A^*(z_{n-1})$.

We now consider the preimage of B_{n-1}. It is easy to check (by looking at the image of vertical lines with increasing real part) that this preimage consists of infinitely many disjoint fingers of width less than 2π which are $2\pi i$-translates of each other. We define B_{n-2} to be the component of $A^*(z_{n-2})$ for which $z_{n-2} \in B_{n-2}$. Then E_λ maps B_{n-2} conformally onto B_{n-1}.

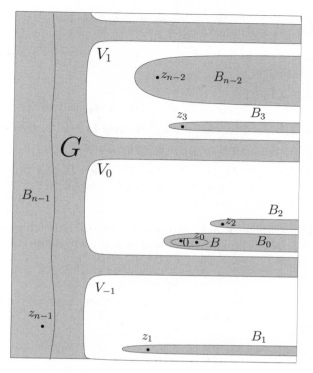

FIGURE 7. Sketch of the sets B_0 to B_{n-1}, G and V_j for $j \in \mathbb{Z}$. Points in grey belong to the basin of attraction of the periodic orbit.

Similarly, we define the sets $B_{n-3}, \ldots, B_0$, by setting B_i to be the connected component of $E_\lambda^{-1}(B_{i+1})$ that contains the point z_i. These inverses are all well defined and the map E_λ sends B_i conformally onto B_{i+1}. Each B_i belongs to the immediate basin $A^*(z_i)$. The following characterization of the sets $B_i, i = 0, \ldots, n - 2$ is proved in [13].

Proposition 6.9. *Let $n > 2$. For $i = 0, \ldots, n - 2$, B_i is a finger of width $c_i < 2\pi$.*

It follows immediately from the above construction that the width of the finger B_{n-2} that is mapped by E_λ conformally onto B_{n-1} is π, while the widths of the other fingers is 0. So we will refer to B_{n-2} as the *big finger*.

We proceed to the final step, by defining the set

$$G = \{z \in \mathbb{C} \mid E_\lambda(z) \in B_0\},$$

which we call the *glove*. We observe from the above construction that G is a connected set and $B_{n-1} \subset G \subset A^*(z_{n-1})$. See Figure 7. Moreover, the complement of G consists of infinitely many fingers, each of which are $2\pi i$

translates of each other. We index these infinitely many connected components by V_j, $j \in \mathbb{Z}$, so that $2\pi i j \in V_j$.

In fact, these V_j form a set of fundamental domains for the Julia set of E_λ in the following sense:

- $J(E_\lambda) \subset \bigcup_{j \in \mathbb{Z}} V_j$;
- E_λ maps each V_j conformally onto $\mathbb{C} \setminus B_0$, and so $E_\lambda(V_j) \supset J(E_\lambda)$.

Hence, for each $j \in \mathbb{Z}$ we have a well defined inverse branch of E_λ:

$$L_j = L_{\lambda,j} : \mathbb{C} \setminus B_0 \longrightarrow V_j.$$

Note that B_0 lies inside V_0 since $0 \in B_0$. The other fingers $B_1, \ldots, B_{n-2}$ may lie inside any of the fundamental domains V_j, depending on the value of λ. In particular, several B_i may lie in the same V_j.

6.3.2. *Kneading sequence and itineraries.* We first introduce the kneading sequence given by the fundamental domains V_j. We define the *kneading sequence* of λ to be

$$K(\lambda) = 0\, k_1\, k_2\, k_3\, \ldots\, k_{n-2}$$

where $B_j \subset V_{k_j}$ for all $0 \leq j \leq n-2$. Notice that there is no entry for B_{n-1}. Observe that, by construction, $k_0 = 0$.

We define the K-*itinerary* of any point $z \in J(E_\lambda)$ to be

$$K(z) = s_0\, s_1\, s_2\, s_3\, \ldots$$

where $E_\lambda^j(z) \in V_{s_j}$ for any $j \geq 0$.

Notice that if γ is a hair of the Julia set, all points in γ must share the same itinerary, since the curve cannot cross the boundaries of the fundamental domains V_j given that these belong to the Fatou set.

One can then use these itineraries together with the kneading sequence to give a complete description of the structure of the Julia set for E_λ in terms of symbolic dynamics. See [13].

6.3.3. *Conclusion of the proof of Proposition 6.8.* We first observe that the connected components of the immediate basin of attraction satisfy $\overline{A * (z_i)} \subset V_j$ where $0 \leq i \leq n-2$ and j is such that $B_i \subset V_j$. In other words, the boundary of each of these components is entirely contained in one and only one of the fundamental domains, more precisely the one whose index provides the corresponding entry in the kneading sequence. The only exception is the connected component $A * (z_{n-1})$, whose boundary intersects all the fundamental domains V_j for all $j \in \mathbb{Z}$. This means that any point z in the boundary of the immediate basin of attraction, say on the boundary of $A * (z_0)$ for example, must have an itinerary equal to

$$K(z) = \overline{0\, k_1\, k_2, \ldots\, k_{n-2} *}$$

where $*$ stands for any integer (which may be different at every round).

On the other hand, any connected component of the Fatou set will eventually map to the immediate basin of attraction. Hence any point on the boundary of a component of the Fatou set must have an itinerary which, after a finite number of entries, ends up exactly as $K(z)$. This concludes the proof of Proposition 6.8.

REFERENCES

[1] Abikoff W., Some remarks on Kleinian groups, *Ann. of Math. Studies* **66** (1971), 1–5.
[2] Abikoff W., The residual limit sets of Kleinian groups, *Acta Math.* **130** (1973), 127–144.
[3] Ahlfors L.V., Sur les domains dans lesquels une fonction méromorphe prend des valeurs appartenant à une region donnée, *Acta Soc. Sci. Fenn. Nova Ser. A. II.* **2** (1933), 1–17.
[4] Baker I.N., Completely invariant domains of entire functions, *Mathematical essays dedicated to A.J. Macintyre*, Ohio University Press, Athens, Ohio (1970), 33–35.
[5] Baker I.N., Wandering domains in the iteration of entire functions, *Proc. London Math. Soc.* **49** (1984), 563–573.
[6] Baker I.N., Kotus J. and Yinian Lü., Iterates of meromorphic functions III: Preperiodic domains, *Ergod. Th. and Dynam. Sys.* **11** (1991), 603–618.
[7] Baker I.N. and Domínguez P., Residual Julia sets, *J. Analysis* **8** (2000), 121–137.
[8] Baker I.N. and Domínguez P., Boundaries of unbounded Fatou components of entire functions, *Ann. Acad. Sci. Fenn. Ser. A. I. Math.* **24** (1999), 437–464.
[9] Barański K. and Karpińska B., Coding trees and boundaries of attracting basins for some entire maps, *Nonlinearity* **20** (2) (2007), 391-415.
[10] Beardon A.F., *Iteration of rational functions*. Graduate Texts in Math. 132, Springer Verlag, New York, 1991.
[11] Beardon A.F., The components of a Julia set, *Ann. Acad. Sci. Fenn. Ser. A. I. Math.* **16** (1991), 173–177.
[12] Bergweiler W., Iteration of meromorphic functions, *Bull. Amer. Math. Soc.* **29** (1993), 151–188.
[13] Bhattacharjee R. and Devaney R.L., Tying hairs for structurally stable exponentials, *Ergod. Th. and Dynam. Sys.* **20** (2000), 1603–1617.
[14] Bhattacharyya P., Iteration of analytic functions. PhD thesis. University of London (1969).
[15] Bodelón C., Devaney R.L., Hayes M., Goldberg L, Hubbard J. and Roberts G., Hairs for the complex exponential family, *International Journal of Bifurcation and Chaos* **9** (1999), 1517–1534.
[16] Carleson L. and Gamelin T.W., *Complex dynamics*. Springer Verlag, New York, 1993.
[17] Devaney R.L., Fagella N. and Jarque X., Hyperbolic components of the exponential family, *Fundamenta Math.* **174** (2002), 193–215.
[18] Devaney R.L., Goldberg L.R. and Hubbard J., A dynamical approximation to the exponential map by polynomials, Preprint (1990).
[19] Devaney R.L, and Krych M., Dynamics of $\exp(z)$, *Ergod. Th. and Dynam. Sys.* **4** (1984), 35–52.
[20] Devaney R.L. and Tangerman F., Dynamics of entire functions near the essential singularity, *Ergod. Th. and Dynam. Sys.* **6** (1986), 489–503.
[21] Domínguez P., Connectedness properties of the Julia sets of transcendental entire functions, *Complex Variables Th. and Appl.* **32** (1997), 199–215.
[22] Domínguez P., Dynamics of transcendental meromorphic functions, *Ann. Acad. Sci. Fenn. Ser. A. I. Math.* **23** (1998), 225–250.

[23] Domínguez P. and Fagella N., Existence of Herman rings for meromorphic functions, *Complex Variables Th. and Appl.* **49** (12) (2004), 851–870.
[24] Domínguez P. and Sienra G., A study of the dynamics of the family $\lambda \sin z$, *International Journal of Bifurcation and Chaos* **12** (2002), 2869–2883.
[25] Eremenko A.E., On the iteration of entire functions, *Dynamical Systems and Ergodic Theory, Banach Center Publications* **23** (1989), 339–345.
[26] Eremenko A.E. and Lyubich M.Yu., Iterates of entire functions, *Soviet Math. Dokl.* **30** (1984), 592–594; translation from Dokl. Akad. Nauk. SSSR **279** (1984).
[27] Eremenko A.E. and Lyubich M.Yu., The dynamics of analytic transformations, *Leningrad Math. J.* **1** (1990), 563–634.
[28] Goldberg L.R. and Keen L., A finiteness theorem for a dynamical class of entire functions, *Ergod. Th. and Dynam. Sys.* **6** (1986), 183–192.
[29] Hua X-H. and Yang C-C., *Dynamics of transcendental functions*. Asian Mathematics Series, Gordon and Breach Science Publishers, 1998.
[30] Kisaka M. and Shishikura M., On multiply connected wandering domains of entire functions, *Transcendental dynamics and complex analysis*, Cambridge University Press, 2008.
[31] McMullen C., *Automorphisms of rational maps, holomorphic functions and modulii I*. MSRI Publications 10, Springer Verlag, New York, 1988.
[32] Milnor J., *Dynamics in one complex variable: introductory lectures*, Vieweg 1999.
[33] Morosawa S., On the residual Julia sets of rational functions, *Ergod. Th. and Dynam. Sys.* **17** (1997), 205–210.
[34] Morosawa S., Nishimura Y., Taniguchi M. and Ueda T., *Holomorphic dynamics*, Cambridge University Press 2000.
[35] Qiao J.Y., The Julia set of the function $z \to z \exp(z + \mu)$, *Chinese Sci. Bulletin* **39** (7) (1994), 529–533.
[36] Qiao J.Y., The buried points on the Julia sets of rational and entire functions, *Sci. China Ser. A* **38** (1995), 1409–1419.
[37] Qiao J.Y., Topological complexity of Julia sets, *Sci. China Ser. A* **40** (1997), 1158–1165.
[38] Rottenfüsser G., Rückert J., Rempe L. and Schleicher D., Dynamic rays of bounded-type entire functions, Preprint.
[39] Schleicher D. and Zimmer J., Escaping points of exponential maps, *J. London Math. Soc. (2)* **67** (2003), 380–400.
[40] Steinmetz N., *Rational iteration, complex analytic dynamical systems*. Walter de Gruyter, Berlin, New York, 1993.
[41] Sullivan D., Quasiconformal homeomorphisms and dynamics I: Solutions of the Fatou-Julia problem on wandering domains, *Ann. of Math.* **122** (3) (1985), 401–418.
[42] Zhang G., On the dynamics of $e^{2\pi i \theta} \sin(z)$, *Illinois J. Math.* **49** (2005), 1171–1179.
[43] Zheng J.H., Ng T.W and Choi Y.Y., Residual Julia sets of meromorphic functions, *Math. Proc. Camb. Phil. Soc.* **141** (2006), 113–126.

F.C. Físico-Matemáticas, B.U.A.P, Av. San Claudio, Col. San Manuel, C.U., Puebla Pue, 72570, México
E-mail address: pdsoto@fcfm.buap.mx

Dept. de Mat. Aplicada i Análisi, Universitat de Barcelona, Gran Via 585, 08007, Barcelona, Spain
E-mail address: fagella@maia.ub.es

BANK-LAINE FUNCTIONS VIA QUASICONFORMAL SURGERY

D. DRASIN AND J. K. LANGLEY

ABSTRACT. Using quasiconformal surgery we construct new examples of entire functions E such that $E(z) = 0$ implies $E'(z) = \pm 1$, these associated with second order linear differential equations with transcendental coefficients. We also extend some previous results on the zero sequences of such functions.

Dedicated to Noel Baker

1. INTRODUCTION

A Bank-Laine function [20, 21] is an entire function E such that $E'(z) = \pm 1$ whenever $E(z) = 0$. These functions arise from differential equations [2]. Indeed, let A be entire, and f_1, f_2 be linearly independent solutions of the equation

(1) $$w'' + A(z)w = 0,$$

normalized so that the Wronskian $W = W(f_1, f_2) = f_1 f_2' - f_1' f_2$ satisfies $W = 1$. Then $E = f_1 f_2$ is a Bank-Laine function and satisfies

(2) $$4A = (E'/E)^2 - 2E''/E - 1/E^2.$$

Conversely, if E is any Bank-Laine function then the function A defined by (2) is entire [4], and E is the product of linearly independent normalized solutions of (1) [2]. Extensive research in recent years has concerned the exponent of convergence $\lambda(f_j)$ of the zeros of solutions f_j, in connection with the order of growth $\rho(A)$ and lower order $\mu(A)$ of the coefficient A, these defined using standard notation from [14] by

$$\rho(A) = \limsup_{r \to \infty} \frac{\log^+ T(r, A)}{\log r},$$

$$\mu(A) = \liminf_{r \to \infty} \frac{\log^+ T(r, A)}{\log r},$$

(3) $$\lambda(f_j) = \limsup_{r \to \infty} \frac{\log^+ N(r, 1/f_j)}{\log r}.$$

1991 *Mathematics Subject Classification.* 30D35.

This research was partly carried out during a visit by the first author to the University of Nottingham, funded by EPSRC. Both authors thank Walter Bergweiler and John Rossi for helpful discussions.

The *Bank-Laine conjecture* asserts that the condition

(4) A transcendental, $\rho(A) < \infty$, $\max\{\lambda(f_1), \lambda(f_2)\} < \infty$

implies that $\rho(A)$ is a positive integer, and this is known to be true [2] under the stronger condition $\max\{\lambda(f_1), \lambda(f_2)\} < \rho(A) < \infty$. It is known further that (4) implies that $\rho(A) > 1/2$ [26, 27] and that E has finite order [2]. Further results, and analogues for higher order equations, may be found in [2, 3, 4, 5, 6, 7, 15, 16, 17, 18, 19, 24].

Note that if E is a Bank-Laine function of finite order then by (2) the associated coefficient function A satisfies

(5) $T(r, A) = m(r, A) = 2m(r, 1/E) + O(\log r).$

In particular, if $\delta(0, E) > 0$ then A is transcendental.

It appears to be relatively difficult to construct Bank-Laine functions of finite order associated via (2) with transcendental coefficient functions A. The simplest [2, 20, 28] are of the following form: given any polynomial P having only simple zeros, Lagrange interpolation gives a non-constant polynomial Q such that Pe^Q is a Bank-Laine function. A second class arises from equations having periodic coefficients [3], leading to Bank-Laine functions of form $E(z) = P(e^{\alpha z})\exp(\beta z)$, with P a polynomial and α, β constants. Of course, all examples arising in this way have $\lambda(E) \leq 1$. We also note that in [20] quasiconformal modifications, applied to the quotient f_1/f_2, were used to convert a Bank-Laine function $f_1(z)f_2(z) = \exp(q(z))\sin \pi z$, with q a polynomial, into a Bank-Laine function $E(z)$ of finite order, with infinitely many zeros and transcendental associated coefficient function A, but having no representation of the form $E(z) = P(e^{\alpha z})\exp(Q(z))$, with P, Q polynomials and α constant. This construction may be modified to give $\rho(E)$ finite but arbitrarily large, but we still do not know whether there are functions so obtained with $\infty > \lambda(E) > 1$.

Another result [21] shows that the zero set of a Bank-Laine function may be very sparse. Indeed, if (c_n) is a positive sequence tending to $+\infty$ then there exists a Bank-Laine function

$$E(z) = e^z \prod_{n=1}^{\infty}(1 - z/\alpha_n),$$

with $|\alpha_n| > c_n$ for each n. Further, $\rho(E) = 1$ and $\lambda(E) = 0$ and E is the product $f_1 f_2$ of normalized linearly independent solutions of an equation (1), with A transcendental, and f_1 has no zeros. These α_n may be chosen close to, but not on, the positive imaginary axis.

In the present paper we employ the quasiconformal surgery method of Shishikura [29] to produce examples with $\lambda(E) = \rho(E)$ finite but arbitrarily large.

Theorem 1.1. *Let n be a positive integer. Then there exists a Bank-Laine function E, associated with a transcendental coefficient function A, such that,* in the notation (1), the order $\rho(E)$, lower order $\mu(E)$, and exponent of convergence $\lambda(E)$ of the zeros of E, satisfy $n \leq \mu(E) \leq \rho(E) = \lambda(E) < \infty$.

The construction of Theorem 1.1 starts from a function f of integer order $n \geq 2$ which satisfies the Bank-Laine condition at all but one of its zeros. Quasiconformal surgery is applied to $z + f(z)$, leading to a Bank-Laine function E which has $\lambda(E) = \rho(E) \geq \mu(E) \geq n$. It is not clear from our methods whether, as seems likely, the order $\rho(E)$ is preserved by this construction (and so would also be n). This is related to the question of how order is transformed by quasiconformal surgery.

We turn next to some results on the zero sequences of Bank-Laine functions. It was observed by Shen [28] that if (a_n) is a complex sequence tending to infinity without repetition, then there exists a Bank-Laine function F with zero sequence (a_n), the construction based on the Mittag-Leffler theorem. A natural question arising from both this observation and the Bank-Laine conjecture is the following: for which sequences (a_n) with finite exponent of convergence does there exist a Bank-Laine function E of finite order with zero sequence (a_n)? A negative answer for sufficiently sparse sequences lying on a line was given in [21]: if L is a straight line in the complex plane and (a_n) is a sequence of pairwise distinct complex numbers, all lying on L, such that $|a_n| \to \infty$ as $n \to \infty$ and

$$(6) \qquad \sum_{a_n \neq 0} |a_n|^{-1} < \infty,$$

then there cannot exist a Bank-Laine function of finite order with zero sequence (a_n).

The obvious example $\sin z$ shows that the condition (6) is not redundant, but we show here that it may be weakened to minimal type, and to mean type if the a_n all lie on a ray.

Theorem 1.2. *Let (a_n) be a sequence of pairwise distinct real numbers, such that $\lim_{n \to \infty} |a_n| = \infty$, and let $n(r)$ be the number of a_n lying in $|z| \leq r$. Assume that at least one of the following holds:*
(i) we have $n(r) = o(r)$ as $r \to \infty$;
(ii) the a_n are all positive, and $n(r) = O(r)$ as $r \to \infty$.
Then there is no Bank-Laine function of finite order with zero sequence (a_n).

It seems likely that if (a_n) is any strictly increasing positive sequence with limit ∞ then there is no Bank-Laine function of finite order with zero sequence (a_n), but the present methods do not suffice to prove this.

2. Lemmas needed for Theorem 1.1

For $a \in \mathbb{C}$ and $r > 0$ we use the standard notation

$$B(a,r) = \{z \in \mathbb{C} : |z-a| < r\}, \quad C(a,r) = \{z \in \mathbb{C} : |z-a| = r\}.$$

Lemma 2.1. *Let ϕ be a K-quasiconformal homeomorphism of the extended plane, fixing 0 and ∞. Then there exist positive constants d_j with the following properties:*
(i) $|z|^{1/d_1} \leq |\phi(z)| \leq |z|^{d_1}$ as $z \to \infty$;
(ii) $|\phi(z) - \phi(w)| \leq |z - w|^{d_2}|w|^{d_3}$ for large z, w with $|z - w| \leq 1$;
(iii) $|\phi(u)| \geq d_4|\phi(v)|$ for large u, v with $|u| = |v|$.

All assertions of Lemma 2.1 are standard [1, 22]. In particular, (iii) may be proved by considering $\psi(z) = \phi(zv)/\phi(v)$, since the K-quasiconformal mappings fixing $0, 1, \infty$ form an equicontinuous family.

The key tool for our construction is Shishikura's main lemma on quasiconformal surgery [8, 29], in the following form.

Lemma 2.2. *Let $g : \mathbb{C} \to \mathbb{C}$ be quasiregular, and let W be a non-empty open subset of $\mathbb{C}$. Assume that g maps W into W, and that $\overline{\partial}g = 0$ a.e. on W and on $\mathbb{C}\setminus g^{-N}(W)$, for some integer $N > 0$. Then there exist an entire function h and a quasiconformal homeomorphism ϕ of the extended plane fixing $0, 1, \infty$, such that $g \equiv \phi^{-1} \circ h \circ \phi$. Further, ϕ is conformal on W and on the interior of $\mathbb{C} \setminus \bigcup_{m=1}^{\infty} g^{-m}(W)$.*

Proof. We sketch the proof from [8, 29]. Set
$$W_0 = W, \quad W_{m+1} = g^{-m-1}(W) \setminus g^{-m}(W) \quad (m = 0, 1, 2, \ldots),$$
and
$$K = \mathbb{C} \setminus \bigcup_{m=0}^{\infty} g^{-m}(W).$$
Define a Beltrami coefficient $\mu(z)$ on $\mathbb{C}$ as follows. For $z \in W_0 \cup K$ we set $\mu = 0$. Assuming that μ has been defined on W_m, we then define μ for $w \in W_{m+1}$ by

(7) $$\mu(w) = \frac{\mu_g(w) + \mu(g(w))A(w)}{1 + \mu(g(w))\overline{\mu_g(w)}A(w)}, \quad A = \frac{\overline{g_w}}{g_w}.$$

In this way, μ is defined inductively a.e. on $\mathbb{C}$ and, since $\mu_g(w) = 0$ a.e. on W_m for $m > N$, we have $|\mu(w)| \leq \kappa < 1$ a.e. We then define ϕ to be the quasiconformal homeomorphism of the extended plane fixing $0, 1, \infty$, and with complex dilatation μ, and (7) gives $\mu_{\phi \circ g} = \mu_\phi$ a.e., so that $\phi \circ g = h \circ \phi$, with h entire. □

3. Proof of Theorem 1.1

Theorem 1.1 is trivially true when $n = 1$: we may take, for example, the function $E(z) = e^{2z} - e^z$; since $\delta(0, E) = \frac{1}{2}$, we see from (5) that the coefficient function A must be transcendental.

Now let $n \geq 2$ be an integer, and set

(8) $$F(z) = z + f(z), \quad f(z) = \frac{e^{-2z^n} - e^{-z^n}}{nz^{n-1}}.$$

Thus f and F are entire, and f has zeros wherever $e^{z^n} = 1$. At such a zero z with $z \neq 0$ we have $f'(z) = -1$. Near 0 we have
$$f(z) \sim \frac{-2z^n + z^n}{nz^{n-1}} = -\frac{z}{n}.$$
It follows that all nonzero fixpoints of F are superattracting, i.e. have multiplier $F'(z) = 0$, while 0 is attracting with $F'(0) = 1 - 1/n \in (0,1)$. The new idea is to perform quasiconformal surgery so that this fixpoint is also superattracting.

Our first lemma of this section follows at once from (8).

Lemma 3.1. *For small positive δ and $j = 0, 1, \ldots, 2n - 1$ let*
$$(9) \qquad \theta_j = \frac{j\pi}{n}, \quad S_{j,\delta} = \{z : z \neq 0, |\arg z - \theta_j| < \frac{\pi}{2n} - \delta\}.$$
Then there exists $c(\delta) > 0$ such that, for large z in $S_{j,\delta}$,
$$\log|f(z)| > c(\delta)|z|^n \quad (j \text{ odd}), \quad \log|f(z)| < -c(\delta)|z|^n \quad (j \text{ even}).$$

Lemma 3.2. *Let λ be an unbounded path in the complex plane. Then F is unbounded on λ.*

Proof. We use a standard argument based on harmonic measure. Choose small positive constants δ and ε. As $z \to \infty$ in $S_{j,\delta}$ we have $F(z)/z \sim 1$ if j is even, and $F(z)/z \to \infty$ if j is odd.

Assume now that λ is an unbounded path on which $F(z)$ is bounded. Then $p(z) = F(z)/z$ is small for large z on λ. Thus there exists, for arbitrarily large positive R, a simple arc $\lambda_R \subseteq \lambda$ joining $C(0, \sqrt{R})$ to $C(0, R^2)$ and, apart from its endpoints, lying in $\sqrt{R} < |z| < R^2$, and with $|p(z)| < \varepsilon$ on λ_R. It is clear from Lemma 3.1 that λ_R must lie in one of the regions between the $S_{j,\delta}$. Since δ is small we may therefore connect $C(0, \sqrt{R})$ to $C(0, R^2)$ by a radial line segment τ_R lying in one of the sets $S_{j,2\delta} \setminus S_{j,4\delta}$ with j even, and such that the angular distance between τ_R and λ_R is at least δ but at most 5δ. Since j is even and R is large we have $|p(z) - 1| < \varepsilon$ on τ_R.

In this way we obtain a domain D_R bounded by λ_R, τ_R and small arcs $U_{\sqrt{R}}$ and V_{R^2} of $C(0, \sqrt{R})$ and $C(0, R^2)$. Since δ is small standard estimates for harmonic measure [31, p.117] now give that
$$\omega(z, D_R, U_{\sqrt{R}} \cup V_{R^2}) < R^{-4n}, \quad z \in D_R, \quad |z| = R,$$
so that since p has order at most n we obtain, provided R is large enough,
$$|p(z)(p(z) - 1)| < 2\varepsilon, \quad z \in D_R, \quad |z| = R.$$
Choosing $z \in D_R$ with $|z| = R$ and $|p(z)| = |p(z) - 1|$ gives an immediate contradiction. □

We apply next some completely standard facts from iteration theory [8, 9, 30]. The attracting fixpoint 0 of F lies in a simply connected component D of the Fatou set of F, with $F(D) \subseteq D$. Let H map D conformally onto

$\Delta = B(0,1)$, with $H(0) = 0$ and $H'(0) > 0$, and define $G = H \circ F \circ H^{-1}$ on Δ. Thus G maps Δ into itself, with $G(0) = 0$ and $G'(0) = 1 - 1/n$. Further, F has at least one critical point z_1 in D. Were this not the case F^{-1} would admit analytic continuation throughout D, since F has no finite asymptotic values and the boundary of D lies in the Julia set of F, and this would imply that F is a conformal self-map of D and G is a conformal self-map of Δ, contradicting the fact that $|G'(0)| < 1$.

The next step of our construction is based on [10, p.106]. Let $w_1 = H(z_1)$ and choose R such that $|w_1| < R < 1$ and such that G has no critical values on $|u| = R$.

Lemma 3.3. *Let V_1 be that component of the set $\{w : |G(w)| < R\}$ which contains 0. Let $U_1 = H^{-1}(V_1)$ and $U_2 = H^{-1}(B(0,R))$, and set $\gamma_j = \partial U_j$. Then the γ_j are disjoint analytic Jordan curves surrounding the origin, such that γ_2 separates γ_1 from 0. Further, F maps U_1 properly onto U_2, and γ_1 onto γ_2, the mapping q-to-one for some $q > 1$.*

Proof. The component V_1 is simply connected by the maximum principle, and therefore so is U_1. Next, $F(U_1)$ is bounded, being a subset of $H^{-1}(B(0,R))$, and so it follows from Lemma 3.2 that U_1 is bounded. Thus the closure of V_1 lies in Δ, and so G maps V_1 properly onto $B(0,R)$ and F maps U_1 properly onto U_2. By Schwarz' lemma V_1 contains a disc $B(0,R')$, $R' > R$. Thus G has at least one critical point in V_1, and the mapping is q-to-one for some $q > 1$. □

Lemma 3.4. *Let $P(z) = H(z)^q$ on $U_2 \cup \gamma_2$. Then $P(z)$ extends to a function continuous on $U_1 \cup \gamma_1$, quasiregular on U_1, with $P(z) \equiv H(F(z))$ on γ_1. Further, P maps U_1 into $B(0,R)$.*

Proof. Let Ω be the doubly connected domain bounded by γ_1 and γ_2, and let $z = s(\zeta)$ map an annulus A given by $S < |\zeta| < T$ conformally onto Ω, with inner boundary mapped to inner boundary. Since $|H(s(\zeta))| \to R$ as $|\zeta| \to S$ it follows from the reflection principle that $(H \circ s)^q$ has an analytic and univalent extension to a neighbourhood of each point of $C(0,S)$, and so has s. Similarly, since $H \circ F$ maps γ_1 onto $C(0,R)$, we see that $H \circ F \circ s$ extends analytically and univalently to a neighbourhood of each point of $C(0,T)$, and again so does s. For real t we may then write

(10) $\quad P(s(Se^{it})) = R_1 \exp(i\psi_1(t)), \quad H(F(s(Te^{it}))) = R_2 \exp(i\psi_2(t)),$

in each which $\psi_j : [-\pi, \pi] \to \mathbb{R}$ is C^1 with positive derivative, and

(11) $\quad R_1 = R^q < R = R_2, \quad \psi_j(\pi) = \psi_j(-\pi) + 2\pi q.$

Using (10) and (11) we readily extend P to the annulus $A = s^{-1}(\Omega)$ as $P(s(\zeta)) = \exp(Q(\zeta))$, in which, for $S \leq r \leq T$ and $-\pi \leq t \leq \pi$,

$$\left(\log \frac{T}{S}\right) Q(re^{it}) = \left(\log \frac{r}{S}\right)(\log R_2 + i\psi_2(t)) - \left(\log \frac{r}{T}\right)(\log R_1 + i\psi_1(t)). \quad (12)$$

A straightforward computation from (12) shows that $Q(\zeta)$ is a locally quasiconformal function of $\log \zeta$, its dilatation uniformly bounded since the ψ_j are C^1. Since $|P(z)| = R$ for $z \in \gamma_1$, the last assertion follows from the maximum principle. Lemma 3.4 is proved. □

Lemma 3.5. *Define $g(z)$ on $\mathbb{C}$ by $g(z) = H^{-1}(P(z))$ for $z \in U_1$, with P as in Lemma 3.4, and set $g(z) = F(z)$ for $z \in \mathbb{C} \setminus U_1$. Then:*
(i) g is analytic on $W = U_2$ and on the exterior of γ_1;
(ii) g is quasiregular on $\mathbb{C}$;
(iii) g maps U_1 into U_2, and the closure of U_1 into U_1;
(iv) $\bar{\partial}g = 0$ on $\mathbb{C} \setminus g^{-2}(W)$;
(v) g has a fixpoint at 0, and no other fixpoint in the closure of U_1;
(vi) if z^* is a fixpoint of g then g is analytic near z^* and z^* is superattracting.

Proof. We first note that (i) is obvious. To prove (ii) we need only note that g is quasiregular on U_1 and analytic outside γ_1, continuous on $\mathbb{C}$, and that γ_1 has two-dimensional Lebesgue measure 0. Next, (iii) holds since P maps U_1 into $B(0, R)$, and (iv) follows, since if $g(g(z_0)) \notin W$ then z_0 lies outside γ_1 and $g(z) = F(z)$ near z_0. Obviously $g(0) = 0$, since $H(0) = 0$, and for z in the closure of U_1 the iterates g_m of g satisfy, using Lemma 3.4,

$$g_2(z) \in U_2, \quad g_{m+2}(z) = H^{-1}(H(g_2(z))^{q^m}) \to H^{-1}(0) = 0,$$

and this proves (v). Assertion (vi) is now obvious. □

In particular, the fixpoint 0, which was only attracting for F in (8), is now superattracting for g. By Lemma 2.2, there exists a quasiconformal homeomorphism ϕ of the extended plane, fixing $0, 1, \infty$, and an entire function h with $g \equiv \phi^{-1} \circ h \circ \phi$. Further, ϕ is conformal on $W = U_2$ and on the interior of $\mathbb{C} \setminus \bigcup_{m=1}^{\infty} g^{-m}(W)$.

Lemma 3.6. *All fixpoints of h are superattracting, and z^* is a fixpoint of g if and only if $\phi(z^*)$ is a fixpoint of h.*

Proof. The second assertion of the lemma is obvious, and it is clear that 0 is a superattracting fixpoint of h, since ϕ is conformal near 0. Next, let $z^* \neq 0$ be a fixpoint of g. By Lemma 3.5, z^* is superattracting and there exists an invariant neighbourhood U^* of z^*, so that in particular $U^* \cap g^{-m}(W) = \emptyset$ for every $m \geq 1$. Thus ϕ is conformal near z^* and so $h'(\phi(z^*)) = 0$. □

Lemma 3.7. *Let $E(z) = h(z) - z$. Then E is a Bank-Laine function of finite order.*

Proof. E is a Bank-Laine function since all fixpoints of h are superattracting, and E has finite order by Lemma 2.1(*i*). □

Lemma 3.8. *There exist $c_0 > 0$ and a family of discs B_k, having finite sum of radii, such that*

(13) $$|E''(z)/E(z)| + |E'(z)/E(z)| \leq |z|^{c_0}, \quad z \notin B = \bigcup_{k=1}^{\infty} B_k.$$

Further, each image $\psi(B_k)$ under the inverse map ψ of ϕ is contained in a disc B_k^*, these discs also having finite sum of radii.

Proof. Since E has finite order we may take $B_k = B(u_k, |u_k|^{-M_1})$ for $k \geq 2$, with M_1 a large positive constant and the u_k zeros of E, and (13) follows from a standard application of the differentiated Poisson-Jensen formula [14, p.22]. The assertion concerning the images $\psi(B_k)$ follows from Lemma 2.1 (*i*) and (*ii*). □

Lemma 3.9. *Let δ be small and positive, and let θ_j and $S_{j,\delta}$, for $j = 0, 1, \ldots, 2n-1$, be as in (9). Then there exists $c > 0$ such that the following hold.*
(i) *If j is odd then*

(14) $$\log |E(z)| \geq |z|^c, \quad z \to \infty, \quad z \in \phi(S_{j,\delta}).$$

(ii) *If j is even then*

(15) $$\log |E(z)| \leq -|z|^c, \quad z \to \infty, \quad z \in \phi(S_{j,\delta}).$$

Further, each $\phi(S_{j,\delta})$ contains a path σ_j tending to infinity and meeting none of the discs B_k of Lemma 3.8, on which

(16) $$(-1)^{j+1} \frac{\log |E(z)|}{\log |z|} \to +\infty$$

as $z \to \infty$.

Proof. We use $c_1, c_2, \ldots$ to denote positive constants. If j is odd then Lemma 3.1 implies that $g(\psi(z)) = F(\psi(z))$ is large for large z in $\phi(S_{j,\delta})$ so that using Lemma 2.1 (*i*) we get

$$\log |h(z)| = \log |\phi(g(\psi(z)))| \geq c_1 \log |F(\psi(z))| \geq c_2 |\psi(z)|^n \geq c_3 |z|^{n/c_4},$$

which gives (14). Next, let j be even. Then $F(\psi(z)) \sim \psi(z)$ for large z in $\phi(S_{j,\delta})$ and applying Lemma 2.1 (*ii*) and Lemma 3.1 we get

$$|g(\psi(z)) - \psi(z)| = |f(\psi(z))| < |\exp(-c(\delta)|\psi(z)|^n) < \exp(-c(\delta)|z|^{n/c_5}),$$

and so

$$|E(z)| = |h(z) - z| \leq |\psi(z)|^{c_6} |g(\psi(z)) - \psi(z)|^{c_7} < \exp(-c_8 |z|^{n/c_9}).$$

Finally, the existence of a suitable path σ_j follows from Lemma 3.8: we need only ensure that $\psi(\sigma_j)$ lies in $S_{j,\delta}$ and avoids the discs B_k^*. □

Since E is a Bank-Laine function, E is the product of linearly independent solutions of (1), in which A is given by (2) and is entire, with $\rho(A) \leq \rho(E)$.

Lemma 3.10. *The coefficient function A is transcendental, and the lower order τ of E is at least n.*

Proof. Take one of the paths σ_j of Lemma 3.9, with j even. Then E is small and A is large as z tends to infinity on σ_j, by (2), (13) and (16). This proves that A is transcendental.

We prove next that E has lower order $\tau \geq n$, the method being essentially that of [26]. By (2) and (13) there exists a positive integer N_1 such that for large z not in the exceptional set B of Lemma 3.8 we have

(17)
$$|E(z)| \geq |z|^{N_1} \Rightarrow |A(z)| \leq |z|^{N_1-1}, \quad |E(z)| \leq |z|^{-N_1} \Rightarrow |A(z)| \geq |z|^{N_1+1}.$$

Choose polynomials P_1, Q_1 of degree less than N_1 such that

$$E_1(z) = z^{-N_1}(E(z) - P_1(z)), \quad A_1(z) = z^{-N_1}(A(z) - Q_1(z))$$

are entire. Lemma 3.9 gives us $2n$ paths σ_j each tending to infinity and avoiding the exceptional set B, and such that $E_1(z) \to 0$ and $A_1(z) \to \infty$ on σ_j, if j is even, and vice versa if j is odd. With N_2 a large positive constant the set $\{z \in \mathbb{C} : |E_1(z)| > N_2\}$ therefore has at least n components $D_1, \ldots, D_n$, and similarly the set $\{z \in \mathbb{C} : |A_1(z)| > N_2\}$ has at least n components $D_{n+1}, \ldots, D_{2n}$. Further, the intersection $D_j \cap D_{j'}, j \neq j'$, lies in the exceptional set B of (13).

Let $\theta_j(t)$ denote the angular measure of the intersection of D_j with the circle $|z| = t > 0$. Since (5) gives

$$\log M(2r, A_1) \leq 3T(4r, A) + O(\log r) \leq 6T(4r, E) + O(\log r),$$

there is a sequence $r_n \to \infty$ with

$$\log M(2r_n, A_1) + \log M(2r_n, E_1) = O(r_n^{\tau + o(1)}), \quad n \to \infty.$$

Applying a standard estimate for harmonic measure [31, p.117] we get

(18)
$$\pi \int_1^{r_n} \frac{dt}{t \theta_j(t)} \leq (\tau + o(1)) \log r_n, \quad n \to \infty.$$

But a standard application of the Cauchy-Schwarz inequality gives

$$4n^2 \leq \sum_{j=1}^{2n} \theta_j(t) \sum_{j=1}^{2n} \frac{1}{\theta_j(t)} \leq (2\pi + o(1)) \sum_{j=1}^{2n} \frac{1}{\theta_j(t)},$$

which with (18) gives $n \leq \tau$. □

Lemma 3.11. *There exists $K > 0$ such that $\log M(r, E) \leq Kn(Kr, 1/E)$ as $r \to \infty$. In particular, $\lambda(E) = \rho(E)$.*

Proof. Let
$$M(r, \phi) = \max\{|\phi(z)| : |z| = r\}, \quad m_0(r, \phi) = \min\{|\phi(z)| : |z| = r\}.$$
We have, using Lemma 2.1 (i),
(19) $\quad \log M(m_0(r, \phi), h) \leq \log M(r, h \circ \phi) = \log M(\phi \circ F) \leq cr^n$

as $r \to \infty$. On the other hand, since $\phi(z^*)$ is a zero of E for every zero z^* of f with z^* large, we get
$$cr^n \leq n(r, 1/f) \leq n(M(r, \phi), 1/E) + c \leq n(cm_0(r, \phi), 1/E) + c,$$
using (8) and Lemma 2.1 (iii), which with (19) proves the lemma. □

This completes the proof of Theorem 1.1.

4. A RESULT NEEDED FOR THEOREM 1.2

Theorem 1.2 will be deduced from the following result.

Theorem 4.1. *Let E be a Bank-Laine function of finite order such that E has infinitely many zeros, all real, and such that E is real on the real axis. Then $\delta(0, E) < 1$.*

The examples e^z and $(1/\pi) \exp(2\pi i z^2) \sin(\pi z)$ show that the assumptions that E is real and has infinitely many zeros are not redundant in Theorem 4.1.

To prove Theorem 4.1, we assume that E satisfies the hypotheses, but that $\delta(0, E) = 1$. We assume further, without loss of generality, that E has infinitely many zeros on the positive real axis. By a theorem of Pfluger [25], ρ is a positive integer. Results from [11, 12, 13] give continuous functions $L(r), L_2(r)$ such that
(20) $\quad L(r) > 0, \quad L(cr) = L(r)(1 + o(1)), \quad L_2(cr) = L_2(r) + o(1)$

as $r \to \infty$, uniformly for $1 \leq c \leq 2$, and such that
(21) $\quad \log|E(re^{i\theta})| = L(r)r^\rho(\cos \rho(\theta - L_2(r))) + o(1)$

uniformly in θ, as $re^{i\theta}$ tends to infinity outside a C^0 set [23], a union U of open discs $B(z_k, r_k)$ of centre z_k and radius r_k such that $\sum_{|z_k|<r} r_k = o(r)$ as $r \to \infty$.

Lemma 4.1. *We may take $L_2(r) \equiv 0$.*

Proof. We may certainly take $L_2(r) \equiv 0$ or $L_2(r) \equiv \pi/\rho$. To see this, just choose a small positive θ_0 such that the rays $\arg z = \pm \theta_0$ have bounded intersection with the C^0 set U. Since E is real on the real axis, we may apply (21) to each of these rays and obtain a contradiction unless $\sin \rho L_2(r) = o(1)$.

If $L_2(r) \equiv \pi/\rho$ then a standard application of (21) and the Phragmén-Lindelöf principle shows that $E(z)$ is small for large z with $|\arg z| < \pi/8\rho$, and so is $E'(z)$. But this contradicts the assumed existence of infinitely many zeros ζ of E on the positive real axis, at which $E'(\zeta) = \pm 1$. □

Lemma 4.2. *Let ε be small and positive. Then*

$$\frac{1}{E(z)} = o(1),$$

$$\frac{E'(z)}{E(z)} = \rho L(r) z^{\rho-1}(1 + o(1)),$$

(22) $$\left(\frac{E'}{E}\right)'(z) = \rho(\rho-1)L(r)z^{\rho-2} + o(L(r)r^{\rho-2}),$$

as $|z| = r \to \infty$, uniformly for $\varepsilon \leq \arg z \leq 2\varepsilon$.

Proof. Let r be large. Since all zeros of E are real, (20) and (21) and Lemma 4.1 give

(23)
$$\log|E(z)| = \operatorname{Re}(L(r)z^\rho) + o(L(r)r^\rho), \quad r/4 \leq |z| \leq 8r, \quad \varepsilon/4 \leq \arg z \leq 8\varepsilon.$$

Set
$$H(z) = \log E(z) - L(r)z^\rho, \quad h(z) = L(r)^{-1}r^{-\rho}H(z).$$

Then (23) and an application of the Borel-Carathéodory inequality give a constant d_r such that

$$h(z) = d_r + o(1), \quad r/2 \leq |z| \leq 4r, \quad \varepsilon/2 \leq \arg z \leq 4\varepsilon.$$

Thus

$$h'(z) = o(r^{-1}), \quad h''(z) = o(r^{-2}), \quad r \leq |z| \leq 2r, \quad \varepsilon \leq \arg z \leq 2\varepsilon,$$

and this gives (22). □

We also obtain from (2) that

$$4A = -(E'/E)^2 - 2(E'/E)' - 1/E^2$$

and (20) and (22) now give, using the fact that E and A are real on the real axis,

(24) $$4A(z) = -\rho^2 L(r)^2 z^{2\rho-2}(1 + o(1)), \quad \varepsilon \leq |\arg z| \leq 2\varepsilon.$$

Lemma 4.3. *Let r be large and positive. Then $A(r) < 0$.*

Proof. Provided ε is small enough, (20), (24) and the Phragmén-Lindelöf principle imply that $A(z) = O(|z|^{2\rho-1})$ for $|\arg z| \leq \varepsilon$. Apply the two constants theorem to the harmonic function $u(z) = \operatorname{Re}(A(z)z^{2-2\rho})$ on the region Ω_r given by $\sqrt{r} < |z| < r^2$, $|\arg z| < \varepsilon$, and denote by c_j positive constants independent of r and ε. Let $L_0(r) = \min\{L(t) : \sqrt{r} \leq t \leq r^2\}$. Then $L_0(r) \geq r^{-o(1)}$ by (20). Now (24) gives

$$u(z) < -c_1 L_0(r)^2, \quad \sqrt{r} \leq |z| \leq r^2, \quad \arg z = \pm\varepsilon.$$

Applying the same standard estimates for harmonic measure [31, p.117] as in Lemma 3.2 shows that

$$A(r)r^{2-2\rho} = \operatorname{Re}\left(A(r)r^{2-2\rho}\right) = u(r) \leq -c_1 L_0(r)^2 + O(r^{4\rho-2}r^{-c_2/\varepsilon}) < 0$$

provided ε is small enough. □

We complete the proof of Theorem 4.1 using Green's transform as in [21]. Write $E = f_1 f_2$, with f_1, f_2 linearly independent solutions of (1). Suppose with no loss of generality that f_1 has infinitely many zeros on the positive x-axis. This leads to a contradiction since

$$\frac{d}{dx}\left(f_1'(x)\overline{f_1(x)}\right) = |f_1'(x)|^2 - A(x)|f_1(x)|^2 \geq 0 \qquad (x \to +\infty).$$

This proves Theorem 4.1.

5. Proof of Theorem 1.2

Assume that (a_n) is as in the statement and that there exists a Bank-Laine function E of finite order with zero sequence (a_n). Then we may write

$$E(z) = \Pi(z)\exp(P(z) + iQ(z))$$

in which Π is a canonical product, and P, Q are real polynomials. Since E is Bank-Laine and has only real zeros, we have $\exp(iQ(z)) = \pm 1$ at every zero z of E, and so may assume that $Q \equiv 0$.

In case (i), we are assuming that $n(r) = o(r)$, and so $N(r, 1/E) = o(r)$ as $r \to \infty$. But every transcendental Bank-Laine function E has [26]

$$\liminf_{r \to \infty} \frac{T(r, E)}{r} > 0,$$

and so $\delta(0, E) = 1$, which in turn contradicts Theorem 4.1.

Now suppose that (a_n) satisfies (ii) but not (i). Then $\sum a_n^{-1}$ diverges, and we may write

$$E(z) = e^{P(z)}\Pi(z), \quad \Pi(z) = \prod_{k=1}^{\infty}(1 - z/a_k)e^{z/a_k},$$

again with P a real polynomial. We may also assume that P has degree at most 1, for otherwise a contradiction arises on applying Theorem 4.1. Now [14, p.29]

$$\log|\Pi(-r)| = \int_0^\infty \left(\frac{-r}{t} + \log\left(1 + \frac{r}{t}\right)\right)dn(t) = -r^2\int_0^\infty \frac{n(t)}{t^2(r+t)}dt.$$

But, using [14, p.25],

$$r\int_0^\infty \frac{n(t)}{t^2(r+t)}dt \geq \frac{1}{2}\int_0^r \frac{n(t)}{t^2}dt \to \infty$$

as $r \to \infty$, since $\sum a_n^{-1}$ diverges by assumption. It follows at once that

$$\lim_{r \to \infty}\frac{T(r, E)}{r} = \lim_{r \to \infty}\frac{T(r, \Pi)}{r} = \infty$$

and again we get $\delta(0, E) = 1$, so that applying Theorem 4.1 gives a contradiction.

References

[1] L.V. Ahlfors, Lectures on quasiconformal mappings, Van Nostrand, Toronto, New York, London 1966.
[2] S. Bank and I. Laine, On the oscillation theory of $f'' + Af = 0$ where A is entire, Trans. Amer. Math. Soc. 273 (1982), 351-363.
[3] S. Bank and I. Laine, Representations of solutions of periodic second order linear differential equations, J. reine angew. Math. 344 (1983), 1-21.
[4] S. Bank and I. Laine, On the zeros of meromorphic solutions of second-order linear differential equations, Comment. Math. Helv. 58 (1983), 656-677.
[5] S. Bank, I. Laine and J.K. Langley, On the frequency of zeros of solutions of second order linear differential equations, Result. Math. 10 (1986), 8-24.
[6] S. Bank, I. Laine and J.K. Langley, Oscillation results for solutions of linear differential equations in the complex domain, Result. Math. 16 (1989), 3-15.
[7] S. Bank and J.K. Langley, Oscillation theory for higher order linear differential equations with entire coefficients, Complex Variables 16 (1991), 163-175.
[8] A. F. Beardon, Iteration of rational functions, Springer, New York, Berlin, Heidelberg, 1991.
[9] W. Bergweiler, Iteration of meromorphic functions, Bull. Amer. Math. Soc. 29 (1993), 151-188.
[10] L. Carleson and T.W. Gamelin, Complex dynamics, Springer, New York, Berlin, Heidelberg, 1993.
[11] D. Drasin, Proof of a conjecture of F. Nevanlinna concerning functions which have deficiency sum two, Acta. Math. 158 (1987), 1-94.
[12] A. Edrei and W.H.J. Fuchs, Valeurs déficientes et valeurs asymptotiques des fonctions méromorphes, Comment. Math. Helv. 33 (1959), 258-295.
[13] A. Eremenko, Meromorphic functions with small ramification, Indiana Univ. Math. Journal 42, no. 4 (1994), 1193-1218.
[14] W.K. Hayman, Meromorphic functions, Oxford at the Clarendon Press, 1964.
[15] S. Hellerstein, J. Miles and J. Rossi, On the growth of solutions of $f'' + gf' + hf = 0$, Trans. Amer. Math. Soc. 324 (1991), 693-706.
[16] S. Hellerstein, J. Miles and J. Rossi, On the growth of solutions of certain linear differential equations, Ann. Acad. Sci. Fenn. Ser. A. I. Math. 17 (1992), 343-365.
[17] I. Laine, Nevanlinna theory and complex differential equations, de Gruyter Studies in Math. 15, Walter de Gruyter, Berlin/New York 1993.
[18] J.K. Langley, Some oscillation theorems for higher order linear differential equations with entire coefficients of small growth, Result. Math. 20 (1991), 517-529.
[19] J.K. Langley, On entire solutions of linear differential equations with one dominant coefficient, Analysis 15 (1995), 187-204.
[20] J.K. Langley, Quasiconformal modifications and Bank-Laine functions, Archiv der Math. 71 (1998), 233-239.
[21] J.K. Langley, Bank-Laine functions with sparse zeros, Proc. Amer. Math. Soc. 129 (2001), 1969-1978.
[22] O. Lehto and K. Virtanen, Quasiconformal mappings in the plane, 2nd edn., Springer, Berlin, 1973.
[23] B.Ja. Levin, Distribution of zeros of entire functions, Amer. Math. Soc., Providence RI, 1980.
[24] J. Miles and J. Rossi, Linear combinations of logarithmic derivatives of entire functions with applications to differential equations, Pacific J. Math. 174 (1996), 195-214.
[25] A. Pfluger, Zur Defektrelation ganzer Funktionen endlicher Ordnung, Comment. Math. Helv. 19 (1946), 91-104.

[26] J. Rossi, Second order differential equations with transcendental coefficients, Proc. Amer. Math. Soc. 97 (1986), 61-66.
[27] L.C. Shen, Solution to a problem of S. Bank regarding the exponent of convergence of the solutions of a differential equation $f'' + Af = 0$, Kexue Tongbao 30 (1985), 1581-1585.
[28] L.C. Shen, Construction of a differential equation $y'' + Ay = 0$ with solutions having prescribed zeros, Proc. Amer. Math. Soc. 95 (1985), 544-546.
[29] M. Shishikura, On the quasi-conformal surgery of rational functions, Ann. Sci. École Norm. Sup (4) 20 (1987), 1-29.
[30] N. Steinmetz, Rational iteration, de Gruyter Studies in Mathematics 16, Walter de Gruyter, Berlin/New York, 1993.
[31] M. Tsuji, Potential theory in modern function theory, Maruzen, Tokyo, 1959.

DEPARTMENT OF MATHEMATICS, PURDUE UNIVERSITY, WEST LAFAYETTE 47907 USA
E-mail address: drasin@math.purdue.edu

SCHOOL OF MATHEMATICAL SCIENCES, UNIVERSITY OF NOTTINGHAM, NG7 2RD, UK
E-mail address: jkl@maths.nott.ac.uk

GENERALISATIONS OF UNIFORMLY NORMAL FAMILIES

W.K. HAYMAN AND A. HINKKANEN

In memory of Noel Baker

ABSTRACT. We obtain separation and growth results for meromorphic functions f in the unit disk such that for some positive integer k, $f \neq 0$ and $f^{(k)} \neq 1$ in the disk, or such that $f'f^k \neq 1$ in the disk. These results are only slightly weaker than those for functions f such that $|f| > \delta$ in the disk.

1. INTRODUCTION

In this paper, we prove the following result. We denote the unit disk by $\mathbb{D}$.

Theorem 1. *Suppose that f is analytic in $\mathbb{D}$ and that there exist positive functions $\delta(r)$ and $\lambda(r)$ defined in $(0,1)$ such that $\delta(r)$ decreases, $0 < \delta(r) < 1$, $\lambda(r)$ increases, and for all r, $0 < r < 1$, the conditions*

(1) $$|z_1| \leq r, \quad |z_2| \leq r,$$

and

(2) $$|f(z_1)| \leq e^{-\lambda(r)}, \quad |f(z_2)| \geq e^{\lambda(r)}$$

imply that

(3) $$\left| \frac{z_2 - z_1}{1 - \overline{z_1} z_2} \right| \geq \delta(r).$$

Then if $|z_0| < r$ and

(4) $$\log |f(z_0)| > \frac{8\lambda(r)}{\delta(r)},$$

we have

(5) $$|f'(z_0)| < \frac{2r|f(z_0)|}{r^2 - |z_0|^2} \left\{ \log |f(z_0)| + \frac{8\lambda(r)}{\delta(r)} \right\}.$$

If $\lambda(r)$ and $\delta(r)$ are constant, then the functions f satisfying the hypotheses of Theorem 1 form a uniformly normal family in the sense of [1].

AMS (2000) *Mathematics Subject Classification.* Primary 30D45. Secondary 30D35.
This material is based upon work supported by the National Science Foundation under Grant No. 0200752.

In most of the applications we have

(6) $$\frac{8\lambda(r)}{\delta(r)} = a + b\log\frac{1}{1-r},$$

where a and b are non-negative constants.

Corollary 1. *Let f, $\lambda(r)$ and $\delta(r)$ satisfy the hypotheses of Theorem 1, and also (6). If $|z| = t$ and $r = 1 - \frac{1}{2}(1-t)^2$, then either $\log|f(z)| \leq \lambda^*(t)$ or*

(7) $$\frac{|f'(z)|}{|f(z)|} < \left\{\frac{2}{1-t^2} + 8\right\}\{\log|f(z)| + \lambda^*(t)\},$$

where $\lambda^*(t) = a + b\log 2 + 2b\log\frac{1}{1-t} = a + b\log\frac{1}{1-r}$. Further we have in all cases

(8) $$\log|f(z)| \leq e^8\left\{3a + 8b + \log^+|f(0)|\right\}\frac{1+|z|}{1-|z|}.$$

2. A SPECIAL CASE

In this section we deal with the case when $\delta(r)$ and $\lambda(r)$ are constant.

Lemma 1. *Suppose that $\lambda(r) = \lambda$ and $\delta(r) = \delta$, where λ and δ are constants which satisfy the hypotheses of Theorem 1, and that (4) holds so that*

(9) $$\alpha = \log|f(z_0)| > \frac{8\lambda}{\delta}.$$

Then

(10) $$|f(z)| > e^\lambda, \quad \text{when } \left|\frac{z-z_0}{1-\overline{z_0}z}\right| < r_1,$$

where

(11) $$1 - r_1 = \frac{4\lambda}{\alpha\delta} < \frac{1}{2}.$$

Thus

(12) $$|f'(z_0)| < \frac{2|f(z_0)|}{1-|z_0|^2}\left\{\log|f(z_0)| + \frac{8\lambda}{\delta}\right\}.$$

We proceed to prove (10) subject to (11). Since the hypotheses (1) to (4) are now conformally invariant, we assume without loss of generality that $z_0 = 0$. Let r_2 be maximal subject to $|f(z)| > e^{-\lambda}$ if $|z| < r_2$.

We suppose first that $r_2 < 1$. Let r_1 be the largest number such that

(13) $$|f(z)| > e^\lambda \quad \text{for } |z| < r_1.$$

Then $0 < r_1 < r_2 < 1$ and there exists $z_2 = r_2 e^{i\theta}$ such that
$$|f(z_2)| = e^{-\lambda}.$$
We set $z_1 = r_1 e^{i\theta}$. Then (3) yields
$$\left|\frac{z_2 - z_1}{1 - \bar{z}_1 z_2}\right| = \frac{r_2 - r_1}{1 - r_1 r_2} \geq \delta,$$
i.e.,

(14)
$$r_2 \geq \frac{r_1 + \delta}{1 + \delta r_1}.$$

We now apply Harnack's inequality to $\log|\phi|$ where
$$\phi(z) = e^\lambda f(r_2 z)$$
and choose z so that $|z| = r_1/r_2$ and $|f(r_2 z)| = e^\lambda$. This is possible since r_1 is maximal subject to (13). Then $|\phi(z)| > 1$ for $|z| < 1$, and so

(15)
$$\log|\phi(z)| \geq \frac{1 - |z|}{1 + |z|} \log|\phi(0)|,$$

i.e.,

(16)
$$2\lambda \geq \frac{r_2 - r_1}{r_2 + r_1}(\lambda + \alpha).$$

If $r_2 = 1$ and r_1 is maximal subject to (13), then either $r_1 = 1$, in which case (14) and (16) are trivial, or we can choose z, such that $|z| = r_1$ and $|f(z_1)| = e^\lambda$. Now (15) still holds and we obtain (16) as before with $r_2 = 1$. Thus (16) is always true. Also (14) is true if $r_2 = 1$ and $r_1 < 1$, since $\delta < 1$. Thus (14) and (16) always hold.

We substitute r_2 from (14) in (16) and obtain
$$\frac{2\lambda}{\lambda + \alpha} \geq \left(\frac{r_1 + \delta}{1 + \delta r_1} - r_1\right) \bigg/ \left(\frac{r_1 + \delta}{1 + \delta r_1} + r_1\right) = \frac{\delta(1 - r_1^2)}{\delta(1 + r_1^2) + 2r_1}$$
$$\geq \frac{\delta(1 - r_1^2)}{(1 + r_1)^2} = \delta \frac{1 - r_1}{1 + r_1} \geq \frac{\delta}{2}(1 - r_1).$$

Thus
$$1 - r_1 \leq \frac{4\lambda}{\delta(\lambda + \alpha)} \leq \frac{4\lambda}{\delta \alpha} < \frac{1}{2}$$

by (9). This proves (10) subject to (11).

The function
$$\Psi(z) = e^{-\lambda} f(r_1 z)$$
satisfies $|\Psi(z)| > 1$ if $|z| < 1$. Thus by Borel's inequality
$$|\Psi'(0)| \leq 2|\Psi(0)| \log|\Psi(0)|,$$

i.e.,

$$\frac{|f'(0)|}{|f(0)|} \leq \frac{2}{r_1}(\log|f(0)| - \lambda) = \frac{2}{r_1}(\alpha - \lambda) \leq 2(\alpha - \lambda)\frac{1}{1 - \frac{4\lambda}{\delta\alpha}}$$
$$= 2(\alpha - \lambda) + 2(\alpha - \lambda)\frac{4\lambda}{\delta\alpha - 4\lambda}$$
$$< 2\alpha + \frac{16\alpha\lambda}{\delta\alpha} = 2\left(\alpha + \frac{8\lambda}{\delta}\right)$$

since $\delta\alpha - 4\lambda > \delta\alpha/2$ by (11). This proves (12) if $z_0 = 0$. If $z_0 \neq 0$, we set $F(z) = f((z + z_0)/(1 + \overline{z_0}z))$ and apply the above result to F instead of f at the origin. This yields (12) in general.

3. Proof of Theorem 1

To prove Theorem 1, we fix r with $0 < r < 1$, and apply Lemma 1 with $F(z) = f(rz)$ instead of $f(z)$ and with $\lambda = \lambda(r)$ and $\delta = \delta(r)$.

We write $z_1 = rZ_1$, $z_2 = rZ_2$ and suppose that (1) and (2) imply (3). Thus if $|Z_1| < 1$, $|Z_2| < 1$ and

(17) $\qquad |F(Z_1)| \leq e^{-\lambda}, \quad |F(Z_2)| \geq e^{\lambda},$

we deduce that

(18) $\qquad \left|\dfrac{r(Z_2 - Z_1)}{1 - r^2\overline{Z_1}Z_2}\right| \geq \delta,$

with $\delta = \delta(r)$ and $\lambda = \lambda(r)$.

We next prove that

(19) $\qquad \dfrac{|Z_2 - Z_1|}{|1 - \overline{Z_1}Z_2|} > \dfrac{r|Z_2 - Z_1|}{|1 - r^2\overline{Z_1}Z_2|}.$

To see this, note that $Z_1 \neq Z_2$ and

$$|1 - r^2\overline{Z_1}Z_2|^2 - r^2|1 - \overline{Z_1}Z_2|^2 = (1 - r^2)(1 - r^2|Z_1|^2|Z_2|^2) > 0,$$

which yields (19). Thus (18) implies that

(20) $\qquad \left|\dfrac{Z_2 - Z_1}{1 - \overline{Z_1}Z_2}\right| > \delta.$

So we can apply Lemma 1 to $F(z) = f(rz)$ instead of $f(z)$, and with $Z_0 = z_0/r$ instead of z_0. This yields (5).

We next prove Corollary 1. We have, with the notation of Corollary 1,

$$\frac{1}{r^2 - t^2} - \frac{1}{1 - t^2} = \frac{(1-r)(1+r)}{(r-t)(r+t)(1-t)(1+t)} \leq \frac{(1-t)^2}{\frac{1}{2}(1-t)^2 \cdot \frac{1}{2}} = 4,$$

since $r = 1 - \frac{1}{2}(1-t)^2 \geq \frac{1}{2}$ and $r - t \geq \frac{1}{2}(1-t)$.

Thus (5) yields, with $\alpha = \log|f(z_0)|$,

$$|f'(z_0)| < \frac{2|f(z_0)|}{r^2 - t^2}\{\alpha + \lambda^*(t)\} < 2|f(z_0)|\left\{\frac{1}{1-t^2} + 4\right\}\{\alpha + \lambda^*(t)\}.$$

This proves (7).

To deduce (8), we fix θ with $\theta \in [0, 2\pi)$ and write $y(t) = \log|f(te^{i\theta})|$ for $0 \le t < 1$. If $y(t) \le \lambda^*(t)$ or if $t = 0$, (8) clearly holds for $z = te^{i\theta}$. So we suppose that for some t with $0 < t < 1$, we have $y(t) > \lambda^*(t)$.

We choose t_0 to be maximal, subject to $0 \le t_0 < t$ and $y(t_0) \le \lambda^*(t_0)$. If $y(\tau) > \lambda^*(\tau)$ for all τ with $0 \le \tau < t$, we set $t_0 = 0$. Then

$$y(\tau) > \lambda^*(\tau) \quad \text{for } t_0 < \tau < t.$$

Thus we can apply (7) with τ instead of t in this range and obtain

$$y'(\tau) - \left\{\frac{2}{1-\tau^2} + 8\right\}y(\tau) < \left\{\frac{2}{1-\tau^2} + 8\right\}\lambda^*(\tau), \qquad \text{for } t_0 < \tau < t.$$

Multiplying by

$$P(\tau) = e^{-8\tau}\frac{1-\tau}{1+\tau}$$

and integrating with respect to τ from t_0 to t we obtain

(21) $$y(t) \le \frac{1}{P(t)}\left\{y(t_0)P(t_0) + \int_{t_0}^t \left(\frac{2}{1-\tau^2} + 8\right)P(\tau)\lambda^*(\tau)\,d\tau\right\}.$$

If $t_0 = 0$, we get $y(t_0)P(t_0) = y(0) \le \log^+|f(0)|$.

If $t_0 > 0$, we have $y(t_0)P(t_0) = \lambda^*(t_0)P(t_0)$.

We write

$$\lambda^*(t) = a^* + b^*\log\frac{1}{1-t}$$

with $a^* = a + b\log 2$ and $b^* = 2b$. Then

$$\lambda^*(t)P(t) = \left(a^* + b^*\log\frac{1}{1-t}\right)\frac{1-t}{1+t}e^{-8t}$$

$$\le a^* + b^*\sup_{0 \le t < 1}(1-t)\log\frac{1}{1-t} = a^* + \frac{b^*}{e}.$$

Again

$$\int_0^1 \left(\frac{2}{1-t^2} + 8\right)\frac{1-t}{1+t}e^{-8t}\,dt < \int_0^1 10e^{-8t}\,dt < \frac{5}{4}$$

while

$$\int_0^1 \left(\frac{2}{1-t^2} + 8\right)\left(\frac{1-t}{1+t}\log\frac{1}{1-t}\right)e^{-8t}\,dt$$

$$< 2\int_0^1 \log\frac{1}{1-t}\,dt + \frac{8}{e}\int_0^1 e^{-8t}\,dt < 2 + \frac{1}{e}.$$

Thus
$$\int_0^1 \left(\frac{2}{1-t^2}+8\right) P(t)\lambda^*(t)\,dt \le \frac{5}{4}a^* + \left(2+\frac{1}{e}\right) b^*.$$

Hence (21) yields finally
$$y(t) \le \frac{1}{P(t)}\left\{a^* + \frac{b^*}{e} + \frac{5}{4}a^* + \left(2+\frac{1}{e}\right)b^* + \log^+|f(0)|\right\}$$
$$\le \frac{1}{P(t)}\left\{3a + 8b + \log^+|f(0)|\right\}$$

which implies (8).

This proves Corollary 1.

4. Applications

Suppose that M_k and A_k are respectively the families of meromorphic functions f in $\mathbb{D}$ and analytic functions f in $\mathbb{D}$, such that $f(z) \ne 0$ and $f^{(k)}(z) \ne 1$ for all $z \in \mathbb{D}$. Then M_k and in particular A_k are normal families in $\mathbb{D}$ when $k \ge 1$, see, e.g., [3, Corollary 4.5.9, p. 150]. It follows that there exist positive constants λ_k and δ_k depending only on k such that if $f \in M_k$ and $|z_j| \le \delta_k$, for $j = 1, 2$, then we cannot have $|f(z_1)| \le e^{-\lambda_k}$ and $|f(z_2)| \ge e^{\lambda_k}$. In fact the conclusion must hold for every sufficiently small δ_k, and λ_k depending on δ_k and k only.

We apply the above conclusion with $|z_1| < 1$ and

(22) $$F(z) = (1-|z_1|)^{-k} f(z_1 + (1-|z_1|)z)$$

instead of $f(z)$. Then clearly F belongs to M_k or A_k if f does.

We deduce the following result.

Corollary 2. *If k is a positive integer and $f \in M_k$, then $1/f$ satisfies the hypotheses and hence the conclusion of Theorem 1 and Corollary 1 with $\delta(r) = \frac{1}{3}\delta_k$ for $0 < r < 1$, and*

(23) $$\frac{8\lambda(r)}{\delta(r)} = \frac{24}{\delta_k}\left\{\lambda_k + k\log\frac{1}{1-r}\right\},$$

(24) $$\lambda^*(t) = \frac{24}{\delta_k}\left\{\lambda_k + k\log 2 + 2k\log\frac{1}{1-t}\right\},$$

where δ_k and λ_k are positive constants depending only on k. If f is also analytic then f satisfies the same conclusions.

We suppose that $|z_1| \leq |z_2| = r$ and that $f \in M_k$. We consider
$$F(Z) = \frac{1}{(1-|z_1|)^k} f(z_1 + (1-|z_1|)Z).$$
Then clearly $F(Z) \neq 0$ and $F^{(k)}(Z) = f^{(k)}(z_1 + (1-|z_1|)Z) \neq 1$ when $Z \in \mathbb{D}$, so that $F \in M_k$. In particular if $Z_1 = 0$ and $Z_2 = (z_2 - z_1)/(1-|z_1|)$ and

(25) $$|Z_2| < \delta_k$$

we cannot have

(26) $$|F(Z_j)| \leq e^{-\lambda_k}, \qquad |F(Z_{j'})| \geq e^{\lambda_k},$$

where (j, j') is a permutation of $(1, 2)$.

Returning to f, we see that

(27) $$|f(z_j)| < e^{-\lambda_k}(1-|z_1|)^k, \quad \text{and} \quad |f(z_{j'})| > e^{\lambda_k}(1-|z_1|)^k$$

imply that

(28) $$|z_{j'} - z_j| > \delta_k(1-|z_1|),$$

i.e.,

(29) $$\frac{|z_2 - z_1|}{1 - |z_1|} > \delta_k.$$

Now if $\eta = |z_2 - z_1|/(1-|z_1|) \leq 1$, we have
$$\begin{aligned} |1 - \overline{z_1}z_2| &= |1 - \overline{z_1}z_1 + \overline{z_1}z_1 - \overline{z_1}z_2| \leq 1 - |z_1|^2 + |z_2 - z_1| \\ &= (1-|z_1|)(1+|z_1|+\eta) < 3(1-|z_1|). \end{aligned}$$
Thus if $|z_2 - z_1| \leq 1 - |z_1|$, we have

(30) $$\left|\frac{z_2 - z_1}{1 - \overline{z_1}z_2}\right| > \frac{|z_2 - z_1|}{3(1-|z_1|)} > \frac{\delta_k}{3}$$

if (29) holds. We have assumed that $|z_2 - z_1| \leq 1 - |z_1|$. But if this is false, (30) is still true by the maximum principle. Thus (28) always implies (30), and so does
$$|f(z_j)| < (1-r)^k e^{-\lambda_k}, \qquad |f(z_{j'})| > (1-r)^{-k} e^{\lambda_k}$$
since this implies (27) and hence (28), because $r = \max\{|z_j|, |z_{j'}|\}$. This proves Corollary 2.

We note that the hypotheses (1) to (3) in Theorem 1 are the same for f and $1/f$. Hence if f satisfies (1) to (3) and $1/f$ is analytic, then $1/f$ satisfies the conclusions of Theorem 1. In particular if $f \in M_k$, so that $f \neq 0$ in $\mathbb{D}$, we can apply the conclusion of Theorem 1 to $1/f$ instead of f, with $\lambda(r)$ given by (23) and hence $\lambda^*(r)$ by (24).

As another example we have the following result. We denote by $\mathcal{M}_k$ and $\mathcal{A}_k$ respectively the family of meromorphic functions f and analytic functions f

in $\mathbb{D}$ such that $f'(z)f(z)^k \neq 1$ in $\mathbb{D}$. By [3, Theorem 4.4.18, p. 132], the family $\mathcal{A}_k$ is normal. For the result that also the family $\mathcal{M}_k$ is normal we refer the reader to [4, p. 226] and the references given therein.

Corollary 3. *If k is a positive integer and $f \in \mathcal{A}_k$, then f satisfies the hypotheses and hence the conclusion of Theorem 1 and Corollary 1 with $\delta(r) = \eta_k$, and*
$$\lambda(r) = \mu_k + \frac{1}{k+1} \log \frac{1}{1-r},$$
where η_k and μ_k are positive constants depending only on k.

We write $F(z) = (1 - |z_1|)^{-1/(k+1)} f(z_1 + (1 - |z_1|)z)$ and proceed as in the proof of Corollary 2.

The functions in M_k and those in $\mathcal{M}_k$ also have locally bounded characteristic. This follows directly from Theorem 1 and Theorem 3 of [2]. In fact if
$$\mu(r) = \sup_{|z| \leq r} f^\#(z) = \sup_{|z| \leq r} \frac{|f'(z)|}{1 + |f(z)|^2},$$
then even for meromorphic f we have under the hypotheses of Corollaries 2 and 3 that
$$\mu(r) = O\left(\frac{1}{1-r}\right) \log \frac{1}{1-r},$$
so that
$$\int_0^1 (1-r)^{1/2} \mu(r)\, dr < \infty,$$
which is the hypothesis of Theorem 3 in [2].

For the definition of locally bounded characteristic and other applications we refer the reader to [2].

References

[1] W.K. Hayman, Uniformly normal families, pp. 199–212 in Lectures on functions of a complex variable, The University of Michigan Press, Ann Arbor, 1955.
[2] W.K. Hayman and Ch. Pommerenke, On meromorphic functions with growth conditions, pp. 61–80 in Complex Analysis, Birkhäuser, Basel, 1988.
[3] Joel Schiff, Normal families, Springer, New York, 1993.
[4] Lawrence Zalcman, Normal families: new perspectives, Bull. Amer. Math. Soc. (N.S.) 35 (1998), 215–230.

DEPARTMENT OF MATHEMATICS, IMPERIAL COLLEGE, LONDON SW7 2BZ, U.K.

UNIVERSITY OF ILLINOIS AT URBANA–CHAMPAIGN, DEPARTMENT OF MATHEMATICS, 1409 WEST GREEN STREET, URBANA, IL 61801 USA

E-mail address: aimo@uiuc.edu

ENTIRE FUNCTIONS WITH BOUNDED FATOU COMPONENTS

AIMO HINKKANEN

In memory of Noel Baker

ABSTRACT. Starting with the work of I.N. Baker that appeared in 1981, many authors have studied the question of under what circumstances every component of the Fatou set of a transcendental entire function must be bounded. In particular, such functions have no domains now known as Baker domains, and no completely invariant domains. There may be wandering domains but not the familiar and more easily constructed unbounded ones that often appear for functions defined by simple explicit formulas.

Two types of criteria are involved in the partial answers obtained for this question: the order of growth, and the regularity of the growth of the function.

Baker himself showed that a function of sufficiently slow growth has only bounded Fatou components and noted that order 1/2 minimal type is the best condition one could hope for. Subsequently his results have been extended to order less than 1/2 except for wandering domains, and also to wandering domains if the growth satisfies, in addition, a mild regularity condition. Similar results have been obtained also for certain functions of faster growth provided that the growth is sufficiently regular.

In this paper we review the results achieved and the methods involved in this area.

1. INTRODUCTION

Let f be a transcendental entire function. We consider the question, initiated by I.N. Baker in 1981, of under what circumstances all the components of the Fatou set of f are bounded. The conjecture that this is the case whenever the growth of f does not exceed order 1/2, minimal type, is still open even though a lot of progress has been made.

We first review some definitions and notations. We write $f^1 = f$, and $f^n = f \circ f^{n-1}$ for $n \geq 2$ for the iterates of f. The Fatou set or the set of normality $\mathcal{F}(f)$ of f consists of all z in the complex plane $\mathbb{C}$ that have a neighbourhood U such that the family $\{f^n|U : n \geq 1\}$ of the restrictions

---

AMS (2000) *Mathematics Subject Classification.* Primary 30D45. Secondary 37F50.
This material is based upon work supported by the National Science Foundation under Grants No. 0200752 and 0457291.

of the iterates of f to U is a normal family. The Julia set $\mathcal{J}(f)$ of f is $\mathcal{J}(f) = \mathbb{C} \setminus \mathcal{F}(f)$. The set $\mathcal{F}(f)$ is trivially open, while by the results of Fatou and of Julia, $\mathcal{J}(f)$ is a non-empty perfect set which coincides with $\mathbb{C}$, or is nowhere dense in $\mathbb{C}$. For the fundamental results in the iteration theory of rational and entire functions we refer to the original papers of Fatou [12, 13, 14] and of Julia [19] and to the books of Beardon [7], Carleson and Gamelin [10], Milnor [20], and Steinmetz [23]. Fatou's paper [14] was the first to address the dynamics of transcendental entire functions.

We say that a set E is (forward) *invariant* under f if $f(E) \subset E$, and *backward invariant* under f if $f^{-1}(E) \subset E$. We say that E is *completely invariant* under f if E is both invariant and backward invariant under f. Each of the sets $\mathcal{F}(f)$ and $\mathcal{J}(f)$ is completely invariant under f.

We next consider the components of $\mathcal{F}(f)$. Let D be a component of $\mathcal{F}(f)$. Then $f^n(D) \subset D_n$ where D_n is a component of $\mathcal{F}(f)$. Incidentally, equality holds if f is rational, and Herring [16] has shown that if f is entire, then $D_n \setminus f^n(D)$ contains at most one point. If all the domains D_n are disjoint, then D is called a *wandering domain*. If there is a smallest positive integer p such that $D_p = D$ then D is *periodic* of period p. In particular, if $p = 1$, then D is called *invariant*. Otherwise, D is called *preperiodic*, and then $D_n = D_{n+p}$ for some $n, p \geq 1$, while for all $p \geq 1$ we have $D_p \neq D$.

We have $\mathcal{F}(f^p) = \mathcal{F}(f)$ for all $p \geq 1$. Hence a periodic domain of period p is an invariant domain for f^p. There is a standard *classification of invariant components* D of $\mathcal{F}(f)$ into the following five types.

There may be a fixed point α of f in D such that $\lim_{n \to \infty} f^n(z) = \alpha$, locally uniformly for $z \in D$. Then the *multiplier* $\lambda = f'(\alpha)$ satisfies $|\lambda| < 1$. If $\lambda = 0$, we call α a superattracting fixed point of f and call D a *superattracting* domain. If $0 < |\lambda| < 1$, we call α an attracting fixed point of f and call D an *attracting* domain.

There may be a fixed point α of f on ∂D such that $\lim_{n \to \infty} f^n(z) = \alpha$, locally uniformly for $z \in D$. Then $f'(\alpha) = 1$. We call α a parabolic fixed point of f and call D a *parabolic* domain.

It may be that D is simply connected, containing a fixed point α of f, such that if φ is a conformal mapping of D onto the unit disk with $\varphi(\alpha) = 0$, then $(\varphi \circ f \circ \varphi^{-1})(z) \equiv e^{2\pi i \beta} z$ for some real irrational number β. In this case D is called a *Siegel disk* with centrum α.

Finally, it may be that $\lim_{n \to \infty} f^n(z) = \infty$, locally uniformly for $z \in D$. Then D is called a *Baker domain*.

We note that this classification differs from that for rational functions in two ways. Rational functions have no Baker domains, whose possible existence stems from the impossibility to define f at the essential singularity at infinity. A component of $\mathcal{F}(f)$ for a rational f that behaves like a Baker domain would have to be a parabolic domain with the parabolic fixed point at infinity. Also, as Baker has shown, it follows from the maximum principle that entire functions (including polynomials) have no Herman rings. A *Herman ring* is a doubly connected invariant component D of $\mathcal{F}(f)$ such that there is a conformal mapping φ of D onto the annulus $\{z : 1 < |z| < R < \infty\}$ for which the conjugated map $(\varphi \circ f \circ \varphi^{-1})(z) \equiv e^{2\pi i \beta} z$ for some real irrational number β. If an entire function f had a Herman ring D, then all points on the inner component of ∂D would have to lie in $\mathcal{J}(f)$. On the other hand, all the iterates of f remain uniformly bounded on an invariant Jordan curve in D and hence, by the maximum principle, in the bounded component of the complement of D, which implies that the iterates of f form a uniformly bounded family and hence a normal family in the union of D and the bounded component of the complement of D. Hence this larger set must lie in $\mathcal{F}(f)$ and so cannot intersect $\mathcal{J}(f)$, which is a contradiction.

Clearly a Baker domain must be unbounded. It is easily seen that for a transcendental entire f, any completely invariant component of $\mathcal{F}(f)$ is unbounded.

For a polynomial f, the set $\mathcal{F}(f)$ contains a neighbourhood of infinity, corresponding to the superattracting fixed point at infinity, and hence has a unique unbounded component, which, furthermore, is completely invariant. One can ask what can be said about the boundedness or unboundedness properties of the components of the Fatou set of a transcendental entire function. Many examples of transcendental entire functions have unbounded components in their Fatou sets, for example, the function $z + 1 + e^{-z}$ considered by Fatou in 1926 [14]. This particular function, however, has order 1.

We now review some concepts related to the growth of an entire function. We use the following notations for the *maximum* and *minimum modulus* of f:

$$M(r, f) = \max\{|f(z)| : |z| = r\}, \qquad m(r, f) = \min\{|f(z)| : |z| = r\}.$$

(The notation $m(r, f)$ is also the standard notation for the proximity function of f in the Nevanlinna theory but we will not need that concept.) Recall that the *order* $\rho(f)$ and *lower order* $\lambda(f)$ of f are defined by

$$\rho(f) = \limsup_{r \to \infty} \frac{\log \log M(r, f)}{\log r}, \qquad \lambda(f) = \liminf_{r \to \infty} \frac{\log \log M(r, f)}{\log r}.$$

If $0 < \rho(f) = \rho < +\infty$, we define the *type* of f by

$$\tau(f) = \limsup_{r \to \infty} \frac{\log M(r, f)}{r^\rho}.$$

If $\tau(f) = 0$, we say that f is of *minimal type*. If $0 < \tau(f) < +\infty$, we say that f is of *mean type*. If $\tau(f) = +\infty$, we say that f is of *maximal type*.

I.N. Baker [2] asked in 1981 whether every component of $\mathcal{F}(f)$ is bounded if the growth of f is sufficiently small. The appropriate growth condition would appear to be of order 1/2, minimal type at most. Baker ([2], p. 489) observed that this condition would be best possible in the following sense.

For any sufficiently large positive a, the function $f(z) = z^{-1/2} \sin \sqrt{z} + z + a$ is of order $1/2$, mean type, and has an unbounded component D of $\mathcal{F}(f)$ containing a segment $[x_0, \infty)$ of the positive real axis, such that $f^n(z) \to \infty$ as $n \to \infty$, locally uniformly in D.

As Baker noted ([2], p. 484), it is possible to have a function of order $1/2$ and of arbitrarily small type with the same properties. He considered $f(z) = \cos\left\{(\varepsilon^2 z + (9/4)\pi^2)^{1/2}\right\}$ where $0 < \varepsilon < \sqrt{3}\pi$. This function is of order $1/2$ and type ε. Now $f(0) = \cos(3\pi/2) = 0$ and $0 < f'(0) = \varepsilon^2/(3\pi) < 1$ so that f has an attracting fixed point at the origin. We have $f([0, \infty)) = [-1, 1]$. Furthermore,

$$0 < f'(x) = -\varepsilon^2 \sin\sqrt{(9/4)\pi^2 + \varepsilon^2 x}/(2\sqrt{(9/4)\pi^2 + \varepsilon^2 x}) < 1$$

for $-1 \leq x \leq 1$, which implies first together with $f(0) = 0$ that $|f(x)| < |x|$ for $x \in [-1, 1] \setminus \{0\}$, and then that $\lim_{n \to \infty} f^n(x) = 0$ for $-1 \leq x \leq 1$. It follows that the component of $\mathcal{F}(f)$ containing the attracting fixed point at the origin contains the positive real axis and is therefore unbounded.

The conjecture that every component of $\mathcal{F}(f)$ is bounded if the growth of f is of order $1/2$, minimal type at most, remains open in general. However, numerous cases have been settled and there is an open problem only for wandering domains, and even then only for functions whose growth is quite irregular, in a sense to be made more precise later on. Several techniques have been developed to attack various instances of this problem. In this paper we survey the results obtained and the techniques developed. We also reproduce a number of proofs related to this subject.

The author would like to extend his sincere thanks to Phil Rippon for reading the manuscript carefully and for making several very helpful comments that resulted in the rewriting of Section 8.

2. Notation

We denote the set of complex numbers by $\mathbb{C}$, and write $\overline{\mathbb{C}} = \mathbb{C} \cup \{\infty\}$ for the extended complex plane, that is, the Riemann sphere. If $z \in \mathbb{C}$ and $r > 0$, we write $B(z, r) = \{w \in \mathbb{C} : |w - z| < r\}$, $S(z, r) = \{w \in \mathbb{C} : |w - z| = r\}$, and $\overline{B}(z, r) = \{w \in \mathbb{C} : |w - z| \leq r\}$. We denote the unit disk $B(0, 1)$ by $\mathbb{D}$.

If γ is a Jordan curve in the complex plane $\mathbb{C}$, and hence bounded, we denote the bounded component of the complement of γ by $\mathrm{int}\,\gamma$ and call it the *interior* of γ. We denote the unbounded component of $\overline{\mathbb{C}} \setminus \gamma$ by $\mathrm{ext}\,\gamma$ and call it the *exterior* of γ.

3. Hyperbolic geometry and Schottky's theorem

Hyperbolic geometry is an indispensable tool in complex analysis, and complex dynamics is no exception. We review a few concepts and results that we shall need.

In the unit disk $\mathbb{D} = \{z \in \mathbb{C} : |z| < 1\}$ we define the density of the hyperbolic or Poincaré metric $\lambda(z) = \lambda_{\mathbb{D}}(z)$ by

$$\lambda(z) = \frac{1}{1-|z|^2}.$$

If U is a simply connected domain in $\overline{\mathbb{C}}$ whose complement contains at least 2 points, there is a conformal mapping φ of $\mathbb{D}$ onto U, and we define the density of the hyperbolic metric $\lambda_U(z)$ in U by

(1) $$\lambda_U(\varphi(z))|\varphi'(z)| = \frac{1}{1-|z|^2}.$$

Then $\lambda_U(z)$ does not depend on the choice of φ.

If U is any domain in $\overline{\mathbb{C}}$ for which there exists an analytic universal covering map φ of $\mathbb{D}$ onto U, we define the density of the hyperbolic metric $\lambda_U(z)$ in U by (1). Then $\lambda_U(z)$ does not depend on the choice of φ, nor, for a fixed φ and for a given $w \in U$, on the choice of $z \in \mathbb{D}$ such that $\varphi(z) = w$.

We write $h_U(z, w)$ for the hyperbolic distance between $z, w \in U$. Thus $h_U(z,w) = \inf_\gamma \int_\gamma \lambda_U(\zeta)\,|d\zeta|$ where γ runs over all rectifiable paths joining z to w in U.

If U is the punctured unit disk $U = \{z : 0 < |z| < 1\}$, then a calculation shows that

(2) $$\lambda_U(z) = -\frac{1}{2|z|\log|z|}$$

since the map $z \mapsto \exp\{(z-1)/(z+1)\}$ is the universal covering map of the unit disk onto U.

Consider the thrice punctured sphere $V = \overline{\mathbb{C}} \setminus \{0, 1, \infty\} = \mathbb{C} \setminus \{0, 1\}$. Suppose that f is analytic in $\mathbb{D}$ and that f omits the values 0 and 1 there. Then f maps $\mathbb{D}$ into V, and by the distance decreasing property of analytic functions, we have

$$h_V(f(z), f(w)) \le h_{\mathbb{D}}(z, w)$$

for all $z, w \in \mathbb{D}$. If $f(0) = a$ and $|z| \le r < 1$, then

$$h_V(f(z), a) \le \frac{1}{2}\log\frac{1+r}{1-r},$$

which imposes an upper bound of $|f(z)|$ as well as a lower bound on each of $|f(z)|$ and $|f(z) - 1|$. This result is known as Schottky's theorem. Explicit bounds have been obtained by several authors; for our purposes, any explicit bounds will do. Therefore we only mention the result ([15], Theorem 5.6, p. 60) that

(3) $$\log^+|f(z)| \le (\pi/2 + 2\log 3 + \log^+|f(0)|)\frac{1+|z|}{1-|z|}.$$

Applying (3) to $1/f$ and to $1/(1-f)$ instead of f, one obtains lower bounds for $|f(z)|$ and $|f(z) - 1|$.

In complex dynamics, one often normalizes a situation by conjugating an entire function by a linear polynomial. Since such a conjugation does not change dynamics, it is a natural thing to do. In that way, one can, for example, move certain two points to two simpler points such as 0 and 1. However, we will often deal with conditions involving circles centred at the origin, and since such a conjugation may change those circles, it might not be clear whether the conjugated function satisfies similar conditions any more. Therefore, it will often be convenient for us not to conjugate the function but to consider the original points. For this purpose, we need a version of Schottky's theorem involving general finite distinct points α and β instead of 0 and 1. This can be achieved as follows. If f is analytic in $\mathbb{D}$ and omits finite distinct points α and β there, then define $g(z) = (f(z) - \alpha)/(\beta - \alpha)$. Now g omits 0 and 1 and satisfies the assumptions of Schottky's theorem. We conclude after a computation that

$$(4) \qquad \log^+ |f(z)| \leq A \frac{1 + |z|}{1 - |z|} + B,$$

where

$$(5) \qquad A = \pi/2 + 2\log 3 + \log^+ |(f(0) - \alpha)/(\beta - \alpha)|,$$

$$(6) \qquad B = \log^+ |\alpha| + \log^+ |\beta - \alpha| + \log 2,$$

whenever $|z| < 1$.

Using these estimates, and the estimates of the hyperbolic metric of the thrice punctured sphere that they are based on, one can prove the following consequence of Schottky's theorem, which is frequently used in complex dynamics.

Theorem 3.1. *Let D be a plane domain whose complement in the sphere contains at least three points. Let f be an analytic function omitting the finite distinct points α and β in D. Let K be a compact subset of D. Then there are constants $B > 1$ and $C > 1$ depending only on K, α, and β such that for all $z, w \in K$ we have*

$$(7) \qquad |f(w) - \alpha| \leq B \max\{1, |f(z) - \alpha|\}^C.$$

4. Growth of entire functions and the $\cos \pi \rho$-theorem

Let f be an entire function. To discuss and properly understand even those results that are not directly in terms of the growth of the function, it is necessary to review a few notions concerning the growth of an entire function. We recall the $\cos \pi \rho$-theorem in the form given by P. Barry in [5], p. 294 (we note in passing that Barry has also obtained an analogous result involving lower order [6]).

If $E \subset [1,\infty)$, the *lower logarithmic density* of the set E is defined by

$$\underline{\log \text{ dens }} E = \liminf_{R \to \infty} \frac{1}{\log R} \int_{E \cap (1,R)} \frac{dt}{t}.$$

Theorem 4.1. *Let f be a transcendental entire function of order $\rho < 1/2$, and suppose that $\rho < \alpha < 1/2$. Then if*

$$E = \{\, r \geq 1 \,:\, \log m(r,f) > (\cos \pi \alpha) \log M(r,f)\,\}$$

we have

$$\underline{\log \text{ dens }} E \geq 1 - \frac{\rho}{\alpha}.$$

In particular, $\limsup_{r \to \infty} m(r,f) = \infty$.

If f is of order $1/2$, minimal type, then $m(r,f)$ is unbounded, and in fact, for each positive integer n, the quantity $m(r,f)/r^n$ is unbounded.

To see that for each positive integer n, the quantity $m(r,f)/r^n$ is unbounded, write the Taylor series of f at the origin as $f(z) = \sum_{k=0}^{\infty} a_k z^k$ and set $P_n(z) = \sum_{k=0}^{n} a_k z^k$. Then $g(z) = (f(z) - P_n(z))/z^{n+1}$ is a transcendental entire function of the same order as f, so that $m(r,g)$ is unbounded. This easily implies that $m(r,f)/r^{n+1}$ is unbounded.

The following example illustrates what Theorem 4.1 actually implies. When $\rho = \rho(f) < 1/2$, set

$$(8) \qquad \alpha = (1+2\rho)/4, \qquad \beta = \cos\left(\pi \frac{1+2\rho}{4}\right), \qquad \sigma = \frac{2}{1-2\rho}.$$

Now it is easily seen that there is a number R_0 such that for each $R \geq R_0$, we have

$$\log m(r,f) > \beta \log M(r,f)$$

for some r with $R \leq r \leq R^\sigma$.

The following observation will be useful for us.

Lemma 4.2. *Let f be a transcendental entire function such that*

$$\limsup_{r \to \infty} m(r,f) = +\infty.$$

If U is an unbounded component of $\mathcal{F}(f)$ and if U_n is the component of $\mathcal{F}(f)$ containing $f^n(U)$, then U_n is unbounded.

Proof of Lemma 4.2. If we pick $w \in U$ and $R > 0$, there exists $r > |w|$ such that $m(r,f) > R$. Since U is unbounded, there exists $z \in U$ with $|z| > r$. Since U is connected, we may join z to w by a path in U and find a point $\zeta \in U$ with $|\zeta| = r$. Thus $|f(\zeta)| > R$. Since R was arbitrary, we conclude that U_1 is unbounded. The same argument combined with induction now shows that each U_n is unbounded. This proves Lemma 4.2.

5. Baker's original results

In the paper [2] where he stated his problem, Baker himself obtained the first results on it. Apart from posing the problem in the first place, the importance of his original contributions extends in two directions. Firstly, he obtained such a good result on components of the Fatou set where the iterates of the function have a finite limit function, that this result cannot be improved in terms of the rate of growth of $M(r, f)$ alone. Secondly, he obtained the first results based on the growth of the maximum modulus of the function alone and not on any other considerations. In this connection, he formulated and proved a lemma ([2], Lemma 9, p. 492) that has since been used in almost all papers to obtain results on the basis of growth.

Further, Baker [2] dealt with the case of what are now known as Baker domains, but only of period one. Since this requires somewhat different techniques from the case of domains where there is a finite limit function, we consider these cases separately.

We begin with the case of finite limit functions ([2], Theorem 1, p. 484).

Theorem 5.1. *Let f be a transcendental entire function whose growth is at most order $1/2$ minimal type. Let U be a component of the Fatou set of f such that every convergent subsequence f^{n_k} of the iterates of f converges locally uniformly in U to a finite limit function. Then U is bounded.*

About ten years earlier, Baker's student Bhattacharyya [9] had proved more specialized results in this direction. For example, he proved that if f is a transcendental entire function whose growth is at most order $1/2$ minimal type, and if U is an invariant attracting or superattracting component of $\mathcal{F}(f)$, then U is bounded.

For Baker domains, Baker [2] obtained the following result ([2], Theorem 1, p. 484).

Theorem 5.2. *Let f be a transcendental entire function whose growth is at most order $1/2$ minimal type. Then there does not exist a Baker domain of period one for f, that is, there is no component U of the Fatou set of f such that $f(U) \subset U$ and f^n converges locally uniformly in U to the constant infinity.*

Concerning growth, Baker proved the following result ([2], Theorem 2, p. 484).

Theorem 5.3. *Let f be a transcendental entire function such that for some real p with $1 < p < 3$, we have*

$$\log M(r, f) = O((\log r)^p) \tag{9}$$

as $r \to \infty$. Then every component of the Fatou set of f is bounded.

We now proceed to give Baker's proofs of Theorems 5.1 and 5.2.

Proof of Theorem 5.1. Let f be a transcendental entire function whose growth is at most order $1/2$ minimal type. Let U be a component of the Fatou set of f such that a subsequence f^{n_k} of the iterates of f converges locally uniformly in U to a finite limit function φ. Then there is a component V of $\mathcal{F}(f)$ such that for some integers $m \geq 0$ and $p \geq 1$, we have $f^m(U) \subset V$ and $f^p(V) \subset V$. By Theorem 4.1 and Lemma 4.2, U is bounded if, and only if, V is bounded. So it suffices to show that V is bounded. Note that our assumption that in U, every convergent subsequence of iterates of f converges locally uniformly in U to a finite limit function, implies that the same is true in V.

To get a contradiction, we assume that V is unbounded.

Choose $\alpha \in V$. Then the sequence $f^n(\alpha)$ is bounded, for otherwise there is a subsequence $f^{m_j}(\alpha) \to \infty$, and by passing to a subsequence without changing notation, we may assume that $f^{m_j} \to \psi$ locally uniformly in V. Now either ψ is analytic or $\psi \equiv \infty$. Since $f^{m_j}(\alpha) \to \infty$, we have $\psi \equiv \infty$, which contradicts our assumption that ψ is finite.

The next observation contains the seed of the generalization of Theorem 5.1 by Zheng (Theorem 6.2 in Section 6). Let A be a constant such that $A > 1$ and such that $|f^n(\alpha)| < A$ for all $n \geq 0$. Let γ be a path in V joining α to a point β with $|\beta| > 2A$ and

$$m(|\beta|, f) > |\beta|^2.$$

This is possible ([2], p. 490, line 12 from the bottom, says "This is impossible" but obviously means "This is possible") since $m(r, f)/r^2$ is unbounded by Theorem 4.1 and since V is unbounded. We may and will replace γ, without changing notation, by its subpath from α to its first point of modulus $= |\beta|$. Hence $|z| \leq |\beta|$ for all $z \in \gamma$.

The path $f(\gamma)$ joins $f(\alpha)$ to a point of modulus $> |\beta|$. By induction on n, we see that the path $f^n(\gamma)$ joins the point $f^n(\alpha)$ of modulus $< A$ to a point of modulus $> |\beta|^2 > |\beta| > 2A$. Since the convergence of any convergent subsequence f^{n_k} to its limit function in V would be uniform of γ, it follows that no limit function can be constant. This implies that V belongs to a Siegel disk cycle of domains. Hence the identity mapping is among the limit functions, that is, $f^{n_k}(z) \to z$ locally uniformly in V for some sequence f^{n_k}. But now $|z| \leq |\beta|$ for all $z \in \gamma$ while every $f^n(\gamma)$ contains a point of modulus $> |\beta|^2$. Since $|\beta|^2$ is a fixed number $> |\beta|$, we get a contradiction. This completes the proof of Theorem 5.1.

Remark. Analysing the proof of Theorem 5.1 we see that it depended only on two things: that a point has bounded orbit under f, and that $m(r, f)/r^2$ is unbounded, or just infinitely often > 1. A closer look reveals that instead of making $m(r, f)/r^2$ large, it would have been enough to make $m(r, f)/r$ large. The theorem of Zheng (Theorem 6.2 in Section 6) is based on this observation.

Proof of Theorem 5.2. Let f be a transcendental entire function whose growth is at most order 1/2 minimal type. Suppose that U is a component of the Fatou set of f such that $f(U) \subset U$ and f^n converges locally uniformly in U to the constant infinity. Then U is necessarily unbounded. Now [2], Theorem 6, p. 485, shows that there is a path γ in U tending to infinity on which $|f(z)| = O(|z|^k)$ for some constant k. Hence $m(r, f) \leq Ar^k$ for some positive constant A, for all large r. This contradicts Theorem 4.1. The proof of Theorem 5.2 is complete.

Remark. By referring to [2], Theorem 6, p. 485 above, we have followed the historically accurate path. A better result, namely, that $|f(z)| = O(|z|)$ along a path going to infinity in a Baker domain was established later by Baker himself in [4]. The proof of [2], Theorem 6, p. 485, is based on Theorem 3.1. After Baker proved that all multiply connected components of the Fatou set of a transcendental entire function are wandering domains, it became clear that Baker domains are simply connected, so that a better estimate of the hyperbolic distance became available, to be used in place of Theorem 3.1. This then leads to the sharper result $|f(z)| = O(|z|)$.

6. FURTHER RESULTS FOR PERIODIC COMPONENTS OF THE FATOU SET

Instead of asking for which functions f all the components of the Fatou set of f are bounded, one may ask for which functions f particular types of components of $\mathcal{F}(f)$ are bounded. Experience shows that wandering domains are the most difficult to deal with, followed by Baker domains, while components of $\mathcal{F}(f)$ where all limit functions of the iterates are finite are the easiest to handle.

If U and V are components of $\mathcal{F}(f)$ with $f(U) \subset V$ and if $m(r, f)$ is unbounded (for example, if the growth of f is at most of order 1/2, minimal type), then by Lemma 4.2, U is unbounded if, and only if, V is unbounded. Therefore, when considering the question of whether a periodic or preperiodic component of the Fatou set of such a function f is unbounded, it is sufficient to consider periodic components.

Unlike the general study of complex dynamics, where one can further restrict one's consideration to invariant components on the grounds that a periodic component of period p for f is an invariant component of $\mathcal{F}(f^p)$ and $\mathcal{F}(f^p) = \mathcal{F}(f)$, one cannot usually perform the same reduction when studying Baker's problem. The reason is that even if something is assumed about the rate or regularity of growth of f, the iterates of f normally no longer possess those qualities.

The first improvement of Baker's original results for periodic components was obtained by Stallard [21, 22] who proved the following theorem ([21], Theorem 3A, p. 49).

Theorem 6.1. *Let f be a transcendental entire function of order $< 1/2$. Then there does not exist a Baker domain of any period for f, that is, there* is no component U of the Fatou set of f such that $f^p(U) \subset U$ for some $p \geq 1$ and such that f^n converges locally uniformly in U to the constant infinity.

Stallard has not published her proof of Theorem 6.1, and the result is not mentioned in [22]. A different proof of Theorem 6.1 was provided by Anderson and the author in [1]. Here we omit the proof but we note that the proof in [1] bears some similarity to the proofs of Theorems 8.3 and 9.1 below, which we will give in Section 10.

Note the small gap left by the fact that in Theorem 6.1, the order of f must be $< 1/2$, while in Theorem 5.2, the growth of f is at most order $1/2$ minimal type. This gap was filled by Zheng Jian-Hua [24] who proved the following results.

Theorem 6.2. ([24], Theorem 1, p. 355) *Let f be a transcendental entire function such that*

(10) $$\limsup_{r \to \infty} \frac{m(r,f)}{r} = +\infty.$$

Let U be a periodic or preperiodic component of the Fatou set of f. Then U is bounded.

In particular, f has no Baker domains or Baker domain cycles.

This applies, in particular, if the growth of f is at most of order $1/2$, minimal type.

Theorem 6.3. ([24], Theorem 2, p. 356) *Let f be a transcendental entire function such that for all r in a sequence tending to ∞, we have*

(11) $$m(r,f) > r.$$

Let U be a component of $\mathcal{F}(f)$. If there is a point $z_0 \in U$ such that the sequence $f^n(z_0)$ is bounded, then U is bounded.

Theorem 6.4. ([24], Theorem 3, p. 357) *Let f be a transcendental entire function of order $< 1/2$ and let U be a wandering domain of $\mathcal{F}(f)$. If there is a point $z_0 \in U$ such that*

$$\log^+ \log^+ |f^n(z_0)| = O(n)$$

as $n \to \infty$, then U is bounded.

We omit the proof of Theorem 6.4. We proceed to prove Theorems 6.2 and 6.3. We reproduce Zheng's arguments, beginning with his lemmas.

Lemma 6.5. ([24], Lemma 2.1, p. 357) *Let γ_1 and γ_2 be disjoint Jordan curves in $\mathbb{C}$ with $z_0 \in \operatorname{int} \gamma_2 \subset \operatorname{int} \gamma_1$. Suppose that f is analytic in $(\operatorname{ext} \gamma_2 \cap \operatorname{int} \gamma_1) \cup \gamma_1 \cup \gamma_2$, that $|f| > R$ on γ_1, and that $|f| \leq R$ on γ_2. Then there exists a Jordan curve $\Gamma \subset (\operatorname{ext} \gamma_2 \cap \operatorname{int} \gamma_1) \cup \gamma_2$ with $z_0 \in \operatorname{int} \Gamma$ such that $|f| = R$ on Γ.*

Lemma 6.6. ([24], Lemma 2.2, p. 357) *Let f be a transcendental entire function. Suppose that $z_0 \in \mathbb{C}$, $R > 0$, and $|f^n(z_0)| < R$ for all $n \geq 0$. Suppose further that $A \geq 1$ and that $m(R,f) > AR$. Then, for each positive integer m, there exists a Jordan curve Γ_m in $\{z : |z| < R\}$ such that $z_0 \in \operatorname{int} \Gamma_m$ and*
$$|f^m(z)| \geq m(R,f) > AR$$
for all $z \in \Gamma_m$.

Proof of Lemma 6.5. We write $\alpha = f(\gamma_1)$. Since $f(\gamma_2) \subset \{w : |w| \leq R\}$ and $\alpha \subset \{w : |w| > R\}$, we have $0 \in \operatorname{int} \alpha$, for otherwise $f(\operatorname{ext} \gamma_2 \cap \operatorname{int} \gamma_1)$ is unbounded, which is impossible. Here we have used the fact that $\partial f(G) \subset f(\partial G)$ when f is analytic in the closure of a domain G. Hence
$$f(\operatorname{ext} \gamma_2 \cap \operatorname{int} \gamma_1) \supset \{w : |w| = R\}.$$
Since f takes each value in $f(\operatorname{ext} \gamma_2 \cap \operatorname{int} \gamma_1)$ only finitely many times in the set $\operatorname{ext} \gamma_2 \cap \operatorname{int} \gamma_1$, there exists at least one component β of $f^{-1}(\{w : |w| = R\})$ such that $z_0 \in \operatorname{int} \beta$, for otherwise there is a Jordan curve L from z_0 to ∞ which does not intersect $f^{-1}(\{w : |w| = R\})$, and this contradicts the fact that $|f|$ is continuous on the part of L lying in $\operatorname{ext} \gamma_2 \cap \operatorname{int} \gamma_1$. Therefore we may extract from β a Jordan curve Γ which has the required properties. This completes the proof of Lemma 6.5.

Proof of Lemma 6.6. The assumption of Lemma 6.6 implies that
$$|f| \geq m(R,f) > AR \tag{12}$$
on $\Gamma \equiv \{z : |z| = R\}$. Since $|f(z_0)| < R$ and $|z_0| < R$, there exists $R_1 > 0$ such that
$$R = \max\{|f(z)| : |z - z_0| = R_1\} > |f(z_0)|.$$
Applying the maximum modulus principle to f, in view of (12), we find that
$$\{z : |z - z_0| \leq R_1\} \subset \operatorname{int} \Gamma = \{z : |z| < R\}$$
and that $R > R_1 > 0$. Since $|f(z)| \leq R$ whenever $|z - z_0| = R$, it follows from Lemma 6.5 that there exists a Jordan curve
$$\Gamma_1 \subset \{z : R_1 < |z - z_0|\} \cap \{z : |z| < R\}$$
such that $z_0 \in \operatorname{int} \Gamma_1$ and such that $|f| = R$ on Γ_1. Furthermore, by (12), we have
$$|f^2| \geq m(R,f) > AR$$
on Γ_1.

To continue by induction on k, suppose that for a positive integer k there exists a Jordan curve $\Gamma_{k-1} \subset \{z : |z| < R\}$ with $z_0 \in \operatorname{int} \Gamma_{k-1}$ such that
$$|f^k| \geq m(R,f) > AR \tag{13}$$
on Γ_{k-1}. Since $|f^k(z_0)| < R$, there exists $R_k > 0$ such that
$$R = \max\{|f^k(z)| : |z - z_0| = R_k\} > |f^k(z_0)|.$$

Applying the maximum modulus principle to f^k, in view of (13), we find that

$$\{z : |z - z_0| \leq R_k\} \subset \text{int}\, \Gamma_{k-1}$$

and that $R > R_k > 0$. Thus, whenever $|z - z_0| = R_k$, we have $|f^k(z)| \leq R$. By Lemma 6.5, there exists a Jordan curve $\Gamma_k \subset \{z : R_k < |z - z_0|\} \cap \text{int}\, \Gamma_{k-1}$ such that $z_0 \in \text{int}\, \Gamma_k$ and such that $|f^k| = R$ on Γ_k. Furthermore, by (13), we have

$$|f^{k+1}| \geq m(R, f) > AR$$

on Γ_k.

By induction on k, we thus obtain the conclusion of Lemma 6.6. This completes the proof of Lemma 6.6.

After this preparation, we are ready to prove Zheng's theorems 6.2 and 6.3.

Proof of Theorem 6.2. Let f be as in Theorem 6.2. To get a contradiction, suppose that U is an unbounded periodic or preperiodic component of $\mathcal{F}(f)$. By (10) and Lemma 4.2, the domain $f^n(U)$ is unbounded for each $n \geq 1$. So we may assume without loss of generality that U is periodic of period $p \geq 1$, so that $f^p(U) \subset U$.

The function f^2 must have at least one fixed point in $\mathbb{C}$, so there exists $z_0 \in \mathbb{C}$ with $f^2(z_0) = z_0$, by, e.g., [8]. By (10), if $A > 1$ is given, there exists a large R such that $m(R, f) > AR$ and $R > |z_0| + |f(z_0)| \geq |f^n(z_0)|$ for all $n \geq 0$.

By Lemma 6.6, for an arbitrary positive integer j, there exists a Jordan curve Γ_j with $\Gamma_j \subset \{z : |z| \leq R\}$ with $z_0 \in \text{int}\, \Gamma_j$ such that

(14) $$|f^j(z)| \geq m(R, f) > AR$$

for all $z \in \Gamma_j$.

Suppose first that $f^n(z) \to \infty$ as $n \to \infty$ locally uniformly for $z \in U$. Then U belongs to a cycle of Baker domains. Since U is a Baker domain of period p, it follows from [4], Corollary, that there exist an arc γ_1 in U going from a finite point to infinity, and a positive constant L_1 such that

$$|f^p(z)| < L_1|z|$$

for all $z \in \gamma_1$.

Let γ_2 be a line segment connecting z_0 and the finite end point of γ_1 (so γ_2 need not be contained in U). Since f^p is entire, there exists a positive constant L_2 such that

$$|f^p(z)| \leq L_2$$

for all $z \in \gamma_2$.
Thus

$$|f^p(z)| \leq L_1|z| + L_2$$

for all $z \in \gamma = \gamma_1 \cup \gamma_2$.

Let z_p be a point of intersection of γ and Γ_p. Clearly $|z_p| \leq R$. From (14) with $j = p$ it follows that
$$AR < |f^p(z_p)| \leq L_1|z| + L_2 \leq L_1 R + L_2.$$
We obtain $A < L_1 + L_2/R \leq L_1 + L_2$. In this inequality, L_1 and L_2 are fixed numbers while A can be chosen to be arbitrarily large. This is a contradiction. It follows that a function f satisfying the assumptions of Theorem 6.2 cannot have a Baker domain cycle.

By the classification of dynamics, it now follows that there exists a point $\alpha \in U$ with bounded orbit, so there is a positive number B such that $|f^n(\alpha)| < B$ for all $n \geq 0$. We may apply Lemma 6.6, taking $A > 2$ and taking R, in addition to other requirements, so large that $R > B + |\alpha|$, and taking $z_0 = \alpha$. This is possible by the assumptions of Theorem 6.2. It follows that there is a Jordan arc γ in U connecting α and a point of $U \cap \{z : |z| = R\}$. It follows from (14) that for each $n \geq 0$, the set $f^n(\gamma)$ contains points in both $\{z : |z| < B\}$ and $\{z : |z| > R\}$. Hence no subsequence of f^n can have a constant limit in U. Thus U belongs to a Siegel disk cycle. Hence there exists a sequence f^{n_k} tending to the identity mapping locally uniformly in U as $k \to \infty$. Then for all large n_k, and for all $z \in \gamma$, we have $|f^{n_k}(z)| < 2|z|$. Choosing z to be a point of intersection of γ and Γ_{n_k}, we find by (14) that also $AR < |f^{n_k}(z)| < 2|z| \leq 2R$. Since $A > 2$, this gives a contradiction. The proof of Theorem 6.2 is now complete.

Proof of Theorem 6.3. Let f satisfy the assumptions of Theorem 6.3. Suppose that $\mathcal{F}(f)$ has an unbounded component U such that for some $z_0 \in U$, the orbit $\{f^n(z_0) : n \geq 0\}$ is bounded. Pick $B > 1$ such that $|f^n(z_0)| < B$ for all $n \geq 0$. We apply Lemma 6.6, taking $A = 1$ and choosing R so large that in addition to all other conditions, we have $R > B + |z_0|$. It follows that for each $j \geq 1$ there is a Jordan curve Γ_j contained in $\{z : |z| \leq R\}$ and containing z_0 in its interior, on which

(15) $$|f^j(z)| \geq m(R, f) > R.$$

As in the proof of Theorem 6.2, when considering a point α with a bounded orbit, we deduce that there exists a positive integer q such that the component V of $\mathcal{F}(f)$ containing $f^q(U)$ belongs to a Siegel disk cycle. We may now assume that $U = V$. There exists a sequence $n_k \to \infty$ such that $f^{n_k}(z) \to z$ locally uniformly for $z \in U$ as $k \to \infty$. Let γ be a Jordan arc in $U \cap \{z : |z| \leq R\}$ joining z_0 to a point of $U \cap \{z : |z| = R\}$. Then for each j there is a point of intersection z_j of γ and Γ_j. Since $m(R, f) > R$, we can choose $\varepsilon > 0$ such that $m(R, f) > (1 + \varepsilon)R$. Set $d = (m(R, f)/R) - \varepsilon > 1$. Then for sufficiently large n_k, due to the uniform convergence of f^{n_k} to the identity map on γ, we have $|f^{n_k}(z_{n_k})| < d|z_{n_k}|$, which contradicts (15). This completes the proof of Theorem 6.3.

Remark. After investigating the above proofs due to Zheng, it might be said that their essence is contained in two ideas: carefully refining Baker's proof of Theorem 5.1, and making use of the Jordan curves Γ_j on which $|f^j|$ is bounded below.

7. Results based on growth alone

A number of authors have obtained results to the effect that all components of the Fatou set of a transcendental entire function f are bounded provided that the maximum modulus of f has a given upper bound on its growth, regardless of the regularity of the growth. The first such result is Theorem 5.3 due to Baker. The next author to extend these results was Baker's student Stallard [22]. She obtained the following theorem ([22], Theorem B, p. 43).

Theorem 7.1. *Let f be a transcendental entire function such that for some $\varepsilon > 0$, we have*

$$\log \log M(r, f) = O\left(\frac{(\log r)^{1/2}}{(\log \log r)^\varepsilon}\right)$$

as $r \to \infty$. Then every component of the Fatou set of f is bounded.

While there have been further results involving the growth of f, they have all included conditions on the regularity of growth, in addition to the rate of growth. Therefore it seems that Theorem 7.1 still yields the best known result involving the rate of growth only, allowing one to conclude that all components, and not only all components of a particular type, of the Fatou set of f are bounded.

The methods employed by Baker, Stallard, and others to obtain results on the basis of the rate of growth have been very clever. The proof by Baker of Theorem 5.3 ([2], Theorem 2, p. 484) was based on careful estimates of P.D. Barry on the minimum modulus of slowly growing functions, a much improved version of the $\cos \pi \rho$–theorem for functions satisfying (9) for some finite p. The proof by Stallard for Theorem 7.1 is considerably more complicated, and starts with growth estimates for functions of zero order that can be found in Cartwright's monograph ([11], Theorems 52 and 53, pp. 83, 84). We omit these proofs here, since while they are ingenious, they are quite technical and lengthy.

However, we shall discuss one of the facts that forms an intellectual basis for those proofs. It is the lemma now often known as the Baker–Stallard lemma. In Baker's paper [2], it appears for all intents and purposes as Lemma 9, p. 492, restricted to the situation considered there. In its present slightly more general form, it was formulated and proved by Stallard ([22], Lemma 2.7, p. 45). For technical reasons, we prefer not to use the notation $(R_n)^{c(n)}$ as given there, where $c(n) > 1$, but denote these numbers by t_n, since they really form a new sequence of radii $> R_n$.

Lemma 7.2. *Suppose that f is a transcendental entire function and that there exist sequences R_n, ρ_n, and t_n, tending to infinity with n, such that*
(1) $R_{n+1} = M(R_n, f)$ *for all n;*
(2) $R_n \leq \rho_n \leq t_n$ *for all n; and*
(3) $m(\rho_n, f) > t_{n+1}$ *for all sufficiently large n.*
Then all the components of the Fatou set of f are bounded.

Proof of Lemma 7.2. Let f satisfy the assumptions of Lemma 7.2. To get a contradiction, suppose that U is an unbounded component of $\mathcal{F}(f)$. Choose distinct points α and β in $\mathcal{J}(f)$ such that $f(\alpha) = \beta$.

Choose a positive integer m_1 such that assumption (3) of Lemma 7.2 holds for all $n \geq m_1$. Since U is unbounded and connected, there is $m_2 \geq m_1$ such that U intersects all circles $\gamma_n = \{z : |z| = R_n\}$, $\gamma'_n = \{z : |z| = t_n\}$, and $\gamma''_n = \{z : |z| = \rho_n\}$ for all $n \geq m_2$. We choose an integer $N \geq m_2$ and note that U contains a Jordan arc C joining a point $w_N \in \gamma_N$ to a point $w'_{N+1} \in \gamma'_{N+1}$. It is clear by assumption (2) of Lemma 7.2 that C must contain a point $w''_{N+1} \in \gamma''_{N+1}$.

Now $f(U)$ is contained in a component of $\mathcal{F}(f)$ containing $f(C)$. By assumption (1) of Lemma 7.2, we have $R_{N+1} = M(R_N, f) \geq |f(w_N)|$. Since $m(\rho_{N+1}, f) > t_{N+2}$, we have $|f(w''_{N+1})| > t_{N+2}$. Hence $f(C)$ must contain an arc joining a point $w_{N+1} \in \gamma_{N+1}$ to a point $w'_{N+2} \in \gamma'_{N+2}$.

We continue this process inductively and find that $f^k(U)$ is contained in a component of $\mathcal{F}(f)$ containing an arc of $f^k(C)$ which joins a point $w_{N+k} \in \gamma_{N+k}$ to a point $w'_{N+k+1} \in \gamma'_{N+k+1}$.

Thus, on C, the function f^k takes a value of modulus at least R_{N+k}. Since $R_{N+k} \to \infty$ as $k \to \infty$, and since U is a component of $\mathcal{F}(f)$, we deduce that $f^k \to \infty$ as $k \to \infty$, locally uniformly in U. Since each f^k omits α and β in U, it follows that there exists a positive integer L such that for all $k \geq L$ and for all $z \in C$, we have $|f^k(z)| > 2\max\{|\alpha|, |\beta|\}$. It now follows from Theorem 3.1 that there exist constants A and B such that

$$|f^k(z)| < A|f^k(w)|^B$$

whenever $z, w \in C$ and $k \geq L$.

For each $k \geq L$, we can choose $z_k, \zeta_k \in C$ such that $f^k(z_k) = w_{N+k} \in \gamma_{N+k}$ and $f^k(\zeta_k) = w'_{N+k+1} \in \gamma'_{N+k+1}$. Hence for each $k \geq L$, we have

$$M(R_{N+k}, f) = R_{N+k+1} \leq t_{N+k+1} = |f^k(\zeta_k)| < A|f^k(z_k)|^B = A(R_{N+k})^B.$$

Since $R_{N+k} \to \infty$ as $k \to \infty$, we obtain a contradiction with the assumption that f is transcendental, so its maximum modulus cannot grow polynomially in any sequence. This completes the proof of Lemma 7.2.

We see that Schottky's theorem is the fundamental reason why Lemma 7.2 works. Other than that, the assumptions of Lemma 7.2 are geared towards assuring that the image of an arc, which itself covers a sufficiently long radial variation, under f stretches so far in the radial direction that this stretch is retained under iteration. Such long stretches are then incompatible with Schottky's theorem.

8. Survey of results based on the regularity of growth

The first results based on the regularity of the growth of the maximum modulus of f were obtained by Stallard [22]. She proved the following theorem ([22], Theorem C, p. 44).

Theorem 8.1. *Let f be a transcendental entire function of order $\rho < 1/2$. Suppose that there exists a real number $c \in [1, \infty)$ such that*

$$\lim_{r \to \infty} \frac{\log M(2r, f)}{\log M(r, f)} = c. \tag{16}$$

Then every component of the Fatou set of f is bounded.

Stallard ([22], p. 44) notes that as Baker observed in [2], p. 489, if

$$f(z) = \frac{\sin \sqrt{z}}{\sqrt{z}} + z + a,$$

where a is a sufficiently large positive constant, then $\mathcal{F}(f)$ has an unbounded component while f is of order $1/2$ with

$$\lim_{r \to \infty} \frac{\log M(2r, f)}{\log M(r, f)} = \sqrt{2}.$$

This shows that the condition $\rho < 1/2$ in Theorem 8.1 cannot be relaxed to include $\rho = 1/2$.

To prove her result, Theorem 8.1, Stallard showed that there exist sequences of radii required by the assumptions of Lemma 7.2. For functions of positive order of such regular growth, this turned out to be reasonably easy. For functions of zero order, she again made use of the estimates found in Cartwright's monograph mentioned before. We omit the details as they are quite technical.

Hua and Yang also studied the question of what can be proved on the basis of the regularity of growth. Theorem 1 in [18], p. 1283 states the following.

Theorem 8.2. *Let f be a transcendental entire function. If the order of f is $< 1/2$, then all components of the Fatou set of f are bounded provided that there exist real numbers C_1 and C_2 with $C_2 > 1$ and $C_1 > C_2^2$ such that for all sufficiently large r, we have*

$$\log M(C_1 r, f) < C_2 \log M(r, f).$$

This means that the increasing function $M(r, f)$ does not jump higher too fast anywhere, for large r.

We omit the proof of Theorem 8.2. Unfortunately, the author is not able to follow certain arguments in the proof of Lemma 2 in [18], which is required for the proof of [18], Theorem 1 (that is, our Theorem 8.2), particularly equation

(13) in [18], p. 1287, where the assumed regularity condition on $M(r, f)$ is applied to an auxiliary function $G(r)$, which is the maximum modulus function of an entire function related to but different from f.

Theorems 2 and 3 in the paper by Hua and Yang ([18], pp. 1283–1284) contain the following claims:

Let f be a transcendental entire function. If the lower order $\mu(f) < 1/2$ then every preperiodic component of $\mathcal{F}(f)$ is bounded. If the lower order $\mu(f) < 1/2$ and if for each real number $m > 1$, we have

$$\log M(r^m, f) \geq m^2 \log M(r, f)$$

for all sufficiently large r, then every component of $\mathcal{F}(f)$ is bounded.

The proofs of both theorems are eventually based on [18], Lemma 4, p. 1288 in an essential way. Here again the author is not able to follow the details of the proof of [18], Lemma 4, which seems to require an inequality for the lower logarithmic density of the set E where the minimum modulus of f is sufficiently large, rather than for the upper logarithmic density, which is what the assumption that the lower order of f is less than $1/2$ would yield.

J.M. Anderson and the author introduced another regularity condition. We set $\varphi(x) = \log M(e^x, f)$ so that, by the Hadamard three-circles theorem, $\varphi(x)$ is an increasing convex function of x. The function $\varphi'(x)$ may fail to exist at a countable set of points x. At such points $\varphi'(x)$ is defined to be the right-hand derivative. We proved the following result ([1], Theorem 2, p. 3245).

Theorem 8.3. *Let f be a transcendental entire function of order $\rho < 1/2$ such that for some positive constant c*

(17) $$\frac{\varphi'(x)}{\varphi(x)} \geq \frac{1+c}{x}$$

for all sufficiently large x, where $\varphi(x) = \log M(e^x, f)$. Then every component of $\mathcal{F}(f)$ is bounded.

The one-sided condition (17) may be compared to Stallard's condition (16), which is essentially the same as

$$\varphi(x+1) \sim c\varphi(x) \qquad \text{as } x \to \infty.$$

The assumption (17) is related to growth as well as regularity. It is equivalent to the condition

$$\liminf_{r \to \infty} \frac{d \log \log M(r, f)}{d(\log \log r)} > 1$$

which implies, in particular, that

$$\log M(r, f) > (\log r)^{1+\delta}$$

for some $\delta > 0$ and all $r > r_0(\delta)$. This guarantees a certain minimal growth rate for $M(r, f)$ from some point onwards. It prevents the existence of large annuli which are almost zero free, where the growth of the function f becomes like that of a polynomial. If the radial spread in such an annulus is large, then a function of order $< 1/2$ (or even of order < 1), which is given by an infinite product, will eventually behave like a polynomial in the annulus. If we have a sequence of such annuli, and if the degree of the polynomial that f resembles in the n-th annulus is p_n, then $p_n \to \infty$ as $n \to \infty$. However, regardless of how large p_n is, the condition (17) will eventually be violated, for any fixed $c > 0$, if the annuli get large enough.

We will prove Theorem 8.3 in Section 10.

Remark. Since the condition (17) essentially amounts to the requirement for a lower bound for the rate of growth of the maximum modulus of f, it is perhaps not surprising that it is satisfied by any function of finite order and positive lower order. To see this, suppose that f has finite order $< \tau$ and lower order $> \delta > 0$. Then for all sufficiently large r we have

$$0 < \delta < \frac{\log \log M(r,f)}{\log r} < \tau < +\infty,$$

that is,

(18) $$e^{\delta x} < \varphi(x) < e^{\tau x}$$

for all sufficiently large x. Choose ε with $0 < \varepsilon = \delta/(2\tau)$ and then choose $c > 0$ so small that $(1+c)(1-\varepsilon) < 1$. It follows that for all sufficiently large x we have

(19) $$(1+c)(1-\varepsilon) \leq 1 - e^{-\tau x} = 1 - \exp(x(\tau - \delta/\varepsilon)).$$

Suppose that $x_0 > 1$ is so large that (18) and (19) are both valid for all $x \geq x_0$. Then pick any $x_2 > x_0/\varepsilon$ and set $x_1 = \varepsilon x_2$, so that $x_0 < x_1 < x_2$. Since $\varphi'(x)$ is increasing, we have

$$\varphi'(x_2)(x_2 - x_1) \geq \int_{x_1}^{x_2} \varphi'(t)\,dt = \varphi(x_2) - \varphi(x_1)$$

and hence

$$x_2 \frac{\varphi'(x_2)}{\varphi(x_2)} \geq \frac{1 - \varphi(x_1)/\varphi(x_2)}{1 - x_1/x_2} \geq \frac{1 - e^{\tau x_1 - \delta x_2}}{1 - \varepsilon} = \frac{1 - e^{x_1(\tau - \delta/\varepsilon)}}{1 - \varepsilon} \geq 1 + c.$$

This yields (17). Of course, even though this argument only required f to be of finite order, in Theorem 8.3 the order of f must be $< 1/2$ for other reasons.

9. A MINIMUM MODULUS PROBLEM FOR FUNCTIONS OF ORDER $< 1/2$

Let us summarize the results that have been obtained concerning Baker's problem; for this purpose, we limit ourselves to the growth considered by Baker in his conjecture. If the growth of the transcendental entire function f does not exceed order $1/2$, minimal type, then every periodic and preperiodic component of the Fatou set of f is known to be bounded (Theorem 6.2). So it remains to study wandering domains, which may be assumed to be simply connected since any multiply connected component of $\mathcal{F}(f)$ is known to be bounded (and, incidentally, a wandering domain) by a result of Baker [3]. If f is exactly of order $1/2$, minimal type, then it appears that nothing is known. If the order of f is $< 1/2$, then wandering domains are bounded if $M(r, f)$ has sufficiently regular growth, as is seen from Theorems 8.1, 8.2, and 8.3. As we have seen, the case where a problem remains can at least heuristically be characterized as the case where there is a sequence of annuli of increasing radial spread in which f has few or no zeros, so that in each of those annuli, f behaves eventually like a polynomial, of degree that tends to infinity in the sequence of annuli.

It does not seem plausible that any fixed regularity condition of the type (17) could capture all transcendental entire functions of order $< 1/2$. Therefore, instead of referring to a prescribed sort of regularity, one might try to develop a condition that simply refers back to the function itself. Equivalently, one can try to distill from the known proofs the property of f minimally required for the proofs to work out. In this connection, the author proved a theorem involving such a condition, and asked whether the condition holds for all transcendental entire functions of order $< 1/2$.

The result is the following ([17], Theorem 1).

Theorem 9.1. *Let f be a transcendental entire function of order $< 1/2$. Suppose that there exist positive numbers R_0, L, δ, and C with $R_0 > e$, $M(R_0, f) > e$, $L > 1$, and $0 < \delta \leq 1$ such that for every $r > R_0$ there exists $t \in (r, r^L]$ with*

$$(20) \qquad \frac{\log m(t, f)}{\log M(r, f)} \geq L \left(1 - \frac{C}{(\log r)^\delta}\right).$$

Then all the components of the Fatou set of f are bounded.

We will prove Theorem 9.1 in Section 10. The corresponding question is then the following ([17]).

Minimum modulus problem. *Does every transcendental entire function of order $< 1/2$ satisfy the assumptions of Theorem 9.1 ?*

As a motivation for this question, the author noted the following in [17]. As we have discussed, the issue is to guarantee that if there is at least a certain radial spread in a component U of $\mathcal{F}(f)$, in the sense that the component contains points z and w such that $\log|w|/\log|z|$ is large, then the same spread is retained in the image components containing $f^n(U)$. The novelty of the condition of Theorem 9.1 is that, unlike any of its predecessors, it allows the radial spread to be reduced, due to the presence of the factor $1 - \frac{C}{(\log r)^\delta}$ on the right hand side of (20). "Nonetheless, it turns out that if we start with a sufficiently large spread, and in an unbounded domain we may start with as large a spread as we like, we are able to keep the propagated spreads large enough even if they are reduced from the original value.

Even though we are allowed to use any $L > 1$ in Theorem 9.1, and in particular we may take L close to 1 if that works, it seems likely that, particularly if the order of f is close to $1/2$, we may need to take L very large, depending on $\rho(f)$. If we are not close to or inside an annulus containing very few zeros of f, it would seem plausible that the condition (20) should be easy to satisfy, with a wide margin, by taking t to be a value arising from the consequences of Theorem 4.1. This is because then $\log m(t,f)/\log M(t,f)$ is greater than a fixed constant while $\log M(t,f)/\log M(r,f)$ should be quite large. So there should be a potential problem at most if we are in an annulus where f behaves like a polynomial. But in that case we should be able to take t close to r^L, and then the three numbers $\log m(t,f)$, $\log M(t,f)$, and $L\log M(r,f)$, should be close together. There may be some error term required to estimate $\log m(t,f)/(L\log M(r,f))$ from below, but (20) allows for such a term. For this reason, it seems sensible to ask whether every transcendental entire function f of order $< 1/2$ satisfies the assumptions of Theorem 9.1, and perhaps even conjecture that this is so." [17]

It might be of some interest to study whether the question can be answered for those functions f for which Baker and Stallard have already proved that all components of the Fatou set are bounded. For example, to start with the strongest reasonable assumptions, does f satisfy the assumptions of Theorem 9.1 provided that f satisfies (9) for some real $p > 1$, for example, for $p = 2$? One can ask whether the careful estimates used by Baker and by Stallard for these functions might be helpful also in the study of this minimum modulus problem.

10. Proofs of results based on the regularity of growth: self-sustaining spread

In this section we give a unified proof for Theorems 8.3 and 9.1. Both are based on what Anderson and the author called self-sustaining spread in [1]. Only the precise cause of the spread differs: in Theorem 8.3 it arises from the assumption (17), and in Theorem 9.1 it arises from the assumption (20).

Thus we assume throughout that f is a transcendental entire function of order $< 1/2$. To get a contradiction, we assume that the Fatou set of f contains an unbounded component D, which is a simply connected wandering domain. As mentioned, all other cases have already been settled.

Roughly speaking, the proof is divided into two very different parts. The first part, and the only part that uses dynamics, amounts to showing that since we are dealing with a component of the Fatou set, the radial spreads that we get cannot get too large in a certain sense. The second part uses (17) or (20), and shows that the spread must be larger than that after all. These two facts are incompatible, so we get a contradiction, which then completes the proofs of these theorems.

Suppose that K is a compact subset of D. In this first part of the proof, our aim is to show that for a certain complex constant a depending only on

D and for a possibly large positive number $C > 1$ depending on K, we have

$$\text{(21)} \qquad \frac{1}{C} \leq \frac{|f^j(z) - a|}{|f^j(w) - a|} \leq C$$

for all $z, w \in K$ and for all $j \geq 0$. This is basically a consequence of standard estimates for the hyperbolic metric in simply connected domains.

To find a, note that since D is a wandering domain, it is disjoint from any of its inverse images. Thus there is a disk $B(a, \rho)$ such that

$$\text{(22)} \qquad B(a, \rho) \cap \cup_{j=0}^{\infty} f^j(D) = \emptyset.$$

Let D_j be the component of the Fatou set of f containing $f^j(D)$. Note that also D_j is an unbounded wandering domain of f and hence is simply connected.

Let $L > 1$ be a large constant, to be determined soon. Pick $j \geq 0$ and $z, w \in K$. Suppose that $|f^j(z) - a|/|f^j(w) - a| > L$. Let $\zeta \in \partial D_j$ be the point closest to a, so that in particular, $|\zeta - a| < |f^j(w) - a|$. Recall that $h_\Omega(z_1, z_2)$ denotes the hyperbolic distance between the points z_1, z_2 of the domain Ω, and that $\lambda_\Omega(z)$ denotes the density of the hyperbolic metric of Ω at $z \in \Omega$. Thus

$$h_{D_j}(f^j(z), f^j(w)) \leq h_D(z, w) \leq L_0 \equiv \max\{h_D(z_1, z_2) : z_1, z_2 \in K\}.$$

Since D_j is simply connected, it follows from Koebe's one-quarter theorem that

$$\lambda_{D_j}(z) \geq \frac{1}{4 \operatorname{dist}(z, \partial D_j)} \geq \frac{1}{4|z - \zeta|} \geq \frac{1}{4(|z - a| + |\zeta - a|)}$$

for all $z \in D_j$, where $\operatorname{dist}(z, \partial D_j)$ denotes the Euclidean distance of z from ∂D_j. Hence

$$L_0 \geq h_{D_j}(f^j(z), f^j(w)) \geq \int_{|f^j(w)-a|}^{|f^j(z)-a|} \frac{dr}{4(r + |\zeta - a|)}$$

$$= \frac{1}{4} \log \frac{|f^j(z) - a| + |\zeta - a|}{|f^j(w) - a| + |\zeta - a|}$$

$$\geq \frac{1}{4} \log \frac{|f^j(z) - a| + |f^j(w) - a|}{2|f^j(w) - a|}$$

$$= \frac{1}{4} \left(\log \left(1 + \frac{|f^j(z) - a|}{|f^j(w) - a|} \right) - \log 2 \right)$$

$$\geq \frac{1}{4} (\log(1 + L) - \log 2),$$

which gives a contradiction if L is sufficiently large if compared to L_0. This proves (21).

Next we observe that even though the constant C depends on K and may be large, we can control the radial spread of the set $f^j(K)$ better by using the logarithmic scale.

Suppose that C_0 is a preassigned constant subject only to $C_0 > 1$. Next we show that by (21) and (22), we have

(23) $$\frac{1}{C_0} \leq \frac{\log(2|f^j(z) - a|/\rho)}{\log(2|f^j(w) - a|/\rho)} \leq C_0$$

for all $z, w \in K$ and for all sufficiently large $j \geq j_0$, say. Having to restrict ourselves to $j \geq j_0$ is one cost that we pay in order to get an estimate involving an arbitrary $C_0 > 1$. For if (23) does not hold, then there are sequences $z_j, w_j \in K$ and integers $n_j \to \infty$ such that

$$\frac{\log(2|f^{n_j}(z_j) - a|/\rho)}{\log(2|f^{n_j}(w_j) - a|/\rho)} > C_0,$$

that is,

(24) $$\frac{2|f^{n_j}(z_j) - a|}{\rho} > \left(\frac{2|f^{n_j}(w_j) - a|}{\rho}\right)^{C_0}.$$

By passing to a subsequence, we may assume that $|f^{n_j}(z_j) - a| \to R_2$ and $|f^{n_j}(w_j) - a| \to R_1$, say, where $\rho \leq R_1 < R_2 < \infty$ or $R_1 = R_2 = \infty$. In the former case, we do not have $f^{n_j} \to \infty$ locally uniformly in D, so that by passing to a further subsequence, we may assume that $f^{n_j} \to \omega$ locally uniformly in D, where ω is a complex number with $|\omega - a| \geq \rho$ (by (22)). Hence $f^{n_j}(z_j) \to \omega$ and $f^{n_j}(w_j) \to \omega$ as $j \to \infty$, which contradicts (24). Thus $R_1 = R_2 = \infty$. But now, by (21),

$$|f^{n_j}(z_j) - a| \leq C|f^{n_j}(w_j) - a| < (2/\rho)^{C_0-1}|f^{n_j}(w_j) - a|^{C_0}$$

when $|f^{n_j}(w_j)|$ is large enough, which is a contradiction. This completes the proof of (23). This also finishes off the first part of the proofs of Theorems 8.3 and 9.1. We have now established an upper bound for radial spread, which is effective since the number $C_0 > 1$ is still at our disposal and so we may choose C_0 to be very close to 1.

We proceed to show that if we choose K to be of large radial spread, as we may since we are choosing a compact subset of an unbounded domain D, then the large radial spread will have to persist, to the extent that we will end up contradicting (23). This contradiction then shows that the domain D with its defining properties could not exist at all, and the proofs of Theorems 8.3 and 9.1 will be complete.

First we deal with the situation in Theorem 8.3, so we assume that (17) holds.

We will explain in a moment how K is to be chosen. Before that, we make a number of general observations. We choose a real number C_0 with $1 < C_0 < 4$. We may assume that (23) holds for $j \geq j_0$, say, for all $z, w \in K$. If $\beta > 0$ is given, we have $\beta \log M(r, f) > \log r$ for all $r \geq R_0$, say. Suppose that (17) holds for all $x \geq x_0 > 1$, with a certain positive constant c. Then, if $x_0 < x_1 < x_2$, we have

$$\log \frac{\varphi(x_2)}{\varphi(x_1)} = \int_{x_1}^{x_2} \frac{\varphi'(x)}{\varphi(x)} dx \geq \int_{x_1}^{x_2} \frac{1+c}{x} dx = (1+c) \log \frac{x_2}{x_1}$$

and so

$$\frac{\varphi(x_2)}{\varphi(x_1)} \geq \left(\frac{x_2}{x_1}\right)^{1+c}.$$

Recall the definition (8) of σ and β, in terms of the order $\rho = \rho(f) < 1/2$ of f. Let $L > \sigma^2 > 4$ be a large constant satisfying also $\beta L^c > \sigma^{2(1+c)}$. Now β determines R_0 as above. We further choose R_0 so large that in addition, we have

(25) $\quad \log(2|w-a|/\rho) \geq \sqrt{4/L} \log|w|, \ \log(2|z-a|/\rho) \leq \sqrt{L/4} \log|z|$

whenever $|z| \geq R_0$ and $|w| \geq R_0$. Suppose that the compact connected subset K of D is chosen so that $\log|\zeta| > \max\{x_0, \log R_0\}$ for all $\zeta \in K$ and so that there are $z, w \in K$ with $\log|w| > L \log|z|$. This is possible since D is unbounded. Write $K_j = f^j(K)$ so that K_j is a compact connected subset of D_j.

It follows from Theorem 4.1 that there are r_1, r_2 with $|z| \leq r_1 \leq |z|^\sigma$ and $|w|^{1/\sigma} \leq r_2 \leq |w|$ such that $\log m(r_j, f) > \beta \log M(r_j, f)$ for $j = 1, 2$. Since K is connected, there are points $\zeta_1, \zeta_2 \in K$ with $|\zeta_j| = r_j$ for $j = 1, 2$. Write $x_j = \log r_j$ for $j = 1, 2$, so that

$$\log|z| \leq x_1 \leq \sigma \log|z| \ \text{ and } \ \sigma^{-1} \log|w| \leq x_2 \leq \log|w|.$$

We obtain

$$\frac{\log|f(\zeta_2)|}{\log|f(\zeta_1)|} \geq \frac{\log m(r_2, f)}{\log M(r_1, f)} \geq \beta \frac{\log M(r_2, f)}{\log M(r_1, f)}$$

$$= \beta \frac{\varphi(x_2)}{\varphi(x_1)} \geq \beta \left(\frac{x_2}{x_1}\right)^{1+c} \geq \beta \left(\frac{\sigma^{-1} \log|w|}{\sigma \log|z|}\right)^{1+c}$$

$$> \beta \left(\frac{L}{\sigma^2}\right)^{1+c} > L.$$

Also for $j = 1, 2$, we have

$$\log|f(\zeta_j)| \geq \log m(r_j, f) > \beta \log M(r_j, f)$$
$$> \log r_j = \log|\zeta_j| > \max\{x_0, \log R_0\}.$$

Note that $\log r_1 = x_1 \geq \log|z| > \max\{x_0, \log R_0\}$ while, since $L > \sigma^2$, we have

$$\log r_2 = x_2 \geq \sigma^{-1} \log|w| > L\sigma^{-1} \log|z|$$
$$> \log|z| > \max\{x_0, \log R_0\}.$$

We find that there are points $\zeta_3 = f(\zeta_1)$ and $\zeta_4 = f(\zeta_2)$ in $f(K)$ with $\log|\zeta_j| > \max\{x_0, \log R_0\}$ for $j = 3, 4$ and with

$$\frac{\log|\zeta_4|}{\log|\zeta_3|} > L.$$

Now we may repeat the above argument and deduce that for all $j \geq 1$, there are $z, w \in f^j(K)$ with $\log|w| > L\log|z|$. Since $C_0 < 4 < L$, we get a contradiction with (23) when $j \geq j_0$, using also (25). This completes the proof of Theorem 8.3.

Under the assumptions of Theorem 9.1, we proceed in much the same way. Only certain technicalities differ, but a close examination of the argument shows that the underlying ideas are quite similar to what was done above.

Suppose that (20) holds for the values of the parameters as given in the statement of Theorem 9.1. Note that the symbols C and L now mean the quantities in (20). Set

$$C_1 = C + \log^+ M(1, f) \geq C > 0.$$

Choose $C_2 > 2$ so large that

$$\frac{1}{L} < \prod_{n=1}^{\infty}\left(1 - \frac{1}{C_2^{n\delta}}\right).$$

Next pick R' so that

(26) $$r < r^{C_2} < M(r, f)$$

for all $r \geq R'$. Choose $R_1 = \max\{R_0, R', 2\exp(C_1^{1/\delta} C_2)\}$. Let K be a compact connected subset of D containing points z_0 and w_0 with

$$|w_0| > |z_0| > R_1$$

and

(27) $$\frac{\log|w_0|}{\log|z_0|} > L^2.$$

Set $K_n = f^n(K)$. We seek to prove that for each $n \geq 1$, there are points $z_n, w_n \in K_n$ with

$$|w_n| > |z_n| > R_1$$

and

$$\frac{\log|w_n|}{\log|z_n|} > L^2 \prod_{k=1}^{n}\left(1 - \frac{1}{C_2^{k\delta}}\right) > L.$$

Since K is connected and (27) holds, there is $\zeta_0 \in K$ with $|w_0| = |\zeta_0|^L$. Thus $|\zeta_0| > |z_0|$. By (20), there is $t \in (|\zeta_0|, |w_0|]$ with

$$\frac{\log m(t, f)}{\log M(|\zeta_0|, f)} \geq L\left(1 - \frac{C}{(\log|\zeta_0|)^\delta}\right).$$

We have
$$|f(z_0)| \leq M(|z_0|, f).$$

Take any point $u_0 \in K$ with $|u_0| = t$. This is possible since K is connected. We have
$$\frac{\log|f(u_0)|}{\log|f(z_0)|} \geq \frac{\log m(t,f)}{\log M(|z_0|,f)} = \frac{\log m(t,f)}{\log M(|\zeta_0|,f)} \frac{\log M(|\zeta_0|,f)}{\log M(|z_0|,f)}.$$

We next find a lower bound for
$$\frac{\log M(|\zeta_0|,f)}{\log M(|z_0|,f)}.$$

If $1 < r_1 < r_2$ and $x_j = \log r_j$ for $j = 1, 2$, and if $r_1 = |z_0|$ and $r_2 = |\zeta_0|$, we have
$$\frac{\log M(|\zeta_0|,f)}{\log M(|z_0|,f)} = \frac{\varphi(x_2)}{\varphi(x_1)}.$$

Since φ is convex, we have
$$\varphi(x_1) \leq \frac{x_2 - x_1}{x_2}\varphi(0) + \frac{x_1}{x_2}\varphi(x_2)$$

so that
$$\varphi(x_2) \geq \frac{x_2}{x_1}\varphi(x_1) - \frac{x_2 - x_1}{x_1}\varphi(0),$$

hence
$$\frac{\varphi(x_2)}{\varphi(x_1)} \geq \frac{x_2}{x_1} - \left(\frac{x_2}{x_1} - 1\right)\frac{\varphi(0)}{\varphi(x_1)}.$$

If $\varphi(0) \leq 0$, we get
$$\frac{\varphi(x_2)}{\varphi(x_1)} \geq \frac{x_2}{x_1}.$$

In general, if $\varphi(0) > 0$,
$$\frac{\varphi(x_2)}{\varphi(x_1)} \geq \frac{x_2}{x_1}\left\{1 - \left(1 - \frac{x_1}{x_2}\right)\frac{\varphi(0)}{\varphi(x_1)}\right\} \geq \frac{x_2}{x_1}\left\{1 - \frac{\varphi(0)}{\varphi(x_1)}\right\}.$$

In particular,
$$\frac{\log M(|\zeta_0|,f)}{\log M(|z_0|,f)} = \frac{\varphi(x_2)}{\varphi(x_1)} \geq \frac{\log|\zeta_0|}{\log|z_0|}\left\{1 - \frac{\log M(1,f)}{\log M(|z_0|,f)}\right\}.$$

We conclude that

$$\frac{\log|f(u_0)|}{\log M(|z_0|,f)} \geq L\frac{\log|\zeta_0|}{\log|z_0|}\left(1 - \frac{C}{(\log|\zeta_0|)^\delta}\right)\left(1 - \frac{\log M(1,f)}{\log M(|z_0|,f)}\right)$$

$$= \frac{\log|w_0|}{\log|z_0|}\left(1 - \frac{C}{(\log|\zeta_0|)^\delta}\right)\left(1 - \frac{\log M(1,f)}{\log M(|z_0|,f)}\right)$$

$$\geq \frac{\log|w_0|}{\log|z_0|}\left(1 - \frac{C}{(\log|\zeta_0|)^\delta} - \frac{\log M(1,f)}{\log M(|z_0|,f)}\right)$$

$$\geq \frac{\log|w_0|}{\log|z_0|}\left(1 - \frac{C}{(\log|z_0|)^\delta} - \frac{\log M(1,f)}{\log M(|z_0|,f)}\right)$$

$$\geq \frac{\log|w_0|}{\log|z_0|}\left(1 - \frac{C + \log^+ M(1,f)}{(\log|z_0|)^\delta}\right)$$

since $\log M(|z_0|,f) \geq \log|z_0| \geq (\log|z_0|)^\delta$. So, whether $\varphi(0) \leq 0$ or $\varphi(0) > 0$, we have

$$\frac{\log|f(u_0)|}{\log M(|z_0|,f)} \geq L^2\left(1 - \frac{C_1}{(\log|z_0|)^\delta}\right) > L^2\left(1 - \frac{1}{C_2^\delta}\right) > L > 1$$

since $(\log|z_0|)^\delta > (\log R_0)^\delta \geq C_1 C_2^\delta$.

Since $|f(z_0)| \leq M(|z_0|,f)$ and $f(K) = K_1$ is connected, the set K_1 contains a point z_1 with $|z_1| = M(|z_0|,f)$. We set $f(u_0) = w_1$ and note that

$$\frac{\log|w_1|}{\log|z_1|} > L^2\left(1 - \frac{1}{C_2^\delta}\right) > L.$$

Lemma 10.1. *Suppose that $n \geq 1$ and that for all m with $1 \leq m \leq n$, there exist $z_m, w_m \in K_m$ with*

$$|w_m| > |z_m| > R_1,$$

$$|z_m| \geq |z_{m-1}|^{C_2} \geq |z_0|^{C_2^m}$$

for $1 \leq m \leq n$, and

$$\kappa_m \equiv \frac{\log|w_m|}{\log|z_m|} > L^2\prod_{k=1}^{m}\left(1 - \frac{1}{C_2^{k\delta}}\right) > L.$$

Then K_{n+1} contains points z_{n+1} and w_{n+1} such that

$$|w_{n+1}| > |z_{n+1}| > R_1,$$

$$|z_{n+1}| \geq |z_n|^{C_2} \geq |z_0|^{C_2^{n+1}},$$

and

$$\kappa_{n+1} \equiv \frac{\log|w_{n+1}|}{\log|z_{n+1}|} > L^2\prod_{k=1}^{n+1}\left(1 - \frac{1}{C_2^{k\delta}}\right) > L.$$

Proof of Lemma 10.1. Since K_n is connected and

$$\frac{\log |w_n|}{\log |z_n|} > L,$$

there is $\zeta_n \in K_n$ with

$$\frac{\log |w_n|}{\log |\zeta_n|} = L.$$

Now by (20), find $t \in (|\zeta_n|, |w_n|] = (|\zeta_n|, |\zeta_n|^L]$ with

$$(28) \qquad \frac{\log m(t,f)}{\log M(|\zeta_n|,f)} \geq L\left(1 - \frac{C}{(\log |\zeta_n|)^\delta}\right) > 1.$$

Then choose $u_n \in K_n$ with $|u_n| = t$. Note that $|z_n| < |\zeta_n|$. We have, as before,

$$\frac{\log |f(u_n)|}{\log M(|z_n|,f)} \geq \frac{\log m(t,f)}{\log M(|\zeta_n|,f)} \frac{\log M(|\zeta_n|,f)}{\log M(|z_n|,f)}$$

$$> L\left(1 - \frac{C}{(\log |\zeta_n|)^\delta}\right) \frac{\log |\zeta_n|}{\log |z_n|} \left(1 - \frac{\log^+ M(1,f)}{\log M(|z_n|,f)}\right)$$

$$\geq \frac{\log |w_n|}{\log |z_n|} \left(1 - \frac{C}{(\log |\zeta_n|)^\delta} - \frac{\log^+ M(1,f)}{\log M(|z_n|,f)}\right)$$

$$\geq \frac{\log |w_n|}{\log |z_n|} \left(1 - \frac{C_1}{(\log |z_n|)^\delta}\right).$$

Now

$$\log |z_n| \geq C_2^n \log |z_0|,$$

so

$$\frac{C_1}{(\log |z_n|)^\delta} \leq \frac{1}{C_2^{n\delta}} \frac{C_1}{(\log |z_0|)^\delta} \leq \frac{1}{C_2^{(n+1)\delta}}.$$

Choose $z_{n+1} \in K_{n+1}$ with $|z_{n+1}| = M(|z_n|,f)$. This is possible since K_{n+1} is connected, $f(z_n) \in K_{n+1}$, $|f(z_n)| \leq M(|z_n|,f)$, and $f(u_n) \in K_{n+1}$ while by (28),

$$|f(u_n)| \geq m(t,f) \geq M(|\zeta_n|,f) > M(|z_n|,f).$$

We get, with $w_{n+1} = f(u_n)$, that

$$\kappa_{n+1} = \frac{\log |w_{n+1}|}{\log |z_{n+1}|} > \kappa_n \left(1 - \frac{1}{C_2^{(n+1)\delta}}\right) > L^2 \prod_{k=1}^{n+1}\left(1 - \frac{1}{C_2^{k\delta}}\right) > L.$$

Also $|z_{n+1}| = M(|z_n|,f) \geq |z_n|^{C_2} > |z_n| > R_1$ by (26). This completes the proof of Lemma 10.1.

We continue with the proof of Theorem 9.1. We have previously shown that the hypothesis of Lemma 10.1 holds for $n = 1$. Hence induction on n

together with Lemma 10.1 shows that for every $n \geq 1$ there are $z_n, w_n \in K_n$ with

$$\frac{\log |w_n|}{\log |z_n|} > L.$$

As in the proof of Theorem 8.3 above, the distinction between $\log |w_n|$ and $\log(2|w_n - a|/\rho)$ is immaterial and is easily handled by taking R_1 even larger; we omit the details. Taking C_0 in (23) with $1 < C_0 < L$ and choosing the appropriate j_0, we obtain a contradiction as soon as $n \geq j_0$. This completes the proof of Theorem 9.1.

References

[1] J.M. Anderson and A. Hinkkanen, Unbounded domains of normality, Proc. Amer. Math. Soc. 126 (1998), 3243–3252.

[2] I.N. Baker, The iteration of polynomials and transcendental entire functions, J. Austral. Math. Soc. (Series A) 30 (1981), 483–495.

[3] I.N. Baker, Wandering domains in the iteration of entire functions, Proc. London Math. Soc. (Series 3) 49 (1984), 563–576.

[4] I.N. Baker, Infinite limits in the iteration of entire functions, Ergodic Theory Dynam. Systems 8 (1988), 503–507.

[5] P.D. Barry, On a theorem of Besicovitch, Quart. J. Math. Oxford (Series 2) 14 (1963), 292–302.

[6] P.D. Barry, On a theorem of Kjellberg, Quart. J. Math. Oxford (Series 2) 15 (1964), 179–191.

[7] A.F. Beardon, Iteration of rational functions, Springer-Verlag, Berlin–New York (1991).

[8] W. Bergweiler, Periodic points of entire functions: proof of a conjecture of Baker, Complex Variables Theory Appl. 17 (1991), 57–72.

[9] P. Bhattacharyya, On the domain of normality of an attractive fixpoint, Trans. Amer. Math. Soc. 153 (1971), 89–98.

[10] L. Carleson and T. Gamelin, Complex dynamics, Universitext: Tracts in Mathematics, Springer, New York, 1993.

[11] M.L. Cartwright, Integral functions, Cambridge University Press, 1956.

[12] P. Fatou, Sur les équations fonctionnelles, Bull. Soc. Math. France 47 (1919), 161–271.

[13] P. Fatou, Sur les équations fonctionnelles, Bull. Soc. Math. France 48 (1920), 33–94, 208–314.

[14] P. Fatou, Sur l'itération des fonctions transcendantes entières, Acta Math. 47 (1926), 337–370.

[15] W.H.J. Fuchs, Topics in the theory of functions of one complex variable, van Nostrand, 1967.

[16] M. E. Herring, Mapping properties of Fatou components, Ann. Acad. Sci. Fenn. Math. Ser. AI 23 (1998), 263–274.

[17] A. Hinkkanen, Entire functions with no unbounded Fatou components. Complex analysis and dynamical systems II, 217-226. Contemp. Math., 382, Amer. Math. Soc., Providence, 2005.

[18] Xinhou Hua and Chung-Chun Yang, Fatou components of entire functions of small growth, Ergodic Theory Dynam. Systems 19 (1999), 1281–1293.

[19] G. Julia, Mémoire sur l'itération des fonctions rationnelles, J. Math. Pures Appl. 8 (1918), 47–245.

[20] J. Milnor, Dynamics in one complex variable, Introductory lectures, Friedr. Vieweg & Sohn, Braunschweig, 1999.
[21] G. Stallard, *Some problems in the iteration of meromorphic functions*, PhD thesis, Imperial College, London, 1991.
[22] G. Stallard, The iteration of entire functions of small growth, Math. Proc. Cambridge Phil. Soc. 114 (1993), 43–55.
[23] N. Steinmetz, Rational iteration, Complex analytic dynamical systems, de Gruyter Studies in Mathematics, 16, Walter de Gruyter & Co., Berlin, 1993.
[24] Jian-Hua Zheng, Unbounded domains of normality of entire functions of small growth, Math. Proc. Cambridge Philos. Soc. 128 (2000), 355–361.

UNIVERSITY OF ILLINOIS, DEPARTMENT OF MATHEMATICS, 273 ALTGELD HALL, 1409 WEST GREEN STREET, URBANA, IL 61801 - USA
E-mail address: aimo@uiuc.edu

ON MULTIPLY CONNECTED WANDERING DOMAINS OF ENTIRE FUNCTIONS

MASASHI KISAKA AND MITSUHIRO SHISHIKURA

Dedicated to the memory of I. N. Baker

ABSTRACT. We investigate the connectivity $\text{conn}(D)$ of a wandering domain D of a transcendental entire function f. First we show that, for large n, $\text{conn}(f^n(D))$ is constant, equal to either $1, 2$ or ∞ (Theorem A). Next we construct an example of a transcendental entire function f with a doubly connected wandering domain (Theorem B), which is the main result of this paper. For this purpose we establish a slightly different version of quasiconformal surgery (Theorem 3.1). Also we construct the following examples by similar methods:
- An entire function f having a wandering domain D with $\text{conn}(D) = p$ for a given $p \in \mathbb{N}$ (Theorem C).
- An entire function f having a doubly connected wandering domain and all its singular values are contained in preperiodic Fatou components (Theorem D).
- An entire function f such that the set $\overline{f(\text{sing}(f^{-1}))}$ is equal to the whole plane $\mathbb{C}$ but f has a wandering domain, hence $J_f \neq \mathbb{C}$ (Theorem E).
- An entire function f with infinitely many grand orbits of doubly connected wandering domains. Furthermore, this f can be constructed so that the Lebesgue measure of the Julia set J_f is positive and there exist countably many measurable invariant line fields supported on mutually disjoint subsets of J_f (Theorem F).

1. INTRODUCTION

Let f be a transcendental entire function and f^n denote the n-th iterate of f. Recall that the *Fatou set* F_f and the *Julia set* J_f of f are defined as follows:

$$F_f = \{z \in \mathbb{C} : \{f^n\}_{n=1}^{\infty} \text{ is a normal family in a neighborhood of } z\},$$
$$J_f = \mathbb{C} \setminus F_f.$$

A connected component D of F_f is called a *Fatou component of f*. A Fatou component D is called a *wandering domain* if $f^m(D) \cap f^n(D) = \emptyset$ for every

---

1991 *Mathematics Subject Classification*. Primary 58F23; Secondary 30D05.
Key words and phrases. complex dynamics, wandering domain, entire functions, quasiconformal maps, quasiconformal surgery.

$m, n \in \mathbb{N}$ $(m \neq n)$. If there exists a $p \in \mathbb{N}$ with $f^p(D) \subseteq D$, then D is called a *periodic component of period p* and it is either an *attracting basin*, a *parabolic basin*, a *Siegel disk* or a *Baker domain*. In particular, if $p = 1$, U is called *an invariant component*.

Here we briefly explain the history of wandering domains. It was I. N. Baker who proved the existence of wandering domains for the first time. In 1963 he proved the following result; see [3, Theorem 1].

Theorem 1.1. *There is an entire function $g(z)$ given by the canonical product*

$$g(z) = Cz^2 \prod_{n=1}^{\infty} \left(1 + \frac{z}{r_n}\right)$$

such that $g(z)$ has at least one multiply connected Fatou component, where $C > 0$ is a constant and r_n is defined by some recursive formula and satisfies $1 < r_1 < r_2 < \cdots$. More precisely, let A_n be the annulus

$$A_n : r_n^2 < |z| < r_{n+1}^{\frac{1}{2}},$$

then there is an integer $N > 0$ such that for all $n > N$ the mapping $z \to g(z)$ maps A_n into A_{n+1} and $g^n(z) \to \infty$ uniformly in A_n. For each $n > N$, A_n belongs to a multiply connected component G_n of F_g.

At this moment, he did not assert that the above Fatou component was a wandering domain, because there was a possibility that G_n were equal for any $n > N$ and hence it was an invariant component. That is, it might be a *Baker domain*, in which, by definition, every point goes to ∞ under iteration. But about ten years later, he proved the following result; see [5, Theorem].

Theorem 1.2. *For $n > N$ the components G_n of F_g described above are all different and each is a wandering domain of g.*

More generally, he proved the following result; see [4, Theorem 1].

Theorem 1.3. *If f is a transcendental entire function, then F_f has no unbounded multiply connected component. That is, any unbounded Fatou component is simply connected*

Thus the first example of a wandering domain was multiply connected. On the other hand, examples of simply connected wandering domains are also known; an example due to Herman is given in [6, Example 5.1]. In this paper we consider the connectivity of a Fatou component, which is defined as follows:

Definition. For a domain D of $\mathbb{C}$, the *connectivity* $\operatorname{conn}(D)$ is defined to be the number of connected components of $\widehat{\mathbb{C}} \setminus D$, which may be ∞.

Note that $\mathrm{conn}(D) = 1$ if and only if D is simply connected, and $\mathrm{conn}(D) = 2$ if and only if D is doubly connected and conformally equivalent to a round annulus
$$\{z \ : \ 0 \leq r_1 < |z| < r_2 \leq \infty\}.$$
By the density of periodic points in the Julia set and Theorem 1.3, it is easily shown that if a Fatou component D is multiply connected, then it must be a wandering domain and $f^n|_D \to \infty$ $(n \to \infty)$. In 1985, Baker constructed an example of a transcendental entire function g with a wandering domain D of infinite connectivity; see [7, Theorem 2].

Theorem 1.4. *There is an entire function $g(z)$ given by the canonical product*
$$g(z) = Cz^2 \prod_{j=1}^{\infty} \left(1 + \frac{z}{r_j}\right)^2, \quad 1 < r_1 < r_2 < \cdots, \ C > 0$$
such that $g(z)$ has a wandering domain with infinite connectivity.

So the following is a natural question to ask:

Question: Is there a wandering domain D with finite connectivity, or more precisely, with $\mathrm{conn}(D) = p$, $p \in \mathbb{N}$?

This question was raised by Baker in [7] and is also explicitly stated as "Question 7" in [10, p.167]. The main purpose of this paper is to construct such an example. Incidentally, the connectivity of the wandering domain discussed in Theorem 1.1 and 1.2 is still unknown.

In this paper we first show the following:

Theorem A. *For a wandering domain D of a transcendental entire function f, the connectivity $\mathrm{conn}(f^n(D))$ is constant for large n and it is either 1, 2 or ∞. If it is 1, then $\mathrm{conn}(D) = 1$. If it is 2, then $f : f^n(D) \to f^{n+1}(D)$ is a covering of annuli for every sufficiently large n.*

According to this theorem, we make the following:

Definition. We define the *eventual connectivity* of a wandering domain D to be $\mathrm{conn}(f^n(D))$ for sufficiently large n.

The main result of this paper is as follows:

Theorem B. *There exists a transcendental entire function f with a wandering domain D such that $f^n(D)$ is doubly connected for each $n \geq 0$, i.e. the eventual connectivity of D is 2. Moreover f has no asymptotic values and all critical values are mapped to 0 which is a repelling fixed point.*

Theorem B gives a negative answer to the following open problem raised by W. Bergweiler [17, p.354].

Problem Let f be a transcendental entire function. Suppose that $f^n|_U \to \infty$ as $n \to \infty$ for some connected component U of the Fatou set of f. Does there exist $\zeta \in \text{sing}(f^{-1})$ such that $f^n(\zeta) \to \infty$? If not, does there exist at least $\zeta \in \text{sing}(f^{-1})$ such that $f^n(\zeta)$ is unbounded?

The main technique used to construct this kind of example is quasiconformal surgery. By using the same technique and some additional arguments, we can also show the following:

Theorem C. *For every $p \in \mathbb{N}$ with $p \geq 3$ there exists a transcendental entire function f with a wandering domain D with $\text{conn}(D) = p$ and $\text{conn}(f^n(D)) = 2$ for every $n \geq 1$.*

Theorem D. *There exists a transcendental entire function f with a wandering domain D such that the eventual connectivity of D is 2. Moreover f has no asymptotic values and all critical values are mapped to 0 which is an attracting fixed point.*

Theorem E. *There exists a transcendental entire function f such that the set $\overline{f(\text{sing}(f^{-1}))}$ is equal to the whole plane $\mathbb{C}$ but f has a wandering domain, hence $J_f \neq \mathbb{C}$.*

Theorem F. *There exists a transcendental entire function f with infinitely many grand orbits of doubly connected wandering domains. That is, there exist doubly connected wandering domains D_i ($i \in \mathbb{N}$) such that if $i \neq j$, then $f^m(D_i) \cap f^n(D_j) = \emptyset$ for any $m, n \in \mathbb{N}$. Furthermore, this f can be constructed so that the Lebesgue measure of the Julia set J_f is positive and there exist countably many measurable invariant line fields supported on mutually disjoint subsets of J_f. In this case f has infinite dimensional qc-deformation supported on J_f.*

Theorem D answers the following question, also by W. Bergweiler [10, Question 10].

Question Can a meromorphic function f have wandering domains if all (or all but finitely many) points of $\text{sing}(f^{-1})$ are contained in preperiodic domains?

For the definitions of an invariant line field and a qc-deformation, see [14]. The first example of a transcendental entire function with countably many measurable invariant line fields was constructed by Eremenko and Lyubich by using approximation theory ([11, Example 5]).

Incidentally, Baker constructed an example of an entire function with infinitely many grand orbits of simply connected wandering domains in [6, p.567, Theorem 5.2]. Also Baker, Kotus and Lü considered the related problem of the existence of multiply connected Fatou components for transcendental meromorphic functions with at least one pole ([8], [9]).

In §2, we prove Theorem A and §3 is devoted to the explanation of the quasiconformal surgery, which is the main tool for the proof of the main Theorem B. We give a proof of Theorem B in §4. Finally we give proofs of Theorems C, D, E and F in §5. In §6, as an Appendix, we prove the main proposition (Proposition 4.1) for the proof of Theorem B.

Acknowledgment: The results presented here have been inspired by Noel Baker's series of papers on wandering domains. We thank him for his pioneering works on the dynamics of transcendental functions. We thank Walter Bergweiler for valuable discussions and drawing our attention to "Question 10" in [10, p.170].

2. Proof of Theorem A

We need some lemmas. These were proved by Baker [6, Theorem 3.1 and its Corollary].

Lemma 2.1. *Let D be a multiply connected wandering domain of an entire function f and $\gamma \subset D$ be a nontrivial curve in D. Then $f^n \to \infty$ $(n \to \infty)$ in D and for every sufficiently large n the winding number of $f^n(\gamma)$ with respect to the origin is positive.*

Lemma 2.2. *If f has an asymptotic value, then every Fatou component of F_f is simply connected.*

Proof of Theorem A. By Lemma 2.1 and Lemma 2.2, we may assume that f has no asymptotic values and D is bounded. Then $f : D \to f(D)$ is a branched covering. If $\text{conn}(D) = \infty$, then $\text{conn}(f^n(D)) = \infty$ for every $n \in \mathbb{N}$. So we assume that $\text{conn}(D) < \infty$. By the Riemann-Hurwitz Theorem, we have

$$2 - \text{conn}(D) = (\deg f|_D)(2 - \text{conn}(f(D))) - ^{\#}\{\text{critical points in } D\}. \quad (2.1)$$

Then it easily follows that $\text{conn}(D) \geq \text{conn}(f(D))$ and hence $\text{conn}(f^n(D))$ is constant for large n. Let us denote it by p. Suppose that $3 \leq p < \infty$, then by replacing D with $f^n(D)$ in (2.1) it follows that $\deg f|_{f^n(D)} = 1$ and hence $f : f^n(D) \to f^{n+1}(D)$ is conformal. By the Argument Principle, f is also 1 to 1 on the bounded components of $\mathbb{C} \setminus f^n(D)$. Then from Lemma 2.1, f must be 1 to 1 on $\mathbb{C}$, which is a contradiction, since f is transcendental. Therefore if p is finite, then $p = 1$ or 2. If $p = 1$, then it is easy to see that $\text{conn}(D) = 1$. If $p = 2$, then from (2.1) we have $\#\{\text{critical points in } f^n(D)\} = 0$ and hence the result follows. □

3. SURGERY AND CONFORMAL STRUCTURE

In this section, we recall the definition of a quasiconformal map and explain quasiconformal surgery (Theorem 3.1).

Definition 3.1. An orientation preserving homeomorphism $\varphi : D \to D'$ between two domains D and D' is called a *quasiconformal map* if it is absolutely continuous on lines on any rectangle $R = \{x + iy : a \leq x \leq b, c \leq y \leq d\}$ contained in D, that is,

(i) $\varphi(x + iy)$ is absolutely continuous as a function of $x \in [a, b]$ for almost every y, and $\varphi(x + iy)$ is absolutely continuous as a function of $y \in [c, d]$ for almost every x

and moreover,

(ii) $|\mu_\varphi(z)| \leq k < 1$ a.e. $z \in D$,

where $\mu_\varphi = \varphi_{\bar{z}}/\varphi_z$ and k is some constant with $0 \leq k < 1$. If $k = 0$, then φ is conformal. If $k \neq 0$, we set $K = (1+k)/(1-k)$ and call φ a *K-quasiconformal* (*K-qc* for short) *map*. The constant

$$K_\varphi = \inf\{K : \varphi \text{ is } K\text{-qc}\}$$

is called the *maximal dilatation of φ*. A map $g : D \to D'$ is called a *K-quasiregular map* if g can be written as $g = f \circ \varphi$ with a K-quasiconformal map φ and an analytic map f.

For properties of quasiconformal maps, see [1].

In order to construct an entire function with doubly-connected wandering domains, we first construct a quasiregular map g with similar properties to what we really want by gluing suitable polynomials together by using interpolation. Then we choose a suitable quasiconformal map φ so that $\varphi \circ g \circ \varphi^{-1}$ is the desired entire function. We call this procedure *quasiconformal surgery*. More precisely, we can formulate this procedure as follows, which is slightly different from the one discussed in [15]:

Theorem 3.1 (quasiconformal surgery). *Let g be a quasiregular mapping from $\mathbb{C}$ to $\mathbb{C}$. Suppose that there are (disjoint) measurable sets $E_j \subset \mathbb{C}$ ($j = 1, 2, \dots$) satisfying:*

(i) *for almost every $z \in \mathbb{C}$, the g-orbit of z passes E_j at most once for every j;*

(ii) *g is K_j-quasiregular on E_j;*

(iii) $K_\infty = \prod_{j=1}^\infty K_j < \infty$;

(iv) g is holomorphic a.e. outside $\bigcup_{j=1}^\infty E_j$ (that is, $\frac{\partial g}{\partial \bar{z}} = 0$ a.e. on $\mathbb{C} \setminus \bigcup_{j=1}^\infty E_j$).

Then there exists a K_∞-quasiconformal map φ such that $f = \varphi \circ g \circ \varphi^{-1}$ is an entire function.

Proof. A *measurable conformal structure* is the measurable conformal equivalence of measurable Riemannian metrics, and can be represented by the metric of the form
$$ds = |dz + \mu(z)d\bar{z}|,$$
where $\mu(z)$ is a $\mathbb{C}$-valued measurable function with
$$||\mu||_\infty = \text{ess.}\sup |\mu(z)| < 1.$$
The distance between two measurable conformal structures $\sigma = [|dz+\mu(z)d\bar{z}|]$ and $\sigma' = [|dz + \mu'(z)d\bar{z}|]$ is defined by
$$d(\sigma, \sigma') = \text{ess.}\sup d_\mathbb{D}(\mu(z), \mu'(z)),$$
where $d_\mathbb{D}$ denotes the Poincaré distance on the unit disk $\mathbb{D}$. A quasiregular map defines the pull-back $g^*(\sigma)$ of the measurable conformal structure σ, and the pull-back preserves the above distance. Let $\sigma_0 = [|dz|]$ denote the standard conformal structure. If g is K-quasiregular, then we have $d(g^*(\sigma_0), \sigma_0) \leq \log K$.

Now define the conformal structures
$$\sigma_n(z) = (g^n)^*(\sigma_0(g^n(z))),$$
which are defined almost everywhere. The pointwise distance (when defined) satisfies
$$d_\mathbb{D}(\sigma_{n+1}(z), \sigma_n(z)) = d_\mathbb{D}(g^*(\sigma_0(g^{n+1}(z))), \sigma_0(g^n(z))) \leq \log K_m$$
if $g^n(z)$ is in some E_m and it is 0 otherwise.

By the hypotheses (i) and (iii), $\{\sigma_n(z)\}_{n=0}^\infty$ is defined and a Cauchy sequence for almost all z. Therefore the pointwise limit $\sigma(z) = \lim_{n\to\infty} \sigma_n(z)$ exists a.e. and satisfies
$$d(\sigma, \sigma_0) \leq \sum_{j=1}^\infty \log K_j = \log K_\infty.$$
Then σ can be written as
$$\sigma(z) = [|dz + \mu(z)d\bar{z}|]$$
with
$$|\mu(z)| \leq \frac{K_\infty - 1}{K_\infty + 1} \quad \text{a.e.}$$

By the Measurable Riemann Mapping Theorem [1, p.98, Theorem 3], there exists a K_∞-quasiconformal mapping $\varphi : \mathbb{C} \to \mathbb{C}$ such that $\frac{\partial \varphi}{\partial \bar{z}} / \frac{\partial \varphi}{\partial z} = \mu(z)$ a.e., in other words $\varphi^*(\sigma_0) = \sigma$. Then $f = \varphi \circ g \circ \varphi^{-1}$ is quasiregular and satisfies $f^*(\sigma_0) = \sigma_0$. This implies that f is locally conformal except at critical points, hence it is analytic. □

Remark. Theorem 3.1 also follows from an idea of Sullivan [16, Theorem 9].

4. Construction (Proof of Theorem B)

Part I : Construction of a model map f_0.

Definition 4.1. For a closed concentric annulus A with center 0, we use the notation
$$A = A(r_1, r_2) = \{z \ : \ r_1 \le |z| \le r_2\}, \quad (0 < r_1 < r_2)$$
and
$$\partial_{\text{inner}} A = \{z \ : \ |z| = r_1\}, \quad \partial_{\text{outer}} A = \{z \ : \ |z| = r_2\},$$
which denote the *inner boundary* and the *outer boundary* of A, respectively. We define the *modulus* of A by
$$\mod(A) = \frac{1}{2\pi} \log \frac{r_2}{r_1}.$$
The *core curve* Core (A) is the unique closed geodesic of A and given by
$$\text{Core}(A) = \{z \ : \ |z| = \sqrt{r_1 r_2}\}.$$

We first construct a model map f_0 which roughly describes the dynamics of what we really want to construct. Let $k_n \in \mathbb{N}$ be given integers with $k_0 \le k_1 \le \cdots \le k_n \le \cdots$. In what follows we choose suitable $R_n \in \mathbb{R}$ with $0 = R_0 < R_1 < R_2 < \cdots$ and set
$$A_n = A(R_n, R_{n+1}) \quad (n \ge 0).$$
(Note that here we abuse the notation — A_0 is a disk, not an annulus.) Then we want to construct a map $f_0 : \mathbb{C} \to \mathbb{C}$ with the following dynamical properties:
$$f_0(z) = a_0 z^{k_0}, \quad z \in A_0 \setminus \partial_{\text{outer}} A_0$$
such that $f_0(A_0) = A_0 \cup A_1$ and
$$f_0(z) = a_n z^{k_n}, \quad z \in A_n \setminus \partial_{\text{outer}} A_n$$
such that $f_0 : A_n \to A_{n+1}$ is a covering map of degree k_n. (See Figure 1, where we describe the annuli A_n as subsets of an infinite cylinder, instead of round annuli in the complex plane.)

For this purpose, we have to choose appropriate $a_n \in \mathbb{C}^*$ and $R_n > 0$. So first we take a_0 and R_1, such that $R_2 = |a_0| R_1^{k_0} > R_1$ holds and also

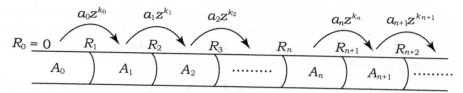

FIGURE 1. The model map f_0. Note that this is only a schematic picture and in reality, $\mod(A_n)$ rapidly increases as n tends to ∞. The same is also true for the following figures.

$M_1 = \exp(2\pi \mod(A_1))$ is large enough to be able to apply Proposition 4.1 in the next part. (Actually we have to ensure that $\mod(A_1) > m_0$, where m_0 is the constant in Proposition 4.1.) Once the constants a_0, R_1 and k_n ($n \in \mathbb{N}$) are chosen, then the constants $a_n \in \mathbb{C}^*$ ($n \geq 1$) and $R_n > 0$ ($n \geq 2$) are determined inductively as follows: define $M_n > 0$ by

$$M_1 = \exp(2\pi \mod(A_1)), \quad M_{n+1} = M_n^{k_n} \ (n \geq 1)$$

and set

$$R_{n+1} = M_n R_n \ (n \geq 1).$$

Also take $a_n \in \mathbb{C}^*$ with the condition

$$R_{n+1} = |a_n| R_n^{k_n}.$$

Note that only $|a_n|$ is determined by the condition above and we can choose $\arg a_n$ freely. Then it is easy to see that

$$\lim_{|z| \nearrow R_n} |f_0(z)| = \lim_{|z| \searrow R_n} |f_0(z)|,$$

because

$$|f_0(z)| = |a_{n-1} z^{k_{n-1}}| \to |a_{n-1}| R_n^{k_{n-1}} \quad (|z| \nearrow R_n)$$

and

$$|a_{n-1}| R_n^{k_{n-1}} = |a_{n-1}| R_{n-1}^{k_{n-1}} M_{n-1}^{k_{n-1}} = R_n M_n = R_{n+1}.$$

On the other hand we have

$$|f_0(z)| = |a_n z^{k_n}| \to |a_n| R_n^{k_n} = R_{n+1} \quad (|z| \searrow R_n).$$

Hence f_0 itself is discontinuous on $|z| = R_n$ but the map $|f_0| : \mathbb{C} \to \mathbb{R}$ is continuous. According to Lemma 2.1, in general, f^n goes to ∞ on a multiply connected wandering domain D and $f^n(D)$ is mapped to an "outer" region by f. So our f_0 indeed satisfies this situation.

Part II : Construction of a quasiregular map f_1 from the model map f_0.

Now we modify the map f_0 to construct a new quasiregular map f_1 and then perform quasiconformal surgery to obtain the desired map f. First we put $k_n = n+1$. For each n we replace f_0 with some different polynomial around some annulus containing the circle $|z| = R_n$ and glue this polynomial and the original map f_0 together by interpolation. More precisely we apply the following proposition.

Proposition 4.1. (1) *Let A, A' and $\widehat{A}$, $\widehat{A}'$ be two pairs of concentric round annuli with center 0 which satisfy*

$$\partial_{\text{outer}} A = \partial_{\text{inner}} A' = \{z \; : \; |z| = R\}, \quad \partial_{\text{outer}} \widehat{A} = \partial_{\text{inner}} \widehat{A}' = \{z \; : \; |z| = \widehat{R}\},$$

and

$$\operatorname{mod}(\widehat{A}) = k \cdot \operatorname{mod}(A), \quad \operatorname{mod}(\widehat{A}') = (k+1) \cdot \operatorname{mod}(A').$$

Let $F_A : A \to \widehat{A}$, $F_A(z) = c_A z^k$, $(k \geq 2)$ be a covering map of degree k which maps A onto $\widehat{A}$. Similarly let $F_{A'} : A' \to \widehat{A}'$, $F_{A'}(z) = c_{A'} z^{k+1}$, $(k \geq 2)$ be a covering map of degree $k+1$ which maps A' onto $\widehat{A}'$. For the annulus A, take annuli $B^\sharp(A)$, $E^\sharp(A)$, $E^\flat(A)$ and $B^\flat(A)$ as in Figure 2 such that

$$\operatorname{mod}(B^\sharp(A)) = \operatorname{mod}(E^\sharp(A)) = \operatorname{mod}(E^\flat(A)) = \operatorname{mod}(B^\flat(A)) = \sqrt{\operatorname{mod}(A)} \tag{4.1}$$

and define

$$A^- = A \setminus (B^\sharp(A) \cup E^\sharp(A) \cup E^\flat(A) \cup B^\flat(A)).$$

Take similar annuli for each A', $\widehat{A}$ and $\widehat{A}'$. Then there exists a constant $m_0 > 0$ such that if $\operatorname{mod}(A) > m_0$ and $\operatorname{mod}(A') > m_0$, then there exists a quasiregular map

$$g : A^- \cup E^\flat(A) \cup B^\flat(A) \cup B^\sharp(A') \cup E^\sharp(A') \cup A'^- \to \mathbb{C}$$

which satisfies the following conditions (I) $\sim$ (III):

(I-a) $g = F_A$ on A^- and $g = F_{A'}$ on A'^-.

(I-b) g is holomorphic on $\operatorname{int} B = \operatorname{int}(B^\flat(A) \cup B^\sharp(A'))$ with a unique critical point $\zeta \in B^\flat(A)$. Also g satisfies $g(\zeta) = \widehat{R}$ and $g(R) = 0$.

(I-c) g is K-quasiregular on $\operatorname{int} E = \operatorname{int}(E^\flat(A) \cup E^\sharp(A'))$ and the maximal dilatation K_g satisfies

$$K = K_g \leq \max\left(1 + \frac{2}{\sqrt{k \cdot \operatorname{mod}(A)}}, 1 + \frac{2}{\sqrt{(k+1) \cdot \operatorname{mod}(A')}}\right) \tag{4.2}$$

$$= \max\left(1 + \frac{2}{\sqrt{\operatorname{mod}(\widehat{A})}}, 1 + \frac{2}{\sqrt{\operatorname{mod}(\widehat{A}')}}\right).$$

(II-a) $g(\operatorname{Core}(A^-)) = \operatorname{Core}(\widehat{A}^-)$. Similarly, $g(\operatorname{Core}(A'^-)) = \operatorname{Core}(\widehat{A}'^-)$.

(II-b) $g(A^-) \subset \widehat{A}^-$ and this inclusion is essential. That is, $g(A^-)$ is an annulus in $\widehat{A}^-$ and its core curve is not 0-homotopic in $\widehat{A}^-$. Similarly, $g(A'^-) \subset \widehat{A}'^-$ essentially.

(III-a) $g(E^\sharp(A')) \subset E^\sharp(\widehat{A'}) \cup \widehat{A'}^-$ essentially.

(III-b) $g(E^\flat(A)) \subset \widehat{A}^- \cup E^\flat(\widehat{A})$ essentially.

(2) *In the case of $k = 1$, the same conclusion holds if we replace the condition* (4.1) *for the annulus A with*

$$\mathrm{mod}\,(E^\flat(A)) = \mathrm{mod}\,(B^\flat(A)) = \mathrm{mod}\,(B^\sharp(A)) = \mathrm{mod}\,(E^\sharp(A)) = 2\sqrt{\mathrm{mod}\,(A)}. \tag{4.3}$$

(*Note that we need not change the conditions* (4.1) *for the annuli A', $\widehat{A}$ and $\widehat{A'}$.*)

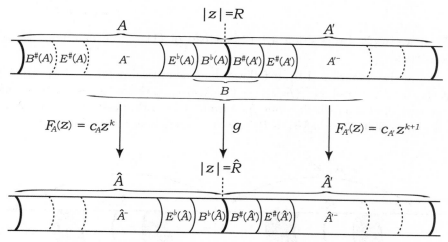

FIGURE 2. Interpolation between the two maps F_A and $F_{A'}$. We glue these two maps together in a neighborhood of the circle $\{z : |z| = R\}$.

We prove this Proposition in the Appendix.

Remark. Note that $g(B)$ covers not only a neighborhood of $\{|z| = \widehat{R}\}$ but also both $\widehat{A}$ and the bounded component of $\mathbb{C} \setminus \widehat{A}$.

Now we apply Proposition 4.1 (1) to each pair of annuli $(A, A') = (A_{n-1}, A_n)$ and maps $F_A(z) = a_{n-1}z^n$, $F_{A'}(z) = a_n z^{n+1}$ $(n = 2, 3, \cdots)$ to obtain a new map $g(z) = g_n(z)$. In this case, of course, $\widehat{A} = A_n$ and $\widehat{A'} = A_{n+1}$. We use the following notation:

$$B_n^\sharp = B^\sharp(A_n), \quad E_n^\sharp = E^\sharp(A_n), \quad E_{n+1}^\flat = E^\flat(A_n), \quad B_{n+1}^\flat = B^\flat(A_n).$$

Also we define

$$B_n = B_n^\flat \cup B_n^\sharp = B^\flat(A_{n-1}) \cup B^\sharp(A_n);$$

see Figure 3. Note that this notation is somehow different to what we defined in Proposition 4.1. Here we use "$\sharp$" and "$\flat$" with respect to the circle $\{z : |z| = R_n\}$ so, for example, the annuli $E_n^\flat$, B_n and $E_n^\sharp$ are located in this order as in Figure 3.

For $n = 1$, we consider the pair $(A_0^\Diamond, A_1)$ rather than (A_0, A_1). More precisely, we take $A_0^\Diamond$ to be a preimage of A_1 by the map $a_0 z$. Then we have $\operatorname{mod}(A_0^\Diamond) = \operatorname{mod}(A_1)$. Define subannuli $B_0^\sharp$, $E_0^\sharp$, A_0^-, $E_1^\flat$ and $B_1^\flat$ such that

$$A_0^\Diamond = B_0^\sharp \cup E_0^\sharp \cup A_0^- \cup E_1^\flat \cup B_1^\flat,$$

$$\operatorname{mod}(E_1^\flat) = \operatorname{mod}(B_1^\flat) = \operatorname{mod}(B_0^\sharp) = \operatorname{mod}(E_0^\sharp) = 2\sqrt{\operatorname{mod}(A_0^\Diamond)}$$
$$(= 2\sqrt{\operatorname{mod}(A_1)}).$$

Then we apply Proposition 4.1 (2) instead of (1) to the pair $(A_0^\Diamond, A_1)$ to construct $g_1(z)$.

From the condition (I-b), it follows that the critical point ζ_n of g_n satisfies $g_n(\zeta_n) = R_{n+1}$ and $g_{n+1}(R_{n+1}) = 0$. Also g_n satisfies an estimate on its maximal dilatation which is obtained from (4.2) in Proposition 4.1. Since we take a_0 so that $R_2 = |a_0|R_1 > R_1$, $z = 0$ is a repelling fixed point of f_0.

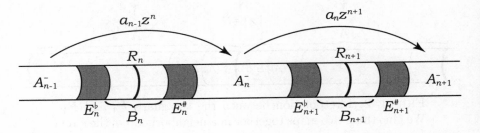

FIGURE 3. Construction of f_1 from f_0 by interpolation.

Then define a new map f_1 by

$$f_1(z) = \begin{cases} f_0(z), & z \in A_0 \setminus (E_1^\flat \cup B_1^\flat) \\ f_0(z), & z \in A_n^- \quad n = 1, 2, \cdots \\ g_n(z), & z \in E_n^\flat \cup B_n \cup E_n^\sharp \quad n = 1, 2, \cdots \end{cases}.$$

Part III : Application of quasiconformal surgery to f_1.

The new map f_1 is a quasiregular map with the desired dynamical properties. Hence we can apply quasiconformal surgery (Theorem 3.1) to obtain a transcendental entire function f with the desired properties. More precisely, the following holds:

Proposition 4.2. *The new map f_1 satisfies the following conditions* (I) $\sim$ (IV):

(I-a) $f_1(z) = a_n z^{n+1}$ on A_n^-.

(I-b) f_1 is holomorphic on B_n.

(I-c) f_1 is K_n-quasiregular on $E_n = E_n^\flat \cup E_n^\sharp$ with
$$K_n \leq 1 + \frac{2}{\sqrt{n! \cdot \mathrm{mod}\,(A_1)}}.$$

(I-d) f_1 has a critical point $\zeta_n \in B_n^\flat$ which satisfies $f_1(\zeta_n) = R_{n+1}$ and $f_1^2(\zeta_n) = 0$ $(n = 1, 2, \cdots)$. $\{\zeta_n\}_{n=1}^\infty$ is the set of all critical points of f_1.

(II-a) $f_1(\mathrm{Core}\,(A_n^-)) = \mathrm{Core}\,(A_{n+1}^-)$.

(II-b) $f_1(A_n^-) \subset A_{n+1}^-$ and this inclusion is essential.

(III-a) $f_1(E_n^\sharp) \subset E_{n+1}^\sharp \cup A_{n+1}^-$ essentially.

(III-b) $f_1(E_n^\flat) \subset A_n^- \cup E_{n+1}^\flat$ essentially.

(IV) $f_1(B_n) \subset \bigcup_{j=0}^{n+1} A_j$.

Hence there exists a quasiconformal mapping φ such that $f = \varphi \circ f_1 \circ \varphi^{-1}$ is holomorphic and entire.

Proof. All the conditions (I) $\sim$ (III) are obtained by applying Proposition 4.1 to each pair of annuli $(A, A') = (A_{n-1}, A_n)$ and the maps $F_A(z) = a_{n-1}z^n$, $F_{A'}(z) = a_n z^{n+1}$ $(n = 1, 2, \cdots)$. Note that
$$K_n \leq 1 + \frac{2}{\sqrt{\mathrm{mod}\,(A_n)}} = 1 + \frac{2}{\sqrt{n! \cdot \mathrm{mod}\,(A_1)}}.$$

Condition (IV) holds from the construction. Then (II-b), (III-a) and (III-b) show that for any $z \in \mathbb{C}$ the f_1-orbit of z passes E_n at most once for every n. Also from (I-c), f_1 is K_n-quasiregular on E_n with
$$K_\infty = \prod_{n=1}^\infty K_n \leq \prod_{n=1}^\infty \left(1 + \frac{2}{\sqrt{n! \cdot \mathrm{mod}\,(A_1)}}\right) < \infty.$$

Finally f_1 is holomorphic outside $\bigcup_{n=1}^\infty E_n$ by (I-a) and (I-b). Therefore we can apply Theorem 3.1 to the map f_1 and hence there exists a K_∞-quasiconformal map φ such that $f = \varphi \circ g \circ \varphi^{-1}$ is a transcendental entire function. □

Part IV : The map f has the desired properties.

Let $\widetilde{A}_n = \varphi(A_n)$, $\widetilde{B}_n = \varphi(B_n)$, $\cdots$ etc. Then f satisfies exactly the same conditions for $\widetilde{A}_n$, $\widetilde{B}_n$ etc. in Proposition 4.2 as f_1 satisfies for A_n, B_n, etc.

Lemma 4.3. *The annuli $\widetilde{A}_n^-$ ($n = 1, 2, \cdots$) are contained in the Fatou set F_f.*

Proof. By the construction, we have $f(\widetilde{A}_n^-) \subset \widetilde{A}_{n+1}^-$ and the iterates tend to ∞ uniformly on $\widetilde{A}_n^-$, hence $\widetilde{A}_n^-$ is contained in F_f. □

Let us denote by D_n the Fatou component containing $\widetilde{A}_n^-$ ($n \geq 1$).

Lemma 4.4. $D_n \neq D_{n+1}$.

Proof. Suppose the sets $\widetilde{A}_n^-$ and $\widetilde{A}_{n+1}^-$ belong to the same Fatou component $D = D_n = D_{n+1}$. Take $z_1 \in \text{Core}\,(\widetilde{A}_n^-)$ and $z_2 \in \text{Core}\,(\widetilde{A}_{n+1}^-)$; see Figure 4.

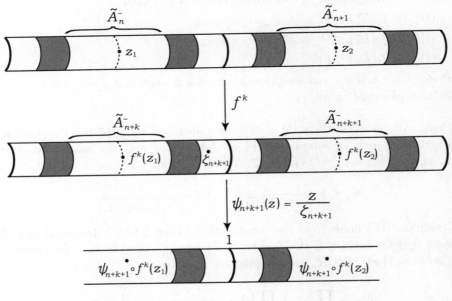

FIGURE 4

Then $f^k(z_1) \in \text{Core}\,(\widetilde{A}_{n+k}^-)$ and $f^k(z_2) \in \text{Core}\,(\widetilde{A}_{n+k+1}^-)$ from Proposition 4.2 (II-a). By the construction, $0 \notin D$ since 0 is a repelling fixed point. Also, for $m \geq 1$, the critical point ζ_m of f satisfies $\zeta_m \in B_m \setminus D$ since $f^2(\zeta_m) = 0$. Let $\psi_m(z) = z/\zeta_m$ then
$$\psi_{n+k+1} \circ f^k(D) \subset \Omega \equiv \widehat{\mathbb{C}} \setminus \{0, 1, \infty\}.$$
Therefore
$$d_\Omega(\psi_{n+k+1} \circ f^k(z_1), \psi_{n+k+1} \circ f^k(z_2)) \leq d_D(z_1, z_2),$$

where d_Ω and d_D are the Poincaré distances of Ω and D respectively. By the construction we have

$$\psi_{n+k+1} \circ f^k(z_1) \to 0 \quad (k \to \infty).$$

In fact, $\{0, f^k(z_1)\}$ and $\{\zeta_{n+k+1}, \infty\}$ are separated by an annulus which is the outer half of $\widetilde{A}^-_{n+k} \setminus \mathrm{Core}\,(\widetilde{A}^-_{n+k})$, and its modulus tends to ∞ as $k \to \infty$. Similarly $\psi_{n+k+1} \circ f^k(z_2) \to \infty$ holds. Hence it follows that

$$d_\Omega(\psi_{n+k+1} \circ f^k(z_1), \psi_{n+k+1} \circ f^k(z_2)) \to \infty.$$

This contradicts the previous statement. □

Remark. This Lemma also follows immediately from Baker's result, Theorem 1.3. Baker's proof of Theorem 1.3 is based on the construction of the hyperbolic metric and so the main idea of our proof of Lemma 4.4 is very similar to his.

Proposition 4.5. *The Fatou component D_n containing $\widetilde{A}^-_n$ can be written as*

$$D_n = \bigcup_{k=0}^{\infty} \widetilde{A}^-_{n,k}, \tag{4.4}$$

where $\widetilde{A}^-_{n,k}$ is the component of $f^{-k}(\widetilde{A}^-_{n+k})$ containing $\widetilde{A}^-_n$. Moreover if all the D_n do not contain critical points, then they are doubly connected, i.e. the eventual connectivity of D_n is 2.

Note that (4.4) is an increasing union, since $f(\widetilde{A}^-_{n+k}) \subset \widetilde{A}^-_{n+k+1}$. In order to prove Proposition 4.5, we need some lemmas.

Lemma 4.6. *Let $a, b > 0$ and $A = \{z \in \mathbb{C} : 0 < \mathrm{Re}\, z < a\}/\sim$, where $z \sim z + nbi$ ($n \in \mathbb{Z}$). Suppose that φ is a quasiconformal mapping from A onto another annulus A'. Denote $\mu = \frac{\partial \varphi}{\partial \bar z} / \frac{\partial \varphi}{\partial z}$. (In other words, A' can be considered as an annulus A with the conformal structure $|dz + \mu(z)\,d\bar z|$.) Then the moduli of A and A' satisfy*

$$\frac{\iint_A 1\,dx\,dy}{\iint_A K_\mu(z)\,dx\,dy} \leq \frac{\mathrm{mod}\,(A')}{\mathrm{mod}\,(A)} \leq \frac{\iint_A K_\mu(z)\,dx\,dy}{\iint_A 1\,dx\,dy},$$

where $K_\mu(z) = \dfrac{1 + |\mu(z)|}{1 - |\mu(z)|}$.

In particular, if $K_\mu(z) = 1$ outside a measurable set $X \subset A$ and $K_\mu(z) \leq K$ on X, then

$$\frac{\mathrm{mod}\,(A')}{\mathrm{mod}\,(A)} \leq K\frac{|X|}{|A|} + \left(1 - \frac{|X|}{|A|}\right),$$

where $|X|$ (resp. $|A|$) denotes the Lebesgue measure of X (resp. A).

The first half is called the Grötzsch inequality and the second half is an easy consequence; see [12, proof of Proposition 3]. (The proof was for a rectangle but it can easily be adapted for annuli of the above form.)

In the construction in Part III, we had

$$\frac{|A_n^+|}{|A_n^-|} \to 1 \quad (n \to \infty),$$

where

$$A_n^+ = B_n^\flat \cup A_n \cup B_{n+1}^\sharp$$

and $|\cdot|$ denotes the Lebesgue measure in the cylinder model $\mathbb{C}/2\pi i\mathbb{Z}$. Since φ is conformal on A_n^-, it follows from Lemma 4.6 that

$$\frac{\mathrm{mod}\,(\widetilde{A}_n^+)}{\mathrm{mod}\,(A_n^+)} \to 1 \quad (n \to \infty).$$

Combining this with $\mathrm{mod}\,(\widetilde{A}_n^-) = \mathrm{mod}\,(A_n^-)$ and $\mathrm{mod}\,(A_n^+)/\mathrm{mod}\,(A_n^-) \to 1$ ($n \to \infty$), we have:

Corollary 4.7.

$$\frac{\mathrm{mod}\,(\widetilde{A}_n^+)}{\mathrm{mod}\,(\widetilde{A}_n^-)} \to 1 \quad (n \to \infty).$$

Lemma 4.8. *For $m > 0$ and $L > 0$, there exists an $\varepsilon = \varepsilon(m, L) > 0$ such that if A_1 is an essential subannulus of an annulus A_2 with $m \leq \mathrm{mod}\,(A_1) \leq \infty$ and $\mathrm{mod}\,(A_2)/\mathrm{mod}\,(A_1) < 1 + \varepsilon$, then any point $z \in A_2$ with the property that $d_{A_2}(z, \mathrm{Core}\,(A_1)) \leq L$ belongs to A_1.*

Proof. Fix constants $m > 0$ and $L > 0$. Suppose that A_1 is an essential subannulus of an annulus A_2 with $m \leq \mathrm{mod}\,(A_1) \leq \infty$ and that there exists a point $z_0 \in A_2 \setminus A_1$ with $d_{A_2}(z_0, \mathrm{Core}\,(A_1)) \leq L$. We want to show that $\mathrm{mod}\,(A_2)/\mathrm{mod}\,(A_1)$ cannot be arbitrarily close to 1.

Choose $z_1 \in \mathrm{Core}\,(A_1)$ such that $d_{A_2}(z_0, z_1) = d_{A_2}(z_0, \mathrm{Core}\,(A_1))$. There exist universal covering maps $\pi_j : \mathbb{D} \to A_j$ with $\pi_j(0) = z_1$ ($j = 1, 2$). Since A_1 is essential in A_2, there exists a lift $\psi : \mathbb{D} \to \mathbb{D}$ of the inclusion map $\iota : A_1 \hookrightarrow A_2$ such that $\pi_2 \circ \psi = \iota \circ \pi_1 = \pi_1$ and $\psi(0) = 0$. There exists a point $\zeta_1 \in \mathbb{D}$ such that the segment $[0, \zeta_1]$ maps onto $\mathrm{Core}\,(A_1)$ by π_1, $d_\mathbb{D}(0, \zeta_1) = \mathrm{length}_{A_1}(\mathrm{Core}\,(A_1))$ and $\pi_1(\zeta_1) = z_1$. Let $\zeta_2 = \psi(\zeta_1)$, then $\pi_2(\zeta_2) = z_1$ and $|\zeta_2| \leq |\zeta_1|$. There is also a point $\zeta_0 \in \mathbb{D}$ such that $\pi_2(\zeta_0) = z_0$, $d_\mathbb{D}(0, \zeta_0) \leq L$ and $\zeta_0 \notin \mathrm{Image}\,\pi_2 \circ \psi$.

It is well known [13, p.12] that

$$\mathrm{length}_{A_j}(\mathrm{Core}\,(A_j)) = \frac{\pi}{\mathrm{mod}\,(A_j)} \quad (j = 1, 2).$$

It follows from the Schwarz-Pick Theorem ([2, Theorem 1-1]) and the definition of geodesics that

$$\frac{\pi}{\mathrm{mod}\,(A_2)} = \mathrm{length}_{A_2}(\mathrm{Core}\,(A_2)) \leq d_{\mathbb{D}}(0,\zeta_2) \leq \mathrm{length}_{A_2}(\mathrm{Core}\,(A_1))$$
$$\leq \mathrm{length}_{A_1}(\mathrm{Core}\,(A_1)) = d_{\mathbb{D}}(0,\zeta_1) = \frac{\pi}{\mathrm{mod}\,(A_1)}.$$

Hence we have

$$\frac{\mathrm{mod}\,(A_1)}{\mathrm{mod}\,(A_2)} \leq \frac{d_{\mathbb{D}}(0,\zeta_2)}{d_{\mathbb{D}}(0,\zeta_1)} \leq 1 \quad \text{and} \quad d_{\mathbb{D}}(0,\zeta_1) \leq \frac{\pi}{m}.$$

Define $\psi_0(0) = \psi'(0)$ and $\psi_0(z) = \psi(z)/z$ ($0 \neq z \in \mathbb{D}$). The Schwarz Lemma applied to ψ implies $|\psi_0(z)| < 1$ since ψ is not surjective. We have

$$|\psi_0(\zeta_1)| = \frac{|\zeta_2|}{|\zeta_1|} \geq \frac{d_{\mathbb{D}}(0,\zeta_2)}{d_{\mathbb{D}}(0,\zeta_1)} \geq \frac{\mathrm{mod}\,(A_1)}{\mathrm{mod}\,(A_2)},$$

where the left inequality follows from the fact that the coefficient $\frac{2}{1-|z|^2}$ of the Poincaré metric in $\mathbb{D}$ is increasing in $[0,1)$. Since

$$d_{\mathbb{D}}(\psi_0(0), \psi_0(\zeta_1)) \leq d_{\mathbb{D}}(0,\zeta_1) \leq \frac{\pi}{m},$$

there exists a function $\delta(\varepsilon, m) > 0$ such that if $\mathrm{mod}\,(A_2)/\mathrm{mod}\,(A_1) < 1 + \varepsilon$ then $|\psi'(0)| = |\psi_0(0)| > 1 - \delta(\varepsilon, m)$ and $\delta(\varepsilon, m) \to 0$ as $\varepsilon \to 0$.

Now decompose ψ as $\psi = \psi_3 \circ \psi_2 \circ \psi_1$, where

$$\psi_3(z) = \frac{z + \zeta_0}{1 + \overline{\zeta_0} z} : \mathbb{D} \to \mathbb{D}$$

is a Möbius transformation sending $-\zeta_0, 0$ to $0, \zeta_0$,

$$\psi_2 : \mathbb{D}^* \equiv \mathbb{D} - \{0\} \hookrightarrow \mathbb{D}$$

is the inclusion and $\psi_1 : \mathbb{D} \to \mathbb{D}^*$ is a holomorphic map sending 0 to $-\zeta_0$ and its image avoids 0. By the Schwarz-Pick Theorem, we have

$$\begin{aligned}
|\psi'(0)| &= \|\psi'(0)\|_{\mathbb{D},\mathbb{D}} \\
&= \|\psi_3'(-\zeta_0)\|_{\mathbb{D},\mathbb{D}} \cdot \|\psi_2'(-\zeta_0)\|_{\mathbb{D}^*,\mathbb{D}} \cdot \|\psi_1'(0)\|_{\mathbb{D},\mathbb{D}^*} \\
&\leq \|\psi_2'(-\zeta_0)\|_{\mathbb{D}^*,\mathbb{D}},
\end{aligned}$$

where $\|\cdot\|_{X,Y}$ denotes the norm of the derivative with respect to the Poincaré metric of the domain X and that of the range Y. Since the Poincaré metric of $\mathbb{D}^*$ is $\frac{|dz|}{|z|\log(1/|z|)}$, we can write

$$\|\psi_2'(-\zeta_0)\|_{\mathbb{D}^*,\mathbb{D}} = \frac{2|\zeta_0|\log(1/|\zeta_0|)}{1 - |\zeta_0|^2} = \frac{t}{\sinh t} \quad \text{with} \quad t = \log(1/|\zeta_0|).$$

Hence there exists $\lambda(L) < 1$ such that if $d(0,\zeta_0) \leq L$, then we have

$$|\psi'(0)| \leq \|\psi_2'(-\zeta_0)\|_{\mathbb{D}^*,\mathbb{D}} \leq \lambda(L).$$

Finally, choose $\varepsilon > 0$ so that $1 - \delta(\varepsilon, m) > \lambda(L)$. If $\mod(A_2)/\mod(A_1) < 1 + \varepsilon$, we have a contradiction, and therefore we have proved the lemma. □

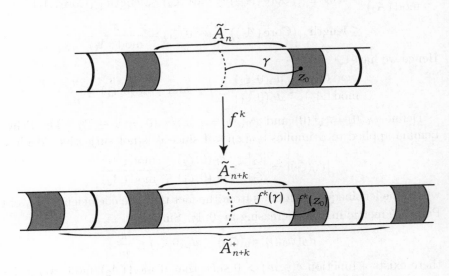

FIGURE 5

Proof of Proposition 4.5. The connected component of $f^{-k}(\widetilde{A}_{n+k}^-)$ containing $\widetilde{A}_n^-$ must be contained in D_n. Hence the right hand side is contained in the left hand side.

In order to show the converse, take any point $z_0 \in D_n$. Join z_0 with $\mathrm{Core}(\widetilde{A}_n^-)$ by a smooth curve γ in D_n; see Figure 5. Let $L = \mathrm{length}_{D_n}(\gamma)$. Note that $f^k(\mathrm{Core}(\widetilde{A}_n^-)) = \mathrm{Core}(\widetilde{A}_{n+k}^-)$ by Proposition 4.2 (II)(i) and that $D_{n+k} \subset \widetilde{A}_{n+k}^+$ by Lemma 4.4. Then by the Schwarz-Pick Theorem again, for $z_0 \in \gamma$, we have

$$d_{\widetilde{A}_{n+k}^+}(f^k(z_0), \mathrm{Core}(\widetilde{A}_{n+k}^-)) \leq d_{D_{n+k}}(f^k(z_0), \mathrm{Core}(\widetilde{A}_{n+k}^-)) \leq L, \quad (k \geq 0).$$

Since $\mod(\widetilde{A}_{n+k}^-) \to \infty$ $(k \to \infty)$, we can apply Lemma 4.8 with $A_1 = \widetilde{A}_{n+k}^-$ and $A_2 = \widetilde{A}_{n+k}^+$ together with Corollary 4.7 and conclude that there exists $k_0 \geq 0$ such that $f^k(\gamma) \subset \widetilde{A}_{n+k}^-$ for $k \geq k_0$. This implies that for large k, γ (and hence z_0) is contained in $\widetilde{A}_{n,k}^-$. Thus D_n is contained in $\bigcup_{k=0}^\infty \widetilde{A}_{n,k}^-$.

Moreover, if all the D_n do not contain critical points, then $\widetilde{A}_{n,k}^-$ is doubly connected. Since $\widetilde{A}_{n,k}^- \subset \widetilde{A}_{n,k+1}^-$ essentially, D_n is also doubly connected as an increasing union of annuli. □

By the construction, all the critical points of f are mapped to 0 by f^2. Since 0 is a repelling fixed point, which is in J_f, all the critical points are in J_f and hence all the D_n do not contain critical points. Therefore D_n is doubly connected for every n by Proposition 4.5. This completes the proof of Theorem B. □

5. Proofs of Theorem C, D, E and F

Proof of Theorem C. We modify the quasiregular map f_1 in the proof of Theorem B to another quasiregular map f_2 which has the required dynamical properties. For this purpose let $p \geq 3$, $p \in \mathbb{N}$ and $h(z)$ be a holomorphic map on the bounded connected component of the complement of $E_{p-2}^\sharp$ which satisfies the following conditions:

(I) h is a $p - 1$ to 1 map;

(II) the origin 0 is a repelling fixed point of h;

(III) h has a unique critical point c in $\{z : |z| < R_{p-2}\}$ which satisfies $h(c) \in A_{p-2}^-$;

(IV) $h(\partial_{\text{inner}} E_{p-2}^\sharp) = \partial_{\text{inner}} E_{p-1}^\sharp$.

More precisely, we construct h as follows: $E_{p-2}^\sharp$ and $E_{p-1}^\sharp$ are given by

$$E_{p-2}^\sharp = A(\alpha, \beta), \quad E_{p-1}^\sharp = A(\alpha', \beta'),$$

where

$$\alpha = R_{p-2} \exp\left(2\pi \sqrt{\operatorname{mod}(A_{p-2})}\right), \quad \beta = R_{p-2} \exp\left(4\pi \sqrt{\operatorname{mod}(A_{p-2})}\right),$$

$$\alpha' = R_{p-1} \exp\left(2\pi \sqrt{\operatorname{mod}(A_{p-1})}\right), \quad \beta' = R_{p-1} \exp\left(4\pi \sqrt{\operatorname{mod}(A_{p-1})}\right).$$

Define h on $A(0, \alpha)$ (the disk with radius α and center 0) by

$$h(z) = \alpha' \cdot \widetilde{h}\left(\frac{z}{\alpha}\right),$$

where

$$\widetilde{h}(z) = M_1 \circ z^{p-1} \circ M_2(z), \quad M_1(z) = \frac{z - a^{p-1}}{1 - a^{p-1}z}, \quad M_2(z) = \frac{z + a}{1 + az}, \quad 0 < a < 1.$$

Obviously h is a $p-1$ to 1 map, satisfies (IV) and has a fixed point at 0 with

$$h'(0) = \frac{\alpha'}{\alpha} \cdot \frac{a^{p-2}(p-1)(1-a^2)}{1 - a^{2(p-1)}}.$$

Also h has a unique critical point $c = -a\alpha$ and critical value $h(c) = -\alpha' a^{p-1}$. For the condition (III) we adjust $0 < a < 1$ so that

$$h(c) \in \operatorname{Core}(A_{p-2}^-) \subset A_{p-2}^-,$$

that is,
$$\alpha' a^{p-1} = \sqrt{R_{p-2}R_{p-1}}$$
and hence we put
$$a = \left(\frac{\sqrt{R_{p-2}R_{p-1}}}{\alpha'}\right)^{\frac{1}{p-1}} = \exp\left(-\frac{\pi}{p-1}\left\{\operatorname{mod}(A_{p-2}) + 2\sqrt{\operatorname{mod}(A_{p-1})}\right\}\right).$$

Thus h has been defined. Note that we can take a as small as we want by taking $\operatorname{mod}(A_1)$ large enough, since $\operatorname{mod}(A_{p-2})$ also gets larger by its definition. Now we have
$$h'(0) = \frac{\sqrt{R_{p-2}R_{p-1}}}{\alpha} \cdot \frac{(p-1)(1-a^2)}{a(1-a^{2(p-1)})}$$
and we have only to check that 0 is repelling. Since
$$\log \frac{\sqrt{R_{p-2}R_{p-1}}}{\alpha} = \pi\left(\operatorname{mod}(A_{p-2}) - 2\sqrt{\operatorname{mod}(A_{p-2})}\right) > 0,$$
we have
$$\frac{\sqrt{R_{p-2}R_{p-1}}}{\alpha} > 1.$$
Also
$$\frac{(p-1)(1-a^2)}{a(1-a^{2(p-1)})} > 1,$$
since a is small enough. Hence it follows that
$$h'(0) > 1.$$

Since the degree of $h|_{|z|=\alpha}$ is equal to that of $f_1|_{|z|=\beta}$, we can glue h and f_1 naturally on $E_{p-2}^\sharp$ so that the map is C^1 covering on this set and satisfies
$$h(E_{p-2}^\sharp) \subset E_{p-1}^\sharp \cup A_{p-1}^-$$
by canonical interpolation (for example, by a similar method to that used in the proof of Lemma 6.1). Hence we obtain a quasiregular map f_2. Of course, we construct f_2 so that f_2 has the same properties as in (I) $\sim$ (IV) above and as $f_1(z)$ satisfies for z in the unbounded component of the complement of $E_{p-2}^\sharp$. Then quasiconformal surgery (Theorem 3.1) gives a new entire function f.

This f has a doubly connected wandering domain U which contains $\widetilde{A}_{p-2}^-$ by exactly the same reason as in the proof of Theorem B. Consider the connected component D of $f^{-1}(U)$ containing $\widetilde{c}$, the critical point of f corresponding to c of f_2. Then we have $D \neq U$, since otherwise U would be an invariant component, which is absurd. So D is also a wandering domain as an inverse image of U. Then by applying the Riemann-Hurwitz formula to $f|_D : D \to U$ we have

$$2 - \mathrm{conn}(D) = (\deg f|_D)(2 - \mathrm{conn}(U)) - \#\{\text{critical points in } D\}.$$

Since

$$\mathrm{conn}(U) = 2, \quad \deg f|_D = p - 1, \quad \#\{\text{critical points in } D\} = p - 2,$$

it follows that $\mathrm{conn}(D) = p$. This completes the proof. □

Proof of Theorem D. In the proof of Theorem B, we put $k_n = n + 1$ for the model map f_0. This time we put $k_n = n + 2$. Then all the estimates follow as above. Note that Lemma 4.4 holds also in this case, because 0 is an attracting fixed point and does not belong to $\bigcup_{n=1}^{\infty} D_n$. □

Proof of Theorem E. Let f_1 and g_n be as in the proof of Theorem B. Then g_n is analytic in B_n and has a unique critical point in B_n, which is mapped to the critical value $v_{n+1} \in B_{n+1}$. By the construction, g_n maps v_n to 0, whereas $g_n(\partial B_n)$ and v_{n+1} are in the unbounded component of $\mathbb{C} \setminus A_n^-$. Hence there exists a closed topological disk U_n with

$$v_n \in \mathrm{int}\, U_n \subset B_n$$

such that

$$g_n|_{U_n} : U_n \to A_n^- \cup Q_n^-$$

is conformal, where Q_n^- is the bounded component of $\mathbb{C} \setminus A_n^-$. Let

$$V_n = (g_n|_{U_n})^{-1}(Q_n^-) \subset U_n$$

and take topological disks W_n with

$$V_n \subset W_n \subset U_n$$

such that

$$\mathrm{mod}\,(U_n \setminus W_n) \geq \mathrm{const.} > 0$$

holds for a constant independent of n. Pick any $w_n \in Q_n^-$. Then there is a unique point $z_n = (g_n|_{U_n})^{-1}(w_n) \in V_n$. Define a K-qc map

$$h_n : U_n \to U_n$$

with a uniform $K \geq 1$ (i.e. independent of n) which satisfies the following:

(I) $h_n = \mathrm{id}$ near ∂U_n,
(II) h_n is conformal on $V_n \cap h_n^{-1}(V_n)$,
(III) $h_n(v_n) = z_n$.

This can be done by a similar method to that used in the proof of Theorem C by using a suitable Möbius transformation. Then define a new map $\widehat{f}_1$ by

$$\widehat{f}_1(z) = \begin{cases} f_1 \circ h_n(z) & \text{if } z \in U_n \\ f_1(z) & \text{otherwise} \end{cases}.$$

This $\widehat{f}_1$ is holomorphic outside

$$\widehat{E} = \Big(\bigcup_{n=1}^{\infty} E_n\Big) \cup \Big(\bigcup_{n=1}^{\infty} (U_n \setminus W_n)\Big),$$

where $\bigcup_{n=1}^{\infty} E_n$ is the set defined before and every $\widehat{f}_1$ orbit of z passes $\widehat{E}$ at most once. Then we can obtain an entire function f from $\widehat{f}_1$ by qc-surgery. This map f has the the same properties as the map in Theorem B satisfies and hence has a wandering domain. Take $w_n \in Q_n^-$ in advance so that the set $\{w_n\}_{n=1}^{\infty}$ forms a dense subset of $\mathbb{C}$. Then f satisfies

$$\overline{f(\operatorname{sing}(f^{-1}))} = \mathbb{C}$$

which completes the proof. $\square$

Remark. In particular, the post singular set $P(f) = \overline{\bigcup_{n=0}^{\infty} f^n(\operatorname{sing}(f^{-1}))}$ is equal to the whole plane $\mathbb{C}$ in this case. This never happens when f is a rational map with $J_f \neq \widehat{\mathbb{C}}$.

Proof of Theorem F. As in the proof of Theorem B, we first construct a model map f_0 for f. We will divide the complex plane into a countable union of concentric round annuli $A_0, A_1, \cdots, A_n, \cdots$ with mutually disjoint interiors satisfying $\partial_{\text{outer}} A_n = \partial_{\text{inner}} A_{n+1}$, that is, $\mathbb{C} = \bigcup_{n=0}^{\infty} A_n$. (Note that A_0 is not an annulus but a round disk). We will choose A_0 and A_1 first and then A_n ($n \geq 2$) will be defined inductively. Moreover for each annulus A_n we choose a countable union of concentric round subannuli

$$A_n \supseteq \bigcup_{\lambda \in \Lambda} A_n^{(\lambda)}$$

with mutually disjoint interiors so that if $\lambda, \lambda' \in \Lambda$ satisfies $\lambda \prec \lambda'$, then $A_n^{(\lambda)}$ is contained in the bounded component of the complement of $A_n^{(\lambda')}$. Here the index set Λ is a countable totally ordered set $(\Lambda, \prec)$ such that every $\lambda \in \Lambda$ has a unique λ' which is right next to λ. That is, there exists a $\lambda' \in \Lambda$ with $\lambda \prec \lambda'$ and there is no λ'' which satisfies $\lambda \prec \lambda'' \prec \lambda'$. We denote such a λ' by $\lambda' = \lambda + 1$. Similarly we require that for every $\lambda \in \Lambda$ there is a unique λ' which is right before λ and denote it by $\lambda' = \lambda - 1$. For example, $\mathbb{Z}$ is the most simple example of Λ with the above properties. We will use a much more complicated Λ later when we construct a Julia set with positive measure.

In order to define the model map f_0, A_n ($n \geq 0$) and $A_n^{(\lambda)}$ ($n \geq 1$, $\lambda \in \Lambda$), we first construct a map $F : \mathbb{R} \to \mathbb{R}$ corresponding to f_0 which satisfies the following commutative diagram:

$$\begin{array}{ccc} \mathbb{C}^* & \xrightarrow{f_0} & \mathbb{C}^* \\ L \downarrow & & \downarrow L \\ \mathbb{R} & \xrightarrow{F} & \mathbb{R} \end{array} \quad (5.1)$$

where

$$L(z) = \frac{1}{2\pi} \log |z| : \mathbb{C}^* \to \mathbb{R},$$

and intervals I_n and $I_n^{(\lambda)} \subset I_n$ corresponding to A_n and $A_n^{(\lambda)} \subset A_n$, respectively. So first let

$$I_0 = (-\infty, r_1], \quad (r_1 > 0).$$

Next take $I_1 = [r_1, r_2]$ and choose $I_1^{(\lambda)} \subset I_1$ so that $I_1^{(\lambda)}$ ($\lambda \in \Lambda$) have mutually disjoint interiors and

$$I_1 \setminus \bigcup_{\lambda \in \Lambda} \operatorname{int} I_1^{(\lambda)}$$

has no interior. Also we require that they are ordered so that

$$\lambda \prec \lambda' \implies \max I_1^{(\lambda)} \leq \min I_1^{(\lambda')}$$

holds, where the equality holds if and only if $\lambda' = \lambda + 1$. Moreover we assume that there is at least one $\lambda \in \Lambda$ such that

$$|I_1^{(\lambda)}| > m_0, \quad |I_1^{(\lambda+1)}| > m_0$$

are satisfied, where $m_0 > 0$ is the constant in Proposition 4.1.

In order to construct $F : \mathbb{R} \to \mathbb{R}$, we define $k : \mathbb{R} \to \mathbb{N}$ as follows: First we put

$$k(x) \equiv 1, \quad x \in I_0$$

on I_0 and define

$$F(x) = x + r_2 - r_1, \quad x \in I_0.$$

Next if $I_n = [r_n, r_{n+1}]$ and $I_n^{(\lambda)} \subset I_n$ ($n \geq 1$) are defined, then define $k(x)$ on I_n by

$$k(x) = k(r_n) + {}^{\#}\left\{ \lambda : I_n^{(\lambda)} \subset [r_n, x], \text{ and } |I_n^{(\lambda)}|, |I_n^{(\lambda+1)}| > m_0 \right\}, \quad x \in I_n.$$

Here note that $k(x)$ is finite, because we have $|I_n| < \infty$. Also note that

$$k(r_1) = 1 = \lim_{\substack{x \in I_1^{(\lambda)} \\ \lambda \to -\infty}} k(x)$$

and in general
$$k(r_n) = \lim_{\substack{x \in I_{n-1}^{(\lambda)} \\ \lambda \to \infty}} k(x) = \lim_{\substack{x \in I_n^{(\lambda)} \\ \lambda \to -\infty}} k(x), \quad n \geq 2,$$
which means that for fixed n there exist $\lambda_+, \lambda_- \in \Lambda$ such that $k(x)$ ($x \in I_{n-1}^{(\lambda)}$) is constant for every $\lambda_+ \prec \lambda$ and so is $k(x)$ ($x \in I_n^{(\lambda)}$) for every $\lambda \prec \lambda_-$ and these constants are both equal to $k(r_n)$.

Then let
$$F(x) = F(r_n) + \int_{r_n}^x k(t)dt, \quad x \in I_n$$
and define I_{n+1} by
$$I_{n+1} = F(I_n)$$
and also
$$I_{n+1}^{(\lambda)} = F(I_n^{(\lambda)}).$$

Thus we have defined all I_n and $I_n^{(\lambda)} \subset I_n$ for every $n \in \mathbb{N}$ and $\lambda \in \Lambda$. Then A_n and $A_n^{(\lambda)}$ are defined by
$$A_n = L^{-1}(I_n) = A(e^{2\pi r_n}, e^{2\pi r_{n+1}}), \quad A_n^{(\lambda)} = L^{-1}(I_n^{(\lambda)}) \subset A_n$$
and finally a model map f_0 is defined by
$$f_0(z) = \exp\left\{2\pi F(L(z)) + ik(L(z))\arg z\right\}$$
so that the commutative diagram (5.1) holds.

By definition, $k(x)$ is constant on $I_n^{(\lambda)}$ for every $n \in \mathbb{N}$ and $\lambda \in \Lambda$ and this value is equal to the degree of the model map $f_0 : A_n^{(\lambda)} \to A_{n+1}^{(\lambda)}$. Actually f_0 is a covering map of the form
$$f_0(z) = a_n^{(\lambda)} z^{k_n^{(\lambda)}}, \quad z \in \text{int } A_n^{(\lambda)}$$
on each int $A_n^{(\lambda)}$ and hence it is piecewise holomorphic. Thus the model map f_0 has been constructed.

Then the following is almost trivial by the definition of $k(x)$.

Proposition 5.1. (1) *The function $k(x)$ is non-decreasing and continuous (in fact, it is constant) on each int $I_n^{(\lambda)}$. If it is discontinuous at $x = x_0$, then there exist an $n_0 \in \mathbb{N}$ and a $\lambda_0 \in \Lambda$ such that*
$$x_0 = \max I_{n_0}^{(\lambda_0)} = \min I_{n_0}^{(\lambda_0+1)}$$
and $k(x)$ satisfies
$$k(x_0 + 0) = k(x_0 - 0) + 1.$$
Moreover we have
$$|I_{n_0}^{(\lambda_0)}| > m_0, \quad |I_{n_0}^{(\lambda_0+1)}| > m_0.$$

(2) For a fixed λ, there exists an $n(\lambda) \in \mathbb{N}$ such that $k(x)$ is continuous at $x = \max I_n^{(\lambda)} = \min I_n^{(\lambda+1)}$ for every $n < n(\lambda)$ and discontinuous at $x = \max I_n^{(\lambda)} = \min I_n^{(\lambda+1)}$ for every $n \geq n(\lambda)$.

Next we modify f_0 to a quasiregular map f_1 by interpolation. We do it around a neighborhood of every circle $L^{-1}(x_0)$, where $x_0 = \max I_{n_0}^{(\lambda_0)} = \min I_{n_0}^{(\lambda_0+1)}$ and $k(x)$ has discontinuity at $x = x_0$. This means that we can apply Proposition 4.1 for the pairs $(A_{n_0}^{(\lambda_0)}, A_{n_0}^{(\lambda_0+1)})$, because

$$\mathrm{mod}\,(A_{n_0}^{(\lambda_0)}) = |I_{n_0}^{(\lambda_0)}| > m_0, \quad \mathrm{mod}\,(A_{n_0}^{(\lambda_0+1)}) = |I_{n_0}^{(\lambda_0+1)}| > m_0$$

are satisfied. Also the above Proposition means that for every fixed λ, there exists an $n(\lambda) \in \mathbb{N}$ such that we can do interpolation between $A_n^{(\lambda)}$ and $A_n^{(\lambda+1)}$ for every $n \geq n(\lambda)$.

In order to apply qc-surgery to the map f_1, we have to modify Theorem 3.1 a little. Actually by using Proposition 4.1 repeatedly, we can construct f_1 so that f_1 satisfies similar properties to those in Proposition 4.2. More precisely, for every $\lambda \in \Lambda$ and $j \geq n(\lambda)$ we can define $A_j^{(\lambda)-}, B_j^{(\lambda)\flat}, B_j^{(\lambda)\sharp}, E_j^{(\lambda)\flat}$ and $E_j^{(\lambda)\sharp}$ which satisfy

$$f_1(\mathrm{Core}\,(A_j^{(\lambda)-})) = \mathrm{Core}\,(A_{j+1}^{(\lambda)-}), \quad f_1(A_j^{(\lambda)-}) \subset A_{j+1}^{(\lambda)-},$$
$$f_1(E_j^{(\lambda)\sharp}) \subset E_{j+1}^{(\lambda)\sharp} \cup A_{j+1}^{(\lambda)-}, \quad f_1(E_j^{(\lambda)\flat}) \subset A_j^{(\lambda)-} \cup E_{j+1}^{(\lambda)\flat}.$$

Also if we put $E_j^{(\lambda)} = E_j^{(\lambda)\flat} \cup E_j^{(\lambda)\sharp}$ ($j \geq n(\lambda)$), then the new map f_1 satisfies the following:

(i) for almost every $z \in \mathbb{C}$, the f_1-orbit of z passes $E_j^{(\lambda)}$ at most once for every $j \geq n(\lambda)$ and λ. Also if the f_1-orbit of z passes $E_{j_0}^{(\lambda_0)}$ for some λ_0, then the orbit never visits other $E_j^{(\lambda)}$ with $\lambda \neq \lambda_0$ and $j \in \mathbb{N}$;

(ii) f_1 is $K_j^{(\lambda)}$-quasiregular on $E_j^{(\lambda)}$;

(iii) $K_\infty^{(\lambda)} = \prod_{j=n(\lambda)}^\infty K_j^{(\lambda)} < \infty$ and $K_\infty^{(\lambda)}$ is uniformly bounded with respect to λ;

(iv) f_1 is holomorphic a.e. outside $\bigcup_{\lambda \in \Lambda} \bigcup_{j \geq n(\lambda)} E_j^{(\lambda)}$.

Then it is not difficult to see that the conclusion of Theorem 3.1 holds also under the above conditions, and therefore we can obtain an entire function $f = \varphi \circ f_1 \circ \varphi^{-1}$ by quasiconformal surgery. By the construction, we can show that f has a doubly connected wandering domain D_λ ($\lambda \in \Lambda$) whose grand orbit corresponds to the sequence of annuli $A_{n(\lambda)}^{(\lambda)}, A_{n(\lambda)+1}^{(\lambda)}, \ldots$. Obviously these D_λ are mutually disjoint.

In order to construct a Julia set of positive area, we choose I_1 and $I_1^{(\lambda)}$ so that $I_1 \setminus \bigcup_{\lambda \in \Lambda} I_1^{(\lambda)}$ has no interior but $|I_1 \setminus \bigcup_{\lambda \in \Lambda} I_1^{(\lambda)}| > 0$. This is done by the following method: Consider the complement of a Cantor set C in I_1 with positive measure which is obtained by the similar construction as for the Cantor Middle-Thirds set in $[0,1]$. Divide each connected component of $I_1 \setminus C$, which is an open interval, into a countable union of closed intervals with disjoint interiors which are labeled by $\mathbb{Z}$ in the natural way. Now let $\{I_1^{(\lambda)} : \lambda \in \Lambda\}$ be the collection of all these intervals. Then the order $I_1^{(\lambda)} \prec I_1^{(\lambda')}$ for $I_1^{(\lambda)}$ and $I_1^{(\lambda')}$ is naturally defined by the order of intervals $I_1^{(\lambda)}, I_1^{(\lambda')} \subset \mathbb{R}$, that is, $I_1^{(\lambda)} \prec I_1^{(\lambda')}$ if and only if $I_1^{(\lambda)}$ is located on the left to $I_1^{(\lambda')}$ in $\mathbb{R}$. Then it is almost obvious that this index set Λ satisfies the properties we mentioned at the beginning of the proof.

Then it holds that $\left|I_n \setminus \bigcup_{\lambda \in \Lambda} I_n^{(\lambda)}\right| > 0$ for every n by the construction and hence

$$A_n \setminus \bigcup_{\lambda \in \Lambda} A_n^{(\lambda)}$$

has positive Lebesgue measure. Note that $f_1 = f_0$ on $A_n \setminus \bigcup_{\lambda \in \Lambda} A_n^{(\lambda)}$ and the union of the boundaries of these doubly connected wandering domains, which is a subset of J_f, accumulates on every point in $\varphi(A_n \setminus \bigcup_{\lambda \in \Lambda} A_n^{(\lambda)})$. Hence we have $\varphi(A_n \setminus \bigcup_{\lambda \in \Lambda} A_n^{(\lambda)}) \subset J_f$. Since we have $|\varphi(A_n \setminus \bigcup_{\lambda \in \Lambda} A_n^{(\lambda)})| > 0$, it follows that $|J_f| > 0$.

By the construction of A_n, $A_n^{(\lambda)}$ and the map f_1, it is now easy to construct countably many measurable invariant line fields supported on mutually disjoint subsets of $\varphi(A_n \setminus \bigcup_{\lambda \in \Lambda} A_n^{(\lambda)}) \subset J_f$. This completes the proof. □

6. Appendix

In this section, as an appendix, we prove Proposition 4.1. For this purpose, we need the following three lemmas:

Lemma 6.1. *Let A, a, b, c, d, δ_0 and δ_1 be positive and $\eta_j : [c,d] \to \mathbb{C}$ be C^1-functions ($j = 1, 2$) which satisfy*

$$|\eta_2(y) - \eta_1(y)| \le \delta_0, \quad |\eta_j'(y) - A| \le \delta_1$$

on $[c,d]$. For $(x,y) \in [a,b] \times [c,d]$ put

$$h(x + iy) = Ax + i\left(\frac{b-x}{b-a}\eta_1(y) + \frac{x-a}{b-a}\eta_2(y)\right).$$

If A, a, b, δ_0 and δ_1 satisfy the inequality

$$C = 1 - \frac{1}{A}\left(\frac{\delta_0}{b-a} + \delta_1\right) > 0,$$

then h is a quasiconformal map on $[a,b] \times [c,d]$ with

$$K_h \leq \frac{1}{C}. \tag{6.1}$$

Proof. Since

$$\frac{\partial h}{\partial x} = A + \frac{i}{b-a}(\eta_2(y) - \eta_1(y)), \quad \frac{\partial h}{\partial y} = i\left(\frac{b-x}{b-a}\eta_1'(y) + \frac{x-a}{b-a}\eta_2'(y)\right),$$

we have

$$\begin{aligned}\frac{\partial h}{\partial z} &= \frac{1}{2}\left\{A + \frac{i}{b-a}(\eta_2(y) - \eta_1(y)) + \left(\frac{b-x}{b-a}\eta_1'(y) + \frac{x-a}{b-a}\eta_2'(y)\right)\right\} \\ &= \frac{1}{2}\left\{2A + \frac{i}{b-a}(\eta_2(y) - \eta_1(y)) + \frac{b-x}{b-a}(\eta_1'(y) - A) \right. \\ &\qquad\qquad \left. + \frac{x-a}{b-a}(\eta_2'(y) - A)\right\}\end{aligned}$$

and similarly

$$\frac{\partial h}{\partial \bar{z}} = \frac{1}{2}\left\{\frac{i}{b-a}(\eta_2(y) - \eta_1(y)) - \frac{b-x}{b-a}(\eta_1'(y) - A) + \frac{x-a}{b-a}(\eta_2'(y) - A)\right\}.$$

Then the following estimates hold:

$$\left|\frac{\partial h}{\partial z}\right| \geq A - \frac{\delta_0}{2(b-a)} - \frac{\delta_1}{2}, \quad \left|\frac{\partial h}{\partial \bar{z}}\right| \leq \frac{\delta_0}{2(b-a)} + \frac{\delta_1}{2}.$$

So if we have

$$\frac{\delta_0}{2(b-a)} + \frac{\delta_1}{2} \leq A - \frac{\delta_0}{2(b-a)} - \frac{\delta_1}{2},$$

that is,

$$C = 1 - \frac{1}{A}\left(\frac{\delta_0}{b-a} + \delta_1\right) > 0,$$

then

$$\left|\frac{\partial h}{\partial \bar{z}}\right| \leq \left|\frac{\partial h}{\partial z}\right|$$

holds and hence h is a quasiconformal map which satisfies (6.1). (Recall that $K = \frac{1+k}{1-k}$; see Definition 3.1.) □

Lemma 6.2. *Let $k \in \mathbb{N}$, $0 < R_1 < R_2$ and $\varphi_j(z)$ be analytic on a neighborhood of $|z| = R_j$ such that $\varphi_j|_{|z|=R_j}$ goes around the origin k-times ($j = 1, 2$).*

If
$$\left|\log\left(\frac{\varphi_2(R_2 e^{iy})}{R_2^k} \frac{R_1^k}{\varphi_1(R_1 e^{iy})}\right)\right| \le \delta_0, \quad \text{for every } y \in [0, 2\pi], \tag{6.2}$$

$$\left|z\frac{d}{dz}\left(\log\frac{\varphi_j(z)}{z^k}\right)\right| \le \delta_1, \quad z = R_j e^{iy}, \ j = 1, 2, \tag{6.3}$$

hold for every $y \in [0, 2\pi]$ and for some positive constants δ_0 and δ_1 satisfying

$$C = 1 - \frac{1}{k}\left(\frac{\delta_0}{\log(R_2/R_1)} + \delta_1\right),$$

then there exists a quasiregular map

$$H : \{z \ : \ R_1 \le |z| \le R_2\} \to \mathbb{C}^*$$

without critical points such that $H = \varphi_j$ on $|z| = R_j$ ($j = 1, 2$) and satisfies

$$K_H \le \frac{1}{C}. \tag{6.4}$$

Proof. By the assumptions, φ_1 and φ_2 can be written as follows:

$$\varphi_j(R_j e^{iy}) = R_j^k \exp(i\eta_j(y)), \quad y \in [0, 2\pi], \ j = 1, 2,$$

where $\eta_j(y) \in \mathbb{C}$ is a C^ω-function with

$$\eta_j(2\pi) - \eta_j(0) = 2k\pi.$$

We apply Lemma 6.1 by taking $[a, b] = [0, \log(R_2/R_1)]$, $[c, d] = [0, 2\pi]$ and $A = k$. From the conditions (6.2) and (6.3), it is easy to see that

$$|\eta_2(y) - \eta_1(y)| \le \delta_0, \quad |\eta'_j(y) - A| \le \delta_1$$

on $[0, 2\pi]$. Then define H as follows:

$$H(R_1 e^{x+iy}) = R_1^k \exp(h(x+iy)),$$

where $h(x + iy)$ is defined as in Lemma 6.1. Note that for $y = 2\pi$, we have

$$H(R_1 e^{x+2\pi i}) = R_1^k \exp(h(x + 2\pi i)) = R_1^k \exp(h(x) + 2\pi i) = H(R_1 e^x).$$

Then it follows that H is a quasiregular map and we have the estimate (6.4) from Lemma 6.1. □

Lemma 6.3. Let $k \in \mathbb{N}$, $k \ge 1$, b, $\omega \in \mathbb{C}^*$ and $\rho^\sharp$, $\lambda^\sharp$, $\rho^\flat$, $\lambda^\flat \in \mathbb{R}$ with $0 < \lambda^\flat < \rho^\flat < 1 < \rho^\sharp < \lambda^\sharp$.

(1) *Suppose that these constants satisfy*

$$\rho^\sharp \ge 2|\omega|, \quad \lambda^\sharp \ge e\rho^\sharp, \quad C^\sharp = 1 - \frac{1}{k+1}\left(\frac{|\log b|}{\log(\lambda^\sharp/\rho^\sharp)} + \frac{4|\omega|}{\rho^\sharp}\right) > 0, \tag{6.5}$$

where log is a branch with $\log 1 = 0$. Then the map $bz^k(z-\omega)$ on $|z| = \rho^\sharp$ and z^{k+1} on $|z| = \lambda^\sharp$ can be interpolated on $\rho^\sharp \leq |z| \leq \lambda^\sharp$ with a quasiregular map g by Lemma 6.2 and g satisfies

$$K_g \leq \frac{1}{C^\sharp}. \tag{6.6}$$

(2) Suppose that these constants satisfy

$$|\omega| \geq 2\rho^b, \quad \rho^b \geq e\lambda^b, \quad C^b = 1 - \frac{1}{k}\left(\frac{|\log(-b\omega)|}{\log(\rho^b/\lambda^b)} + \frac{4\rho^b}{|\omega|}\right) > 0. \tag{6.7}$$

Then the map $bz^k(z-\omega)$ on $|z| = \rho^b$ and z^k on $|z| = \lambda^b$ can be interpolated on $\lambda^b \leq |z| \leq \rho^b$ with a quasiregular map g by Lemma 6.2 and g satisfies

$$K_g \leq \frac{1}{C^b}. \tag{6.8}$$

Proof. We prove only (1), since (2) can be proved similarly. We apply Lemma 6.2 by taking

$$R_1 = \rho^\sharp, \quad R_2 = \lambda^\sharp, \quad \varphi_1(z) = bz^k(z-\omega), \quad \varphi_2(z) = z^{k+1},$$

and changing k to $k+1$. Also we take constants δ_0 and δ_1 as follows:

$$\left|\log\left(\frac{\varphi_2(R_2 e^{iy})}{R_2^{k+1}} \frac{R_1^{k+1}}{\varphi_1(R_1 e^{iy})}\right)\right| = \left|\log b\left(1 - \frac{\omega}{\rho^\sharp e^{iy}}\right)\right| \leq |\log b| + \frac{2|\omega|}{\rho^\sharp} =: \delta_0$$

and

$$\left|z\frac{d}{dz}\left(\log\frac{\varphi_1(z)}{z^{k+1}}\right)\right| = \left|\frac{1}{z/\omega - 1}\right| \leq \frac{1}{\rho^\sharp/|\omega| - 1} \leq \frac{2|\omega|}{\rho^\sharp} =: \delta_1,$$

where $z = R_1 e^{iy} = \rho^\sharp e^{iy}$. Here we used the assumption $\rho^\sharp \geq 2|\omega|$ in the second estimate. Then it is easy to see that these δ_0 and δ_1 satisfy the assumption in Lemma 6.2. So from Lemma 6.2, we conclude that there exists a quasiregular map g on $\rho^\sharp \leq |z| \leq \lambda^\sharp$ such that $g|_{|z|=\rho^\sharp} = bz^k(z-\omega)$ and $g|_{|z|=\lambda^\sharp} = z^{k+1}$. It is easy to see that

$$\frac{1}{k+1}\left(\frac{\delta_0}{\log(R_2/R_1)} + \delta_1\right) = \frac{1}{k+1}\left(\frac{|\log b| + 2|\omega|/\rho^\sharp}{\log(\lambda^\sharp/\rho^\sharp)} + \frac{2|\omega|}{\rho^\sharp}\right)$$

$$\leq \frac{1}{k+1}\left(\frac{|\log b|}{\log(\lambda^\sharp/\rho^\sharp)} + \frac{4|\omega|}{\rho^\sharp}\right)$$

by using the assumption $\lambda^\sharp \geq e\rho^\sharp$ and hence the estimate (6.6) follows. □

Now we are ready to prove Proposition 4.1

Proof of Proposition 4.1. (1) The condition $\operatorname{mod}(A) > 4\sqrt{\operatorname{mod}(A)}$, that is,
$$\operatorname{mod}(A) > 16$$
is necessary to be able to take annuli which satisfy (4.1). Similarly we must have $\operatorname{mod}(A') > 16$. To use Lemma 6.3, we apply $\psi_A(z) = z/R$ and $\psi_{\widehat{A}}(z) = z/\widehat{R}$ around $\{|z| = R\}$ and $\{|z| = R'\}$, respectively, and take $0 < \lambda^{\flat} < \rho^{\flat} < 1 < \rho^{\sharp} < \lambda^{\sharp}$. Then we construct $G(z)$ in this normalized coordinate by using Lemma 6.3 and then define $g(z) = \psi_{\widehat{A}}^{-1} \circ G \circ \psi_A(z) = \widehat{R} \cdot G(z/R)$; see Figure 6. From (4.1), we have to take $\rho^{\sharp}$, $\lambda^{\sharp}$, $\rho^{\flat}$ and $\lambda^{\flat}$ as follows:

$$\rho^{\sharp} = \exp(2\pi\sqrt{\operatorname{mod}(A')}), \quad \lambda^{\sharp} = (\rho^{\sharp})^2,$$
$$\rho^{\flat} = \exp(-2\pi\sqrt{\operatorname{mod}(A)}), \quad \lambda^{\flat} = (\rho^{\flat})^2.$$

Next we put $\omega = 1$ in Lemma 6.3. Then $bz^k(z-1)$ has a unique critical point $z = \frac{k}{k+1}$ other than the origin, which lies in $A(\rho^{\flat}, 1)$.

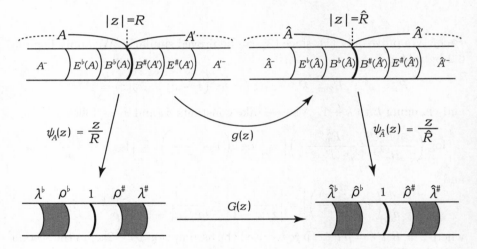

FIGURE 6. How to construct the new map g via the normalized coordinate.

Then the conditions (6.5) for "$\sharp$" in Lemma 6.3 (1) become
$$\exp(2\pi\sqrt{\operatorname{mod}(A')}) \geq 2,$$
$$\exp(4\pi\sqrt{\operatorname{mod}(A')}) \geq e \exp(2\pi\sqrt{\operatorname{mod}(A')}),$$
$$C^{\sharp} = 1 - \frac{1}{k+1}\left(\frac{|\log b|}{2\pi\sqrt{\operatorname{mod}(A')}} + \frac{4}{\exp(2\pi\sqrt{\operatorname{mod}(A')})}\right) > 0.$$

These conditions are apparently satisfied if $\operatorname{mod}(A')$ is large enough. Similarly the conditions (6.7) for "$\flat$" in Lemma 6.3 (2) are satisfied if $\operatorname{mod}(A)$ is large enough. (Actually, $\operatorname{mod}(A) > 16$ and $\operatorname{mod}(A') > 16$ are sufficient.) Hence by Lemma 6.3 the map which is equal to z^k on $|z| = \lambda^{\flat}$ and $bz^k(z-1)$

on $\rho^\flat \leq |z| \leq \rho^\sharp$ and z^{k+1} on $|z| = \lambda^\sharp$ can be interpolated on $\lambda^\flat \leq |z| \leq \lambda^\sharp$ with a quasiregular map $G(z)$. Also we define $G(z) = z^k$ for $|z| \leq \lambda^\flat$ and $G(z) = z^{k+1}$ for $\lambda^\sharp \leq |z|$. Thus the map $g(z) = \widehat{R} \cdot G(z/R)$ has been constructed for $z \in A^- \cup E^\flat(A) \cup B^\flat(A) \cup B^\sharp(A') \cup E^\sharp(A') \cup A'^-$.

Next we check if this map g satisfies the conditions (I) $\sim$ (III).

(I-a): This is trivial from the construction.

(I-b): g is holomorphic on int B by the construction. Also g has a unique critical point $\zeta = \frac{k}{k+1} R \in B^\flat(A)$ corresponding to the critical point $\frac{k}{k+1}$ of $bz^k(z-1)$. In order to adjust so that $g(\zeta) = \widehat{R}$, we put

$$b = -(k+1)\left(\frac{k}{k+1}\right)^{-k}$$

so that the critical point $\frac{k}{k+1}$ is mapped to 1 by $bz^k(z-1)$. On the other hand, we have $g(R) = \widehat{R}G(1) = 0$.

(I-c): From Lemma 6.3, we have

$$K_g \leq \max\left(\frac{1}{C^\sharp}, \frac{1}{C^\flat}\right).$$

With some more calculation, it is not hard to prove the estimate (4.2).

(II-a) $g(\text{Core}(A^-)) = \text{Core}(\widehat{A}^-)$:

Since $\text{Core}(A^-) = \text{Core}(A)$ by the construction, we have

$$g(\text{Core}(A^-)) = g(\text{Core}(A)) = F_A(\text{Core}(A)) = \text{Core}(\widehat{A}) = \text{Core}(\widehat{A}^-).$$

Also $g(\text{Core}(A'^-)) = \text{Core}(\widehat{A}'^-)$ holds similarly.

(II-b) $g(A^-) \subset \widehat{A}^-$:

In general the following holds: if A and B are concentric annuli with

$$\text{Core}(A) = \text{Core}(B) \quad \text{and} \quad \text{mod}(A) < \text{mod}(B),$$

then we have $A \subset B$. So in order to show that $g(A^-) \subset \widehat{A}^-$, it is enough to show that

$$\text{mod}(g(A^-)) < \text{mod}(\widehat{A}^-),$$

since it is easy to see that $\text{Core}(g(A^-)) = \text{Core}(\widehat{A}^-)$ by using $\text{Core}(A^-) = \text{Core}(A)$ and the fact that g maps $\text{Core}(A^-)$ to $\text{Core}(g(A^-))$. This is shown by a simple calculation. Also $g(A'^-) \subset \widehat{A}'^-$ follows similarly.

(III-a) $g(E^\sharp(A')) \subset E^\sharp(\widehat{A}') \cup \widehat{A}'^-$:

Since $g(z) = \widehat{R}G(z/R)$ on $\partial_{\text{inner}} E^\sharp(A')$ and

$$|G(z)| = |bz^k(z-1)| > \frac{1}{2}|b||z|^{k+1} \quad (|z| \geq 2),$$

we have
$$|g(z)| > \frac{1}{2}|b|(\rho^\sharp)^{k+1}\widehat{R}, \quad z \in \partial_{\text{inner}} E^\sharp(A').$$
Also we have
$$|g(z)| = (\lambda^\sharp)^{k+1}\widehat{R}, \quad z \in \partial_{\text{outer}} E^\sharp(A')$$
so it follows that
$$g(E^\sharp(A')) \subset A\Big(\frac{1}{2}|b|(\rho^\sharp)^{k+1}\widehat{R}, (\lambda^\sharp)^{k+1}\widehat{R}\Big).$$
Therefore
$$A\Big(\frac{1}{2}|b|(\rho^\sharp)^{k+1}\widehat{R}, (\lambda^\sharp)^{k+1}\widehat{R}\Big) \subset E^\sharp(\widehat{A}') \cup \widehat{A}'^{-} \tag{6.9}$$
is sufficient for $g(E^\sharp(A')) \subset E^\sharp(\widehat{A}') \cup \widehat{A}'^{-}$. By comparing moduli, it is easy to see that the following condition is equivalent to (6.9) (see Figure 7):
$$\mod\Big(A\Big(\widehat{R}, \frac{1}{2}|b|(\rho^\sharp)^{k+1}\widehat{R}\Big)\Big) > \mod(B^\sharp(\widehat{A}')),$$
$$\mod(A(\widehat{R}, (\lambda^\sharp)^{k+1}\widehat{R})) < \mod(B^\sharp(\widehat{A}') \cup E^\sharp(\widehat{A}') \cup \widehat{A}'^{-}).$$
This is equivalent to
$$(k+1-\sqrt{k+1})\sqrt{\mod(A')} + \frac{1}{2\pi}\log\frac{|b|}{2} > 0$$
$$2 + \frac{2}{\sqrt{k+1}} < \sqrt{\mod(A')}$$
and it is easy to show that both inequalities hold, since we have $k \geq 2$, $|b| > 2$ and $\mod(A') > 16$.

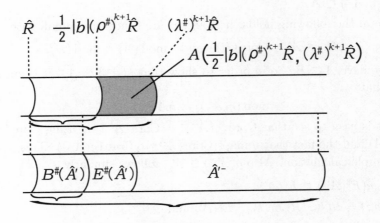

FIGURE 7. Comparison of moduli.

(III-b) $g(E^b(A)) \subset \widehat{A}^- \cup E^b(\widehat{A})$: Similarly in this case, we can show that
$$g(E^b(A)) \subset A((\lambda^b)^k \widehat{R}, 2|b|(\rho^b)^{k+1} \widehat{R}),$$
so
$$A((\lambda^b)^k \widehat{R}, 2|b|(\rho^b)^{k+1} \widehat{R}) \subset \widehat{A}^- \cup E^b(\widehat{A})$$
is sufficient for $g(E^b(A)) \subset \widehat{A}^- \cup E^b(\widehat{A})$. By comparing moduli again, we can show that this is equivalent to the condition
$$\text{mod}\,(A(2|b|(\rho^b)^{k+1}\widehat{R}, \widehat{R})) > \text{mod}\,(B^b(\widehat{A})),$$
$$\text{mod}\,(A((\lambda^b)^k \widehat{R}, \widehat{R})) < \text{mod}\,(\widehat{A}^- \cup E^b(\widehat{A}) \cup B^b(\widehat{A})),$$
which is equivalent to
$$(k+1-\sqrt{k})\sqrt{\text{mod}\,(A)} - \frac{1}{2\pi}\log 2|b| > 0,$$
$$2 + \frac{2}{\sqrt{k}} < \sqrt{\text{mod}\,(A)}.$$

It is not hard to show these inequalities, since we have $k \geq 2$, $|b| \leq 4k$ and $\text{mod}\,(A) > 16$.

(2) In the case of $k = 1$, the property (III-b) does not hold under the condition (4.1). This is why we have to change the condition (4.1) to (4.3). Then the condition $\text{mod}\,(A) > 4 \times 2\sqrt{\text{mod}\,(A)}$, that is,
$$\text{mod}\,(A) > 64$$
is necessary to be able to take annuli which satisfy (4.3). By changing ρ^b and λ^b to
$$\rho^b = \exp(-4\pi\sqrt{\text{mod}\,(A)}), \quad \lambda^b = (\rho^b)^2,$$
respectively, we can apply Lemma 6.3 again provided that $\text{mod}\,(A)$ is large enough (actually, $\text{mod}\,(A) > 64$ is sufficient). Also, the properties (I) $\sim$ (III) can be proved by exactly the same method as in the proof of (1). So we omit the details. □

References

[1] L. Ahlfors, "Lectures on Quasiconformal Mappings", Van Nostrand (1966).
[2] L. Ahlfors, "Conformal Invariants", McGraw-Hill (1973).
[3] I. N. Baker, *Multiply connected domains of normality in iteration theory*, Math. Z. **81** (1963), 206–214.
[4] I. N. Baker, *The domains of normality of an entire function*, Ann. Acad. Sci. Fenn. Ser. A. I. Math. **1** (1975), no.2, 277–283.
[5] I. N. Baker, *An entire function which has a wandering domain*, J. Austral. Math. Soc. Ser. A **22** (1976), 173–176.
[6] I. N. Baker, *Wandering domains in the iteration of entire functions*, Proc. London Math. Soc. (3) **49** (1984), 563–576.

[7] I. N. Baker, *Some entire functions with multiply-connected wandering domains*, Erg. Th. & Dyn. Sys. **5** (1985), 163–169.

[8] I. N. Baker, J. Kotus and Y. Lü, *Iterates of meromorphic functions II: Examples of wandering domains*, J. London Math. Soc. (2) **42** (1990), 267–278.

[9] I. N. Baker, J. Kotus and Y. Lü, *Iterates of meromorphic functions III: Preperiodic domains*, Erg. Th. & Dyn. Sys. **11** (1991), 603–618.

[10] W. Bergweiler, *Iteration of meromorphic functions*, Bull. AMS **29** No. 2 (1993), 151–188.

[11] A. E. Eremenko and M. Ju. Lyubich, *Examples of entire functions with pathological dynamics*, J. London Math. Soc. (2) **36** (1987), 458–468.

[12] F. P. Gardiner and N. Lakic, "Quasiconformal Teichmüller Theory", Mathematical Surveys and Monographs, **76** AMS, (2000).

[13] C. T. McMullen, "Complex Dynamics and Renormalization", Annals of Mathematics Studies, **135** Princeton University Press, (1994).

[14] C. T. McMullen and D. Sullivan, *Quasiconformal homeomorphisms and dynamics. III. The Teichmüller space of a holomorphic dynamical system*, Adv. Math. **135** No. 2 (1998), 351–395.

[15] M. Shishikura, *On the quasiconformal surgery of rational functions*, Ann. Sci. Éc. Norm. Sup. **20** (1987), 1–29.

[16] D. Sullivan, *Conformal dynamical systems*, in Geometric dynamics (Rio de Janeiro, 1981), Lecture Notes in Math. **1007** Springer, Berlin, (1983), 725–752.

[17] C.-C. Young, G.-C. Wen, K.-Y. Li and Y.-M Chiang, "Complex analysis and its applications", Pitman Research Notes in Math. Ser. **305** Longman Scientific & Technical, (1994).

GRADUATE SCHOOL OF HUMAN AND ENVIRONMENTAL STUDIES, KYOTO UNIVERSITY, KYOTO 606-8501, JAPAN

E-mail address: kisaka@math.h.kyoto-u.ac.jp

DEPARTMENT OF MATHEMATICS, FACULTY OF SCIENCE, KYOTO UNIVERSITY, KYOTO 606-8502, JAPAN

E-mail address: mitsu@math.kyoto-u.ac.jp

FRACTAL MEASURES AND ERGODIC THEORY OF TRANSCENDENTAL MEROMORPHIC FUNCTIONS

JANINA KOTUS AND MARIUSZ URBAŃSKI

Dedicated to the memory of Professor I.N. Baker

ABSTRACT. We discuss Hausdorff and packing measures of some significant subsets of Julia sets of large classes of transcendental entire and meromorphic functions. In particular, the class of hyperbolic entire and meromorphic functions satisfying some mild derivative growth conditions is explored in Chapter 5. This class contains hyperbolic exponential and elliptic functions, which are also dealt with in Chapters 2 and 3, respectively. A substantial fraction of Chapter 2 and the whole of Chapter 3 treat functions which are not hyperbolic or not necessarily hyperbolic. Also Walters expanding and non-expanding conformal maps are discussed at length. Other classes of transcendental maps are also mentioned and we prove a number of results about invariant measures, Hausdorff dimension (especially its real analyticity), box and packing dimensions. Special attention is given to the methods developing thermodynamic formalism and conformal measures. Issues concerning the Lebesgue measure of various subsets of Julia sets are addressed in Chapter 7. In Section 3.3 a positive answer to Question 1 posed in [67] is given.

CONTENTS

(1) Introduction (presentation of topics to be dealt with)
(2) Exponential family $\mathcal{E}_\lambda, \lambda \in \mathbb{C} \setminus \{0\}$
   2.1 Preliminaries
   2.2 Hyperbolic maps: measures and dimensions
   2.3 Hyperbolic maps: thermodynamic formalism and multifractal analysis
   2.4 $\mathcal{E}_{1/e}$ and parabolic implosion
   2.5 Non-hyperbolic exponential maps
   2.6 Fatou functions, sine, cosine, and further families
(3) Elliptic functions
   3.1 General facts
   3.2 Gibbs and equilibrium states
   3.3 Critically non-recurrent elliptic functions

2000 *Mathematics Subject Classification.* Primary 37F35. Secondary 37F10, 30D05.
The research of the first author was supported in part by the the Polish KBN Grant No 2 PO3A 034 25 and the Warsaw University of Technology Grant no 504G 11200043000. The second author was supported in part by the NSF Grant DMS 0400481. The research of both authors was supported in part by the NSF/PAN grant INT-0306004.

(4) Walters expanding conformal maps
   4.1 Basic facts and definitions
   4.2 Hausdorff and box dimensions, Hausdorff and packing measures
   4.3 Barański and post-Barański maps, I
   4.4 Jump-like conformal maps
   4.5 Barański and post-Barański maps, II
(5) Hyperbolic entire and meromorphic functions
   5.1 Thermodynamic formalism
   5.2 Real analyticity
(6) Non-hyperbolic Barański maps
   6.1 The class $\mathcal{R}$
   6.2 The class $\mathcal{H}$
(7) Transcendental entire and meromorphic functions – the Lebesgue measure viewpoint
   7.1 The Lebesgue measure of the set of points escaping to ∞
   7.2 Milnor metric attractors
   7.3 The Lebesgue measure of Julia sets
(8) Appendix 1: $K(V)$ method of constructing semi-conformal measures
(9) Appendix 2: Martens' methods of constructing σ-finite invariant measures

1. Introduction (presentation of topics to be dealt with)

The main goal of this survey is to provide an overview of methods and results which have been used in the past five years (plus Barański's paper [7]) to analyse in detail the finer fractal structure of Julia sets of some classes of transcendental entire and meromorphic functions. By finer fractal structure we, roughly speaking, mean any knowledge about vanishing, positivity, finiteness and infiniteness of Hausdorff and packing measures of some significant subsets of Julia sets, including the Julia sets themselves. The methods used take as a starting point the development of appropriate versions of thermodynamic formalism, conformal measures and (infinite) iterated function systems. Apart from discussing Hausdorff and packing measures, we frequently mention the closely related concepts of Hausdorff, box and packing dimension, as well as invariant measures equivalent to either Hausdorff or packing measure. Our primary interest is in the classes of hyperbolic entire and meromorphic functions satisfying some mild derivative growth conditions (see Chapter 5), exponential ($z \mapsto \lambda \exp(z)$), elliptic and Walters expanding conformal maps, although we also discuss other classes in Section 2.5, Section 3.3, and separately in Chapter 4, where various subclasses of Walters expanding conformal maps are defined and explored. We devote Chapter 7 to issues related to the Lebesgue measure of the Julia sets and some of their subsets, for instance the set of points escaping to ∞. Results concerning Lebesgue measure are also scattered in previous sections; in Chapter 5 however they are treated more systematically and with greater generality. As an immediate application of results proved in [44] and stated in Chapter 5, we give in Section 3.3 a positive answer to Question 1 posed in [67]. For the background concerning the topological dynamics of transcendental entire and meromorphic functions the reader is referred for example to [1]-[5], [8] and [57]. We would like to make it clear that this article has been written over a relatively long period of time, during which new ideas and directions of development have been emerging. Wishing to include them, we have tried to present the material in a way which respects the historical development of the field and the existing classes of entire and meromorphic functions.

Acknowledgement. We thank the referee whose valuable remarks and suggestions influenced the final form of our paper.

2. EXPONENTIAL FAMILY

2.1. Preliminaries. In this chapter, except for the last two sections, we deal with the family $\{\mathcal{E}_\lambda\}_{\lambda \in \mathbb{C} \setminus \{0\}}$ of entire maps $\mathcal{E}_\lambda : \mathbb{C} \mapsto \mathbb{C}$ given by the formula

$$\mathcal{E}_\lambda(z) = \lambda e^z.$$

All these maps are called exponential and $\{\mathcal{E}_\lambda\}_{\lambda \in \mathbb{C} \setminus \{0\}}$ is called the exponential family. This family was investigated for the first time from the dynamical point of view by P. Fatou [28]. He conjectured that the Julia set of the map $\mathcal{E}_1$ is the entire complex plane $\mathbb{C}$. M. Misiurewicz proved this conjecture about sixty years later in [56] by ingenious but rather elementary methods. It was this event which revived anew the work on exponential family. Two other pioneering works in this area are due to M. Rees [61] and M. Lyubich [48] who proved that the map $\mathcal{E}_1$ is not ergodic with respect to the Lebesgue measure and the ω-limit set of Lebesgue almost every point coincides with the orbit of zero, $\{\mathcal{E}_1(0)\}_{n=0}^\infty$. The third one is due to R. Devaney and M. Krych who introduced in [24] a symbolic representation of exponential maps. This approach was the key point to clarify the topological picture of the Julia sets of hyperbolic exponential maps (see [20]-[25]), and was developed in the paper [1] by classifying hyperbolic Julia sets as Cantor bouquets and straight brushes. One should also mention at this point the paper [6] by I. Baker and P. Rippon who provided the first proof of Sullivan's Non-Wandering Theorem in the class of exponential functions. We should also mention the work of C. McMullen [55] who proved that the Hausdorff dimension of the Julia set of each exponential map is 2. In fact he proved more, that the Hausdorff dimension of the set of points escaping to infinity is equal to 2. Although interesting, this was not the end of the story since the set of points escaping to infinity is actually not significant from an ergodic point of view: for example, it cannot support any invariant Borel probability measure, although it does exhibit some interesting geometrical features (see [33], [34] and [35]). This situation made M. Urbański and A. Zdunik (see [76]-[79]) ask whether the complement of points escaping to infinity is dynamically and geometrically more interesting. The answer to this question is provided in the sections of this chapter. The basic construction used in all these sections, introduced for the first time in a slightly different form in [7] (see also Sections 3.2 and 4.3), whose analogs are applied to all maps considered in this article, is this. Let $\sim$ be the equivalence relation on $\mathbb{C}$ defined by declaring that $w \sim z$ if and only if $w - z \in 2\pi i \mathbb{Z}$, where $\mathbb{Z}$ is the set of all integers. Let $Q = \mathbb{C}/\sim$ be the quotient space of $\mathbb{C}$ by the relation $\sim$. Then Q is an infinite cylinder, conformally equivalent to the punctured plane $\mathbb{C}\setminus\{0\}$. Let $\Pi : \mathbb{C} \to Q$ be the canonical projection. Since each map $\mathcal{E}_\lambda : \mathbb{C} \to \mathbb{C}$ is constant on equivalence classes of the relation $\sim$, it can be treated as a map from Q to $\mathbb{C}$. The object we are after here, the map $E_\lambda : Q \to Q$, is defined as follows

$$E_\lambda = \Pi \circ \mathcal{E}_\lambda.$$

Notice that $E_\lambda \circ \Pi = \Pi \circ \mathcal{E}_\lambda$, that is, the following diagram commutes.

$$\begin{array}{ccc} \mathbb{C} & \xrightarrow{\mathcal{E}_\lambda} & \mathbb{C} \\ \Pi \downarrow & & \downarrow \Pi \\ Q & \xrightarrow{E_\lambda} & Q \end{array} \qquad (2.1)$$

So E_λ is a factor, via a conformal, locally isometric semiconjugacy Π. It is clear that all local geometric fractal features of $\mathcal{E}_\lambda$ are the same as those of E_λ, and this is also true of most of the dynamical features. Note that, since Π is a local isometry,

$$J(E_\lambda) = \Pi(J(\mathcal{E}_\lambda)) \quad \text{and} \quad J(\mathcal{E}_\lambda) = \Pi^{-1}(J(E_\lambda)).$$

Let $J_{bd}(\mathcal{E}_\lambda)$ be the set of all points $z \in J(\mathcal{E}_\lambda)$, whose orbit $\{\mathcal{E}_\lambda^n(z)\}_{n=0}^\infty$ is bounded. The following general result, often needed in the metric theory of exponential maps, has been proved in [76].

Proposition 2.1. *For every* $\lambda \in \mathbb{C}\setminus\{0\}$, $\operatorname{HD}(J_{bd}(\mathcal{E}_\lambda)) > 1$.

2.2. Hyperbolic maps: measures and dimensions.

An exponential map $\mathcal{E}_\lambda$ is called hyperbolic if it has an attracting periodic orbit. This equivalently means (see [77]) that there are two constants $c > 0$ and $\kappa > 1$ such that

$$|(\mathcal{E}_\lambda^n)'(z)| \geq c\kappa^n$$

for all $z \in J(\mathcal{E}_\lambda)$ and all $n \geq 0$. From now on, unless otherwise stated, we fix a hyperbolic exponential map $\mathcal{E}_\lambda$, denote it by $\mathcal{E}$, and denote E_λ by E. The subfamily of all hyperbolic exponential maps is very big, in particular it contains all parameters $\lambda \in (0, 1/e) \subset \mathbb{R}$. The reader may wish to consider only the case in which the map E corresponds to a parameter from this segment $(0, 1/e)$. In this section we describe the results and methods worked out in [76] and [77], although we closely follow the more mature exposition from [45]. The main objects of our focus are the radial Julia sets $J_r(\mathcal{E})$ and $J_r(E)$ introduced in [76]. The first one is defined as the complement in the Julia set of the set of points escaping to infinity, that is, the set of points $z \in J(\mathcal{E})$ that have a finite ω-limit point. Since $|\mathcal{E}(z)| = |\lambda|e^{\operatorname{Re}(z)}$, $J_r(\mathcal{E})$ is the set of points $z \in J(\mathcal{E})$ such that the sequence $\{\operatorname{Re}(\mathcal{E}^n(z))\}_{n=0}^{\infty}$ has a finite accumulation point. The set $J_r(E)$ is also defined as the set of points in the Julia set $J(E)$ that do not escape to infinity under the action of iterates of E. A point $z \in J_r(E)$ if and only if $z \in J(E)$ and there exists a point $y \in J(E)$ and an unbounded increasing sequence $\{n_k\}_{k=1}^{\infty}$ such that

$$y = \lim_{k \to \infty} E^{n_k}(z) \quad \text{and} \quad E^{n_k}(z) \in B(y, \delta/4)$$

for all $k \geq 1$, where

$$\delta = \frac{1}{4}\operatorname{dist}(J(E), \{E^n(\Pi(0))\}_{n=1}^{\infty}) \tag{2.2}$$

is positive since the map E is hyperbolic. The most important feature of points z from $J_r(E)$ is that for every $k \geq 1$ there exists a unique holomorphic branch $E_z^{-n_k} : B(y, 2\delta) \to Q$ of E^{-n_k} sending the point $E^{n_k}(z)$ to z. What distinguishes here the points of $J_r(E)$ from other points escaping to ∞ is that the inverse branches $E_z^{-n_k}$ are all defined on the same ball.

2.2.1. Pressure, Perron–Frobenius operators and generalised conformal measures.
In this subsection we gather material which will be needed to formulate and to sketch the proof of Bowen's formula, in order to discuss geometric measures and the proof of real-analytic dependence on λ of the Hausdorff dimension of the Julia sets $J_r(\mathcal{E}_\lambda)$. We begin with the notion of topological pressure. The problem here is that the phase space $J(E)$ is no longer compact and the classical approach using covers, (n, ϵ)-separated sets or (n, ϵ)-spanning sets fails in the context of exponential functions. We therefore adapt the pointwise approach, equivalent in the case of open expanding maps to the classical ones (see [59]), which for exponential maps works very well. Given $t \geq 0$, the topological pressure of the potential $-t \log |E'|$ is given by the formula

$$\operatorname{P}(t) = \lim_{n \to \infty} \frac{1}{n} \log \sum_{x \in E^{-n}(z)} |(E^n)'(z)|^{-t},$$

where z is an arbitrary point in $Q \setminus PC(E)$ and

$$PC(E) = \overline{\bigcup_{n \geq 0} E^n(\Pi(0))}.$$

It is not at all obvious that the limit defining topological pressure $\operatorname{P}(t)$ must exist. It can be proved using finer properties of the corresponding Perron–Frobenius operator. Independence of z is much easier; it almost immediately follows from Koebe's Distortion Theorem. In order to get closer to the meaning of this topological pressure $P(t)$, let us look at the term corresponding to $n = 1$, namely,

$$P_z(1,t) = \sum_{x \in E^{-1}(z)} |E'(x)|^t = \sum_{n \in \mathbb{Z}} |E'(z_n)|^{-t} = \sum_{n \in \mathbb{Z}} |E(z_n)|^{-t} = \sum_{n \in \mathbb{Z}} |\tilde{z} + 2\pi i n|^{-t},$$

where $\tilde{z}$ is an arbitrary point from $\Pi^{-1}(z)$ and $z_n \in Q$ is the only point such that $\lambda e^{z_n} = \tilde{z} + 2\pi i n$. The series $P_z(1,t)$ converges if and only if $t > 1$. Before moving on we would like to make two remarks. Firstly, it is straightforward to see that without projecting the dynamics to the cylinder Q, the series involved in the definition of the topological pressure would not converge even for $n = 1$. And this is the main reason to project $\mathcal{E}$ down to E. A different method to resolve this 'divergence' difficulty is provided in Chapter 5. Secondly, the dynamical significance of the pressure function and other concepts in the thermodynamic formalism is evident from Subsection 2.2.2. Its geometrical significance comes from Subsection 2.2.3, notably Theorem 2.11 (Bowen's formula).

The basic properties of the topological pressure $P(t)$ are summarised in the following.

Lemma 2.2. *The function $t \mapsto P(t)$, $t \geq 0$, has the following properties.*
  (a) *There exists $t \in (0,1)$ such that $0 \leq P(t) < +\infty$.*
  (b) *$P(t) < +\infty$ for all $t > 1$.*
  (c) *The function $P(t)$ restricted to the interval $(1, +\infty)$ is convex, continuous and strictly decreasing.*
  (d) *$\lim_{t \to +\infty} P(t) = -\infty$.*
  (e) *There exists exactly one $t > 1$ such that $P(t) = 0$.*

The basic concept of any version of thermodynamic formalism is the Perron–Frobenius operator and eigenvector of its dual operator, called in the sequel a generalised conformal measure. Let $C_b(J(E))$ be the Banach space of all bounded complex-valued continuous functions defined on $J(E)$. Given $t > 1$, the (bounded) linear operator $\mathcal{L}_t : C_b(J(E)) \to C_b(J(E))$ is given by the formula

$$\mathcal{L}_t(g)(z) = \sum_{x \in E^{-1}(z)} |E'(x)|^{-t} g(x) = \sum_{n \in \mathbb{Z}} |\tilde{z} + 2\pi i n|^{-t} g(z_n).$$

Let $\mathcal{L}_t^* : C_b^*(J(E)) \to C_b^*(J(E))$ be the operator dual to $\mathcal{L}_t$, that is,

$$\mathcal{L}_t^*(\nu)(g) = \nu(\mathcal{L}_t(g)).$$

Following the classical thermodynamic formalism, and what is extremely important for a geometrical purposes, one would like to find an eigenvector of the operator $\mathcal{L}_t^*$ corresponding to a positive eigenvalue. The classical approach (see [12], comp. [59]) is to consider the map

$$\nu \mapsto \frac{\mathcal{L}_t^*(\nu)}{\mathcal{L}_t^*(\nu)(\mathbb{1})} \qquad (2.3)$$

defined on the convex space of Borel probability measures on $J(E)$. This map is easily seen to be continuous and, in the classical case of open distance expanding maps, ones applies the Schauder-Tichonov fixed point theorem to get a fixed point of the map (2.3) and, consequently, the required eigenvector of the operator $\mathcal{L}_t^*$. In our case however this method fails since the space of all Borel probability measures on $J(E)$ is not compact. So, one must proceed in a different way. A Borel probability measure m_t is called (t, α_t)-conformal (with $t > 1$ and $\alpha_t \geq 0$) if $m_t(J(E)) = 1$ and for any Borel set $A \subset Q$, restricted to which E is injective, one has

$$m_t(E(A)) = \int_A \alpha_t |E'|^t dm_t. \qquad (2.4)$$

Note that a measure is (t, α_t)-conformal if and only if it is an eigenmeasure of the dual operator $\mathcal{L}_t^*$ corresponding to the eigenvalue α_t. One proceeds to construct an m_t measure by using the $K(V)$-methods described in Appendix 1. Namely, for every $n \geq 1$, put

$$K_n = \bigcap_{j=0}^{\infty} E^{-j}(\{z \in J(E) : \text{Re} z \leq n\}).$$

Then each set K_n is a compact subset of the cylinder Q and $E(K_n) \subset K_n$. So, Lemma 8.1 applies to produce for every $t \geq 0$, a number $\text{P}_n(t)$ and a measure m_n supported on K_n with the following two properties. If $A \subset K_n$ is a Borel set and $E|_A$ is one-to-one, then

$$m_n(E(A)) \geq \int_A e^{\text{P}_n(t)} |E'|^t m_n. \qquad (2.5)$$

If in addition $A \cap \{z \in Q : \text{Re} z = n\} = \emptyset$, then

$$m_n(E(A)) = \int_A e^{\text{P}_n(t)} |E'|^t m_n. \qquad (2.6)$$

It is easy to see that in our context one can replace the inclusion $A \subset K_n$ by $A \subset \{z \in Q : \text{Re} z \leq n\}$ and the two formulae above still hold. It follows from the definition of the numbers $\text{P}_n(t)$ that $\text{P}_n(t) \leq \text{P}(t)$. Making use of the flexibility involved in this definition, one can arrange for the sequence $\{\text{P}_n(t)\}_{n=1}^{\infty}$ to be non-decreasing ($\text{P}_n(t) \leq \text{P}_{n+1}(t)$). If the cylinder Q were compact, the next step would be obvious and rather straightforward. Take as a candidate for a (t, α_t)-conformal measure any weak limit of the sequence $\{m_n\}_{n=1}^{\infty}$. Since however the cylinder Q and the Julia set $J(E)$ are not compact, one needs to show that the sequence $\{m_n\}_{n=1}^{\infty}$ is tight, which in our setting means that the measures of this sequence do not accumulate at $+\infty$. And this has been done in [76] and [77]. Now, taking any weakly converging subsequence of the sequence $\{m_n\}_{n=1}^{\infty}$ and using (2.5) along with (2.6), one relatively easily checks that its weak limit m_t satisfies formula (2.4) with $\alpha_t = \exp(\lim_{n \to \infty} P_n(t))$. This means that m_t is (t, α_t)-conformal and $\mathcal{L}_t^*(m_t) = \alpha_t m_t$. Its basic property, obtained without making use of the properties of the Perron–Frobenius operator, is the following. Put

$$I_\infty(f) = \{z \in \mathbb{C} : \lim_{n \to \infty} f^n(z) = \infty\}.$$

Then we have the following.

Proposition 2.3. *For every $t > 1$ there exists $M > 0$ such that for m_t-a.e. $x \in J(E)$*

$$\liminf_{n \to \infty} \operatorname{Re}(E^n(x)) \leq M.$$

In particular, $m_t(I_\infty(E)) = 0$ or equivalently $m_t(J_r(E)) = 1$.

This is the first sign that $J_r(E)$ is the right object to deal with. There will be more. In order to study the Perron–Frobenius operator $\mathcal{L}_t$, it is convenient to consider its normalised version $\hat{\mathcal{L}}_t = \alpha_t^{-1} \mathcal{L}_t$. As relatively soft facts (although obtained with some non-obvious tricks) one proves in [77] the following.

Lemma 2.4. $\sup_{n \geq 1}\{||\hat{\mathcal{L}}_t^n(\mathbb{1})||_\infty\} < \infty$.

and

Lemma 2.5. $\inf_{n \geq 1}\{\inf\{\hat{\mathcal{L}}_t^n(\mathbb{1})(z) : z \in \{w \in Q : \operatorname{Re} w \leq x\}\}\} > 0$ *for every $x \geq 0$ large enough.*

These two facts are the main ingredients in the proof linking the topological pressure $P(t)$ and the eigenvalue α_t. Namely

$$\alpha_t = e^{P(t)}.$$

To obtain further and harder properties of the Perron–Frobenius operator $\hat{\mathcal{L}}_t$ we have to define the Banach space H_α of locally α-Hölder bounded continuous functions. Fix $\alpha \in (0,1]$. Given $g : J(E) \to \mathbb{C}$, let

$$v_\alpha(g) = \inf\{L \geq 0 : |g(y) - g(x)| \leq L|y-x|^\alpha \text{ for } x, y \in J(E) \text{ with } |y-x| \leq \delta\}$$

be the α-variation of the function g, where $\delta > 0$ was defined in formula (2.2). Any function with bounded α-variation will be called α-Hölder (or simply Hölder continuous if we do not want to specify the exponent of Hölder continuity). Let $C_b = C_b(J(E))$ be the space of all bounded continuous complex valued functions defined on $J(E)$ and let

$$||g||_\alpha = v_\alpha(g) + ||g||_\infty.$$

Clearly the space
$$H_\alpha = H_\alpha(J(E)) = \{g \in J(E) : ||g||_\alpha < \infty\}$$
endowed with the norm $||\cdot||_\alpha$ is a Banach space densely contained in C_b with respect to the $||\cdot||_\infty$ norm. Any member of the space H_α will be called a bounded Hölder continuous function with exponent α and any member of the space $\bigcup_{\alpha>0} H_\alpha$ will be simply called a bounded Hölder continuous function. Using the inequality $||\hat{\mathcal{L}}_t^n||_\alpha \leq \frac{1}{2} v_\alpha(g) + c||g||_\infty$ (for all $n \geq 1$ large enough), noting that images of bounded subsets of H_α under $\hat{\mathcal{L}}_t$ are relatively compact as subsets of $C_b(J(E))$ and applying the Ionescu-Tulcea and Marinescu Theorem, we obtain the final (at least for our purposes) properties of the Perron–Frobenius operator, stated in the following theorem.

Theorem 2.6. *If $t > 1$ then we have the following.*
  (a) *The operator $\hat{\mathcal{L}}_t : H_\alpha \to H_\alpha$ has a simple isolated eigenvalue at 1.*
  (b) *The eigenspace of the eigenvalue 1 is generated by a nowhere vanishing function $\psi_t \in H_\alpha$ such that $\int \psi_t dm_t = 1$ and $\lim_{\mathrm{Re} z \to +\infty} \psi_t(z) = 0$.*
  (c) *The number 1 is the only eigenvalue of modulus 1.*
  (d) *There exists $S : H_\alpha \to H_\alpha$ such that*
$$\hat{\mathcal{L}}_t = Q_1 + S,$$
where $Q_1 : H_\alpha \to \mathbb{C}\psi_t$ is a projection on the eigenspace $\mathbb{C}\psi_t$,
$$Q_1 \circ S = S \circ Q_1 = 0 \quad \text{and} \quad ||S^n||_\alpha \leq C\xi^n$$
for some constant $C > 0$, some constant $\xi \in (0,1)$ and all $n \geq 1$.

2.2.2. Invariant measures. Mainly as a consequence of Theorem 2.6, we get the following results.

Theorem 2.7. *If $t > 1$, then the measure $\mu = \mu_t = \psi_t m_t$ is E-invariant, ergodic with respect to each iterate of E and equivalent to the measure m_t. In particular $\mu(J_r(E)) = 1$.*

Due to Theorem 2.6 the E-invariant measure μ has much finer stochastic properties than ergodicity of all iterates of E, as follows.

Theorem 2.8. *The dynamical system (E, μ_t) is metrically exact, that is, its Rokhlin natural extension is a K-system.*

The proof of this fact is the same as the proof of Corollary 3.7 in [18]. The next two theorems are standard consequences of Theorem 2.6 (see [16] and

[59] for example). Let g_1 and g_2 be real square-μ_t-integrable functions on $J_r(E)$. For every positive integer n the n-th correlation of the pair g_1, g_2, is given by

$$C_n(g_1, g_2) = \int g_1 \cdot (g_2 \circ E^n) \, d\mu_t - \int g_1 \, d\mu_t \int g_2 \, d\mu_t,$$

provided the above integrals exist. Notice that due to the E-invariance of μ we can also write

$$C_n(g_1, g_2) = \int (g_1 - \int g_1 d\mu_t)\big((g_2 - \int g_2 d\mu_t) \circ E^n\big) d\mu_t.$$

We now have the following.

Theorem 2.9. *There exist $C \geq 1$ and $\rho < 1$ such that for all $g_1 \in H_\alpha(Q)$ and $g_2 \in L^1(\mu_t)$*

$$C_n(g_1, g_2) \leq C\rho^n \|g_1 - \int g_1 d\mu_t\|_\alpha \|g_2 - \int g_2 d\mu_t\|_{L^1}.$$

Let $g : J_r(E) \to \mathbb{R}$ be a square-integrable function. The limit

$$\sigma^2(g) = \lim_{n \to \infty} \frac{1}{n} \int \Big(\sum_{j=0}^{n-1} g \circ E^j - n \int g d\mu_t\Big)^2 d\mu_t$$

is called the asymptotic variance or dispersion of g, provided it exists.

Theorem 2.10. *If $g \in H_\alpha(Q)$, $\alpha \in (0,1)$, then $\sigma^2(g)$ exists and, if we have $\sigma^2(g) > 0$, then the sequence of random variables $\{g \circ E^n\}_{n=0}^\infty$ with respect to the probability measure μ_t satisfies the Central Limit Theorem, that is,*

$$\mu_t\Big(\Big\{x \in J_r(E) : \frac{\sum_{j=0}^{n-1} g \circ E^j - n \int g d\mu_t}{\sqrt{n}} < r\Big\}\Big) \to \frac{1}{\sigma\sqrt{2\pi}} \int_{-\infty}^r e^{-t^2/2\sigma^2} \, dt.$$

2.2.3. Bowen's formula, Hausdorff and packing measures. It was stated in Lemma 2.2 that there exists exactly one value $t > 1$ such that $P(t) = 0$, and we call this value h. Then the measure m_h is h-conformal in the sense of Sullivan, meaning that

$$m_h(E(A)) = \int_A |E'|^h dm_h \qquad (2.7)$$

for every Borel set $A \subset J(E)$ such that $E|_A$ is one-to-one. By the analogy with the classical situation of hyperbolic rational functions (see [75]), one may guess that h is the Hausdorff dimension of $J_r(E)$ and m_h, up to a multiplicative constant, the Hausdorff and packing measure on $J_r(E)$. The first guess turns out to be entirely true, the latter one only partially.

Theorem 2.11. *(Bowen's formula) The number h, the unique zero of the pressure function $t \to \mathrm{P}(t), t > 1$, is equal to $\mathrm{HD}(J_r(E))$.*

Since the definition of the pressure function $\mathrm{P}(t)$ has a priori nothing to do with the set $J_r(E)$, this theorem in particular indicates that $J_r(E)$ is the right object to deal with.

Theorem 2.12. $0 < \mathrm{H}^h(J_r(E)) < \infty$.

This theorem indicates that the h-dimensional Hausdorff measure H^h on $J_r(E)$ is the right geometric measure. In particular, the purely dynamically defined, h-conformal measure m_h gets its geometrical interpretation as the normalised Hausdorff measure.

Concerning the h-dimensional packing measure P^h, the situation seems to be much worse. We have the following.

Proposition 2.13. *We have $\mathrm{P}^h(J_r(E)) = \infty$. In fact $\mathrm{P}^h(G) = \infty$ for every open nonempty subset of $J_r(E)$.*

It might seem that a measure with such strange properties is completely useless. However since the 2-dimensional packing measure on $\mathbb{C}$ is proportional to the 2-dimensional Lebesgue measure and this latter one is not locally infinite, we immediately get from Proposition 2.13 the following.

Corollary 2.14. *It holds that $h = \mathrm{HD}(J_r(E)) < 2$.*

Since $J_r(E) \supset J_{bd}(E)$, as an immediate consequence of Proposition 2.1, we get the following.

Corollary 2.15. *It holds that $h = \mathrm{HD}(J_r(E)) > 1$.*

In our case every compact E-invariant set is hyperbolic. So Shishikura's Hausdorff dimension, being by definition the supremum over all hyperbolic sets, is equal to the supremum over all E-invariant compact sets. Since any E-invariant Borel probability measure μ on $J(E)$ is supported on $J_r(E)$ (that is, $\mu(J_r(E)) = 1$), and since each compact E-invariant subset of $J(E)$ is contained in $J_r(E)$, we get the following result.

Theorem 2.16. *The hyperbolic dimension of $J(E)$ and the supremum of the Hausdorff dimensions of all Borel probability E-measures, both being bounded above by $\mathrm{HD}(J_r(E))$, are strictly less than $\mathrm{HD}(J(E)) = 2$.*

The two suprema appearing in Theorem 2.16 are in fact equal (see [59]). Note that it is still an open question whether there exists a rational function with hyperbolic dimension less than the Hausdorff dimension of its Julia set.

2.2.4. Real analyticity of the Hausdorff dimension function.

Since this article primarily concerns measures, and dimensions are treated more briefly, we shall provide a very short description of how to prove that the Hausdorff dimension of the radial Julia set $J_r(E_\lambda)$ depends in real-analytic fashion on hyperbolic parameters λ. The full proof, technically rather complicated, can be found in [77] and [45]. One starts with the trivial observation that we can restrict our attention to an arbitrary fixed component W of the set of all hyperbolic parameters. Fix one parameter $\lambda_0 \in W$. It is known (see[27], comp.[78]) that each function $E_\lambda : \mathbb{C} \to \mathbb{C}$ is topologically conjugate to E_{λ_0} via a quasi-conformal homeomorphism $h_\lambda : \mathbb{C} \to \mathbb{C}$. This allows us to define new Perron–Frobenius operators

$$\mathcal{L}^0_{\lambda,t} : \mathrm{H}_\alpha(J(E_{\lambda_0})) \to \mathrm{H}_\alpha(J(E_{\lambda_0})), \quad (\lambda,t) \in W \times (1,+\infty),$$

as follows

$$\mathcal{L}^0_{\lambda,t} g(z) = \sum_{x \in E^{-1}_{\lambda_0,t}(z)} |E'_\lambda(h_\lambda(x))|^{-t} g(x).$$

The most technically involved task now is to demonstrate that, given any $t_0 > 1$, there exists a polydisc $\mathbb{D}((\lambda_0, t_0); R) \subset \mathbb{C}^2 \times \mathbb{C}$ such that the operator-valued function $(\lambda, t) \mapsto \mathcal{L}^0_{\lambda,t} \in L(\mathrm{H}_\alpha(J(E_{\lambda_0})))$, $(\lambda, t) \in W \times (1,+\infty)$, has a holomorphic extension to this polydisc $\mathbb{D}((\lambda_0, t_0), R)$. The rest of the proof is the right combination of Theorem 2.6 (that $e^{P(t)}$ is a simple isolated eigenvalue of $\mathcal{L}^0_{\lambda,t}$), perturbation theory for linear operators, Theorem 2.11 (Bowen's formula) and the inverse function theorem.

2.3. Hyperbolic maps: thermodynamic formalism and multifractal analysis.

In this section our intention is to provide a relatively compact, nevertheless comprehensive summary of the article [74], which provides a systematic account of the thermodynamic formalism and multifractal analysis of hyperbolic exponential maps, and 1^+-tame Hölder continuous potentials, which are natural in this context. We keep the notation and terminology from the previous section.

2.3.1. Pressures, potentials, transfer operators and conformal measures.

Given $n \geq 0$ let

$$S_n \phi = \phi + \phi \circ E + \cdots + \phi \circ E^{n-1}.$$

The following simple distortion fact is needed for nearly all the results presented below.

Lemma 2.17. *For every $\alpha > 0$ there exists $L_\alpha > 0$ such that if $\phi : J(E) \to \mathbb{C}$ is an α-Hölder function, then*
$$|S_n\phi(E_v^{-n}(y)) - S_n\phi(E_v^{-n}(x))| \leq L_\alpha v_\alpha(\phi)|y - x|^\alpha$$
for all $n \geq 1$, all $x, y \in J(E)$ with $|x - y| \leq \delta$, and all $v \in E^{-n}(x)$.

Put $Q_+^c = \{z \in Q : \text{Re} z \geq \inf\{\text{Re}(J(E))\}\}$. Given $\kappa \in \mathbb{R}$ a Hölder continuous function $\phi : Q_+^c \to \mathbb{R}$ is called κ-tame (and we put $\kappa = \kappa(\phi)$) if
$$A_\phi = \sup\{|\phi(z) + \kappa \text{Re} z| : z \in Q_+^c\} < +\infty.$$
A Hölder continuous function $\phi : Q_+^c \to \mathbb{R}$ is called κ^+-tame if it is a-tame with some $a > \kappa$. Especially important will turn out to be the 1^+-tame functions. Any κ-tame function with any $\kappa \in \mathbb{R}$ is called a tame function. The above tameness notions refer to any Hölder continuous complex-valued function if its real part satisfies the respective conditions. Note that $\phi : J(f) \to \mathbb{C}$ is a 0^+-tame or 0-tame function if and only if e^ϕ is a bounded Hölder continuous function with the same exponent.

The topological pressure of the tame potentials is defined by the pointwise method. It can be proved that for $\phi : J(E) \to \mathbb{R}$, a 0^+-tame potential, the following limit
$$P_z(\phi) = \lim_{n\to\infty} \frac{1}{n} \log \sum_{x \in E^{-n}(z)} \exp(S_n\phi(x))$$
exists and is independent of $z \in J(E)$. Its common value is denoted by $\text{P}(\phi)$ and is called the topological pressure of the potential ϕ. Assume now ϕ to be a 1^+-tame potential. The Perron–Frobenius operator $\mathcal{L}_\phi : C_b \to C_b$ is defined by the formula

$$\mathcal{L}_\phi g(z) = \sum_{x \in E^{-1}(z)} \exp(\phi(x))g(x) = \sum_{k=-\infty}^{+\infty} \exp(\phi(z_k))g(z_k), \quad (2.8)$$

where z_k is the only point of the singleton $\pi\big(\mathcal{E}^{-1}(\tilde{z} + 2\pi i k)\big)$ and $\tilde{z}$ is an arbitrary point from $\pi^{-1}(z)$ and Q. The dual operator $\mathcal{L}_\phi^* : C_b^* \to C_b^*$ is given by the formula $\mathcal{L}_\phi^* \mu(g) = \mu(\mathcal{L}_\phi g)$. Applying the $K(V)$ method from [18] (see Appendix 1) and the tightness argument, one can show that there exists m_ϕ, a unique Borel probability measure on $J(E)$ that is an eigenmeasure of $\mathcal{L}_\phi^*$. The corresponding eigenvalue is equal to $e^{\text{P}(\phi)}$. This equivalently means that
$$m_\phi(E(A)) = \int_A \exp(\text{P}(\phi) - \phi) dm_\phi$$

for every Borel set $A \subset Q$ restricted to which E is injective. The measure m_ϕ is called ϕ-conformal. Applying the famous Tulcea–Ionescu and Marinescu Theorem, using Lemma 2.17 and (2.8), one eventually ends up with the following properties of the Perron–Frobenius operator $\mathcal{L}_\phi$.

Theorem 2.18. *If $\phi : J(E) \to (0, \infty)$ is a 1^+-tame potential, then we have the following.*
  (a) *The operator $e^{-\mathrm{P}(\phi)}\hat{\mathcal{L}}_\phi : \mathrm{H}_\alpha \to \mathrm{H}_\alpha$ has a simple isolated eigenvalue at 1.*
  (b) *The eigenspace of the eigenvalue 1 is generated by a nowhere vanishing function $g_\phi \in \mathrm{H}_\alpha$ such that $\int g_\phi dm_\phi = 1$ and $\lim_{\mathrm{Re} z \to +\infty} g_\phi(z) = 0$.*
  (c) *The number 1 is the only eigenvalue of modulus 1.*
  (d) *With $S : \mathrm{H}_\alpha \to \mathrm{H}_\alpha$ as in Theorem 2.6, we have*
$$e^{-\mathrm{P}(\phi)}\hat{\mathcal{L}}_\phi = Q_1 + S,$$
where $Q_1 : \mathrm{H}_\alpha \to \mathbb{C}g_\phi$ is a projector on the eigenspace $\mathbb{C}g_\phi$,
$$Q_1 \circ S = S \circ Q_1 = 0 \text{ and } ||S^n||_\alpha \leq C\xi^n$$
for some constant $C > 0$, some constant $\xi \in (0,1)$ and all $n \geq 1$.

2.3.2. Ergodic theory of invariant measures. As a fairly straightforward consequence of of Theorem 2.18, one links conformal measures with ergodic theory. More precisely:

Theorem 2.19. *If $\phi : J(E) \to (0, \infty)$ is a 1^+-tame potential, then the measure $\mu = \mu_\phi = g_\phi m_\phi$ is E-invariant, ergodic with respect to each iterate of E and equivalent to the measure m_ϕ. In particular $\mu(J_r(E)) = 1$. In addition the dynamical system (E, μ_ϕ) is metrically exact, that is, its Rokhlin natural extension is a K-system.*

The measure μ_ϕ is called the invariant Gibbs state of the potential ϕ. Due to Theorem 2.18 the E-invariant measure μ_ϕ has additional stochastic properties to those listed in the theorem above. Namely, the correlations of Hölder continuous bounded potentials decay (with respect to the dynamical system (E, μ_ϕ)) exponentially fast, and for every $g \in \mathrm{H}_\alpha$, the sequence $\{g \circ E^n\}_{n=1}^\infty$ of identically distributed random variables (with respect to the measure μ_ϕ) converges in distribution to a Gauss distribution provided that the asymptotic variance (dispersion) of g is positive. As it was proved in [15] (and the proof is repeated in [74]), the Gibbs state μ_ϕ is an equilibrium state for the potential ϕ. As a matter of fact, the following, nearly classical version of the Variation Principle holds.

Theorem 2.20. *If $\mathcal{E} : \mathbb{C} \to \mathbb{C}$ is hyperbolic and $\phi : J(E) \to \mathbb{C}$ is a 1^+-tame potential, then the invariant measure μ_ϕ is an equilibrium state of the potential ϕ, that is*

$$P(\phi) = \sup\{h_\mu(E) + \int \phi d\mu\},$$

where the supremum is taken over all Borel probability E-invariant ergodic measures μ with $\int \phi d\mu > -\infty$, and

$$P(\phi) = h_{\mu_\phi} + \int \phi d\mu_\phi.$$

2.3.3. Analytic properties of the pressure function.
Sections 7 and 8 of [74] contain a detailed analysis of differentiability properties of the topological pressure function. Let $\phi, \psi : J(E) \to \mathbb{R}$ be two arbitrary tame functions. Consider the sets

$$\Sigma_1(\phi, \psi) = \{q \in \mathbb{C} : \text{Re}(q)\kappa(\phi) + \kappa(\psi) > 1\}$$

and

$$\Sigma_2(\phi, \psi) = \{(q, t) \in \mathbb{C} \times \mathbb{C} : \text{Re}(q)\kappa + \text{Re}(t)\gamma > 1\}.$$

The key ingredient to understanding all further analytic properties of various 'thermodynamical objects' is the following.

Proposition 2.21. *If $\phi, \psi : J(E) \to \mathbb{R}$ are two arbitrary tame functions, then the function $q \mapsto \mathcal{L}_{q\phi+\psi} \in L(H_\alpha)$, $q \in \Sigma_1(\phi, \psi)$, is holomorphic.*

Using perturbation theory for linear operators (Kato–Rellich Theorem, see Theorem XII.8 in [60]), we obtain the following fairly straightforward consequence of Proposition 2.21.

Lemma 2.22. *If ϕ and ψ are arbitrary tame functions, then the function $q \mapsto P(q\phi + \psi)$, $q \in \Sigma_1(\phi, \psi) \cap \mathbb{R}$, is real-analytic, and likewise, the function $(q, t) \mapsto P(q\phi + t\psi)$, $(q, t) \in \Sigma_2(\phi, \psi)$, is real-analytic.*

In the classical theory of distance expanding maps (see [59]) the first and the second derivative of the pressure function are calculable although the calculations leading to the formula for the second derivative are rather tedious. Even more tedious calculations, performed in Section 8 of [74] led to the following results.

Theorem 2.23. *Suppose that $\phi : J(E) \to \mathbb{R}$ is a 1^+-tame function and $\psi : J(E) \to \mathbb{R}$ is a tame function. Then*

$$\frac{d}{dt}\Big|_{t=0} P(\phi + t\psi) = \int \psi d\mu_\phi.$$

Theorem 2.24. *Suppose that* $\phi : J(E) \to \mathbb{R}$ *is a* 1^+*-tame function and* $\psi, \zeta : J(E) \to \mathbb{R}$ *are tame functions. Then*

$$\frac{\partial^2}{\partial s \partial t}\bigg|_{(0,0)} P(\phi + s\psi + t\zeta) = \sigma^2(\psi, \zeta),$$

where

$$\sigma^2(\psi, \zeta) = \lim_{n \to \infty} \frac{1}{n} \int S_n\big(\psi - \mu_\phi(\psi)\big) S_n\big(\zeta - \mu_\phi(\zeta)\big) d\mu_\phi$$
$$= \int (\psi - \mu_\phi(\psi))(\zeta - \mu_\phi(\zeta)) d\mu_\phi$$
$$+ \sum_{k=1}^\infty \int (\psi - \mu_\phi(\psi))(\zeta - \mu_\phi(\zeta)) \circ E^k d\mu_\phi$$
$$+ \sum_{k=1}^\infty \int (\zeta - \mu_\phi(\zeta))(\psi - \mu_\phi(\psi)) \circ E^k d\mu_\phi.$$

(If $\psi = \zeta$, then we simply write $\sigma^2(\psi)$ for $\sigma^2(\psi, \psi)$.)

2.3.4. Cohomologies. Of course the natural question arises of when the two 1^+-tame potentials have the same Gibbs (equilibrium) states. This problem is completely solved in Theorem 9.1 from [74] and, although the solution is the same as in the classical case, some of its constituents, especially item (3) below, are rather unexpected. Let $\mathcal{F}$ be any class of real-valued functions defined on $J(E)$. Two functions $\phi, \psi : J(E) \to \mathbb{R}$ are said to be cohomologous in the class $\mathcal{F}$ if there exists a function $u \in \mathcal{F}$ such that

$$\phi - \psi = u - u \circ E.$$

Our solution is this.

Theorem 2.25. *If $\phi, \psi : J(E) \to \mathbb{R}$ are two arbitrary 1^+-tame functions, then the following conditions are equivalent:*

(1) $\mu_\phi = \mu_\psi$.
(2) *There exists a constant R such that for each $n \geq 1$, if $E^n(z) = z$ ($z \in J(E)$), then*

$$S_n\phi(z) - S_n\psi(z) = nR.$$

(3) *The difference $\psi - \phi$ is cohomologous to a constant R in the class of bounded Hölder continuous functions.*
(4) *The difference $\psi - \phi$ is cohomologous to a constant in the class of bounded continuous functions.*
(5) *The difference $\psi - \phi$ is cohomologous to a constant in the class of all functions defined everywhere in $J(E)$.*

(6) There exist constants S and T such that for every $z \in J(E)$ and every $n \geq 1$
$$|S_n\phi(z) - S_n\psi(z)| \leq T.$$
If these conditions are satisfied, then $R = S = \mathrm{P}(\phi) - \mathrm{P}(\psi)$.

As its complement we list the following result whose proof uses Rokhlin's natural extension and the concept of canonical conditional measures.

Proposition 2.26. *If $\phi : Q_+^c \to \mathbb{R}$ is a 1^+-tame function and $\psi : Q_+^c \to \mathbb{R}$ is a tame function, then $\sigma^2_{\mu_\phi}(\psi) = 0$ if and only if ψ is cohomologous to a constant function in the class of bounded Hölder continuous functions on $J(E)$.*

2.3.5. Hausdorff dimension of Gibbs states.

Given a Borel probability measure μ on a metric space, we define for a point x in this space, the number (called the local dimension at the point x)
$$d_\mu(x) = \lim_{r \to 0} \frac{\log \mu(B(x,r))}{\log r}$$
provided that the limit exists. Recall that $\mathrm{HD}(\mu)$, the Hausdorff dimension of the measure μ, is the infimum of the numbers $\mathrm{HD}(Y)$ taken over all Borel sets Y such that $\mu(Y) = 1$. If $d_\mu(x)$ is constant a.e. (denote it by d_μ), then $\mathrm{HD}(\mu) = d_\mu$. If μ is a Borel probability E-invariant measure on $J(E)$, then the number
$$\chi_\mu = \int \log |E'| d\mu$$
is called the Lyapunov exponent of the map E with respect to the measure μ. Note that this exponent is always positive. The proof of the following result appeared first in [73] and was repeated in [74].

Theorem 2.27. *If $\mathcal{E} : \mathbb{C} \to \mathbb{C}$ is a hyperbolic exponential map and $\phi : J(E) \to \mathbb{C}$ is a 1^+-tame potential, then the local dimension $d_\mu(x)$ exists for μ_ϕ-a.e. $z \in J(E)$ and is equal to $h_{\mu_\phi}/\chi_{\mu_\phi}$. In particular*
$$\mathrm{HD}(\mu_\phi) = \frac{h_{\mu_\phi}}{\chi_{\mu_\phi}}.$$

2.3.6. Multifractal analysis of 1^+-tame functions.

Fix a 1^+-tame potential $\phi : Q_+^c \to \mathbb{R}$. Subtracting $\mathrm{P}(\phi)$ from ϕ, we can assume without loss of generality that $\mathrm{P}(\phi) = 0$ and call ϕ normalised. Consider the two-parameter family of potentials $\phi_{q,t} : P_+ \to \mathbb{R}$, $q, t \in \mathbb{R}$, defined as follows.
$$\phi_{q,t} = -t \log |E'| + q\phi.$$
Note that $\phi_{q,t}$ is a $t + q\kappa(\phi)$-tame function. We have the following.

Lemma 2.28. *For every $q \in \mathbb{R}$ there exists a unique $t = T(q) \in \mathbb{R}$ such that $P(\phi_{q,t}) = 0$. In addition, $(q, T(q)) \in \Sigma_2(\phi, -\log|E'|)$.*

The mapping $T : \mathbb{R} \to \mathbb{R}$ is called the temperature function. The multifractal analysis of a Borel probability measure μ studies the level sets of the pointwise dimension function d_μ. Although defined in an entirely 'fractal manner', this function frequently exhibits very regular features. This is also the case in our situation, and we will discuss them now in the context of Gibbs (equilibrium) states μ_ϕ. For technical reasons our analysis will be performed on a subset $J_{rr}(E)$ of $J_r(E)$, which will also turn out to be fairly large, and whose (somewhat technical) definition we provide now. Given an integer $s > \inf(\operatorname{Re} J(E))$ and a point $z \in J(E)$ let $\{t_s^n(z)\}_{n=1}^\infty$ denote the sequence of consecutive visits of the point z to $Q_s = \{z \in J(E) : \operatorname{Re} z \leq s\}$ under the action of E, that is, this sequence is strictly increasing (perhaps finite, perhaps empty), $E^{t_s^n(z)}(z) \in Q_s$ for all $n \geq 1$ and $E^j(z) \notin Q_s$ for all $t_s^n(z) < j < t_s^{n+1}(z)$. Now define the set M_s to be

$$\left\{ z \in J_r(E) : \lim_{n \to \infty} \frac{\log\left|\left(E^{t_s^{n+1}(z) - t_s^n(z)}\right)'\left(E^{t_s^n(z)}(z)\right)\right|}{\log\left|\left(E^{t_s^n(z)}\right)'(z)\right|} = 0, \lim_{n \to \infty} \frac{t_s^{n+1}(z)}{t_s^n(z)} = 1 \right\}$$

and put

$$J_{rr}(E) = \bigcup_s M_s,$$

where the union is taken over all integers $s > \inf(\operatorname{Re} J(E))$. The robustness of the set $J_{rr}(E)$ is reflected in the fact that $\mu(J_{rr}(E)) = 1$ for every Borel probability E-invariant ergodic measure on $J(E)$ with finite Lyapunov exponent, in particular if $\mu = \mu_\phi$, where ϕ is a 1^+-potential.

Given a 1^+-tame function $\phi : Q_+^c \to \mathbb{R}$, and a real number $\alpha \geq 0$, we define the following two sets.

$$\mathcal{K}_\phi(\alpha) = \left\{ z \in J_r(E) : \lim_{n \to \infty} \frac{P(\phi)n - S_n\phi(z)}{\log|(E^n)'(z)|} = \alpha \right\}$$

and

$$D_\phi(\alpha) = \left\{ z \in J_{rr}(E) : \lim_{r \to 0} \frac{\log \mu_\phi(B(z,r))}{\log r} = \alpha \right\}.$$

The first relation between the sets $\mathcal{K}_\phi(\alpha)$ and $D_\phi(\alpha)$ is that

$$D_\phi(\alpha) = \mathcal{K}_\phi(\alpha) \cap J_{rr}(E). \tag{2.9}$$

We set

$$k_{\mu_\phi}^+ = \operatorname{HD}(\mathcal{K}_\phi(\alpha)) \text{ and } k_{\mu_\phi}^- = \operatorname{HD}(D_\phi(\alpha)).$$

By (2.9), we have
$$k^-_{\mu_\phi} \leq k^+_{\mu_\phi}.$$
The first result of this section establishes, as its main feature, the real analyticity of the functions $k^-_{\mu_\phi}$ and $k^+_{\mu_\phi}$. As a by-product, a closer relationship between the sets $\mathcal{K}_\phi(\alpha)$ and $D_\phi(\alpha)$ is obtained.

Theorem 2.29. *Suppose that $\phi : Q^c_+ \to \mathbb{R}$ is a 1^+-tame function. Then the functions $k^-_{\mu_\phi}$ and $k^+_{\mu_\phi}$ coincide in their natural domain (α_1, α_2) given in item (c). Denote their common value by k_{μ_ϕ}. Then the following statements are true.*

(a) *The pointwise dimension $d_{\mu_\phi}(x)$ exists for μ_ϕ-almost every $x \in J(E)$ and*
$$d_{\mu_\phi}(x) = \frac{\mathrm{P}(\phi) - \int \phi d\mu_\phi}{\int \log |E'| d\mu_\phi}.$$

(b) *The function $q \mapsto T(q)$, $q \in \mathbb{R}$ is real-analytic, $T(0) = \mathrm{HD}(J_r(E))$, $T'(q) < 0$ and $T''(q) \geq 0$. In addition T'' vanishes at one point if and only if it vanishes at all points, if and only if $\mu_\phi = \mu_{-h \log |E'|}$, if and only if ϕ and $-h \log |E'|$ are cohomologous modulo constant in the class of all (bounded) Hölder continuous functions.*

(c) *The function $-T'(q)$ attains values in an interval (α_1, α_2) where either $0 \leq \alpha_1 < \alpha_2 < \infty$ or this set of values is a singleton $\alpha_1 = \alpha_2$.*

(d) *For every $q \in \mathbb{R}$, $k_{\mu_\phi}(-T'(q)) = T(q) - qT'(q)$.*

(f) *The function k_{μ_ϕ} is real-analytic throughout its whole domain (α_1, α_2).*

(f) *If $\mu_\phi \neq \mu_{-h \log |E'|}$, then the functions $k_{\mu_\phi}(\alpha)$ and $T(q)$ form a Legendre transform pair.*

This theorem has been established assuming only that ϕ is Hölder continuous (and 1^+-tame). Assuming more, that ϕ is harmonic on a half-cylinder containing the Julia set $J(E)$, and performing a more involved analysis of analytic properties of the relevant Perron–Frobenius operators, we were able to show that the function k_{μ_ϕ} also depends in a real-analytic fashion on the parameter λ. This can be regarded as an extension of the real analyticity of the Hausdorff dimension of the Julia set $J_r(E)$ discussed in Subsection 2.2.4. Let us now explain in greater detail what we mean by saying that k_{μ_ϕ} depends real-analytically on λ. Fix $a \in \mathbb{R}$ and set
$$\mathrm{Hyp}(a) = \left\{\lambda \in \mathrm{Hyp} : \mathrm{Re}\bigl(\inf\{J(\mathcal{E}_\lambda)\}\bigr) > a\right\}.$$
Clearly $\mathrm{Hyp}(a)$ is an open set. Put
$$U_{a,\phi} = \bigcup_{\lambda \in \mathrm{Hyp}(a)} \{\lambda\} \times (\alpha_1(\lambda), \alpha_2(\lambda)),$$
where $\alpha_1(\lambda)$ and $\alpha_2(\lambda)$ are the numbers coming from items (c) and (f) of Theorem 2.29. The set $U_{a,\phi}$ is open and vertically connected, the latter meaning

that for every $\lambda \in \mathrm{Hyp}(a)$, the set $(\{\lambda\} \times \mathbb{R}) \cap U_{a,\phi}$ is connected. Given $\lambda \in \mathrm{Hyp}(a)$, let μ_ϕ^λ be the Gibbs state corresponding to the potential ϕ and the dynamical system $E_\lambda : J(E_\lambda) \to J(E_\lambda)$. The function $k_\phi : U_{a,\phi} \to [0,2]$ is defined by the formula

$$k_\phi(\lambda, \alpha) = k_{\mu_\phi^\lambda}(\alpha).$$

The second main theorem of this section is the following.

Theorem 2.30. *If $\phi : Q_a^c \to \mathbb{R}$ is a harmonic 1^+-tame potential, then the function $k_\phi : U_{a,\phi} \to \mathbb{R}$ is real-analytic.*

In view of item (b) of Theorem 2.29 the function k_ϕ is non-degenerate provided that the function ϕ is not cohomologous to $-\mathrm{HD}(J_r(E_\lambda))\log|E_\lambda'|$ for any $\lambda \in \mathrm{Hyp}(a)$ in the class of all (bounded) Hölder continuous functions. The potential ϕ is then called essential. The last result proved in [74] provides an easy sufficient condition for a harmonic 1^+-tame potential to be essential. One requires that $\kappa(\phi) \geq 2$.

2.4. $\mathcal{E}_{1/e}$ and parabolic implosion. In this section we describe the results obtained in [78] and [80]. We first, summarising [78], deal with the map $\mathcal{E}_{1/e} : \mathbb{C} \to \mathbb{C}$ alone, and then, describing [80], we look at what happens if λ converges to $1/e$ from the left and from the right along the real axis. It turns out that in the former case the Hausdorff dimension of the radial Julia set $J_r(E_\lambda)$ behaves continuously at the point $1/e$, and in the latter case it behaves 'highly' discontinuously at the parameter $\lambda = 1/e$.

2.4.1. The map $\mathcal{E}_{1/e}$ alone. We again work with the map $E_{1/e} : Q \to Q$. The set $J_r(E_{1/e})$ must now, however, be modified slightly. First notice that $E_{1/e}(1) = 1, E'_{1/e}(1) = 1$ and $E''_{1/e}(1) = 1 \neq 0$ so that 1 is a parabolic fixed point with one petal. Since the Julia set near the fixed point 1 lies entirely in the repelling petal, there exists $\theta > 0$ such that if $z \in J(E_{1/e}) \cap B(1,\theta) \setminus \{1\}$ then $E_{1/e}^n(z) \notin B(1,\theta)$ for some $n \geq 1$. Since in addition $E_{1/e}^n(0)$ converges to 1 through the attracting petal at 1, we see that there exists $\delta > 0$ such that

$$\bigcup_{z \in J(E_{1/e}) \setminus B(1,\theta)} B(z, 4\delta) \cap \{E_{1/e}^n(0)\}_{n=0}^\infty = \emptyset. \quad (2.10)$$

Now, almost as in the case of hyperbolic parameters, a point $z \in J_r(E_{1/e})$ is said to belong to $J_r(E)$ if and only if there exist $y \in J_r(E_{1/e}) \setminus B(1,\theta)$ (note here the difference with the hyperbolic case) and an unbounded increasing sequence $\{n_k\}_{k=1}^\infty$ such that $y = \lim_{k\to\infty} E_{1/e}^{n_k}(z)$ and $E_{1/e}^{n_k}(z) \in B(y, 4\delta)$ for all $k \geq 1$. Note that because of (2.10), for every $k \geq 1$ there exists a unique holomorphic branch $E_{1/e}^{-n_k} : B(y, 2\delta) \to Q$ of $E_{1/e}^{-n_k}$ sending $E_{1/e}^{n_k}(z)$ to z. Note also that $J_r(E_{1/e})$ coincides with the radial set defined exactly as in the hyperbolic case after removing the inverse image of 1, that is, the set $\bigcup_{n\geq 0} E_{1/e}^{-n_k}(1)$. Given $t \geq 0$, a t-conformal measure for $E_{1/e} : J_r(E_{1/e}) \to J_r(E_{1/e})$ is defined in the same way as in the hyperbolic case (see the beginning of Subsection 2.2.3) with h replaced by t in (2.7). Let $h = \mathrm{HD}(J_r(E_{1/e}))$. Some properties of t-conformal measures are described by the following theorem.

Theorem 2.31. *The following hold.*
  (1) *h is the unique t for which an atomless conformal measure exists.*
  (2) *For $E_{1/e} : J(E_{1/e}) \to J(E_{1/e})$, there exists a unique h-conformal measure m. The measure m is atomless.*
  (3) *The h-conformal measure m is ergodic and conservative.*
  (4) *If ν is a t-conformal measure for $E_{1/e}$ and $t > 1$, $t \neq h$, then $t > h$ and $\nu\left(\bigcup_{n\geq 0} E_{1/e}^{-n}(1)\right) = 1$.*

The geometric, Hausdorff, and packing measures have features analogous to the hyperbolic case (although the proofs about Hausdorff measures are even more involved).

Theorem 2.32. $0 < \mathrm{H}^h(J_r(E_{1/e})) < \infty$.

Proposition 2.33. *We have $\mathrm{P}^h(J_r(E_{1/e})) = \infty$. In fact $\mathrm{P}^h(G) = \infty$ for every open nonempty subset G of $J_r(E_{1/e})$.*

Based on this proposition and Proposition 2.1, the same arguments as in the hyperbolic case give the following.

Corollary 2.34. $1 < \mathrm{HD}(J_r(E_{1/e})) < 2$.

2.4.2. Left-hand side continuity. In this section we present the idea and single out the basic steps of the proof of the following.

Theorem 2.35. $\lim_{\lambda \nearrow 1/e} \mathrm{HD}(J_r(E_\lambda)) = \mathrm{HD}(J_r(E_{1/e}))$.

To start the short discussion of the proof, we take an arbitrary sequence $\{\lambda_n\}_{n=1}^\infty$ converging to $1/e$ from the left. Let $h_n = \mathrm{HD}(J_r(E_{\lambda_n}))$ and let m_n be the h_n conformal measure for E_{λ_n}. The first thing is to prove that the sequence $\{m_n\}_{n=1}^\infty$ is tight. Then let u be an arbitrary accumulation point of the sequence $\{h_n\}_{n=1}^\infty$. We know that $u \in [1, 2]$. Passing to a subsequence we may assume without loss of generality that $u = \lim_{n\to\infty} h_n$. Passing to yet another subsequence we may assume (due to tightness of $\{m_n\}_{n=1}^\infty$) that the sequence $\{m_n\}_{n=1}^\infty$ converges weakly to a measure m. It is not too difficult to show that m is u-conformal and $m(J_r(E_{1/e})) = 1$. In fact, inspecting carefully the proof of Proposition 2.1, we see that there exists $s > 1$ such that $\text{HD}(J_r(E_\lambda)) \geq s$ for all $\lambda \in (0, 1/e)$ sufficiently close to $1/e$. Therefore $1 < s \leq u \leq 2$. It now follows from Theorem 2.31 that $u \geq \text{HD}(J_r(E_{1/e}))$. It also follows from this theorem that in order to conclude the proof, that is, to show that $u \leq \text{HD}(J_r(E_{1/e}))$, it suffices to demonstrate that the u-conformal measure m is atomless. Conjugating the map E_λ by affine transformations so that their attracting fixed point becomes 0 (so, independent of the parameter λ) one carefully checks that the measures m_n (actually their images under these conjugacies) do not charge arbitrarily small neighbourhoods of zero too much.

2.4.3. Parabolic implosion for $E_{1/e}$.

The question we want to discuss in this short section is what happens if $\lambda \searrow 1/e$ along the real axis. Since the Julia set $J(\mathcal{E}_{1/e})$ is a nowhere dense Cantor bouquet and since $J(\mathcal{E}_\lambda) = \mathbb{C}$ for all $\lambda > 1$ (this follows from the fact that the trajectory of 0, $\{\mathcal{E}_\lambda^n(0)\}_{n=0}^\infty$, escapes to infinity (see [6])), we have obvious discontinuity of the Julia set of $\mathcal{E}_\lambda$ if $\lambda \searrow 1/e$ but no discontinuity of Hausdorff dimension, since all these sets are known to have Hausdorff dimension 2 (see [55]). In order to observe more interesting phenomena concerning Hausdorff dimensions, one considers the dynamically more significant sets $J_r(E_\lambda)$. So far, we have defined these sets only in the hyperbolic and parabolic cases. Now, we do this in full generality, using Definition 3.7 from [79].

Definition 2.36. $J_r(\mathcal{E}) \subset \mathbb{C}$ is the set of those points $z \in \mathbb{C}$ for which there exists an unbounded sequence $\{n_k(z)\}_{k=1}^\infty$ such that

$$\text{dist}\left(\{\mathcal{E}^{n_k(z)}(z)\}_{k=1}^\infty, \{\mathcal{E}_\lambda^n(0)\}_{n=0}^\infty\right) > 0$$

and the set $\text{Re}\left(\{\mathcal{E}^{n_k(z)}(z)\}_{k=1}^\infty\right)$ is bounded. The set $J_r = J_r(E) \subset Q$ is defined to be $\Pi(J_r(\mathcal{E}))$.

Note that if the trajectory of 0, $\{\mathcal{E}_\lambda^n(0)\}_{n=0}^\infty$, is dense, then $J_r(E) = \emptyset$. Note also that, as in the hyperbolic and parabolic cases, all points in $J_r(E)$ allow univalent holomorphic pull backs from a bounded region. To be precise, this means that if $z \in J_r(E)$ then there exist $y \in J_r(E)$, $\delta(z) > 0$ and an unbounded increasing sequence $\{s_k\}_{k=1}^\infty$ (which could be extracted from the sequence $\{n_k(\tilde{z})\}_{k=1}^\infty$ guaranteed by Definition 2.36 with some $\tilde{z} \in \Pi^{-1}(z)$) such that

$$\lim_{k\to\infty} E^{s_k}(z) = y, \quad |E^{s_k}(z) - y| < \frac{1}{4}\delta(z), \quad k = 1, 2, \ldots,$$

and for every $k \geq 1$ there exists a unique holomorphic branch

$$E_z^{-s_k} : B(E^{s_k}(z), 2\delta(z)) \to Q$$

of E^{-s_k} sending $E^{s_k}(z)$ to z. Finally note that the radial Julia set defined in Definition 2.36 coincides in the hyperbolic and parabolic cases with the respective sets defined in Section 2.2 and Subsection 2.2.3. The ultimate results obtained in [80], which will be called global dimension-wise parabolic implosion phenomena, are these.

Theorem 2.37. *For every $\sigma \in \mathbb{R}$ there exists a sequence $\{\epsilon_k\}_{k=1}^{\infty}$ of positive reals converging to 0 such that $(-\pi/\sqrt{\epsilon_k}) \to \sigma(\mathrm{mod}\ 1)$ as $k \to \infty$, and*

$$\liminf_{k\to\infty} \mathrm{HD}(J_r(E_{1/e+\epsilon_k})) > \mathrm{HD}(J_r(E_{1/e})).$$

Corollary 2.38. *We have that*

$$\limsup_{\lambda \searrow 1/e} \mathrm{HD}(J_r(E_\lambda)) > \mathrm{HD}(J_r(E_{1/e})).$$

The structure of the proof of Theorem 2.37 is this. First, it was proved in [79] (see Section 2.5) that for every $\lambda > 1/e$ there exists a unique $\mathrm{HD}(J_r(E_\lambda))$-conformal measure m_λ for the map E_λ. Let u be the value of the lower limit appearing in Theorem 2.37. Using Lavaurs maps, which are characteristic of any parabolic implosion, one shows as the first step that the sequence $\{m_{\lambda_k}\}_{k=1}^{\infty}$, $\lambda_k = \frac{1}{e} + \epsilon_k$, is tight, and any of its weak limits is a measure, u-conformal for both $E_{1/e} : \mathbb{C} \to \mathbb{C}$ and $g_\sigma : \mathbb{C} \setminus J(E_{1/e}) \to \mathbb{C}$, where g_σ is the Lavaurs map corresponding to the parameter σ. The second step is to prove that the series

$$\sum_{n=1}^{\infty} \sum_{x \in E_{1/e}^{-n}(1) \setminus \bigcup_{k=0}^{n-1} E_{1/e}^{-k}(1)} |(E_{1/e}^n)'(x)|^{-u}$$

converges. Since, by Theorem 2.31(2), the same series, with u replaced by $\mathrm{HD}(J_r(E_{1/e}))$ diverges (since otherwise, we would get an atomic $\mathrm{HD}(J_r(E_{1/e}))$-conformal measure supported on $\bigcup_{n\geq 0} E_{1/e}^{-n}(1) \subset J(E_{1/e})$), we conclude that $u > \mathrm{HD}(J_r(E_{1/e}))$. So, we are done.

2.5. Non-hyperbolic exponential maps.

In [79] a class of exponential maps was considered whose members all had Julia sets equal to the whole complex plane. In particular, 0 belongs then to the Julia set and the maps are not hyperbolic (see the beginning of Section 2.2). The lack of hyperbolicity was to some extent compensated by extremely fast convergence to ∞ of forward iterates of zero. Namely, a parameter $\lambda \in \mathbb{C} \setminus \{0\}$ is called a super-growing parameter if $\lim_{n\to\infty} \mathrm{Re}(\mathcal{E}^n(0)) = +\infty$, and there exists $c > 0$ such

that for all $n \geq 1$,
$$|\text{Re}(\mathcal{E}_\lambda^{n+1}(0))| \geq c\exp(\text{Re}(\mathcal{E}_\lambda^n(0)) = \frac{c}{|\lambda|}|\mathcal{E}_\lambda^{n+1}(0)|.$$

Now fix an arbitrary super-growing parameter λ. Put $\mathcal{E} = \mathcal{E}_\lambda$ and $E = E_\lambda$. We have observed at the beginning of Subsection 2.2.3 that $J(E) = \mathbb{C}$. The radial Julia set $J_r(E)$ defined in Definition 2.36 turns out to be as 'nice' as in the hyperbolic and parabolic cases. Indeed, although the Perron–Frobenius operator method does not seem to be naturally applicable for a map whose Julia set contains 0, using $K(V)$-methods (see Appendix 1) and the tightness of the sequence of semi-conformal measures this method produces, it was possible to prove the existence and uniqueness of an h-conformal measure m for the map $E : Q \to Q$, where $h = \text{HD}(J_r(E))$. In addition this measure was proved to be ergodic, conservative (meaning that $\sum_{n\geq 0} \mathbb{1}_F \circ E^n(z) = \infty$ for m-a.e. $z \in \mathbb{C}$) and $m(J_r(E)) = 1$. As in the hyperbolic case the conformal measure turns out to be a normalised version of the h-dimensional Hausdorff measure H^h, and in fact we have this result.

Theorem 2.39. $0 < \text{H}^h(J_r(E)) < \infty$.

Proposition 2.40. *We have* $\text{P}^h(J_r(E)) = \infty$. *In fact* $\text{P}^h(G) = \infty$ *for every open nonempty subset G of $J_r(E)$.*

This proposition implies that $h < 2$, in the same way as in Subsection 2.4.3. Since $J_r(E)$ contains $J_{bd}(E)$, using Proposition 2.1, we deduce the following.

Theorem 2.41. $1 < \text{HD}(J_r(E)) < 2$.

This theorem has the following rather unexpected corollary, proved for the first time in [48] by a different method.

Corollary 2.42. *If λ is a super-growing parameter then for Lebesgue almost every point $z \in \mathbb{C}$, $\omega(z) = \{\mathcal{E}_\lambda^n(0)\}_{n=0}^\infty \cup \{\infty\}$.*

We show how to prove Corollary 2.42. Let $\mathcal{E} = \mathcal{E}_\lambda$. Since $\text{HD}(J_r(\mathcal{E})) < 2$, the complement of $J_r(\mathcal{E})$ is a set of full measure. Fix a point $z \notin J_r(\mathcal{E})$. By the definition of $J_r(\mathcal{E})$, this implies that $\omega(z) \subset \{\mathcal{E}_\lambda^n(0)\}_{n=0}^\infty \cup \{\infty\}$. We only have to check that, actually, equality holds for almost every point z. So, assume that $\omega(z) = \infty$. The set of such points has Lebesgue measure 0; actually, this is true for a large class of maps (see, for example, [55] or [27]). Next, assume that $\omega(z) = \{\infty\} \cup \{\mathcal{E}_\lambda^n(0)\}_{n=k}^\infty$ for some $k > 0$. Then, there exists an infinite sequence of integers s_i such that $\mathcal{E}^{s_i}(z) \to \mathcal{E}_\lambda^k(0)$. Then, denoting $n_i = s_i - 1$, we see that $\text{Re}(\mathcal{E}^{n_i}(z)) \to \text{Re}(\mathcal{E}_\lambda^k(0))$ and, moreover, $\text{dist}(\mathcal{E}^{n_i}(z), \{\mathcal{E}_\lambda^n(0)\}_{n=k}^\infty) > 0$. Consequently, $z \in J_r(\mathcal{E})$, a contradiction.

By the tightness argument the following fact was also established in [79].

Theorem 2.43. *The function $\lambda \to \text{HD}(J_r(E_\lambda)), \lambda \in (1/e, +\infty)$ is continuous.*

An open problem is whether the above function is real-analytic.

Leaving geometric measures and dimensions, let us have a closer look at the dynamical properties of the h-conformal measure m for the (super-growing) map $E : \mathbb{C} \to \mathbb{C}$. Ergodicity and conservativity of the measure m along with the fact that the orbit of zero escapes to ∞, make it possible to apply M. Martens' method (see Appendix 2) to construct (up to a multiplicative constant) a σ-finite E-invariant Borel measure μ equivalent to m. The method of M. Martens leaves open a natural procedure of checking whether the measure is finite or infinite. One must carefully control the distortion in the case when orbits of points approach singularities, which in the case of exponential maps (after projecting on the cylinder) are the orbits of zero, $-\infty$ and $+\infty$. This was done in [79, Theorem 4.6] by fairly technical and rather complicated arguments. The result is this.

Theorem 2.44. *There exists a unique Borel probability E-invariant measure μ absolutely continuous with respect to the $\text{HD}(J_r(E))$-conformal measure m. In addition μ is ergodic and equivalent to m.*

2.6. Fatou functions, sine, cosine and further families. In the paper [45] the family $f_\lambda(z) = e^{-z} + z + \lambda, \text{Re}\lambda \geq 1$, of Fatou functions was investigated in great detail. Although all of them have a Baker domain at ∞, this turned out not to preclude the possibility of analysing these maps from the geometrical (fractal) and (measure-theoretical) dynamical point of view. The paper [45] provides a uniform treatment of issues dealt with in [76] and [77] in the setting of the technically more complicated Fatou functions f_λ. The results here are the same as those discussed in Section 2.2. Also the family of transcendental meromorphic functions $f_\lambda(z) = \lambda(1 - e^{2z})^{-1}$ was studied in detail with its own methods in [46]. The appropriate results about conformal, Hausdorff, packing and invariant measures were obtained there.

Also, I. Coiculescu and B. Skorulski solved in [13] and [14] the problems suggested in [76] and [77], by extending the results proved there to the case of the family $\mathcal{H}$ of hyperbolic maps of the form $\sum_{j=0}^n a_j e^{(j-k)z}$, where $0 < k < n$. The Julia sets of these maps contain Cantor bouquets. Note that this family includes the exponential, sine and cosine families.

3. ELLIPTIC FUNCTIONS

3.1. General facts. If $\omega_1, \omega_2 \in \mathbb{C} \setminus \{0\}$ are such that $\text{Im}(\omega_2/\omega_1) \neq 0$, then the set
$$\Lambda = \{m\omega_1 + n\omega_2 : m, n \in \mathbb{Z}\}$$
is called the lattice generated by the numbers ω_1 and ω_2. A non-constant meromorphic function $f : \mathbb{C} \to \overline{\mathbb{C}}$ is called elliptic (with respect to the lattice Λ) if and only if
$$f(z + \omega_1) = f(z + \omega_2) = f(z)$$
for all $z \in \mathbb{C}$. This equivalently means that f is Λ-invariant, $f(z + \omega) = f(z)$ for all $z \in \mathbb{C}$ and all $\omega \in \Lambda$. The Fatou set of a meromorphic function consists of points $z \in \mathbb{C}$ which admit neighbourhoods restricted to which all iterates of our meromorphic functions are well-defined and form a normal family. The Julia set is by definition the complement (in $\mathbb{C}$) of the Fatou set. Since $f^{-1}(\infty)$, the set of poles is an infinite set, $J(f)$ is the closure of all prepoles $\bigcup_{n \geq 0} f^{-n}(\infty)$. For every pole b of f let $q_b \geq 1$ denote its multiplicity. Define
$$q = \max\{q_b : b \in f^{-1}(\infty)\} \geq 1.$$
Associating with the elliptic function f an infinite conformal iterated function system in the sense of [54] and determining its θ number introduced in [54], we were able to prove in [43] the following.

Theorem 3.1. *If $f : \mathbb{C} \to \overline{\mathbb{C}}$ is an elliptic function, then*
$$\text{HD}(J(f)) > \frac{2q}{1+q} \geq 1.$$

The obvious consequence of Theorem 3.1 is that if f has poles of large multiplicities, then the Hausdorff dimension of the Julia set of f is close to 2. This may suggest that this dimension is always equal to 2. In Section 3.3 we will describe a large class of examples with Julia set of dimension less than 2. Here we will give a simple construction showing that for each lattice Λ there exists an elliptic Λ-invariant function whose Julia set is not the entire complex plane $\mathbb{C}$. Indeed, let Λ be a lattice and let $g : \mathbb{C} \to \overline{\mathbb{C}}$ be an elliptic Λ-invariant function for which some zero, call it b, is not a critical point of g. Consider the family of functions $\{g_\lambda = \lambda g + b\}_{\lambda \in \mathbb{C} \setminus \{0\}}$. This family consists of Λ-invariant elliptic functions and $g_\lambda(b) = \lambda g(b) + b = 0 + b = b$, $g'_\lambda(b) = \lambda g'(b)$ for all $\lambda \in \mathbb{C} \setminus \{0\}$. So, if $|\lambda| < 1/|g'(b)|$, then b is an attracting fixed point of g. The basin of attraction of b under g_λ is contained in the Fatou set, and consequently $J(g_\lambda) \neq \mathbb{C}$.

We would also like to mention that, using the same methods as in [43], it was shown in [51] that the Hausdorff dimension of the Julia set of any function of the form $\exp \circ f$, where f is elliptic, is equal to 2. These functions are doubly periodic but have essential singularities. Also, in the papers [29]-[30] elliptic functions, actually Weierstrass elliptic $\wp$-functions, were found with Julia sets of various topological types.

As the reader may recall, we dealt in Chapter 2 with the set of points not escaping to ∞ and, in spite of the fact that the set of points escaping to ∞ was large (of Hausdorff dimension 2), its complement turned out to be geometrically and dynamically sound. For elliptic functions, the situation is in some sense better. We have the following.

Theorem 3.2. *If $f : \mathbb{C} \to \overline{\mathbb{C}}$ is an elliptic function, then*
$$\mathrm{HD}(I_\infty(f)) \leq \frac{2q}{1+q}.$$

As an immediate consequence, we obtain the following.

Corollary 3.3. *If $f : \mathbb{C} \to \overline{\mathbb{C}}$ is an elliptic function, $h = \mathrm{HD}(J(f))$, then $\mathrm{H}^h(I_\infty(f)) = 0$, and consequently $l_2(I_\infty(f)) = 0$.*

3.2. Gibbs and equilibrium states. In this section we continue to assume that $f : \mathbb{C} \to \overline{\mathbb{C}}$ is an arbitrary elliptic function. Our aim is to discuss results concerning Gibbs and equilibrium states obtained in [52]. Let $T = \mathbb{C}/\Lambda$ be the torus generated by the lattice Λ and let $\Pi : \mathbb{C} \to \mathbb{C}/\Lambda$ be the canonical projection. Let $\mathcal{P} = \Pi(f^{-1}(\infty))$ be the set of 'poles' on the torus T. The map $f : \mathbb{C} \setminus f^{-1}(\infty) \to \mathbb{C}$ uniquely projects down to the holomorphic map $F : T \setminus \mathcal{P} \to T$ so that $F \circ \Pi = \Pi \circ f$, that is, the following diagram commutes.

$$\begin{array}{ccc} \mathbb{C} \setminus f^{-1}(\infty) & \xrightarrow{f} & \mathbb{C} \\ \Pi \downarrow & & \downarrow \Pi \\ T \setminus \mathcal{P} & \xrightarrow{F} & T. \end{array} \qquad (3.1)$$

Notice that we have a little bit more. Since the function f is constant on fibres of Π, there exists a unique holomorphic map $\hat{f} : T \to \overline{\mathbb{C}}$ such that

$\hat{f}(\Pi(z)) = f(z)$ for all $z \in \mathbb{C}$. As in Chapter 2, our primary object of interest in this section will be the dynamical system $F : T \setminus \mathcal{P} \to T$.

Following the classical case of a subshift of finite type [12] (more generally the case of open distance expanding maps) or, more appropriately in this context, the approach initiated in [19], one is tempted to develop the theory of Gibbs and equilibrium states for Hölder continuous potentials $\phi : T \to \mathbb{R}$. To be very general, suppose that $\phi : T \to \mathbb{R}$ is an arbitrary function, with no other assumptions. The basic tool of any known version of thermodynamic formalism is an appropriate Perron–Frobenius operator, which in our context would take the form

$$\mathcal{L}_\phi g(x) = \sum_{y \in F^{-1}(x)} e^{\phi(y)} g(y).$$

Notice that the series defining the Perron–Frobenius operator $\mathcal{L}_\phi$ is infinite and in order to make it well-defined and bounded on the Banach space $C(T)$ of continuous functions on T, one should demand that, with a universal constant $C > 0$,

$$\mathcal{L}_\phi(\mathbb{1}) = \sum_{y \in F^{-1}(x)} e^{\phi(y)} \leq C$$

for all $x \in T$. Let us examine what this requirement really means. First, we immediately see that ϕ cannot be uniformly bounded from below. To get a deeper insight fix $\tilde{x} \in \Pi^{-1}(x)$. Then $y \in F^{-1}(x)$ if and only if there exists $\omega \in \Lambda$ such that $\hat{f}(y) = \tilde{x} + \omega$. Therefore

$$\mathcal{L}_\phi(\mathbb{1}) = \sum_{\omega \in \Lambda} \sum_{y \in \hat{f}^{-1}(\tilde{x}+\omega)} e^{\phi(y)}.$$

If $|\omega|$ is large, then $y \in \hat{f}^{-1}(\tilde{x}+\omega)$ is near the pole b of $\hat{f} : T \to \mathbb{C}$, where we can write

$$\tilde{x} + \omega = \hat{f}(y) = \frac{G_b(y)}{(y-b)^{q_b}}.$$

Here G_b is a holomorphic function defined near b, such that $G_b(b) \neq 0$ and, let us recall, $q_b \geq 1$ is the multiplicity of the pole b. Since the set of poles $\mathcal{P} \subset T$ is finite the series

$$\sum_{b \in \mathcal{P}} \sum_{\omega \in \Lambda} |\tilde{x} + \omega|^{-(2+\epsilon_b)}$$

converges with arbitrarily chosen $\epsilon_b > 0$. Trying to apply the comparison test, we would therefore require that, with some constant $L > 0$,

$$\exp(\phi(y)) \leq L|\tilde{x}+\omega|^{-(2+\epsilon_b)} = L \left(\frac{|y-b|^{q_b}}{|G_b(y)|} \right)^{2+\epsilon_b}$$

for all poles $b \in \mathcal{P}$ and all $y \in \hat{f}^{-1}(\tilde{x}+\omega)$ which are close to b or, equivalently,

$$\phi(y) \leq \log L - (2+\epsilon_b) \log |G_b(y)| + (2+\epsilon_b) q_b \log |y-b|$$

near b. This inequality suggests that we deal with the class of potentials $\phi : T \to \mathbb{C}$, called in the sequel summable, which satisfy the following two conditions.

(a) For any open set V containing $\mathcal{P}$, ϕ is Hölder continuous on $T \setminus V$.
(b) For every pole $b \in \mathcal{P}$ there are $\epsilon_b > 0$ and a Hölder continuous function H_b such that $\phi(z) = H_b(z) + (2+\epsilon_b) q_b \log|z-b|$ on a sufficiently small neighbourhood of b.

Of course the most significant potentials are those of the form $-t \log|F'|$, where $t > 0$. If the map F is hyperbolic then these potentials are summable and Bowen's formula holds. If however F is not hyperbolic, in particular if the Julia set contains critical points, then the potentials $-t \log|F'|$ are no longer summable and, as in the case of rational functions, the theory described in this section does not apply to them. However, as long as the critical points are not recurrent a lot can be said about conformal, invariant and geometric measures associated to the potential $-h \log|F'|$ (see Section 3.3).

Given a measurable function $\psi : T \to [0, +\infty]$, a Borel probability measure m on T is said to be ψ-conformal if and only if $m(J(F)) = 1$ and

$$m(F(A)) = \int_A \psi \, dm$$

for every Borel set $A \subset T$ such that $F_{|A}$ is one-to-one. Unlike the case of hyperbolic exponential functions, it is much easier here to construct (generalised) conformal measures, due to the compactness of the torus T. Namely, the map

$$\nu \mapsto \frac{\mathcal{L}_\phi^* \nu}{\mathcal{L}_\phi^* \nu(\mathbb{1})}$$

discussed in Subsection 2.2.1 is, in our space, continuous on the compact convex set of Borel probability measures on T. So, the Schauder–Tichonov theorem applies and we obtain a $\kappa e^{-\phi}$-conformal measure for some constant $\kappa > 0$. The problem of defining pressure is however the same as in Subsection 2.2.1 and we resolve it in the same way as there by employing the pointwise definition.

$$P(\phi, x) = \limsup_{n \to \infty} \frac{1}{n} \log \sum_{y \in F^{-n}(x)} \exp\left(\sum_{j=0}^{n-1} \phi \circ F^j(y)\right) = \limsup_{n \to \infty} \frac{1}{n} \log \mathcal{L}_\phi^n(\mathbb{1})(x).$$

The key to obtaining all the results discussed below is a very detailed analysis of the behaviour of the normalised Perron–Frobenius operator $\hat{\mathcal{L}}_\phi = \kappa^{-1} \mathcal{L}_\phi$. Apart from (a) and (b) the third general assumption is that

$$\sup\{P(\phi, x) : x \in T\} > \sup(\phi).$$

Concerning the operator $\hat{\mathcal{L}}_\phi$ itself, it turns out to be almost periodic and admits a continuous, everywhere positive function $\rho : T \to \mathbb{R}$ such that

$$\hat{\mathcal{L}}_\phi(\rho) = \rho.$$

By an extensive analysis of its behaviour, one gets the following.

Lemma 3.4. *For every $x \in T, P(\phi, x)$ is the same, and the common value $P(\phi)$, called the topological pressure of ϕ, is given by the following formula*

$$P(\phi) = \lim_{n\to\infty} \frac{1}{n} \log \mathcal{L}_\phi^n(x) = \lim_{n\to\infty} \frac{1}{n} \log \|\mathcal{L}_\phi^n(\mathbb{1})\|_\infty.$$

Theorem 3.5. *There exists a unique $\exp(P(\phi) - \phi)$-conformal measure m_ϕ on T and a unique Borel probability F-invariant measure μ_ϕ absolutely continuous with respect to the measure m_ϕ. The measure μ_ϕ is in fact equivalent to m_ϕ, and $\frac{d\mu_\phi}{dm_\phi} = \rho$, the fixed point of the Perron–Frobenius operator $\hat{\mathcal{L}}_\phi$, normalised so that $\int \rho \, d\mu_\phi = 1$. The measure μ_ϕ is called the Gibbs state of the summable potential ϕ.*

Passing to equilibrium states, denote by M_ϕ (M_ϕ^e) the space of all F-invariant (ergodic) Borel probability measures on $J(F)$ for which $\int \phi \, d\mu > -\infty$. Let $\chi_\mu = \int \log |F'| d\mu$ be the characteristic Lyapunov exponent of the measure μ. A simple observation (based on the similar behaviour of ϕ and $\log |F'|$ near poles) is that if $\mu \in M_\phi^e$ then $\chi_\mu < +\infty$. A more involved argument (more difficult than in the case of rational functions since $F : T \setminus \mathcal{P} \to T$ is not Lipschitz continuous) leads to the following theorem.

Theorem 3.6. *(Ruelle's inequality) If $\mu \in M_\phi^e$, then $h_\mu(F) \leq 2\max\{0, \chi_\mu\}$. In particular if $h_\mu(F) > 0$, then $\chi_\mu > 0$.*

Another technical fact needed to establish the variational principle and to identify all equilibrium states (which are defined just below) of ϕ, is this.

Proposition 3.7. *If $\mu \in M_\phi^e$ and $\chi_\mu > 0$, then there exists a countable generating partition α (mod μ) such that its entropy $\mathrm{H}_\mu(\alpha)$ is finite. In particular, $h_\mu(F) = \int \log J_\mu d\mu$, where J_μ is the Jacobian $\left(\frac{d\mu \circ F}{d\mu}\right)$ of F with respect to the measure μ, well-defined on the complement of a set of measure zero.*

Armed with these last two results and Theorem 3.5, one proves the following.

Theorem 3.8. *(Variational Principle) We have*

$$P(\phi) = \sup\{h_\mu(\phi) + \int \phi d\mu : \mu \in M_\phi\}.$$

A measure $\mu \in M_\phi$ is called an equilibrium state of ϕ if $h_\mu(\phi) + \int \phi d\mu = P(\phi)$. The following result can be therefore treated as a completion of Theorem 3.8.

Theorem 3.9. *The Gibbs state μ_ϕ (proved to exist in Theorem 3.5) is the unique equilibrium state of the potential ϕ.*

Addressing a referee's questions we would like to remark that in Theorem 3.6 and Proposition 3.7 the assumption $\mu \in M_\phi$ can be relaxed thus giving, as in the case of rational functions, the following result.

Theorem 3.10. *If μ is a Borel probability F-invariant ergodic measure with positive Lyapunov exponent, then*

$$\mathrm{HD}(\mu) = \frac{h_\mu}{\xi_\mu},$$

and, moreover,

$$\lim_{r \to 0} \frac{\log \mu(B(x,r))}{\log r} = \mathrm{HD}(\mu)$$

for μ_ϕ-a.e. $x \in J(F)$.

3.3. Critically non-recurrent elliptic functions.

3.3.1. Preliminaries. Throughout this section, except at the very end, we discuss the content of [44]. In the previous section an elliptic function was treated exclusively as a dynamical system. In trying to say something finer than in Section 3.1 about the fractal geometry of the Julia set of an elliptic function, some restrictions on the class of functions to be analysed are needed. In [44] this class was defined by analogy with the case of rational functions of the Riemann sphere treated in [71], [72]; comp. [75].

Definition 3.11. *An elliptic function $f : \mathbb{C} \to \overline{\mathbb{C}}$ is called critically non-recurrent if and only if the following three conditions are satisfied:*
  (a) $c \notin \omega(c)$ *for every critical point c of f lying in $J(f)$,*
  (b) $\omega(c)$ *is a compact set for every critical point c of f lying in $J(f)$,*
  (c) *every critical point c in the Fatou set belongs to the basin of attraction of either an attracting or parabolic periodic point.*

Denote by $\Omega(f)$ the set of parabolic periodic points of f. Let

$$\mathrm{Crit}(J(f))$$

be the set of all critical points of f that are contained in the Julia set $J(f)$. The result which permits us to carry out our analysis of critically non-recurrent elliptic function is the following theorem, which is Mañé's theorem from [49] adapted to the context of elliptic functions.

Theorem 3.12. *Let $f : \mathbb{C} \to \overline{\mathbb{C}}$ be a critically non-recurrent elliptic function. If $X \subset J(f) \setminus \Omega(f)$ is a closed subset of $\mathbb{C}$, then for every $\epsilon > 0$ there exists $\delta > 0$ such that for every $x \in X$ and every $n \geq 0$, all the connected components of $f^{-n}(B(x,\delta))$ have Euclidean diameters $\leq \epsilon$.*

For the rest of this section, f will be a critically non-recurrent elliptic function. Let

$$\mathrm{Sing}^-(f) = \bigcup_{n \geq 0} f^{-n}\big(\Omega(f) \cup \mathrm{Crit}(J(f)) \cup f^{-1}(\infty)\big).$$

The following proposition is a consequence of Theorem 3.12. It is important for further applications in [44] and is interesting in itself as it gives a very precise description of radial points.

Proposition 3.13. *If $z \in J(f) \setminus \mathrm{Sing}^-(f)$, then there exist a positive number $\eta(z)$, an increasing sequence of positive integers $\{n_j\}_{j \geq 1}$, and a point*

$$x = x(z) \in \omega(z) \setminus (\Omega(f) \cup \omega(\mathrm{Crit}(z)))$$

such that $x \neq \infty$ if $z \notin I_\infty(f)$, $\lim_{j \to \infty} f^{n_j}(z) = x$ and

$$\mathrm{Comp}(z, f^{n_j}(z), f^{n_j}, \eta(z)) \cap \mathrm{Crit}(f^{n_j}) = \emptyset$$

for every $j \geq 0$, where $\mathrm{Comp}(z, f^{n_j}(z), f^{n_j}, \eta(z))$ denotes the connected component of $f^{-n_j}(B(f^{n_j}(z), \eta(z)))$ that contains z and

$$\mathrm{Crit}(z) = \{c \in \mathrm{Crit}(J(f)) : c \in \omega(z)\}.$$

3.3.2. Hausdorff, packing and conformal measures. As in Chapter 2 the link between dynamics and geometry of elliptic functions is provided by Sullivan's conformal measures. Recall that given $t \geq 0$, a Borel probability measure m_s on $J(f)$ is called t-conformal if and only if

$$m_s(f(A)) = \int_A |f^*|^t dm_s,$$

whenever $A \subset J(f)$ is a Borel set such that $f_{|A}$ is one-to-one and f^* is the derivative of f with respect to the spherical metric on $\mathbb{C}$. If we give up the finiteness assumption of m_s but we replace f^* by the standard Euclidean derivative f' of f, we denote the resulting conformal measure by m_e and call it the Euclidean t-conformal measure for f. The relation between these two measures is that

$$\frac{dm_e}{dm_s}(z) = (1+|z|^2)^t.$$

In fact, several weaker versions of conformal measures are needed in [44], but we will not discuss them all here. Let $h = \mathrm{HD}(J(f))$ be the Hausdorff dimension of the Julia set of f. The h-conformal measure for f is constructed by the $K(V)$-method described in Appendix 1. The bad set here, called crossing in [44], is any finite set

$$Y \subset \{\infty\} \cup \Omega(f) \cup \bigcup_{n=1}^{\infty} f^n(\mathrm{Crit}(J(f))$$

such that the following conditions are satisfied

(y1) $\infty \in Y$.
(y2) $Y \cap \{f^n(x) : n \geq 1\}$ is a singleton for all $x \in \mathrm{Crit}(J(f))$.
(y3) $Y \cap \mathrm{Crit}(f) = \emptyset$.
(y4) $\Omega(f) \subset Y$.

Now, we choose any sequence $\{r_n\}_{n=1}^{\infty}$ of positive reals converging to zero and such that $\partial B_s(Y, r_n) \cap \mathrm{Crit}(f) = \emptyset$, where the subscript '$s$' indicates that the ball is considered with respect to the spherical metric on $\mathbb{C}$. Following the general scheme outlined in Appendix 1, one then defines the compact forward invariant sets

$$K_n = \bigcap_{j=0}^{\infty} J(f) \setminus f^{-n}(B(Y, r_n)), \quad n \geq 1,$$

and one applies Lemma 8.2 to get semi-conformal measures. This is the first step. Introducing a special order in $\mathrm{Crit}(J(f))$, a stratification of the closure of the postcritical set, studying the parabolic periodic points, critical points and poles of f, which are potential candidates for atoms of the conformal measure, we eventually arrive at the following result.

Theorem 3.14. *Let* $f : \mathbb{C} \to \overline{\mathbb{C}}$ *be a critically non-recurrent elliptic function. There exists a unique h-conformal measure m for f. This measure is atomless.*

Using this conformal measure m to gain information about Hausdorff and packing measure we end up, after lengthy and fairly technical considerations, with the following.

Theorem 3.15. *Let $f : \mathbb{C} \to \overline{\mathbb{C}}$ be a critically non-recurrent elliptic function. If $h = \mathrm{HD}(J(f)) = 2$, then $J(f) = \mathbb{C}$. If $h < 2$, then*
  (a) $\mathrm{H}^h(J(f)) = 0$,
  (b) $\mathrm{P}^h(J(f)) > 0$,
  (c) $\mathrm{P}^h(J(f)) = \infty$ *if and only if* $\Omega(f) \neq \emptyset$,

where H^h and P^h are defined using the spherical metric on $\overline{\mathbb{C}}$.

So, if $\mathrm{HD}(J(f)) < 2$, then the Hausdorff measure always vanishes, whereas the packing measure turns out to be the right geometric measure exactly when there are no parabolic periodic points. If $\Omega(f) \neq \emptyset$, then $\mathrm{H}^h(J(f)) = 0$ and $\mathrm{P}^h(J(f)) = \infty$ (even locally). No geometric interpretation of the h-conformal measure has been found so far in this case.

Note that a similar phenomenon has been observed by D. Sullivan in the context of geometrically finite Kleinian groups with cusps of different ranks (see [69]) and in [54] in the case of conformal irregular infinite iterated function systems.

In Question 1 of [67], Stallard asked whether there exists a transcendental meromorphic function f for which the packing and Hausdorff dimensions are both equal to h, where h is not an integer. It follows from Theorem 3.15 and Theorem 3.1 that this is true for any critically non-recurrent elliptic function f with no parabolic periodic points for which $J(f) \neq \mathbb{C}$. The existence of such functions was shown in [29].

3.3.3. Invariant measures equivalent to h-conformal measure.
Since it is not difficult to show that the h-conformal measure m of the postcritical set vanishes, the method of M. Martens (see Appendix 2) applies and leads to the following result.

Theorem 3.16. *There exists a σ-finite f-invariant measure μ that is absolutely continuous with respect to the h-conformal measure m. In addition, μ is ergodic and conservative.*

The most intriguing problem here is to determine whether the σ-finite invariant measure μ is finite or infinite. In order to deal with this problem, it is useful to recall from [72] the concepts of finite and infinite condensations. Namely, z is a point of finite condensation of a Borel measure ν if there is an open neighbourhood U of z such that $\nu(U) < \infty$; otherwise z is said to be a point of infinite condensation of ν. Our strategy to cope with the problem of finiteness of the measure μ was to identify the points of its finite and infinite condensation. To our surprise, careful estimates permitted us to prove the following.

Theorem 3.17. ∞ *is a point of finite condensation of the measure μ.*

We were able to go further to establish the following.

Theorem 3.18. *The set of points of infinite condensation of μ is contained in the set of parabolic points $\Omega(f)$.*

As an immediate consequence of this theorem, we get the following.

Corollary 3.19. *If $\Omega = \emptyset$, then there exists an f-invariant probability measure μ equivalent to m.*

Since the case $J(f) = \mathbb{C}$ rules out parabolic points, as an immediate consequence of this corollary we get

Corollary 3.20. *If $J(f) = \mathbb{C}$, then there is a unique probability measure μ equivalent to the Lebesgue measure on $\mathbb{C}$.*

It follows from the above that in order to understand the problem of finiteness of the σ-finite f-invariant measure μ, one must analyse in detail the parabolic points. Such analysis has been done in [72] and applies unchanged to the case of critically non-recurrent elliptic functions to give the following.

Proposition 3.21. *If $\omega \in \Omega \setminus \overline{O_+(\mathrm{Crit}(J(f)))}$, then μ has infinite condensation at ω if and only if $h \leq \frac{2p(\omega)}{p(\omega)+1}$.*

As an immediate consequence of this proposition and Theorem 3.1, we get the following remarkable corollary.

Corollary 3.22. *Let $\mathcal{R}$ be a fundamental parallelogram for the elliptic function f. If $\Omega \cap \overline{O_+(\mathrm{Crit}(J(f)))} = \emptyset$ and*

$$\max\{q_b : b \in \mathcal{R} \cap f^{-1}(\infty)\} \geq \max\{p(\omega) : \omega \in \Omega(f)\},$$

then the invariant measure μ is finite.

We also deduce the following results.

Proposition 3.23. *If $\omega \in \Omega$ and $h \leq \frac{2p(\omega)}{p(\omega)+1}$, then μ has infinite condensation at ω.*

Theorem 3.24. *If $c \in J(f)$ is a critical point of f of order s, $\omega = f(c) \in \Omega$, and $h \leq \frac{2sp(\omega)}{p(\omega)+1}$, then μ has infinite condensation at ω.*

4. WALTERS EXPANDING CONFORMAL MAPS

In this chapter we present the theory developed in [41]. The chapter begins with a very general setting and then we gradually narrow it down to applications to particular meromorphic functions.

4.1. Basic facts and definitions.
We first define Walters expanding maps and collect their properties needed in the sequel. For a full account of Walters' theory see [81].

Let X_0 be an open and dense subset of a compact metric space X endowed with a metric ρ. We call a continuous map $T : X_0 \to X$ Walters expanding provided that the following conditions are satisfied:

(1a) The set $T^{-1}(x)$ is at most countable for each $x \in X$.
(1b) There exists $\delta > 0$ such that for every $x \in X$ and every $n \geq 0$, $T^{-n}(B(x, 2\delta))$ can be written uniquely as a disjoint union of open sets $\{B_y(x)\}_{y \in T^{-n}(x)}$ such that $y \in B_y(x)$ and $T^n : B_y(x) \to B(x, 2\delta)$ is a homeomorphism from $B_y(x)$ onto $B(x, 2\delta)$. The corresponding inverse map from $B(x, 2\delta)$ to $B_y(x)$, $y \in T^{-n}(x)$, will be denoted by T_y^{-n}.
(1c) There exist $\lambda > 1$ and $n \geq 1$ such that for every $x \in X$, every $y \in T^{-n}(x)$ and all $z_1, z_2 \in B_y(x)$
$$d(T^n(z_1), T^n(z_2)) \geq \lambda d(z_1, z_2).$$
(1d) $\forall \epsilon > 0 \, \exists s \geq 1 \, \forall x \in X \; T^{-s}(x)$ is ϵ-dense in X.

Recall that a function $g : Y \to \mathbb{R}$, where (Y, ρ) is a metric space, is Hölder continuous if there exist $\beta > 0$ and $L > 0$ such that for all $y_1, y_2 \in Y$, $|g(y_1) - g(y_2)| \leq L\rho(y_1, y_2)^\beta$. The parameter β is called the Hölder exponent of the function g and L is called its Hölder constant. A function $\phi : X_0 \to \mathbb{R}$ is called dynamically Hölder if there exists $\beta > 0$ and $L > 0$ such that for every $n \geq 1$, every $x \in X$ and every $y \in T^{-n}(x)$, the restriction $\phi|_{T_y^{-n}(B(x,\delta))}$

is Hölder continuous with exponent $\beta > 0$ and constant L. For $n \geq 1$ put

$$S_n\phi(x) = \sum_{j=0}^{n-1} \phi \circ T^j(x).$$

Using (1c), a standard argument in thermodynamic formalism shows that there exists a constant $C > 0$ such that
$\forall x \in X$, $\forall y, z \in B(x, \delta)$, $\forall n \geq 0$, $\forall u \in T^{-n}(x)$

$$|S_n\phi(T_u^{-n}(y)) - S_n\phi(T_u^{-n}(z))| \leq Cd(y, z)^\beta. \tag{4.1}$$

The function $\phi : X_0 \to X$ is called summable if

$$\sup_{x \in X} \left\{ \sum_{y \in T^{-1}(x)} \exp(\phi(y)) \right\} < \infty.$$

Given $x \in X$ we define, as in the case of elliptic functions,

$$P_x(\phi) = \limsup_{n \to \infty} \frac{1}{n} \log \sum_{y \in T^{-n}(x)} \exp(S_n(\phi(y))).$$

It is not difficult to prove that if $\phi : X_0 \to \mathbb{R}$ is dynamically Hölder, then $P_x(\phi) = P_y(\phi)$ for all $x, y \in X$. The common value is called the topological pressure of ϕ with respect to T and is denoted by $P(\phi)$. We note that $P(\phi) < \infty$ if and only if ϕ is summable. From the results of P. Walters in [81] one can extract the following theorem.

Theorem 4.1. *If $T : X_0 \to X$ is a Walters expanding map and $\phi : X_0 \to X$ is a dynamically Hölder summable function, then there exist m_ϕ and μ_ϕ, Borel probability measures on X such that*

(a) *for all $n \geq 1$, $x \in X$, $y \in T^{-n}(x)$ and for every Borel set $A \subset T_y^{-n}(B(x, \delta))$*

$$m_\phi(T^n(A)) = \int_A e^{P(\phi) - S_n(\phi)} dm_\phi,$$

(b) *μ_ϕ is T-invariant, which means that $\mu_\phi \circ T^{-1} = \mu_\phi$, ergodic and equivalent to m_ϕ with continuous Radon–Nikodym derivative bounded away from zero and infinity.*

The reader familiar with thermodynamic formalism, for example with Subsection 2.2.1 or Section 3.2, may notice that property (a) means that the measure m_ϕ is an eigenmeasure of the operator dual to the appropriate Perron–Frobenius operator with eigenvalue $e^{P(\phi)}$. Many additional stochastic properties of the dynamical system (T, μ_ϕ) can be found in [81].

A Walters expanding map $F : X_0 \to X$ is called conformal if $X \subset \mathbb{C}$ and if for every $x \in X$, every $n \geq 1$ and every $y \in F^{-n}(x)$ the inverse map $F_y^{-n} : B_X(x, 2\delta) \to X_0$ has a (unique) holomorphic extension to the ball

$B_\mathbb{C}(x, 2\delta)$. This extension will be denoted by the same symbol F_y^{-n}. For the rest of this chapter we assume that F is a Walters expanding conformal map. Of special importance will be the functions $g_t : X_0 \to \mathbb{R}, t \geq 0$ given by the formula

$$g_t(x) = -t \log |F'(x)|.$$

It immediately follows from Koebe's distortion theorem that each function g_t is dynamically Hölder with the Hölder exponent $1/3$. Following [54] we define θ_F to be the infimum of all $t \geq 0$ for which the function g_t is summable. Thanks to Proposition 2.4 in [41], we have

$$\theta_F = \inf\{t \geq 0 : P(g_t) < \infty\}.$$

The following proposition is a straightforward and standard consequence of the definition of pressure and property (1c).

Proposition 4.2. *The function $P : (\theta_F, \infty) \to \mathbb{R}$ is convex, continuous, strictly decreasing and $\lim_{t \to +\infty} P(t) = -\infty$.*

We define

$$h_F = h = \inf\{t : P(t) \leq 0\}.$$

Obviously $h_F \geq \theta(F)$. Following the terminology of [54] we call the map F regular if $P(h) = 0$, strongly regular if there exists $t \geq 0$ such that $0 < P(t) < \infty$ and hereditarily regular if $P(\theta_F) = \infty$ (which, by (1c) and (1d), implies that $\lim_{t \searrow \theta_F} P(t) = +\infty$). In view of Proposition 4.2 each strongly regular map is regular and each hereditarily regular map is strongly regular. If F is regular, then $m = m_{g_h}$ is called the h-conformal measure for F. Its F-invariant version will be denoted by μ. Let

$$X_\infty = \bigcap_{n \geq 0} F^{-n}(X_0).$$

The following result is an immediate consequence of Theorem 4.1.

Theorem 4.3. *If F is a regular Walters expanding conformal map, then there exists a unique F-invariant Borel probability measure μ_h absolutely continuous with respect to the h-conformal measure m_h. The measure μ_h is ergodic and its Radon–Nikodym derivative is bounded away from 0 and ∞.*

4.2. Hausdorff and box dimensions, Hausdorff and packing measures. In this section we will be primarily interested in the dynamical system $F : X_\infty \to X_\infty$ and the geometry of the set X_∞. The first result in this direction, a version of Bowen's formula, is this.

Theorem 4.4. *If $F : X_0 \to X$ is a Walters expanding conformal map, then $\mathrm{HD}(X_\infty) \leq h$. If, in addition, F is strongly regular, then $\mathrm{HD}(X_\infty) = h$ and, in particular, $\mathrm{HD}(X_\infty) > \theta_F$.*

Passing to the upper ball-counting dimension (also called the upper box dimension or Minkowski dimension), we let X be an arbitrary metric space and A an arbitrary subset of X. We denote by $N_r(A)$ the minimal number of balls with centres in the set A and of radius $r > 0$ needed to cover A. The upper ball-counting dimension of A is defined to be

$$\mathrm{BD}(A) = \limsup_{r \to 0} \frac{\log N_r(A)}{-\log r}.$$

A formula for the upper ball-counting dimension of the set X is given by the following.

Theorem 4.5. *If F is a regular Walters expanding conformal map and if W is a finite δ-net of X, then*

$$\mathrm{BD}(X) = \max\{\mathrm{HD}(X_\infty), \mathrm{BD}(F^{-1}(W))\}$$
$$= \max\{\mathrm{HD}(X_\infty), \max\{\mathrm{BD}(F^{-1}(w)) : w \in W\}\}.$$

We would like to bring the reader's attention to the fact that although the set $F^{-1}(W)$ is countable, its upper box dimension can be positive. We would also like to emphasise the fact that in Theorem 4.5 only the first inverse iterate $F^{-1}(W)$ is involved and higher inverse iterates are not needed. The problem of determining whether the Hausdorff (H^h) and packing (P^h) measures of the set X_∞ are finite and positive is a more delicate issue. However the following general result holds.

Theorem 4.6. *If F is a regular Walters expanding conformal map, then $\mathrm{H}^h(X_\infty) < \infty$ and $\mathrm{P}^h(X_\infty) > 0$. In addition $\mathrm{H}^h \ll m_h$ and $m_h \ll \mathrm{P}^h$.*

The general tools, applied for example to jump-like conformal maps, to deal with the problem whether $H^h(X_\infty)$ is positive or vanishes, or $P^h(X_\infty)$ is finite or infinite are collected in the following theorems.

Theorem 4.7. *Suppose F is a regular Walters expanding conformal map. Assume that there exist $\gamma \geq 1$ and $L > 0$ such that for every $x \in X_0$ and for every r satisfying the condition $r \geq \gamma \operatorname{diam}(F_x^{-1}(B(F(x), \delta)))$, we have $m(B(x), r)) \leq Lr^h$. Then $\mathrm{H}^h(X_\infty) > 0$.*

Theorem 4.8. *Suppose F is a regular Walters expanding conformal map. If there exist a sequence of points $z_j \in X, j \geq 1$, and a sequence of positive reals $\{r_j\}_{j=1}^\infty$ such that $r_j \leq \delta/2$ and*

$$\overline{\lim}_{j \to \infty} \frac{m_h(B(z_j, r_j))}{r_j^h} = \infty,$$

then $\mathrm{H}^h(X_\infty) = 0$.

Theorem 4.9. *Suppose F is a regular Walters expanding conformal map. Assume that there exist $\gamma \geq 1$ and $0 < \xi \leq \delta$ such that for every $x \in X_0$ and for every r satisfying the condition $\gamma \operatorname{diam}(F_x^{-1}(B(F(x), \delta)))) \leq r \leq \xi$, we have $m_h(B(x,r)) \geq Lr^h$. Then $\mathrm{P}^h(X_\infty) < \infty$.*

Theorem 4.10. *Suppose F is a regular Walters expanding conformal map. If there exists a sequence of points $z_j \in X, j \geq 1$ and a sequence of positive reals $\{r_j\}_{j=1}^\infty$ such that*

$$\varliminf_{j \to \infty} \frac{m_h(B(z_j, r_j))}{r_j^h} = 0,$$

then $\mathrm{P}^h(X_\infty) = \infty$.

Let $\operatorname{Asymp}(f)$ denote from now on the set of asymptotic values of the function f under consideration. Concluding this rather abstract section we mention that if $f : \mathbb{C} \to \overline{\mathbb{C}}$ is a meromorphic map for which

$$J(f) \cap \overline{\bigcup_{n=0}^\infty f^n(\operatorname{Crit}_G(f) \cup \operatorname{Asymp}(f))} = \emptyset,$$

where $\operatorname{Crit}_G(f)$ is the set of general critical points (that is, the set of critical points or multiple poles of f), then it is not difficult to prove that if $M : \overline{\mathbb{C}} \to \overline{\mathbb{C}}$ is a Möbius transformation such that $M(\infty) \notin J(f)$, then the Julia set of the map $\tilde{f} = M^{-1} \circ f \circ M : \overline{\mathbb{C}} \setminus M^{-1}(\infty) \to \overline{\mathbb{C}}$ is a compact subset of $\mathbb{C}$ and $\tilde{f}$ restricted to its Julia set is a conformal Walters expanding map. In particular all the theorems proved in this section apply to $\tilde{f}$.

4.3. Barański and post-Barański Maps, I. In this section based on [41, Section 4] we give a brief account of the class of Barański and post-Barański maps. The latter are all Walters expanding conformal maps and provide a good motivation for dealing in the next section with the larger class of jump-like conformal maps. We consider the class of functions of the form

$$f(z) = H(\exp(Q(z))) \quad \text{and} \quad \tilde{f}(z) = \exp(Q(H(z))),$$

where Q and H are non-constant rational functions. Let

$$Q^{-1}(\infty) = \{d_j : j = 1, \ldots, m\}$$

be the set of poles of Q. Then

$$f(z) = H(\exp(Q(z))) : \overline{\mathbb{C}} \setminus \{d_j : j = 1, \ldots, m\} \to \overline{\mathbb{C}}$$

and

$$\tilde{f}(z) = \exp(Q(H(z))) : \overline{\mathbb{C}} \setminus H^{-1}(\{d_j : j = 1, \ldots, m\}) \to \overline{\mathbb{C}} \setminus \{0, \infty\}.$$

We additionally assume that there is at least one pole d_i of Q such that $d_i \neq H(0), H(\infty)$. We may assume without loss of generality that $d_i = d_1$. Then the set

$$Ess_\infty(f) = \bigcup_{n=0}^\infty f^{-n}(\{d_j : i = 1, \ldots, m\})$$

contains infinitely many points. Since $\{0, \infty\} \cap H^{-1}(d_1) = \emptyset$, the set

$$Ess_\infty(\tilde{f}) = \bigcup_{n=0}^\infty \tilde{f}^{-n}(H^{-1}(\{d_j := 1, \ldots, m\}))$$

contains infinitely many points. Using Montel's criterion it can be easily proved that

$$J(f) = \overline{Ess_\infty(f)} \quad \text{and} \quad J(\tilde{f}) = \overline{Ess_\infty(\tilde{f})}.$$

Notice that $\mathrm{Asymp}(f) = \{H(0), H(\infty)\}$ and $\mathrm{Asymp}(\tilde{f}) = \{0, \infty\}$ are respectively the sets of asymptotic values f and $\tilde{f}$. We say that f is a Barański map if the following conditions are satisfied:

(1) $J(f) \cap \overline{\bigcup_{n=0}^\infty f^n(\mathrm{Crit}(f) \cup \mathrm{Asymp}(f))} = \emptyset$,
(2) if $a \in \mathrm{Crit}(Q)$, then $\exp(Q(a))$ is not a pole of H,
(3) if H has a multiple pole, then $Q(\infty) \neq \infty$.

The map $\tilde{f}$ is then called a post-Barański map. K. Barański himself in his pioneering paper [7] considered the case where Q is the identity map. The maps f and $\tilde{f}$ are closely related, as the following two formulae show:

$$f \circ H(z) = H \circ \tilde{f}(z) \quad z \notin Ess(\tilde{f}); \tag{4.2}$$

$$\exp(Q) \circ f = \tilde{f} \circ \exp(Q) \quad z \notin Ess(f). \tag{4.3}$$

These relations allow us to deduce lots of valuable dynamical and geometrical properties of the map f from the corresponding properties of the map $\tilde{f}$. This is why the rest of this section is devoted to the post-Barański map $\tilde{f}$. The first observation is that there is a $\kappa \in (0, +\infty)$ such that

$$J(\tilde{f}) \subset \{z : e^{-\kappa} < |z| < e^\kappa\}.$$

The second one is that

$$J(\tilde{f}) \cap \overline{\bigcup_{n=0}^\infty \tilde{f}^n \left(\mathrm{Crit}_G(\tilde{f}) \cup \mathrm{Asymp}(\tilde{f})\right)} = \emptyset.$$

Armed with these two observations, one can prove the following first basic results about post-Barański maps.

Theorem 4.11. $\tilde{f} : J(\tilde{f}) \setminus \{b_j : j = 1, \ldots, p\} \to J(\tilde{f})$ *is a Walters expanding conformal map.*

Here
$$\{b_j : j = 1, 2, \ldots, p\} = (Q \circ H)^{-1}(\infty).$$
For every $j = 1, \ldots, p$, let $q_j \geq 1$ be the order of b_j treated as a pole of $Q \circ H$. Note that for every $z \in J(\tilde{f})$, each holomorphic branch of $\tilde{f}^{-1}$ defined on the ball $B(z, 2\delta)$ can be expressed in the form
$$\tilde{f}_{j,a,n}^{-1}(w) = (Q \circ H)_{j,a}^{-1}(\log w + 2\pi i n), \tag{4.4}$$
where $j = 1, \ldots, p$, $a = 1, \ldots, q_j$, $\log w$ is the value of the logarithm of w lying in the rectangle $[-\kappa, \kappa] \times [0, 2\pi]$ and $(Q \circ H)_{j,a}^{-1}$ is a local holomorphic inverse branch of $Q \circ H$. For $n \in \mathbb{Z}$ with sufficiently large modulus, each such inverse branch can be interpreted as a branch of $(Q \circ H)^{-1}$ defined on some vertical strip of the form either $[-\kappa, \kappa] \times [T, +\infty]$ or $[-\kappa, \kappa] \times [-\infty, -T]$, depending on whether n is positive or negative and sending ∞ to the pole b_j of $Q \circ H$. Let (X, d) be a compact metric space. For every $A, B \subset X$ define
$$\text{dist}(A, B) = \inf\{d(a, b) : a \in A, b \in B\}$$
and
$$\text{Dist}(A, B) = \sup\{d(a, b) : a \in A, b \in B\}.$$
Tedious technical calculations lead to the following basic facts about a post-Barański map.

Theorem 4.12. *The map $\tilde{f}_{|J(\tilde{f})}$ is a jump-like conformal map, that is, there exist $C \geq 1$ and $A \geq 2$ such that the following conditions are satisfied:*

(2a) $\{b_j : j = 1, \ldots, b_p\} \cap \tilde{f}^{-1}(J(\tilde{f})) = \emptyset$.

(2b) For every $x \in J(\tilde{f})$ the set $\tilde{f}^{-1}(x)$ can be uniquely represented as
$$\{x_{j,a,n} : n \in \mathbb{Z}, 1 \leq j \leq p, 1 \leq a \leq q_j\}.$$

(2c) $\max_{1 \leq j \leq p} \max_{1 \leq a \leq q_j} \sup_{x \in J(\tilde{f})} \{\lim_{n \to \infty} \text{Dist}(b_j, \tilde{f}_{j,a,n}^{-1}(B(x, \delta)))\} = 0$.

(2d) $\forall z \in J(\tilde{f})$, $\forall 1 \leq j \leq p$, $\forall 1 \leq a \leq q_j$, $\forall n \in \mathbb{Z}$, $|n| \geq A$
$$C^{-1}|n|^{-\frac{q_j+1}{q_j}} \leq |(\tilde{f}_{j,a,n}^{-1})'(z)| \leq C|n|^{-\frac{q_j+1}{q_j}}.$$

(2e) $\forall w, z \in J(\tilde{f})$, $\forall 1 \leq j \leq p$, $\forall a, b \in \{1, \ldots, q_j\}$, $\forall k, n \in \mathbb{Z}$, $||k| - |n|| \geq A$, $|n| \geq A$, $|k| \geq A$
$$\text{dist}(\tilde{f}_{j,a,k}^{-1}(B(w, \delta)), \tilde{f}_{j,b,n}^{-1}(B(z, \delta))) \geq C^{-1}\left||k|^{-\frac{1}{q_j}} - |n|^{-\frac{1}{q_j}}\right|.$$

(2f) $\forall w, z \in J(\tilde{f})$, $\forall 1 \leq j \leq p$, $\forall a \in \{1, \ldots, q_j\}$, $\forall k, n \in \mathbb{Z}$, $kn > 0$, $||k| - |n|| \geq A$, $|n| \geq A$, $|k| \geq A$
$$\text{Dist}(\tilde{f}_{j,a,k}^{-1}(B(w, \delta)), \tilde{f}_{j,b,n}^{-1}(B(z, \delta))) \leq C\left||k|^{-\frac{1}{q_j}} - |n|^{-\frac{1}{q_j}}\right|.$$

As an immediate consequence of (2d), we get the following (perhaps with a larger constant C):

$$\forall x \in J(\tilde{f}), \ \forall 1 \le j \le p, \ \forall 1 \le a \le q_j, \ \forall n \in \mathbb{Z}, \ |n| \ge A$$

$$C^{-1}|n|^{-\frac{q_j+1}{q_j}} \le \operatorname{diam}(\tilde{f}_{j,a,n}^{-1}(B(x,\delta))) \le C|n|^{-\frac{q_j+1}{q_j}}. \tag{4.5}$$

Letting $k \to \infty$, it immediately follows from (2c), (2e) and (2f) that

$$\forall 1 \le j \le p, \ \forall_{x \in J(\tilde{f})}, \ \forall 1 \le a \le q_j, \ \forall |n| \ge 2A$$

$$C^{-1}|n|^{-\frac{1}{q_j}} \le \operatorname{dist}(b_j, \tilde{f}_{j,a,n}^{-1}(B(x,\delta))) \le \operatorname{Dist}(b_j, \tilde{f}_{j,a,n}^{-1}(B(x,\delta))) \le C|n|^{-\frac{1}{q_j}} \tag{4.6}$$

We end this section with large classes of examples of Barański maps. The family $f_\lambda(z) = \lambda \tan(z)$, $0 < |z| < 1$ was explored in [7]. In fact the map $\lambda \tan(z)$ is a Barański map as long as both asymptotic values $\pm \lambda i$ lie in basins of attraction of attracting periodic cycles. If $Q(z)$ is a polynomial different from the identity, we obtain a transcendental meromorphic function f of the complex plane with one essential singularity at ∞. The most transparent class of examples is provided by the following.

Example 1. Let

$$f(z) = \frac{A \exp(z^p) + B \exp(-z^p)}{C \exp(z^p) + D \exp(-z^p)}, \quad p \in \mathbb{N}, \ AD - BC \ne 0.$$

Thus $\operatorname{Crit}(f) = \{0\}$, $\operatorname{Crit}_G(f) = \operatorname{Crit}(f)$ and $\operatorname{Asymp}(f) = \{\frac{A}{C}, \frac{B}{D}\}$. If we have $\frac{A}{C}, \frac{B}{D} \ne \infty$, then f is not entire. Notice that conditions (2) and (3) of the definition of Barański maps are always satisfied for the map f. If, in addition, condition (1) is satisfied, then f is a Barański map and all the results stated in the forthcoming Section 4.5 apply. If Q is not a polynomial, then f has more then one essential singularity. Let us analyse in detail the following concrete example of this type.

Example 2. Let $H(z) = z, Q(z) = \frac{z-1}{z+1}$. Then $f: \mathbb{C} \setminus \{-1\} \to \mathbb{C} \setminus \{0, \infty\}$,

$$f(z) = \exp\left(\frac{z-1}{z+1}\right).$$

and $\tilde{f} = f$. Since the pole of Q is not an omitted value of f, we see that $\bigcup_{n=0}^\infty f^{-n}(-1)$ contains infinitely many points and consequently

$$J(f) = \overline{\bigcup_{n=0}^\infty f^{-n}(-1)}.$$

Since $f^{-1}(S^1) \subset S^1$, we have $f^{-n}(-1) \in S^1$ for all $n \in \mathbb{N}$. Therefore $J(f) \subset S^1$. We shall prove that f is a Barański map and its Julia set $J(f)$ is a topological Cantor set. Note that $\text{Crit}_G(f) = \emptyset$ and $\text{Asymp}(f) = \{0,\infty\}$. One can check that $f(1) = 1$ and $f'(1) = 1/2$, so the number 1 is an attracting fixed point of f. Thus $J(f)$ is a topological Cantor set contained in the circle S^1. In order to conclude the proof it is now sufficient to demonstrate that 1 attracts both asymptotic values 0 and ∞. Since $f'(x) > 0$ for $x \in \mathbb{R} \setminus \{-1\}$, the function f is strictly increasing on $(-\infty, -1)$ and $(-1, +\infty)$. Now, if $x \in (1, +\infty)$, then $f(1) < f(x) < x$. This implies that $\lim_{n\to\infty} f^n(x) = 1$ for all $x \in (1, +\infty)$. In particular $\lim_{n\to\infty} f^n(\infty) = 1$ since $f(\infty) = e \in (1, +\infty)$. If $x \in (-1, 1)$, then $x < f(x) < f(1) = 1$. This implies that $\lim_{n\to\infty} f^n(x) = 1$ for all $x \in (-1, 1)$. In particular $\lim_{n\to\infty} f^n(0) = 1$ since $f(0) = 1/e \in (-1, 1)$. We are done.

4.4. Jump-like conformal maps. It turns out that the properties established in Theorem 4.11 and Theorem 4.12 are themselves sufficient to provide a fairly complete description of the dynamics and geometry of the maps appearing in these theorems. This motivated us to single them out and to introduce the class of jump-like conformal maps. We call a Walters expanding conformal map $F : X_0 \to X$ jump-like if the following requirements are met. There exist $C \geq 1, p \geq 1, A \geq 2, b_j \in X$ and $q_j \geq 1$ for every $j = 1, \ldots, p$ such that the following conditions are satisfied:

(3a) $\{b_1, \ldots, b_p\} \cap F^{-1}(X) = \emptyset$.
(3b) For every $x \in X$, the set $F^{-1}(x)$ can be uniquely represented as
$$\{x_{j,a,n} : n \in \mathbb{Z},\ 1 \leq j \leq p,\ 1 \leq a \leq q_j\}.$$
(3c) $\max_{1\leq j\leq p} \max_{1\leq a\leq q_j} \sup_{x\in X}\{\lim_{n\to\infty} \text{Dist}(b_j, F^{-1}_{x_{j,a,n}}(B(x,\delta)))\} = 0$.
(3d) $\forall x \in X,\ \forall 1 \leq j \leq p,\ \forall 1 \leq a \leq q_j,\ \forall n \in \mathbb{Z},\ |n| \geq A$,
$$C^{-1}|n|^{\frac{q_j+1}{q_j}} \leq |F'(x_{j,a,n})| \leq C|n|^{\frac{q_j+1}{q_j}}.$$
(3e) $\forall y, z \in X,\ \forall 1 \leq j \leq p,\ \forall a, b \in \{1, \ldots, q_j\},\ \forall k, n \in \mathbb{Z}$,
$||k| - |n|| \geq A,\ |n| \geq A,\ |k| \geq A$,
$$\text{dist}(F^{-1}_{y_{j,a,k}}(B(y,\delta)), F^{-1}_{z_{j,b,n}}(B(z,\delta))) \geq C^{-1}\left||k|^{-\frac{1}{q_j}} - |n|^{-\frac{1}{q_j}}\right|.$$
(3f) $\forall y, z \in X,\ \forall 1 \leq j \leq p,\ \forall a \in \{1, \ldots, q_j\},\ \forall k, n \in \mathbb{Z},\ kn > 0$,
$||k| - |n|| \geq A, |n| \geq A,\ |k| \geq A$,
$$\text{Dist}(F^{-1}_{y_{j,a,k}}(B(y,\delta)), F^{-1}_{z_{j,a,n}}(B(z,\delta))) \leq C\left||k|^{-\frac{1}{q_j}} - |n|^{-\frac{1}{q_j}}\right|.$$

The name 'jump-like' is motivated by a fairly strong formal similarity between these maps and the jump maps considered in the theory of parabolic rational functions. By a relatively simple calculation, we can prove the following.

Proposition 4.13. *Suppose that $F : X_0 \to X$ is a jump-like conformal map and let $q = \max\{q_j : 1 \leq j \leq p\}$. Then the map F is hereditarily regular and $\theta_F = \frac{q}{q+1}$.*

As an immediate consequence of this proposition and Theorem 4.3, we get the following.

Theorem 4.14. *If F is a jump-like conformal mapping, then there exists a unique F-invariant Borel probability measure μ_h absolutely continuous with respect to the h-conformal measure m_h. The measure μ_h is ergodic and its Radon–Nikodym derivative is bounded away from 0 and ∞.*

It is easy to see that if F is a jump-like conformal map, then for every $x \in X$, $\mathrm{BD}(F^{-1}(x)) = \frac{q}{q+1}$, where $q = \max\{q_j : 1 \leq j \leq p\}$. Thus, as an immediate consequence of Proposition 4.13 and Theorem 4.5, we get the following.

Theorem 4.15. *If $F : X_0 \to X$ is a jump-like conformal map, then*
$$\mathrm{BD}(X) = \mathrm{HD}(X_\infty).$$

Based on the assumptions (3a)-(3f) one can verify by very technical considerations the sufficient conditions established in Section 4.2. As the result we get the following complete description of geometric measures, and simultaneously, a geometric characterisation of the dynamically defined conformal measure m_h.

Theorem 4.16. *Suppose that $F : X_0 \to X$ is a jump-like conformal map.*
  (a) *If $h < 1$, then $0 < \mathrm{P}^h(X_\infty) < \infty$ and $\mathrm{H}^h(X_\infty) = 0$.*
  (b) *If $h = 1$, then $0 < \mathrm{P}^h(X_\infty), \mathrm{H}^h(X_\infty) < \infty$.*
  (c) *If $h > 1$, then $0 < \mathrm{H}^h(X_\infty) < \infty$ and $\mathrm{P}^h(X_\infty) = \infty$.*

4.5. Barański and post-Barański Maps, II.
In view of Theorem 4.11 and Theorem 4.12 from Section 4.3, all post-Barański maps are jump-like. Therefore, their fractal and dynamical properties proved in [41] can be summarised briefly in the following theorem.

Theorem 4.17. *Proposition 4.13, Theorem 4.15, Theorem 4.14 and Theorem 4.16 are true with jump-like conformal maps replaced by post-Barański maps.*

Since $H(J(\tilde{f})) = J(f)$, where f is a Barański map and H is the rational function involved in the formula defining it, one can deduce relatively easily the following results.

Theorem 4.18. *Let f be a Barański map, and let Hausdorff measure and packing measure be defined using the Euclidean metric.*
  (a) *If $h < 1$, then $\mathrm{P}^h(J(f)) > 0$ and $\mathrm{P}^h|_{J(f)}$ is σ-finite, whereas $\mathrm{H}^h(J(f)) = 0$.*
  (b) *If $h = 1$, then $\mathrm{P}^h(J(f)) > 0$ and $\mathrm{H}^h(J(f)) > 0$, and both measures restricted to $J(f)$ are σ-finite.*
  (c) *If $h > 1$, then $\mathrm{H}^h(J(f)) > 0$ and $\mathrm{H}^h|_{J(f)}$ is σ-finite, whereas $\mathrm{P}^h(J(f)) = \infty$.*

Theorem 4.19. *Let f be a Barański map, and let Hausdorff measure and packing measure be defined using the spherical metric.*
  (a) *If $h < 1$, then $0 < \mathrm{P}^h(J(f)) < \infty$ and $\mathrm{H}^h(J(f)) = 0$.*
  (b) *If $h = 1$, then $0 < \mathrm{P}^h(J(f))$ and $\mathrm{H}^h(J(f)) < \infty$.*
  (c) *If $h > 1$, then $0 < \mathrm{H}^h(J(f)) < \infty$ and $\mathrm{P}^h(J(f)) = \infty$.*

We say that Hausdorff or packing measure on $J(f)$ is geometric if it is finite and positive. Using the semi-conjugacy $H \circ \tilde{f} = f \circ H$ established in Section 4.3, the following result is an immediate consequence of Theorem 4.17.

Theorem 4.20. *If f is a Barański map, then there exists a unique f-invariant probability measure equivalent to a conformal measure.*

5. Hyperbolic Meromorphic and Entire Functions

The paper [53], whose contents we briefly describe in this section, presents a new uniform approach to the theory of thermodynamic formalism for a very wide class of meromorphic functions of finite order. The key point is to associate to a given meromorphic function $f : \mathbb{C} \to \overline{\mathbb{C}}$ a suitable Riemannian metric $d\sigma = \gamma |dz|$. One then uses Nevanlinna theory to construct conformal measures for the potentials $-t \log |f'|_\sigma$ and control the corresponding Perron–Frobenius operators. Here

$$|f'(z)|_\sigma = |f'(z)| \frac{\gamma \circ f(z)}{\gamma(z)}$$

is the norm of the derivative of f with respect to the metric $d\sigma$. With this tool in hand one is able to obtain geometric information on the Julia set $J(f)$ and on the radial (or conical) Julia set

$$J_r(f) = \{z \in J(f) : \liminf_{n \to \infty} |f^n(z)| < \infty\}.$$

In [53], in contrast to the work reported on in the previous sections, neither periodicity nor the Walters expanding property is assumed to hold. We now give a fairly precise description of the results obtained in [53].

5.1. Thermodynamic formalism.
The main idea which allows one to abandon periodicity is to associate to a given meromorphic function f a Riemannian conformal metric $d\sigma = \gamma|dz|$ with respect to which the Perron–Frobenius–Ruelle (or transfer) operator

$$\mathcal{L}_t \phi(w) = \sum_{z \in f^{-1}(w)} |f'(z)|_\sigma^{-t} \phi(z) \tag{5.1}$$

is well-defined and has all the required properties that make the thermodynamic formalism work. Such a good metric can be found for meromorphic functions $f : \mathbb{C} \to \overline{\mathbb{C}}$ that are of finite order ρ and satisfy the following growth condition for the derivative.

Rapid derivative growth: There exist $\alpha_2 > \max\{0, -\alpha_1\}$ and $\kappa > 0$ such that

$$|f'(z)| \geq \kappa^{-1}(1+|z|^{\alpha_1})(1+|f(z)|^{\alpha_2}) \tag{5.2}$$

for all finite $z \in J(f) \setminus f^{-1}(\infty)$. If (5.2) holds, then we write $\alpha = \alpha_1 + \alpha_2$.

This condition is very general and forms the second main idea of [53]. It is relatively easy to verify for a large natural class of functions which include the entire exponential family λe^z, certain other periodic functions ($\sin(az+b)$, $\lambda \tan(z)$, elliptic functions, ...), the cosine-root family $\cos(\sqrt{az+b})$ and the composition of these functions with arbitrary polynomials. The Riemannian metric σ we mentioned above is this:

$$d\sigma(z) = (1+|z|^{\alpha_2})^{-1}|dz|.$$

The third and fourth main ideas in [53] were to revive the old method of construction of conformal measures from [17] (which itself stemmed from the work of D. Sullivan [68], [69], [70] and S. Patterson [58]) and to employ results and methods from Nevanlinna theory. These allow us to perform the construction of conformal measures and to get good control of the Perron–Frobenius–Ruelle operator, resulting in the following key result of [53].

Theorem 5.1. *If $f : \mathbb{C} \to \overline{\mathbb{C}}$ is an arbitrary hyperbolic meromorphic function of finite order ρ that satisfies the rapid derivative growth condition (5.2), then for every $t > \frac{\rho}{\alpha}$ the following are true.*
  (1) *The topological pressure $\mathrm{P}(t) = \lim_{n \to \infty} \frac{1}{n} \log \mathcal{L}_t^n(\mathbb{1})(w)$ exists and is independent of $w \in J(f) \cap \mathbb{C}$.*
  (2) *There exists a unique $\lambda |f'|_\sigma^t$-conformal measure m_t and necessarily $\lambda = e^{\mathrm{P}(t)}$. Also, there exists a unique Gibbs state μ_t, that is, μ_t is f-invariant and equivalent to m_t. Moreover, both measures are ergodic and supported on the radial (or conical) Julia set.*
  (3) *The density $\psi = d\mu_t/dm_t$ is a continuous and bounded function on the Julia set $J(f)$.*

Note that for the existence of $e^{P(t)}|f'|_\sigma^t$-conformal measures the assumption of hyperbolicity is not needed. Note also that, even in the context of exponential functions λe^z and Walters expanding conformal maps, this result is strictly speaking new since it concerns the map f itself and not its projection onto the infinite cylinder. An important special case in Theorem 5.1 is when h is a zero of the pressure function $t \mapsto P(t)$. In this situation, the corresponding measure m_h is $|f'|_\sigma^h$-conformal, also called simply h-conformal. Such a (unique) zero $h > \rho/\alpha$ exists provided the function f satisfies the following two additional conditions.

Divergence type: The series $\Sigma(t,w) = \sum_{z \in f^{-1}(w)} |z|^{-t}$ diverges at the critical exponent (which is the order of the function $t = \rho$); here w is any non-Picard exceptional value.

Balanced growth condition: There exist $\alpha_2 > \max\{0, -\alpha_1\}$ and $\kappa > 0$ such that

$$\kappa^{-1}(1+|z|^{\alpha_1})(1+|f(z)|^{\alpha_2}) \leq |f'(z)| \leq \kappa(1+|z|^{\alpha_1})(1+|f(z)|^{\alpha_2})$$
(5.3)

for all finite $z \in J(f) \setminus f^{-1}(\infty)$.

Indeed, we have the following.

Theorem 5.2. *(Bowen's formula) If $f : \mathbb{C} \to \overline{\mathbb{C}}$ is a hyperbolic meromorphic function that is of finite order $\rho > 0$, of divergence type and of balanced derivative growth with $\alpha_1 \geq 0$, then the pressure function $P(t)$ has a unique zero $h > \rho/\alpha$ and*

$$\mathrm{HD}(J_r(f)) = h \ .$$

In addition, one easily proves that $\mathrm{HD}(J_r(f)) < 2$.

5.2. Real analyticity. In the paper [53] the authors developed a new approach to the problem of real analyticity of the hyperbolic dimension of hyperbolic meromorphic functions. This allowed them to employ the method of holomorphic extensions of generalised Perron–Frobenius operators worked out in [77] (comp. [14] and [45]). As the most transparent outcome of this work, the following theorem (extending results from [76], [45] and [14]) has been proved.

Theorem 5.3. *Let $f : \mathbb{C} \to \overline{\mathbb{C}}$ be either the sine, tangent, exponential or the Weierstrass elliptic function and let $f_\lambda(z) = f(\lambda_d z^d + \lambda_{d-1} z^{d-1} + \cdots + \lambda_0)$, $\lambda = (\lambda_d, \lambda_{d-1}, \ldots, \lambda_0) \in \mathbb{C}^* \times \mathbb{C}^d$. Then the function*
$$\lambda \mapsto \mathrm{HD}(J_r(f_\lambda))$$
is real-analytic in a neighbourhood of each parameter λ^0, giving rise to a hyperbolic function f_{λ^0}.

This result is an example of an application of the general Theorem 5.4 that we present below. Recall that the Speiser class $\mathcal{S}$ is the set of meromorphic functions $f : \mathbb{C} \to \overline{\mathbb{C}}$ that have a finite set of singular values $\mathrm{Sing}(f^{-1})$. We will work in the subclass $\mathcal{S}_0$ which consists of the functions $f \in \mathcal{S}$ that have a strictly positive and finite order $\rho = \rho(f)$, and that are of divergence type. Fix Λ, an open subset of $\mathbb{C}^N$, $N \geq 1$. Let
$$\mathcal{M}_\Lambda = \{f_\lambda \in \mathcal{S}_0 : \lambda \in \Lambda\}, \quad \Lambda \subset \mathbb{C}^N,$$
be a holomorphic family such that the points of $\mathrm{Sing}(f_\lambda^{-1}) = \{a_{1,\lambda}, \ldots, a_{d,\lambda}\}$ depend continuously on $\lambda \in \Lambda$. Now let $\mathcal{H}$ denote the set of hyperbolic functions from $\mathcal{S}_0$, and put
$$\mathcal{H}\mathcal{M}_\Lambda = \mathcal{M}_\Lambda \cap \mathcal{H}.$$
We say that $\mathcal{M}_\Lambda$ is of bounded deformation if there exists $M > 0$ such that, for all $j = 1, \ldots, N$,
$$\left| \frac{\partial f_\lambda(z)}{\partial \lambda_j} \right| \leq M |f_\lambda'(z)|, \quad \lambda \in \Lambda \text{ and } z \in J(f_\lambda). \tag{5.4}$$

We also say that $\mathcal{M}_\Lambda$ is of uniformly balanced growth provided every $f \in \mathcal{M}_\Lambda$ satisfies the condition (5.3) with some fixed constants $\kappa, \alpha_1, \alpha_2$.

Theorem 5.4. *Suppose $\phi_{\lambda^0} \in \mathcal{H}\mathcal{M}_\Lambda$ and that $U \subset \Lambda$ is an open neighbourhood of λ^0 such that $\mathcal{M}_U$ is of uniformly balanced growth and of bounded deformation. Then the map*
$$\lambda \mapsto \mathrm{HD}(J_r(f_\lambda))$$
is real-analytic near λ^0.

6. Non-hyperbolic Barański maps

In this chapter we report on the class $\mathcal{R}$ of function of the form
$$f(z) = R \circ \exp(z), \tag{6.1}$$
where R is a non-constant rational function and the singular set $\mathrm{Sing}(f^{-1})$ of f^{-1} is allowed to intersect the Julia set. The results we describe were

obtained by J. Kotus and B. Skorulski in [40], [37], [38] and [65]. Note that the above function f is of the form of the Barański functions explored in the last three sections of Chapter 4, with $R = H$ and Q being the identity map. The singular set $\mathrm{Sing}(f^{-1})$ clearly consists of (finitely many) critical values of f and two asymptotic values $R(0), R(\infty)$. The class $\mathcal{R}$ is defined to consist of those functions given by (6.1) for which there exists an integer $q \geq 0$ such that
$$\infty \in f^q\big(\{R(0), R(\infty)\}\big).$$
The class $\mathcal{H}$, on the other hand, consists of those functions defined by (6.1) for which $\infty \notin \{R(0), R(\infty)\}$ and the forward orbit of the set $\{R(0), R(\infty)\}$ under iteration of f stays at a positive distance from the Julia set $J(f)$. In what follows, we will consider the classes $\mathcal{R}$ and $\mathcal{H}$ separately.

6.1. The class $\mathcal{R}$. First, we consider the family $\mathcal{P}$ of all entire functions in $\mathcal{R}$; thus $f \in \mathcal{P}$ if and only if $f \in \mathcal{R}$ and $R^{-1}(\infty) \in \{0, \infty\}$. Then set $\mathcal{Q} = \mathcal{R} \setminus \mathcal{P}$. Let $\mathcal{Q}_1$ be the family of all those functions in $\mathcal{Q}$ for which exactly one of the asymptotic values $R(0)$ or $R(\infty)$ is eventually mapped onto ∞. Put $\mathcal{Q}_2 = \mathcal{Q} \setminus \mathcal{Q}_1$, that is, both $R(0)$ and $R(\infty)$ eventually land at ∞. Put
$$P_1(f) = \overline{\Theta^+(\mathrm{Sing}(f^{-1})) \setminus \Theta^+(\{R(\infty)\})} \tag{6.2}$$
and
$$P_2(f) = \overline{\Theta^+(\mathrm{Sing}(f^{-1})) \setminus \Theta^+(\{R(0), R(\infty)\})}. \tag{6.3}$$
Let χ denote the spherical metric on the Riemann sphere $\overline{\mathbb{C}}$. Set
$$\mathcal{P}^* = \{f \in \mathcal{P} : \mathrm{dist}_\chi(P_1(f), J(f)) > 0\},$$
$$\mathcal{Q}_1^* = \{f \in \mathcal{Q}_1 : \mathrm{dist}_\chi(P_1(f), J(f)) > 0\},$$
$$\mathcal{Q}_2^* = \{f \in \mathcal{Q}_2 : \mathrm{dist}_\chi(P_2(f), J(f)) > 0\},$$
and
$$\mathcal{R}^* = \mathcal{P}^* \cup \mathcal{Q}_1^* \cup \mathcal{Q}_2^*. \tag{6.4}$$
Exactly as in Chapter 2, the map f is projected down to a map F of the cylinder $\mathcal{C} = \mathbb{C}/\sim$ and $J_r(F)$ is defined to be the set of all points $z \in J(F)$ whose ω-limit set $\omega(z)$ is not contained in the union of $\{-\infty, +\infty\}$ and the closure of the postsingular set of f. Using the $K(V)$ method (see Appendix 1) the following two theorems were proved in [65] (comp.[64]).

Theorem 6.1. *Let $f \in \mathcal{R}^*$. Then*
   a) *There exists an h-conformal measure m on $J(F)$ for F such that m is atomless and $m(J_r(F)) = 1$.*
   b) *If m' is a probabilistic measure on $J(F)$ which is t-conformal for some $t > 1$, then $m' = m$.*
   c) *F is ergodic with the respect to the measure m.*
   d) *$1 < \mathrm{HD}(J_r(F)) = h < 2$.*

Theorem 6.2. *If $f \in \mathcal{R}^*$ and $t > 1$, then there exist a unique α_t and a unique (t, α_t)-conformal measure m_t for $F : J(F) \to J(F) \cup \{\infty\}$. In addition, $m_t(J_r(F)) = 1$ and the map F is ergodic with respect to the measure m_t.*

Restricting our attention to the class $\mathcal{Q}_2^* \subset \mathcal{R}^*$, we mention that J. Kotus, making use of Theorem 6.1, was able to prove the following.

Theorem 6.3. *If $f \in \mathcal{Q}_2^*$, then $0 < \mathrm{H}^h(J_r(F)) < \infty$, and, in particular, H^t and the conformal measure m (from Theorem 6.1) coincide up to a multiplicative constant.*

Theorem 6.4. *The packing measure P^h, restricted to $J_r(F)$, is locally infinite at every point of $J_r(F)$.*

Theorem 6.5. *For every function $f \in \mathcal{Q}_2^*$ there exists exactly one Borel probability F-invariant measure μ absolutely continuous with respect to the h-conformal measure m. Moreover μ is ergodic and equivalent to m.*

6.2. The class $\mathcal{H}$. Let $B = \Pi(f^{-1}(\infty))$, where Π is the canonical projection from $\mathbb{C}$ onto the cylinder $\mathbb{C}/\sim$. Notice that if $f \in \mathcal{H}$, then there exists $\tilde{K} > 0$ such that

$$-\tilde{K} < \mathrm{Re} J(F) < \tilde{K}. \tag{6.5}$$

Observe that every point in B is a discontinuity point of F. We call the points of B the poles of F. The set of critical points of F we denote by

$$\mathrm{Crit}(F) = \pi(\mathrm{Crit}(f)).$$

Let

$$J_{bd}(F) = \{z \in J(F) : \inf_{n \geq 1}\{|f^n(z) - b_i|\} > 0, \, b_i \in B\,\}.$$

Let c be a critical point of f eventually mapped onto ∞, that is, there exist $k \geq 2$ and a pole $b \in B$ such that $f^{k-1}(c) = b$. Then there exist $A = A(f^{k-1}, c) \geq 1$ and $p \geq 2$ such that we have, in a neighbourhood of c,

$$A^{-1}|z - c|^p \leq |f^{k-1}(z) - f^{k-1}(c)| \leq A|z - c|^p. \tag{6.6}$$

Then we call $p = p(f^{k-1}, c)$ the order of f^{k-1} at the critical point c. Let q_b denote the multiplicity of the pole b. Define

$$p = \sup\{p_c : \, c \in \mathrm{Crit}(f) \text{ s.t. } f^k(c) = \infty \text{ for some } k \in \mathbb{N}\,\}$$
$$q = \sup\{q_b : \, b \in B \text{ s.t. } \exists c \in \mathrm{Crit}(f), \, \exists k \in \mathbb{N} \text{ and } f^k(c) = b\}.$$

Let $J_r(F)$ be the set of the points in $J(F)$ whose ω-limit set is not contained in B. The following results were proved in [37].

Theorem 6.6. *If $f \in \mathcal{H}$, then* $\mathrm{HD}(J_{bd}(F)) > \frac{pq}{pq+1}$.

Theorem 6.7. *If $f \in \mathcal{H}$, then* $\mathrm{HD}(J_r(f)) = \mathrm{HD}(J_r(F)) = h \in (\frac{pq}{pq+1}, 2)$.

Theorem 6.8. *If $f \in \mathcal{H}$, then the h-conformal measure m is a unique t-conformal probability measure, with $t > \frac{pq}{pq+1}$, for $F : J(F) \to J(f) \cup \{\infty\}$. In addition, m is conservative and ergodic.*

Concerning Gibbs and equilibrium states, we would like to end this chapter by bringing to the reader's attention the fact that with the same methods as those developed in [52] one can prove, as was shown in [40], all the same results for the functions in the class $\mathcal{H}$ (and appropriate potentials), as those stated in Section 3.2 for elliptic functions.

7. Transcendental entire and meromorphic functions – the Lebesgue measure viewpoint

7.1. The Lebesgue measure of the set of points escaping to ∞.
In the previous chapter we explored in detail the fractal and dynamical properties of some significant classes of transcendental entire and meromorphic functions. We have frequently supplied the reader with information concerning the Lebesgue measure of the Julia set and points escaping to infinity. In this chapter we would like to deal with much bigger classes ($\mathcal{S}$ and $\mathcal{B}$) of transcendental functions and discuss the Lebesgue measure of the corresponding Julia sets and points escaping to ∞. Recall that

$$\mathcal{S} = \{f : \mathbb{C} \to \overline{\mathbb{C}} : \text{transcendental meromorphic s.t. } \mathrm{Sing}(f^{-1}) \text{ is finite}\}$$

and

$$\mathcal{B} = \{f : \mathbb{C} \to \overline{\mathbb{C}} : \text{transcendental meromorphic s.t. } \mathrm{Sing}(f^{-1}) \text{ is bounded}\},$$

where

$$\mathrm{Sing}(f^{-1}) = \{z \in \mathbb{C} : z \text{ is a finite singularity of } f^{-1}\}.$$

Recall also that

$$P(f) = \{z \in \mathbb{C} : z \text{ is a finite singularity of } f^{-n} \text{ for some } n \geq 1\}$$

and for every $n \geq 1$ define

$$I_n(f) = \{z \in \mathbb{C} : \lim_{m \to \infty} f^{mn}(z) = \infty\}.$$

Let l_2 be the Lebesgue measure on $\mathbb{C}$. For a long time, it was expected in conformal dynamics that either $J(f) = \overline{\mathbb{C}}$ or $l_2(J(f)) = 0$. Whereas this is still an open problem in the class of all rational functions, for transcendental functions this dichotomy fails. This failure was already established in 1987 by C. McMullen who proved in [55] the following remarkable fact.

Theorem 7.1. *Let for all $a, b \in \mathbb{C}$, $f_{a,b}(z) = \sin(az + b)$. If $a \neq 0$, then $l_2(I_1(f_{a,b})) > 0$. Consequently $l_2(J(f_{a,b})) > 0$.*

We would like however to add that I. Coiculescu and B. Skorulski proved in [13] that the set of points of the Julia set not escaping to infinity under the action of $f_{a,b}, a \neq 0$ has Hausdorff dimension less than 2 (see Section 2.5). In the opposite direction A. Eremenko and M. Lyubich in [27] formulated a rather general sufficient condition for the set $I_1(f)$ to have zero Lebesgue measure. Given $r, R > 0$, let $\Theta_R(r, f)$ denote the linear measure of the set $\{\theta \in [0, 2\pi] : |f(re^{i\theta})| < R\}$, and put

$$E(f, R) = \liminf_{r \to \infty} \frac{1}{\ln r} \int_1^r \Theta_R(t, f) \frac{dt}{t}. \quad (7.1)$$

The sufficient condition of Eremenko and Lyubich is the following.

Theorem 7.2. *If $f \in \mathcal{B}$ is a transcendental entire function and $E(f, R) > 0$, then $l_2(I_1(f)) = 0$.*

The assumption of this theorem holds for all entire functions of finite order which have at least one finite logarithmic singularity (see [27]), in particular for all exponential maps $z \in \lambda \exp(z)$, $\lambda \in \mathbb{C} \setminus \{0\}$, studied in Chapter 2.

Theorem 7.2 was extended by L. Keen and J. Kotus in [36] to the class of meromorphic functions in $\mathcal{B}$, under some additional conditions on the orders of poles, their residual and principal parts. In particular, these assumptions are satisfied for all elliptic functions. However in [43] (see Section 3.1) we proved for these maps the much stronger result that $\mathrm{HD}(I_1(f)) < 2$, using different methods. An important class of meromorphic functions fulfilling the assumptions of [36] consists of the maps $\lambda \tan(z)$, $\lambda \in \mathbb{C} \setminus \{0\}$.

By other methods, results of a different kind were obtained by Bock in [11]. First, for every $n \geq 1$, he introduced the class

$$\mathcal{B}_n = \{f : \mathbb{C} \to \overline{\mathbb{C}} : \text{transcendental meromorphic s.t.}$$

$$\bigcup_{i=0}^{n-1} f^i(\mathrm{Sing}(f^{-1})) \cap \mathbb{C} \text{ is bounded}\}.$$

Notice that $\{\mathcal{B}_n\}_{n=1}^{\infty}$ is a decreasing sequence and $\mathcal{B}_1 = \mathcal{B}$. A plane set E is called thin at ∞ if its density is bounded away from 1 in all sufficiently large disks, that is, if there exist a positive R and ϵ such that, for all complex z and every disc $B(z,r)$ of centre z and radius $r > R$,

$$\text{density}(E, B(z,r))) = \frac{l_2(E \cap B(z,r))}{D(z,r)} < 1 - \epsilon.$$

Bock's first result is this.

Proposition 7.3. *Let $\omega \in \mathbb{C} \setminus \{0\}$ and let $f \in \mathcal{B}_n$ be periodic with period ω. If there exists $r > 0$ such that $f^{-n}(B(0,r))$ is thin at ∞, then $l_2(I_n(f)) = 0$.*

It easily follows from this proposition that the Lebesgue measure of the set of points escaping to infinity under any member of the exponential or tangent family is equal to zero.

Bock's next result goes in the opposite direction. For any $s \in (-\pi, \pi)$, $\alpha > 0$ and $K > 0$, let

$$W_{\alpha,K}(s) = \{z \in D(0,1) : \exists v \in \{-1, 0, 1\} \ |\arg(z) - s - 2v\pi| \leq K/(-\log|z|)^{\alpha}\}.$$

Theorem 7.4. *Let $f \in \mathcal{B}_n$, $n \geq 1$. Suppose that there exist $\alpha > 0, t_0 > 0$, $R_0 > 1, N \geq 1$, and angles $s_0, \ldots, s_{N-1} \in [-\pi, \pi)$ such that, for all $t > t_0$,*

$$B(0, R_0) \setminus \bigcup \{W_{\alpha,t}(s_v) : v \in \{0, \ldots, N-1\}\} \subset f^{-n}(B(0, e^t)).$$

Then $l_2(I_n(f)) > 0$.

This theorem along with Proposition 7.3 has rather unexpected consequences for the tangent family. Let $g_\lambda(z) = \lambda \tan(z), \lambda \in \mathbb{C} \setminus \{0\}$. For every $p \geq 1$, let

$$\mathcal{C}_p = \{\lambda \in \mathbb{C} \setminus \{0\} : g_\lambda^p(\pm \lambda i) = \infty\}.$$

It follows from Theorem 7.4 that if $\lambda \in \mathcal{C}_p$, then $l_2(I_{p+1}(g_\lambda)) > 0$, whereas we have already noted that $l_2(I_1(g_\lambda)) = 0$, by Proposition 7.3.

7.2. Milnor metric attractors. Let (M, ρ) be a compact Riemannian manifold, let X be an arbitrary subset of M and let $T : X \to M$ be a continuous map. Put $X_\infty = \bigcap_{n=0}^\infty T^{-n}(M)$. Then $T(X_\infty) \subset X_\infty$. A closed set $A \subset M$ is called a Milnor (metric) attractor of T provided that there exists a Borel set $B \subset X_\infty$ with the same Lebesgue measure as X_∞ and such that $\lim_{n\to\infty} \rho(T^n(z), A) = 0$ or equivalently $\omega(z) \subset A$ for all $z \in B$. Notice that any countable intersection of Milnor attractors is a Milnor attractor. A is called a minimal Milnor attractor if it does not contain any proper subset which is a Milnor attractor.

The structure of Milnor attractors for transcendental meromorphic functions was described fairly completely in [11]. Let us formulate it here. We treat $\overline{\mathbb{C}}$ as a compact Riemannian manifold with the spherical metric. Bock's result is this.

Theorem 7.5. *Let f be a transcendental meromorphic function. Then at least one of the following statements holds:*
  (a) *the set $\overline{P(f)} \cup \{\infty\}$ is a Milnor attractor for $f : J(f) \to \overline{\mathbb{C}}$,*
  (b) *$J(f) = \overline{\mathbb{C}}$ and $\overline{\mathbb{C}}$ is the minimal Milnor attractor for $f : J(f) \to \overline{\mathbb{C}}$. Furthermore, the map $f : \overline{\mathbb{C}} \to \overline{\mathbb{C}}$ is conservative with respect to the Lebesgue measure.*

We would like to note that an analogous result in the class of rational functions has been proved by M. Lyubich in [47]. In [42] we have provided an alternative proof of part (b) of Bock's theorem. The precise formulation of our result is the following.

Proposition 7.6. *Let $f : \mathbb{C} \to \overline{\mathbb{C}}$ be a transcendental meromorphic function such that $J(f) = \overline{\mathbb{C}}$. If $l_2(\{z : \omega(z) \subset \overline{P(f)} \cup \{\infty\}\}) = 0$, then f is ergodic and conservative with respect to the Lebesgue measure.*

As an immediate consequence of Theorem 7.5, we get the following result noted in Bock's paper [11].

Corollary 7.7. *If the set $\mathrm{Sing}(f^{-1})$ is finite and each singularity of f^{-1} is preperiodic but not periodic then part (b) holds for f. In particular, $\overline{\mathbb{C}}$ is the minimal Milnor attractor for f.*

For an entire transcendental function f it is impossible to satisfy the assumption of this criterion because in this case ∞ is a transcendental singularity which does not belong to the domain of f and therefore cannot be a preperiodic point of f. However Bock proved in [10] the following.

Corollary 7.8. *Let f be non-constant entire function. Suppose that $\mathrm{Sing}(f^{-1})$ is finite and each singularity of f^{-1} is preperiodic but not periodic. Then either ∞ or $\overline{\mathbb{C}}$ is a minimal Milnor attractor for $f : \mathbb{C} \to \mathbb{C}$. In the latter case it even holds that $\omega(z) = \mathbb{C}$ for Lebesgue a.e. $z \in \mathbb{C}$.*

The first example of a transcendental function for which $\overline{P(f)}$ is a metric attractor is due to M. Rees who proved in [61] that if $E(z) = e^z$, then $\{E^n(0)\}_{n=0}^{\infty} \cup \{\infty\}$ is a Milnor attractor for f. She also proved that the exponential map is not recurrent, that is, there exists a Borel set $B \subset \overline{\mathbb{C}}$ with positive Lebesgue measure such that $B \cap f^n(B) = \emptyset$ for all $n \geq 1$. M. Lyubich has clarified the situation completely by proving in [48] the following (see also Section 2.1).

Theorem 7.9. $\{E^n(0)\}_{n=0}^{\infty} \cup \{\infty\}$ *is the minimal Milnor attractor for the exponential function* $E(z) = e^z$. *In addition, E is not ergodic with respect to the Lebesgue measure.*

J. Hemke has given several sufficient conditions for the alternative (a) in Theorem 7.5 to hold (see [31] and also [32]). We now present some of his results. The most general of Hemke's sufficient conditions is the following.

Theorem 7.10. *Let $f : \mathbb{C} \to \overline{\mathbb{C}}$ be a meromorphic function, $A \subset \mathbb{C}$ finite and $G \subset \mathbb{C}$, such that*

(a) *there exists $\epsilon > 0$, such that the map*

$$\overline{s} : G \to A \cup \{0\} : z \mapsto \begin{cases} s & \text{if } \exists s \in A : |f(z) - s| \leq \exp(|z|^\epsilon) \\ 0 & \text{if } |f(z)| \geq \exp(|z|^\epsilon) \end{cases}$$

is well-defined and there exist $\delta_1, \delta_2 \in \mathbb{R}$ such that, for all $z \in G$,

$$|z|^{\delta_1} \leq \left|\frac{f'(z)}{f(z) - \overline{s}(z)}\right| \leq |z|^{\delta_2};$$

(b) *there exist $B > 1$ and $\beta \in (-\infty, 1)$ such that, for every measurable set $D \subset \{z : \text{dist}(z, \mathbb{C} \setminus G) \leq 2|z|^{-\delta_1}\}$, we have*

$$l_2(D) \leq B \operatorname{diam}(D) \sup_{z \in D} |z|^\beta;$$

(c) $\lim_{m \to \infty} f^m(s) = \infty$ *and* $B(f^m(s), 2|f^m(s)|^\tau) \subset G$ *for some $\tau > \beta$, almost all $m \in \mathbb{N}$ and all $s \in A$.*

Then the set $T(f) = \{z : \omega(z) \subset \overline{\Theta^+(A)}\}$ has positive measure. In particular, if $A = P(f)$, then $\overline{P(f)}$ is a Milnor attractor for f.

A point $z \in \mathbb{C}$ is said to escape exponentially if $\lim_{n \to \infty} f^n(z) = \infty$ and

$$|f^n(z)| \geq \exp(|f^{n-1}(z)|^\delta)$$

for some $\delta > 0$ and all integers $n \geq 1$. Theorem 7.10 was applied in [31] to the class of function of the form

$$f(z) = \int_0^z P(t) \exp(Q(t))dt + c \qquad (7.2)$$

with polynomials P and Q and $c \in \mathbb{C}$, such that Q is non-constant and P not identically zero. These functions have at most $\deg(Q)$ finite asymptotic values and $\deg(P)$ critical points.

Theorem 7.11. *Let f be a meromorphic function of the form (7.2). Suppose that all the finite asymptotic values escape exponentially. Then the Julia set $J(f)$ has positive Lebesgue measure and $\overline{P(f)}$ is a Milnor attractor for $f : J(f) \to J(f) \cup \{\infty\}$. In addition, if $\deg(Q) \geq 3$, then $l_2(F(f)) < \infty$.*

The function $f(z) = \exp(z^3 + az + b)$, where $a = (27\pi^2/16)^{1/3}$ and $b = \log \sqrt{a/3}$, was proved in [31] to satisfy all the assumptions of Theorem 7.10. Under an additional assumption, Hemke was able to identify the Milnor minimal attractor by proving the following.

Theorem 7.12. *Let f satisfy the assumptions of Theorem 7.11. Suppose also that every critical point either escapes exponentially, is preperiodic or is contained in an attracting Fatou-component. Then $\overline{\Theta^+(A)}$ is the minimal Milnor attractor for f, where A denotes the set of finite asymptotic values of f.*

All the functions defined by (7.2) are entire and have a rational Schwarzian derivative. The asymptotic behaviour of such functions is understood fairly well. Theorem 7.11 continues to be true for all such functions satisfying some natural additional assumptions (see [31]).

It is well-known and easy to see that if f is a transcendental meromorphic function and $\bigcup_{n \geq 0} f^{-n}(\infty)$ contains at least three distinct points, then

$$J(f) = \overline{\bigcup_{n \geq 0} f^{-n}(\infty)}.$$

It may in particular happen that if all the singular values of f^{-1} belong to $J(f)$, then after finitely many iterates all these singular values land on poles. This situation has been thoroughly studied for functions of the form

$$f(z) = \frac{a \exp(z^p) + b \exp(-z^p)}{c \exp(z^p) + d \exp(-z^p)},$$

where $p \in \mathbb{N}$ and $ad - bc \neq 0$, mentioned in Section 4.3, which form the class we call $\mathcal{F}$. Given $z \in \mathbb{C}$ we denote the trajectory of z by

$$\Theta^+(z) = \{f^n(z)\}_{n=1}^{\infty}.$$

We say that $f \in \mathcal{F}$ satisfies condition (C1) if the following statements are true.

(i) The asymptotic values $\frac{a}{c}$ and $\frac{b}{d}$ are finite and eventually mapped onto ∞, that is, there exist $q_1, q_2 \in \mathbb{N}$ such that $f^{q_1-1}(\frac{a}{c}) = f^{q_2-1}(\frac{b}{d}) = \infty$.

(ii) If 0 is a critical point, then either the trajectory of 0 is bounded or 0 is eventually mapped onto ∞ and

$$\Theta^+(0) \cap \left(\Theta^+\left(\frac{a}{c}\right) \cup \Theta^+\left(\frac{b}{d}\right)\right) = \{\infty\}.$$

We say that a function $f \in \mathcal{F}$ satisfies condition (C2) if condition (C1) holds and if 0 is a critical point and it is not eventually mapped onto infinity, then

$$l_2(\{z \in J(f) : \omega(z) \subset \overline{\Theta^+(0)}\}) = 0.$$

B. Skorulski proved in [63] and [64] the following.

Theorem 7.13. *If f satisfies condition (C1), then*
  (i) *there exists a set $E \subset J(f)$ with positive Lebesgue measure such that $\omega(z) = P_{Asymp(f)} = \Theta^+(\frac{a}{c}) \cup \Theta^+(\frac{b}{d})$ for all $z \in E$, and the action of f is not ergodic on $J(f)$ with respect to the Lebesgue measure;*
  (ii) *if $q_1 \neq q_2$, then $l_2(I_n(f)) = 0$ for every $n \geq 1$.*
Moreover, if condition (C2) is satisfied, then in particular $P_{Asymp(f)}$ is the Milnor attractor of f.

Corollary 7.14. *Let $p > 1, k \in \mathbb{Z}$ and*

$$f(z) = \sqrt[p]{\pi i/2 + k\pi i} \, \frac{\exp(z^p) - \exp(-z^p)}{\exp(z^p) + \exp(-z^p)}.$$

Then the Fatou set $F(f)$ is nonempty since 0 is a superattracting fixed point. The asymptotic values $\xi_1 = \sqrt[p]{\pi i/2 + k\pi i}$, $\xi_2 = -\sqrt[p]{\pi i/2 + k\pi i}$ are mapped by f onto ∞. The Julia set $J(f)$ has positive measure and for almost all $z \in J(f)$, $\omega(z) = P_{Asymp(f)}$. In particular, $P_{Asymp(f)}$ is the Milnor attractor of f. Moreover, f is not ergodic with respect to the Lebesgue measure on $J(f)$.

Theorem 7.13 has been recently extended by Skorulski [65] to the large class $\mathcal{R}$ treated in detail in Section 6.1. Note that each function $f \in \mathcal{P}$ ($f \in \mathcal{R}$ and f entire) can be represented in the following form

$$f(z) = \sum_{j=-n_2}^{n_1} a_j e^{jz}.$$

So, if $n_1, n_2 > 0$, then f has no finite asymptotic values. Denote the class of those functions by $\mathcal{P}_2$ and its complement (in $\mathcal{P}$) by $\mathcal{P}_1$. Put also

$$\mathcal{R}_1 = \mathcal{Q}_1 \cup \mathcal{P}_1 \quad \text{and} \quad \mathcal{R}_2 = \mathcal{Q}_2 \cup \mathcal{P}_2.$$

For every $f \in \mathcal{R}$, put

$$P_{Asymp}(f) = \Theta^+(\{R(0), R(\infty)\}).$$

Combining results of Skorulski from [64] and [65], we obtain the following.

Theorem 7.15. *If $f \in \mathcal{R}_2$, then $P_{\text{Asymp}}(f)$ is a Milnor attractor for f. In particular, the Lebesgue measure of $J(f)$ is positive.*

Let
$$\mathcal{R}_2^* = \{f \in \mathcal{R}_2 : \text{dist}_\chi(P_2(f), J(f)) > 0\}.$$
Skorulski also proved the following in [65].

Theorem 7.16. (1) *If $f \in \mathcal{R}_2^*$, then $P_{\text{Asymp}}(f)$ is the minimal Milnor attractor for f.*
(2) *If $q_1 \neq q_2$, then the Lebesgue measure of the set*
$$I_n(f) = \{z : \lim_{k \to \infty} f^{nk}(z) = \infty\}$$
is equal to zero for all $n \in \mathbb{N}$.
(3) *If $f \in \mathcal{R}_2^* \cap \mathcal{Q}$, then there does not exist any f-invariant measure on $J(f)$ which is absolutely continuous with the respect to the Lebesgue measure and finite on all compact subsets of $J(f)$.*
(4) *If $f \in \mathcal{R}_2^*$ and $n_1 = n_2$, then f is not ergodic on $J(f)$ with the respect to the Lebesgue measure.*

7.3. The Lebesgue measure of Julia sets. It is by now a standard fact (see [70] for the first proof) that the Julia set of an expanding rational function has Lebesgue measure zero. As Theorem 7.1 shows, with $a, b \neq 0$ sufficiently small in moduli, this property already fails in the class of expanding sine maps. Let us, however, see what can be said if we assume in addition that the Julia set is thin at ∞. In order to avoid any confusion let us introduce the following two classes of functions.
$$\mathcal{E} = \{f : \mathbb{C} \to \mathbb{C} : f \text{ is entire, dist}(\overline{P(f)}, J(f)) > 0\}$$
and
$$\mathcal{E}_0 = \{f : \mathbb{C} \to \mathbb{C} : f \text{ is entire, } \overline{P(f)} \text{ is compact, } \overline{P(f)} \cap J(f) = \emptyset\}.$$
Obviously $\mathcal{E}_0 \subset \mathcal{E}$. The following theorem was first proved by C. McMullen in [55] for the class $\mathcal{E}_0$ and by G. Stallard in [66] for the class $\mathcal{E}$.

Theorem 7.17. *If $f \in \mathcal{E}$ and B is a measurable completely invariant subset of $J(f)$ such that B is thin at ∞, then $l_2(B) = 0$. In particular if $J(f)$ is thin at ∞, then $l_2(J(f)) = 0$.*

Going beyond hyperbolicity but still keeping a rather general setting, we formulate the following two remarkable results, the first one proved by A. Eremenko and M. Lyubich in [27] and the second proved by B. Skorulski in [63]

Theorem 7.18. *Suppose that $f \in \mathcal{S}$ is a transcendental entire function such that $E(f, R) > 0$ for all $R > 0$. Assume that the orbit of every finite singularity of f^{-1} is either absorbed by a repelling cycle or converges to an attracting or to a neutral rational cycle. Then either $J(f) = \overline{\mathbb{C}}$ or $l_2(J(f)) = 0$.*

Note that the set $E(f, R)$ was defined in (7.1).

Theorem 7.19. *Let $f \in \mathcal{F}$. If one of the two asymptotic values of f is mapped onto ∞, while the second asymptotic value and the critical point of f are in the Fatou set, then $l_2(J(f)) = 0$.*

8. APPENDIX 1: $K(V)$ METHOD OF CONSTRUCTING SEMI-CONFORMAL MEASURES

Given a continuous map $T : X \to X$ from a topological space X into itself, the map T is said to be non-open at the point $x \in X$ if and only if for every open neighbourhood V of x there exists an open set $U \subset V$ such that $T(U)$ is not open. The set of all points in X at which T is non-open, is denoted by $NO(T)$. A point $c \in X$ is said to be a critical point of T if there is no open neighbourhood W of c such that the map $T_{|W}$ is one-to-one. The set of all critical points of T is denoted by $\text{Crit}(T)$.

Now let X be a compact subset of the extended complex plane $\overline{\mathbb{C}}$. We say that $f \in \mathcal{A}(X)$ provided that $f : X \to X$ is a continuous map which can be meromorphically extended to a neighbourhood $U(f) = U(f, X)$ of X in $\overline{\mathbb{C}}$. Denote by $M_e^+(f)$ the set of all Borel probability ergodic f-invariant measures on X with positive entropy and, for any Borel measure μ on X, denote by $\text{HD}(\mu)$ the Hausdorff dimension of the measure μ. Finally, define

$$\text{DD}(X) = \sup\{\text{HD}(\mu) : \mu \in M_e^+(f)\}.$$

Obviously $\text{DD}(X) \leq \text{HD}(X)$. Proceeding as in [18] (comp. [59, Chapter 10] for a more mature exposition) with the set $K(V)$ replaced by X, one can prove the following two useful auxiliary results.

Lemma 8.1. *Suppose that X is a compact subset of $\overline{\mathbb{C}}$, $f \in \mathcal{A}(X)$ and $f : X \to X$ has no critical points. Then for all $t \geq 0$ there exist $\text{P}(t) \in \mathbb{R}$ and a Borel probability measure m_t on X with the following two properties:*

$$m_t(f(A)) \geq \int_A e^{\text{P}(t)} |f'|^t m_t,$$

and, if in addition $A \subset U(f) \setminus NO(f)$, then

$$m_t(f(A)) = \int_A e^{\text{P}(t)} |f'|^t m_t.$$

Lemma 8.2. *Suppose that X is a compact subset of $\overline{\mathbb{C}}$, $f \in \mathcal{A}(X)$ and $f : X \to X$ has no critical points. Then there exists $s(X) \in [0, DD(X)]$ and a Borel probability measure m called $s(X)$-semi-conformal on X with the following two properties:*

$$(a) \qquad m(f(A)) \geq \int_A |f'|^{s(X)} dm$$

for any Borel set $A \subset U(f)$ such that $f_{|A}$ is one-to-one

$$(b) \qquad m(f(A)) = \int_A |f'|^{s(X)} dm$$

for any Borel set $A \subset U(f) \setminus NO(f)$ such that $f_{|A}$ is one-to-one.

Given two compact sets $X \subset Y \subset \mathbb{C}$ and a function $f \in \mathcal{A}(Y)$, we say that X is a branchwise contained in Y provided that the following condition is satisfied. There is $\delta > 0$ such that for every $x \in X$ and for every $n \geq 0$ there exists a holomorphic inverse branch $f_x^{-n} : B(f^n(x), \delta) \to \mathbb{C}$ of f^n sending $f^n(x)$ to x and such that $f^j(f_x^{-n}(B(f^n(x), \delta)) \subset U(f, Y) \setminus NO(f)$ for all $j = 0, 1, \ldots, n$.

Proceeding as in the proof of Lemma 3.2 in [79], with the same obvious modifications, one gets the following result.

Lemma 8.3. *Suppose that Y is a compact subset of $\mathbb{C}$ and $f \in \mathcal{A}(Y)$. If f has no critical points in Y and X is a compact set branchwise contained in Y (notice that we do not assume X to be a forward invariant under f), then $\mathrm{HD}(X) \leq s(Y)$.*

These two lemmas are most frequently applied in the context when $f : S \to \overline{S}$ is a holomorphic map of a Riemann surface S (usually $\mathbb{C}$, $\overline{\mathbb{C}}$ or $\overline{\mathbb{C}} \setminus \{0\}$) into a Riemann surface $\overline{S}$ such that $\bigcup_{n \geq 0} f^{-n}(V) \supset (\overline{S} \setminus S) \cup \mathrm{Crit}(f)$. Next, one defines the set

$$K(V) = \bigcap_{n=0}^{\infty} f^{-n}(\overline{S} \setminus V).$$

Then $f(K(V)) \subset K_V$ and if $\overline{S} \setminus V$ is compact, then so is the set $K(V)$. One can apply Lemma 8.1 and Lemma 8.2 with $X = K(V)$ and $U(f|_{K(V)}) = \overline{S} \setminus V$. Notice that $NO(f|_{K(V)}) \subset \partial V$. In order to get a conformal measure one lets V decrease to a set which is usually finite and one takes an arbitrary weak limit measure of semi-conformal measures produced in Lemma 8.1 or Lemma 8.2. It requires a separate proof (sometimes easy, sometimes difficult) to show that such a limit measure is conformal for $f : S \to \overline{S}$. Finally, one may control the exponent with the help of Lemma 8.2 and Lemma 8.3.

However, the latter lemma is not always applicable and then one must undertake another approach in order to determine the exponent of the conformal measure produced as that weak limit.

9. Appendix 2: Martens' method of constructing σ-finite invariant measures

Suppose that X is a σ-compact metric space, ν is a Borel probability measure on X, positive on open sets, and that a measurable map $f : X \to X$ is given with respect to which the measure ν is quasi-invariant, that is, $\nu \circ f^{-1} \ll \nu$. Moreover, we assume the existence of a countable partition $\alpha = \{A_n : n \geq 0\}$ of subsets of X which are all σ-compact and of positive measure ν. We also assume that $\nu(X \setminus \bigcup_{n \geq 0} A_n) = 0$, and if additionally for all $m, n \geq 1$ there exists $k \geq 0$ such that

$$\nu(f^{-k}(A_m) \cap A_n) > 0,$$

then the partition α is called irreducible. Martens' result, comprising Proposition 2.6 and Theorem 2.9 of [50], reads as follows.

Theorem 9.1. *Suppose that $\alpha = \{A_n : n \geq 0\}$ is an irreducible partition for $T : X \to X$. Suppose that T is conservative and ergodic with respect to the measure ν. If for every $n \geq 1$ there exists $K_n \geq 1$ such that, for all $k \geq 0$ and all Borel subsets A of A_n,*

$$K_n^{-1} \frac{\nu(A)}{\nu(A_n)} \leq \frac{\nu(f^{-k}(A))}{\nu(f^{-k}(A_n))} \leq K_n \frac{\nu(A)}{\nu(A_n)},$$

then T has a σ-finite T-invariant measure μ that is absolutely continuous with respect to ν. In addition, μ is equivalent to ν, conservative and ergodic, and unique up to a multiplicative constant. Moreover, for every Borel set $A \subset X$,

$$\mu(A) = \lim_{n \to \infty} \frac{\sum_{k=0}^{n} \nu(f^{-k}(A))}{\sum_{k=0}^{n} \nu(f^{-k}(A_0))}.$$

This theorem is widely used in conformal dynamics in the context where ν is a conformal measure. The distortion assumption, the higher displayed formula above, is usually derived from Koebe's distortion theorem.

REFERENCES

[1] J.M. Aarts, L.G. Oversteegen, The geometry of Julia sets, Trans. A.M.S. 338(2) (1993), 897–918.
[2] I.N. Baker, J. Kotus, Y. Lü, Iterates of meromorphic functions I, Erg. Th. and Dynam. Sys. 11 (1991), 241–248.
[3] I.N. Baker, J. Kotus, Y. Lü, Iterates of meromorphic functions II: Examples of wandering domains, J. London Math. Soc. 42 (1990), 267–278.
[4] I.N. Baker, J. Kotus, Y. Lü, Iterates of meromorphic functions III: Preperiodic domains, Erg. Th. and Dynam. Sys. 11 (1991), 603–618.
[5] I.N. Baker, J. Kotus, Y. Lü, Iterates of meromorphic functions IV: Critically finite functions, Results in Math. 22 (1992), 651–656.
[6] I.N. Baker, P.J. Rippon, Iteration of exponential functions, Ann. Acad. Sci. Fenn. Ser. A I Math. 9 (1984), 49–77.
[7] K. Barański, Hausdorff dimension and measures on Julia sets of some meromorphic functions, Fund. Math. 147 (1995), 239–260.
[8] W. Bergweiler. Iteration of meromorphic functions, Bull. A.M.S. 29:2 (1993), 151–188.
[9] W. Bergweiler, Comments and corrections to *Iteration of meromorphic functions*, see http://analysis.math.uni-kiel.de/bergweiler/bergweiler.engl.html
[10] H. Bock, On the dynamics of entire functions on the Julia set, Results in Math. 30 (1996), 16–20.
[11] H. Bock, Über das Iterationsverhalten meromorpher Funktionen auf der Juliamenge, PhD Thesis, Aachen, 1998.
[12] R. Bowen, Equilibrium states and the ergodic theory for Anosov diffeomorphisms. Lect. Notes in Math. 470, Springer, 1975.
[13] I. Coiculescu, B. Skorulski, Thermodynamic formalism of transcendental entire maps of finite singular type, to appear in Monatsh. Math.
[14] I. Coiculescu, B. Skorulski, Perturbations in the Speiser class, to appear in Rocky Mountain J. Math.
[15] A. M. Davie, M. Urbański, A. Zdunik, Maximizing measures on metrizable noncompact spaces, Proc. Edinb. Math. Soc. 50 (2007), 123–151.
[16] M. Denker, F. Przytycki, M. Urbański, On the transfer operator for rational functions on the Riemann sphere, Ergod. Th. and Dynam. Sys. 16 (1996), 255–266.
[17] M. Denker, M. Urbański, On the existence of conformal measures, Trans. A.M.S. 328 (1991), 563–587.
[18] M. Denker, M. Urbański, On Sullivan's conformal measures for rational maps of the Riemann sphere, Nonlinearity 4 (1991), 365–384.
[19] M. Denker, M. Urbański, Ergodic theory of equilibrium states for rational maps, Nonlinearity 4 (1991), 103–134.
[20] R.L. Devaney, Julia sets and bifurcation diagram for exponential maps, Bull. A.M.S. 11 (184), 167–181.
[21] R.L. Devaney, Structural instability of exp(z), Proc. A.M.S. 94 (1985), 545–548.
[22] R.L. Devaney, Dynamics of entire maps, Dynamical Systems and Ergodic Theory, Banach Center Publ. vol. 23, Polish Scientific Publishers, Warsaw, 1989, 221–228.
[23] R.L. Devaney, e^z: dynamics and bifurcations, Intern. J. Bifurcation Chaos, 1 (1991), 287–308.
[24] R.L. Devaney, M. Krych, Dynamics of exp(z), Erg. Th. and Dynam. Sys. 4 (1984), 35–52.
[25] R.L. Devaney, F. Tangerman, Dynamics of entire functions near the essential singularity, Erg. Th. and Dynam. Sys. 6 (1986), 498–503.

[26] A.E. Eremenko, On the iteration of entire functions, Dynamical Systems and Ergodic Theory, Banach Center Publ. vol. 23, Polish Scientific Publishers, Warsaw, 1989, 339–345.

[27] A.E. Eremenko, M.Yu. Lyubich, Dynamical properties of some class of entire functions, Ann. Inst. Fourier (Grenoble) 42 (1992), 989–1020.

[28] P. Fatou, Sur l'itération des fonctions transcendantes entèries, Acta Math. 47 (1926), 337–370.

[29] J. Hawkins and L. Koss, Ergodic properties and Julia sets of Weierstrass elliptic functions, Monat. Math. 137, (2002), 273–301.

[30] J. Hawkins, L. Koss, Connectivity properties of Julia sets of Weierstrass elliptic functions, Topology Appl. 152, no. 1-2 (2005), 107–137.

[31] J.M. Hemke, Measurable dynamics of meromorphic maps, PhD Thesis, Kiel, 2005.

[32] J.M. Hemke, Recurrence of entire transcendental functions with simple post-singular sets, Fund. Math. 187, no. 3 (2005), 255–289.

[33] B. Karpińska, Area and the Hausdorff dimension of the set of accessible points of the Julia sets of $\lambda \exp z$ and $\lambda \sin z$, Fund. Math. 159 (1999), 269–287.

[34] B. Karpińska, The Hausdorff dimension of the hairs without endpoints for $\lambda \exp z$, C. R. Acad. Sci. Paris, Série I. t. 328 (1999), 1039–1044.

[35] B. Karpińska, M. Urbański, How points escape to infinity under exponential maps, J. London Math. Soc. 73 (2006), 141–156.

[36] L. Keen, J. Kotus, Ergodicity of some classes of meromorphic functions, Ann. Acad. Sci. Fenn. 24 (1999), 133–145.

[37] J. Kotus, Conformal measures for non-entire functions with critical values eventually mapped onto infinity, Analysis 25 (2005), 333-350.

[38] J. Kotus, Probabilistic invariant measures for non-entire functions with asymptotic values mapped onto ∞, Illinois J. Math. 49 (2005) 1203–1220.

[39] J. Kotus, Elliptic functions with critical points eventually mapped onto infinity, Monatsh. Math. 149 (2006), 103–117.

[40] J. Kotus, Gibbs state for some class of meromorphic functions, Indag. Math., N.S. 17 no. 4 (2006), 583–597.

[41] J. Kotus and M. Urbański, Conformal, geometric and invariant measures for transcendental expanding functions, Math. Ann. 324 (2002), 619–656.

[42] J. Kotus, M. Urbański, Existence of invariant measures for transcendental subexpanding functions, Math. Zeit. 243 (2003), 25–36.

[43] J. Kotus, M. Urbański, Hausdorff dimension and Hausdorff measures of Julia sets of elliptic functions, Bull. London Math. Soc. 35 (2003), 269–275.

[44] J. Kotus, M. Urbański, Geometry and ergodic theory of non-recurrent elliptic functions, Journal d'Analyse Math. 93 (2004), 35–102.

[45] J. Kotus and M. Urbański, The dynamics and geometry of the Fatou functions, Discrete & Continuous Dyn. Sys. 13 (2005), 291–338.

[46] J. Kotus, M. Urbański, Geometry and dynamics of some meromorphic functions, Math. Nachr. 279 no. 13-14 (2006), 1565–1584.

[47] M.Yu. Lyubich, The dynamics of rational transforms: the topological picture, Russian Math. Surveys 41 (1986) 43-117; translation from Uspekhi Math. Nauk 41 (1986), 35–95.

[48] M.Yu. Lyubich, Measurable dynamics of the exponential, Sib. Math. J. 28 (1988), 780-793; translation from Sib. Math. Zh. 28 (1987), 11–127.

[49] R. Mañé, On a theorem of Fatou, Bol. Soc. Bras. Mat. 24 (1993), 1–11.

[50] M. Martens, The existence of σ-finite invariant measures, Applications to real one-dimensional dynamics, Front for the Mathematics ArXiv, http://front.math.ucdavis.edu/; math.DS/9201300.

[51] V. Mayer, M. Urbański, Exponential elliptics give dimension two, Illinois J. Math. 49 (2005), 291–294.
[52] V. Mayer, M. Urbański, Gibbs and equilibrium measures for elliptic functions, Math. Zeit. 250 (2005), 657–683.
[53] V. Mayer, M. Urbański, Geometric thermodynamical formalism and real analyticity for meromorphic functions of finite order, Preprint 2005 (available on Urbanski's webpage).
[54] R.L. Mauldin, M. Urbański, Dimensions and measures in infinite iterated function systems. Proc. London Math. Soc. 73:3 (1996), 105–154.
[55] C. McMullen, Area and Hausdorff dimension of Julia set of entire functions, Trans. A.M.S. 300 (1987), 329–342.
[56] M. Misiurewicz, On the iterates of e^z, Ergod. Th. and Dynam. Sys. 1 (1981), 103–106.
[57] S. Morosawa, Y. Nishimura, M. Taniguchi, T. Ueda, Holomorphic Dynamics, Cambridge University Press, 2000.
[58] S. J. Patterson, The limit set of a Fuchsian group, Acta Math. 136 (1976), 241–273.
[59] F. Przytycki, M. Urbański, Fractals in the plane - Ergodic Theory Methods, To appear (available on Urbanski's webpage).
[60] M. Reed, B. Simon, Methods of Modern Mathematical Physics, IV: Analysis of Operators, Academic Press 1978.
[61] M. Rees, The exponential map is non-recurrent, Math. Zeit. 191 (1986), 593–598.
[62] B. Skorulski, Non-ergodic maps in the tangent family, Indag. Math. N.S. 14 no. 1 (2003), 103–118.
[63] B. Skorulski, Metric properties of the Julia set of some meromorphic functions with an asymptotic value eventually mapped onto a pole, Math. Proc. Cambridge Phil. Soc. 138 (2005), 117–138.
[64] B. Skorulski, Metric properties of the Julia set of meromorphic functions, PhD Thesis, Warsaw, 2004.
[65] B. Skorulski, The existence of conformal measures for some transcendental meromorphic functions, Contemporary Mathematics 396 (2006), 169–201.
[66] G.M. Stallard, Entire functions with Julia sets of zero measure, Math. Proc. Cambridge Philos. Soc. 108 (1990), 551–557.
[67] G.M. Stallard, Dimensions of Julia sets of transcendental meromorphic functions, Transcendental dynamics and complex analysis, Cambridge University Press, 2008.
[68] D. Sullivan, *Seminar on conformal and hyperbolic geometry*, Preprint IHES (1982).
[69] D. Sullivan, Entropy, Hausdorff measures old and new, and the limit set of a geometrically finite Kleinian groups, Acta. Math. 153 (1984), 259–277.
[70] D. Sullivan, Conformal dynamical systems, Lect. Notes in Math. vol. 1007 (Springer 1983), 725–752.
[71] M. Urbański, Rational functions with no recurrent critical points, Ergod. Th. and Dynam. Sys. 14 (1994), 391–414.
[72] M. Urbański, Geometry and ergodic theory of conformal nonrecurrent dynamics, Ergod. Th. and Dynam. Sys. 17 (1997), 1449–1476.
[73] M. Urbański, Recurrence rates for loosely Markov dynamical systems, J. Australian Math. Soc. 82 (2007), 39–57.
[74] M. Urbański, Thermodynamic formalism and multifractal analysis of finer Julia sets of exponential family, Preprint 2005 (available on Urbanski's webpage).
[75] M. Urbański, Measures and dimensions in conformal dynamics, Bull. A.M.S. 40 (2003), 281–321.
[76] M. Urbański, A. Zdunik, The finer geometry and dynamics of exponential family, Michigan Math. J. 51 (2003), 227–250.

[77] M. Urbański, A. Zdunik, Real analyticity of Hausdorff dimension of finer Julia sets of exponential family, Ergod. Th. and Dynam. Sys. 24 (2004), 279–315.
[78] M. Urbański, A. Zdunik, The parabolic map $f_{1/e}(z) = \frac{1}{e}e^z$, Indag. Math. 15 (2004), 419–433.
[79] M. Urbański, A. Zdunik, Geometry and ergodic theory of non-hyperbolic exponential maps, Trans. A.M.S. 359 (2007), 3973–3997.
[80] M. Urbański, M. Zinsmeister, Parabolic implosion and Julia-Lavaurs sets in the exponential family, Monatsh. Math. 149 (2006), 129–140.
[81] P. Walters, An introduction to ergodic theory, Springer-Verlag, 1982.

JANINA KOTUS, FACULTY OF MATHEMATICS AND INFORMATION SCIENCES
WARSAW UNIVERSITY OF TECHNOLOGY
WARSAW 00-661, POLAND
E-MAIL: JANINAK@IMPAN.GOV.PL

MARIUSZ URBAŃSKI, DEPARTMENT OF MATHEMATICS, UNIVERSITY OF NORTH TEXAS,
P.O. BOX 311430, DENTON, TX 76203-1430, USA
E-MAIL:URBANSKI@UNT.EDU
WEB: HTTP://WWW.MATH.UNT.EDU/~URBANSKI

COMBINATORICS OF BIFURCATIONS IN EXPONENTIAL PARAMETER SPACE

LASSE REMPE AND DIERK SCHLEICHER

In memory of Professor I. N. Baker

ABSTRACT. We give a complete combinatorial description of the bifurcation structure in the space of exponential maps $z \mapsto \exp(z) + \kappa$. This combinatorial structure is the basis for a number of important results about exponential parameter space. These include the fact that every hyperbolic component has connected boundary [26, 29], a classification of escaping parameters [14], and the fact that all dynamic and parameter rays at periodic addresses land [23, 27].

CONTENTS

1. Introduction	317
2. Combinatorics of Exponential Maps	323
3. Orbit Portraits and Itineraries	326
4. Hyperbolic components	334
5. Local Bifurcation Results	338
6. Bifurcation from a Hyperbolic Component	342
7. Internal Addresses	349
Appendix A. Further Topics	358
Appendix B. Combinatorial Algorithms	363
List of Symbols	367
Index	368
References	369

1. INTRODUCTION

Ever since Douady and Hubbard's celebrated study of the Mandelbrot set [8], combinatorics has played a fundamental role for the dynamics of complex polynomials. In particular, the concept of *external rays*, both in the dynamical and parameter plane, and the landing behavior of such rays, has

---

2000 *Mathematics Subject Classification.* Primary 37F10; Secondary 30D05.

The first author was supported in part by a postdoctoral fellowship of the German Academic Exchange Service (DAAD) and by the German-Israeli Foundation for Scientific Research and Development (G.I.F.), grant no. G-643-117.6/1999.

helped in the understanding of polynomial Julia sets and bifurcation loci. This program has been particularly successful for the simplest polynomial parameter spaces: the quadratic family $z \mapsto z^2 + c$ and its higher-dimensional cousins, the unicritical families $z \mapsto z^d + c$ [8, 30, 9].

In this article, we consider the space of exponential maps,

$$E_\kappa : \mathbb{C} \to \mathbb{C}; z \mapsto \exp(z) + \kappa.$$

It is well-known that a restriction on the number of *singular values* (i.e., critical and asymptotic values) of an entire function generally limits the amount of different dynamical features that can appear for the same map. Since exponential maps are the only transcendental entire functions which have just one singular value, namely the omitted value κ (see e.g. [18, Appendix D]), exponential maps form the simplest parameter space of entire transcendental functions. In addition, the exponential family can be considered as the limit of the polynomial unicritical families, and thus is an excellent candidate to apply the combinatorial methods which were so successful for Mandel- and Multibrot sets.

For these reasons, the exponential family has been studied extensively over the past twenty-five years. Among the first to do so were Devaney [5], Baker and Rippon [1] and Eremenko and Lyubich [11]. The idea to interpret exponential maps as limit dynamics of unicritical polynomials, and to understand them using the successful combinatorial tools from the polynomial case, appeared in [7]. Recently, some progress has been made in this direction: a complete classification of escaping points for exponential maps in terms of *dynamic rays* was given in [31], and a similar construction was carried out to obtain *parameter rays* [13, 15]. Also, exponential maps with attracting periodic cycles were classified in [28] using combinatorics.

Nonetheless, a basic description of exponential dynamics in analogy to the initial study of the Mandelbrot set should involve at least the following results. (See the end of this section for basic definitions.)

(a) For every hyperbolic component W, there is a homeomorphism of pairs $(W, \overline{W}) \to (\mathbb{H}, \overline{\mathbb{H}})$, where $\mathbb{H}$ is the left half plane. (In particular, ∂W is a Jordan curve.)
(b) Every periodic parameter ray lands at a parabolic parameter.
(c) If the singular value of E_κ does not escape to ∞, then all periodic dynamic rays of E_κ land.
(d) If the singular value of E_κ does not escape to ∞, then every repelling periodic point of E_κ is the landing point of a periodic dynamic ray.

For unicritical polynomials, the analogs of these statements all have relatively short analytic proofs (see e.g. [21] for (b) and [20, Theorems 18.10 and 18.11] for (c) and (d)), but these break down in the exponential case. Nonetheless, it is possible to prove items (a) through (c), using a novel approach based on a thorough study of parameter space. One of the goals of

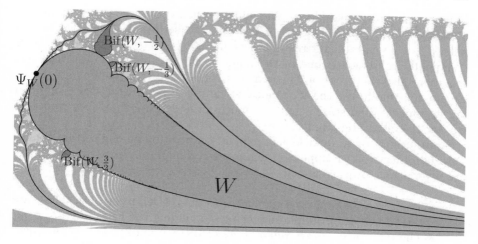

FIGURE 1. Structure of child components bifurcating from a period 3 hyperbolic component in exponential parameter space

this article is to provide the first ingredients in this approach by obtaining a complete description of the combinatorial structure of parameter space (as given by bifurcations of hyperbolic components). In the sequel [26], this description is used to prove (a), which, in turn, leads to proofs of (b) [27] and (c) [23], as well as some progress on (d) [23].

To illustrate the difficulties we face, let us consider the structure of hyperbolic components bifurcating from a given hyperbolic component. If we already knew results (a) and (b) above, it would be quite easy to obtain the following description; compare Figure 1.

Let W be a hyperbolic component of period $n \geq 2$, and let $\mu : W \to \mathbb{D}^*$ be the multiplier map (which maps each attracting parameter to the multiplier of its unique attracting cycle). Then there exists a conformal isomorphism $\Psi_W : \mathbb{H} \to W$ with $\mu \circ \Psi_W = \exp$ which extends continuously to $\partial \mathbb{H}$ and such that $\Psi_W(0)$ (the *root* of W) is the landing point of two periodic parameter rays. The region containing W which is enclosed by these two rays is called the *wake* of W.

For every $h = \frac{p}{q} \in \mathbb{Q} \setminus \mathbb{Z}$, the point $\Psi_W(2\pi i h)$ is the root point of a (unique) hyperbolic component $\mathrm{Bif}(W, h)$ of period qn (called a *child component* of W). The component $\mathrm{Bif}(W, h)$ tends to infinity above or below W depending on whether $h < 0$ or $h > 0$ (respectively). If $0 < h_1 < h_2$ or $h_1 < h_2 < 0$, then $\mathrm{Bif}(W, h_1)$ tends to infinity below $\mathrm{Bif}(W, h_2)$. Any hyperbolic component other than W which lies in the wake of W is contained in the wake of a unique child component $\mathrm{Bif}(W, h)$.

The problem we face is that, without knowing (a), we do not know that all the parabolic parameters $\Psi_W(2\pi i h)$ really exist, and hence that all these bifurcations really take place. We will nonetheless obtain (in Theorem 6.8) a purely *combinatorial* version of the above description. This description allows us to separate different objects in parameter space based on their combinatorics, which is one of the main ideas used in [26] to establish (a) (compare also the proof of Theorem A.1 for a use of this technique). Since we are not able to use the topological structure of parameter space, many of our arguments (and statements) are much more delicate than they would be e.g. for the quadratic family.

Another goal of our article is to explain the relation among certain combinatorial objects which appear in exponential dynamics. In particular, there are several such objects associated to any hyperbolic component.

- The *characteristic external addresses* of W (Definition 3.4). These are the addresses of the parameter rays bounding the wake of W.
- The *intermediate external address* of W (Section 2). This is an object which does not appear in Multibrot sets. It describes the combinatorial position of the singular value within the dynamical plane of a parameter in W. At the same time, it describes the position of the hyperbolic component W itself in the vertical structure of parameter space.
- The *kneading sequence* of W (Definition 3.5). This object describes the itinerary of the singular orbit with respect to a natural *dynamical* partition (as opposed to the *static* partition used to define external addresses).
- The *internal address* of W (Definition 7.8). Introduced for Multibrot sets in [17], this address describes the position of W within the bifurcation structure of hyperbolic components. Its relative, the *angled internal address* (Definition A.6), is decorated with some additional information.

Our study yields algorithms to convert between these different objects (where possible), and also to compute the address of any child component. These algorithms (many of which are nontrivial even in the case of the quadratic family) are collected in Appendix B.

Finally, the combinatorial objects and methods used in this article have applications far beyond the scope of our present investigation, and are likely to play a significant role in further studies of the exponential family (as they did for the Mandelbrot set). Thus, we aim to present a comprehensive exposition of these concepts which may serve as a reference in the future.

We should emphasize that all results of this article — with the exception of the analytical considerations of Section 5 — are completely combinatorial and could be formulated and proved without any reference to the underlying exponential maps. However, we prefer to carry out an argument within an actual dynamical plane whenever possible, as we find this much more intuitive. (Compare for example the definition of orbit portraits in Section 3, as well as the proofs of Lemma 3.10 and Proposition 7.4.)

Since the combinatorial structure of exponential parameter space is a limit of that for unicritical polynomials, it would be possible to infer many of our results from corresponding facts for these families. However, many of these — particularly for Multibrot sets of higher degrees — are themselves still unpublished. Also, there are aspects of exponential dynamics, such as the *intermediate external address* of a hyperbolic exponential map, which would not feature in such an approach. We have thus decided to give a clean self-contained account in the exponential case.

Structure of this article. In the following two sections, we give a comprehensive overview of several combinatorial concepts for exponential maps: external addresses, dynamic rays, intermediate external addresses, orbit portraits, characteristic rays and itineraries. In Section 4, we consider some basic facts about hyperbolic components of exponential maps, and how they are partitioned into *sectors*.

Section 5 is the only part of the article in which analytical considerations are made: we investigate the stability of orbits at a parabolic point, allowing us to understand the structure of bifurcations occurring at such points. This provides the link between our subsequent combinatorial considerations and the exponential parameter plane. While the arguments in this section are very similar to those in the polynomial setting, there are some surprises: the combinatorics of a parent component can be determined with great ease from that of a child component, thanks to the new feature of intermediate external addresses.

With these preliminaries, we will be in a position to prove our main results. Section 6 deals with the structure of (combinatorial) child components of a given hyperbolic component, as discussed above. Section 7 introduces internal addresses, giving a "human-readable" combinatorial structure to parameter space, and shows how they are related to the combinatorical concepts defined before.

In Appendix A, we consider some further concepts. These are not required for the proofs in [26] but follow naturally from our discussion and will be collected for future reference. Appendix B explicitly collects the combinatorial algorithms which are implied by our results.

For the reader's convenience, a list of notation and an index of the relevant combinatorial concepts is provided at the end of the article.

Some remarks on notation. We have chosen to parametrize our exponential maps as $z \mapsto E_\kappa(z) = \exp(z) + \kappa$. Traditionally, they have often been parametrized as $\lambda \exp$, which is conjugate to E_κ if $\lambda = \exp(\kappa)$. We prefer our parametrization mainly because the behavior of exponential maps at ∞, and in particular the asymptotics of external rays, do not depend on the parameter in this parametrization. Note that this is also the case in the usual parametrization of quadratic polynomials as $z \mapsto z^2 + c$. Also, under our parametrization the picture in the parameter plane reflects the situation in the dynamical plane, which is a conceptual advantage. Note that E_κ and $E_{\kappa'}$ are conformally conjugate if and only if $\kappa - \kappa' \in 2\pi i \mathbb{Z}$. This will prove useful in the combinatorial description. When citing known results, we always translate them into our parametrization.

We should also note that both dynamical and parameter rays have often been referred to as "hairs" in the transcendental setting. We chose our terminology to emphasize the analogy with the polynomial case. Similarly, *itinerary* is often also used to denote what we call *external adresses*. We prefer our convention since it not only avoids overloading the former term, but also highlights the relation to external angles of polynomials.

Preliminaries. It is well-known [12, Theorem 4] that an exponential map can have at most one nonrepelling periodic orbit. We call a parameter κ itself *attracting, parabolic, indifferent*, etc. if the map E_κ has a periodic orbit of the corresponding type. Similarly, if the singular orbit of E_κ (that is, the forward orbit of the singular value κ) tends to ∞ or is finite, we call κ an *escaping* or *postsingularly finite* parameter, respectively. A maximal open and connected region in the κ-plane consisting of attracting parameters is called a *hyperbolic component*.

If $\gamma : [0, \infty) \to \mathbb{C}$ is a curve, we shall say that $\lim_{t \to \infty} \gamma(t) = +\infty$ (or, in short, call γ a *curve to* $+\infty$) if $\operatorname{Re} \gamma(t) \to +\infty$ and $\operatorname{Im} \gamma$ is bounded; analogously for $-\infty$. The n-th iterate of any function f will be denoted by f^n. Whenever we write a rational number as a fraction $\frac{p}{q}$, we will assume p and q to be coprime.

We will be concerned with different kinds of rays throughout this article: dynamic rays and attracting dynamic rays in the dynamical plane (Section 2) as well as internal rays (Section 4) in the parameter plane. When there is no danger of confusion, we may, for brevity, refer to any of these simply as "rays".

We conclude any proof and any result which immediately follows from previously proved theorems by the symbol ■. A result which is cited without proof, or whose proof is merely sketched with details provided elsewhere, is concluded by □.

Acknowledgements. We would like to thank Walter Bergweiler, Alex Eremenko, Markus Förster, Misha Lyubich, Jack Milnor, Rodrigo Perez, Phil Rippon and Juan Rivera-Letelier for many helpful discussions, and the referee for useful suggestions. We also thank the Institute of Mathematical Sciences at Stony Brook as well as the Mathematics Institute at the University of Warwick for continued support and hospitality.

2. Combinatorics of Exponential Maps

An important combinatorial tool in the study of polynomials is the structure provided by *dynamic rays*, which foliate the basin of infinity. Similarly, throughout this article, we will assign combinatorics to curves in the dynamical plane of an exponential map, both in the set of escaping points and in Fatou components. This section will review these methods, which were introduced in [31] and [28].

External addresses and dynamic rays. A sequence $\underline{s} = s_1 s_2 \ldots$ of integers is called an (infinite) external address[1]. If $s_1, \ldots, s_n \in \mathbb{Z}$, then the address obtained by periodically repeating this sequence will be denoted by $\overline{s_1 \ldots s_n}$.

Let $\kappa \in \mathbb{C}$ and let $\gamma : [0, \infty) \to \mathbb{C}$ be a curve in the dynamical plane of E_κ. Then we say that γ has *external address* $\underline{s}$ if and only if

$$\lim_{t \to \infty} \operatorname{Re} E_\kappa^{j-1}(\gamma(t)) = +\infty \quad \text{and} \quad \lim_{t \to \infty} \operatorname{Im} E_\kappa^{j-1}(\gamma(t)) = 2\pi s_j$$

for all $j \geq 1$; in this case, we also write $\underline{s} = \operatorname{addr}(\gamma)$. An external address $\underline{s}$ is called *exponentially bounded* if there exists some $x > 0$ with $2\pi|s_k| < F^{k-1}(x)$ for all $k \geq 1$, where $F(t) = \exp(t) - 1$ is used as a model function for exponential growth.

The *set of escaping points* of E_κ is defined to be

$$I := I(E_\kappa) := \{z \in \mathbb{C} : E_\kappa^n(z) \to \infty\}.$$

It is known that the Julia set $J(E_\kappa)$ is the closure of $I(E_\kappa)$ [10, 12]. In [31], the set $I(E_\kappa)$ has been completely classified. In particular, it was shown that it consists of curves to ∞, so-called *dynamic rays*. We will use this result in the following form. (Note that the fact that dynamic rays are the path-connected components of $I(E_\kappa)$ was stated but not proved in [31]; for a proof compare [14].)

2.1. Theorem and Definition (Dynamic Rays).
Let $\kappa \in \mathbb{C}$. Then, for every exponentially bounded address $\underline{s}$, there exists a unique injective curve $g_{\underline{s}} : [0, \infty) \to I(E_\kappa)$ or $g_{\underline{s}} : (0, \infty) \to I(E_\kappa)$ which has external address $\underline{s}$ and whose trace is a path-connected component of $I(E_\kappa)$. The curve $g_{\underline{s}}$ is called the dynamic ray at address $\underline{s}$.

If $\kappa \notin I(\overline{E}_\kappa)$, then every path-connected component of $I(E_\kappa)$ is a dynamic ray. If $\kappa \in I(E_\kappa)$, then every such component is either a dynamic ray or is mapped into a dynamic ray under finitely many iterations.

Remark 1. In order to state this theorem as given, dynamic rays need to be parametrized differently from [31]. In this article, we will only be using

---
[1] For brevity, we will frequently omit the adjective "external"; "address" will always mean "external address" unless explicitly stated otherwise.

dynamic rays at periodic addresses $\underline{s}$, and for these our parametrization agrees with that of [31], provided that the singular orbit does not escape.

Remark 2. As usual, we say that a dynamic ray $g_{\underline{s}}$ *lands* at a point $z_0 \in \hat{\mathbb{C}}$ if $\lim_{t \to 0} g_{\underline{s}}(t) = z_0$.

Intermediate external addresses. We shall also need to assign combinatorics to certain curves in Fatou components of E_κ which, under finitely many iterations, map to a curve to $-\infty$. Let $\gamma : [0, \infty) \to \mathbb{C}$ be a curve in the dynamical plane of E_κ such that, for some $n \geq 1$, $\lim_{t \to \infty} E_\kappa^{n-1}(\gamma(t)) = -\infty$. Then there exist $s_1, \ldots, s_{n-2} \in \mathbb{Z}$ and $s_{n-1} \in \mathbb{Z} + \frac{1}{2}$ such that

$$\lim_{t \to \infty} \mathrm{Re}(E_\kappa^{j-1}(\gamma(t))) = +\infty \quad \text{and} \quad \lim_{t \to \infty} \mathrm{Im}(E_\kappa^{j-1}(\gamma(t))) = 2\pi s_j$$

for $j = 1, \ldots, n - 1$. We call

(1) $$\mathrm{addr}(\gamma) := s_1 s_2 \ldots s_{n-1} \infty$$

the *intermediate external address* of γ. Any sequence of the form (1) with $s_1, \ldots, s_{n-2} \in \mathbb{Z}$ and $s_{n-1} \in \mathbb{Z} + \frac{1}{2}$ is called an intermediate external address (of length n). Note that ∞ is the unique intermediate external address of length 0.

To illustrate the relationship between infinite and intermediate external addresses, consider the following construction. Define

$$f : \mathbb{R} \setminus \{(2k - 1)\pi; k \in \mathbb{Z}\} \to \mathbb{R}, t \mapsto \tan(t/2).$$

Then to any (infinite) external address $\underline{s}$ we can associate a unique point x for which $f^{k-1}(x) \in \big((2s_k - 1)\pi, (2s_k + 1)\pi\big)$ for all $k \geq 1$. However, there are countably many points which are not realized by any external address in this way, namely the preimages of ∞ under the iterates of f. Adding intermediate external addresses corresponds to filling in these points. The space $\overline{\mathcal{S}}$ of all infinite and intermediate external addresses is thus order-isomorphic to the circle $\overline{\mathbb{R}} \cong \mathbb{S}^1$. We also set $\mathcal{S} := \overline{\mathcal{S}} \setminus \{\infty\}$. The *shift map* is the function

$$\sigma : \mathcal{S} \to \overline{\mathcal{S}}; s_1 s_2 \ldots \mapsto s_2 \ldots;$$

note that σ corresponds to the function f in the above model.

Lexicographic and vertical order. The space $\mathcal{S}$ naturally comes equipped with the *lexicographic order* on external addresses. (As seen above, this ordered space is isomorphic to the real line $\mathbb{R}$, and in particular is complete.) Similarly, the space $\overline{\mathcal{S}}$ carries a (complete) circular ordering. In our combinatorial considerations, we will routinely use the following fact.

2.2. Observation (Shift Preserves Order On Small Intervals).
For every $\underline{s} = s_1 s_2 s_3 \ldots$ and $\underline{s}' := (s_1 + 1) s_2 s_3 \ldots$, the map $\sigma : [\underline{s}, \underline{s}') \to \overline{\mathcal{S}}$ preserves the circular order of $\overline{\mathcal{S}}$.

Any family of pairwise disjoint curves to $+\infty$ has a natural *vertical order*: for any two such curves, one is *above* the other. More precisely, suppose that $\gamma : [0, \infty) \to \mathbb{C}$ is a curve to $+\infty$ and define $\mathcal{H}_R := \{z \in \mathbb{C} : \operatorname{Re} z > R\}$ for $R > 0$. If R is large enough, then the set $\mathcal{H}_R \setminus \gamma$ has exactly two unbounded components, one above and one below γ. Any curve $\widetilde{\gamma}$ to $+\infty$ which is disjoint from γ must (eventually) tend to ∞ within one of these.

It is an immediate consequence of the definitions that, if γ and $\widetilde{\gamma}$ have (infinite or intermediate) external addresses $\operatorname{addr}(\gamma) \neq \operatorname{addr}(\widetilde{\gamma})$, then γ is above $\widetilde{\gamma}$ if and only if $\operatorname{addr}(\gamma) > \operatorname{addr}(\widetilde{\gamma})$.

Intermediate address of attracting and parabolic dynamics. Let us suppose that E_κ has an attracting or parabolic periodic point. Then the singular value κ is contained in some periodic Fatou component; we call this the *characteristic Fatou component*. Let $U_0 \mapsto U_1 \mapsto \ldots \mapsto U_n = U_0$ be the cycle of periodic Fatou components, labeled such that U_1 is the characteristic component. (This will be our convention for the remainder of the paper.) Since U_1 contains a neighborhood of the singular value, U_0 contains a left half plane. In particular, U_0 contains a horizontal curve along which $\operatorname{Re}(z) \to -\infty$. Its pullback to U_1 under E_κ^{n-1} has an intermediate external address $\underline{s}$ of length n. (The address $\underline{s}$ does not depend on the initial choice of the curve to $-\infty$, since the latter is unique up to homotopy in U_0.)

We call $\underline{s}$ the *intermediate external address of* κ and denote it by $\operatorname{addr}(\kappa)$; it will play a special role throughout this article. The following was proved independently in [28] and [6]; the idea of the proof goes back to [1, Section 7].

2.3. Proposition (Existence of Attracting Maps with Prescribed Combinatorics).
Let $\underline{s}$ be an intermediate external address. Then there exists an attracting parameter κ with $\operatorname{addr}(\kappa) = \underline{s}$. □

A converse result was also proved in [28]: the external address $\operatorname{addr}(\kappa)$ determines E_κ up to quasiconformal conjugacy (see Proposition 4.2).

Attracting dynamic rays. We shall frequently have need for a canonical choice of certain curves in a Fatou component. Let E_κ have an attracting orbit, which we label $a_0 \mapsto a_1 \mapsto \ldots \mapsto a_n = a_0$ such that $a_i \in U_i$. Note that we can connect κ to a_1 in U_1 by a straight line in linearizing coordinates. The pullback of this curve under E_κ^n along the orbit of a_1 is then a curve $\gamma \subset U_1$ which connects a_1 to ∞ and has $\operatorname{addr}(\gamma) = \operatorname{addr}(\kappa)$. We call γ the *principal attracting ray* of E_κ. More generally, any maximal curve in U_1 which starts at a_1 and is mapped into a radial line by the extended Kœnigs map $\varphi : U_1 \to \mathbb{C}$ (see [20, Corollary 8.4]) is called an *attracting dynamic ray* of E_κ. Apart from the principal attracting ray, the attracting dynamic ray which contains κ will be particularly important. Those attracting dynamic rays which map

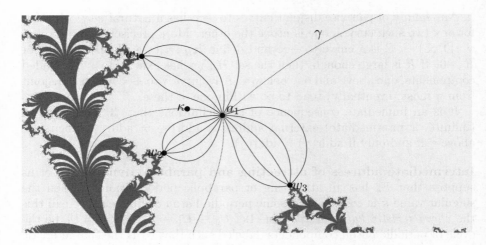

FIGURE 2. Attracting dynamic rays for a parameter where the attracting multiplier has angle $1/3$. All unbroken attracting rays land at the distinguished boundary orbit $\{w_i\}$, while the three broken attracting rays (shown as dotted lines) contain the singular orbit.

to a curve to $-\infty$ in U_0 under an iterate E_κ^{nj-1}, $j > 0$, are called *broken*; the principal attracting ray is one of these. (Other attracting dynamic rays are called *unbroken*.) Note that such a broken attracting ray is mapped to a proper subcurve of the attracting ray containing κ by E_κ^{nj}.

It can be shown that every unbroken attracting dynamic ray lands at a point in $\mathbb{C}$ [22, Theorem 4.2.7]. We shall only need this fact in the case of rational multipliers (illustrated in Figure 2), where it is simpler to prove.

2.4. Definition and Lemma (Distinguished Boundary Orbit).
Consider an attracting exponential map E_κ of period n whose attracting multiplier has rational angle $\frac{p}{q}$. Then the orbit of the principal attracting ray of E_κ under E_κ^n consists of q attracting dynamic rays and contains all points of the singular orbit in U_1.

Every other attracting dynamic ray starting at a_1 is periodic of period q and lands at one of the points of a unique period q orbit of E_κ^n on ∂U_1. This orbit is called the distinguished boundary orbit *on ∂U_1.*

Proof. Analogous to the case $q = 1$ [28, Lemma 6.1]. □

3. ORBIT PORTRAITS AND ITINERARIES

Orbit Portraits. Following Milnor [19], we will use the notion of *orbit portraits* to encode the dynamics of periodic rays landing at common points. As in the case of quadratic polynomials, this is important for understanding the structure of parameter space.

3.1. Definition (Orbit Portrait).
Let $\kappa \in \mathbb{C}$ and let $\{z_1, \ldots, z_n\}$ be a repelling or parabolic periodic orbit for E_κ. Define

$$A_k := \{\underline{r} \in \mathcal{S} : \underline{r} \text{ is periodic and the dynamic ray } g_{\underline{r}} \text{ lands at } z_k\}.$$

Then $\mathcal{O} := \{A_1, \ldots, A_n\}$ is called the orbit portrait of $\{z_k\}$. The orbit (and the orbit portrait) is called essential if every A_k contains at least two addresses. An essential orbit portrait is called of satellite type if it contains only one cycle of addresses; otherwise it is called primitive.

3.2. Lemma (Basic Properties of Orbit Portraits).
Let $\mathcal{O} = \{A_1, \ldots, A_n\}$ be an orbit portrait. Then all A_k are finite, and the shift map carries A_k bijectively onto A_{k+1}. Furthermore, all addresses in the portrait share the same period qn (called the ray period of the orbit) for some integer $q \geq 1$.

Proof. The proof that all rays share the same period and are transformed bijectively by the shift is completely analogous to the polynomial setting [20, Lemma 18.12]. Let qn be the common period of the rays in $\mathcal{O}$ and let $\underline{s} = \overline{s_1 s_2 \ldots s_{qn}} \in A_1$. It is easy to see that

$$A_1 \subset \{\overline{s'_1 s'_2 \ldots s'_{qn}} : |s'_k - s_k| \leq 1\}$$

(see e.g. [24, Lemma 5.2]). The set on the right hand side is finite, as required. ∎

An orbit portrait can also be defined as an abstract combinatorial object, without reference to any parameter. We will not do this here, but we will often suppress the actual choice of parameter present in its definition.

Characteristic rays. For quadratic polynomials, every orbit portrait has two distinguished rays, which are exactly the two rays which separate the critical value from all other rays in the portrait [19, Lemma 2.6]. A corresponding statement for exponential maps is given by the following result.

3.3. Definition and Lemma (Characteristic Rays [28, Lemma 5.2]).
Let $\kappa \in \mathbb{C}$. Suppose that $\{z_k\}$ is a repelling or parabolic periodic orbit with essential orbit portrait $\mathcal{O}$. Then there exist ℓ and two periodic rays $g_{\underline{r}}$ and $g_{\underline{\tilde{r}}}$ landing at z_ℓ (the characteristic rays of the orbit $\{z_k\}$) such that the curve $g_{\underline{r}} \cup \{z_\ell\} \cup g_{\underline{\tilde{r}}}$ separates the singular value from all other rays of the orbit portrait. The addresses $\underline{r}$ and $\underline{\tilde{r}}$ are called the characteristic addresses of $\mathcal{O}$; they depend on $\mathcal{O}$ but not on κ. The interval in $\mathcal{S}$ bounded by $\underline{r}$ and $\underline{\tilde{r}}$ is called the characteristic sector of $\mathcal{O}$.

Furthermore, if there are at least three rays landing at each z_k, then the orbit portrait of $\{z_k\}$ is of satellite type.

A pair $\langle \underline{r}, \underline{\tilde{r}} \rangle$ with $\underline{r} < \underline{\tilde{r}}$ is called a characteristic ray pair for E_κ if E_κ has an orbit portrait whose characteristic rays are $\underline{r}$ and $\underline{\tilde{r}}$. More generally, we call $\langle \underline{r}, \underline{\tilde{r}} \rangle$ a characteristic ray pair *if there is any* $\kappa \in \mathbb{C}$ *with such an orbit portrait.* If $\langle \underline{r}, \underline{\tilde{r}} \rangle$ is a characteristic ray pair of period n, then $\sigma^{n-1}(\underline{\tilde{r}}) < \sigma^{n-1}(\underline{r})$. □

Remark. The final claim does not appear in the statement of [28, Lemma 5.2], but is immediate from its proof.

3.4. Theorem and Definition (Characteristic Rays [28, Lemma 5.2]).
Let κ be an attracting or parabolic parameter and suppose that $\underline{s} := \mathrm{addr}(\kappa)$ has length n. Then there exists a unique characteristic ray pair $\langle \underline{s}^-, \underline{s}^+ \rangle$ for E_κ such that the common landing point of $g_{\underline{s}^-}$ and $g_{\underline{s}^+}$ lies on the boundary of the characteristic Fatou component U_1; both rays have period n. This ray pair separates κ from all other periodic points with essential orbit portraits.

The addresses $\underline{s}^-$ and $\underline{s}^+$ depend only on $\underline{s}$, and are called the characteristic addresses *of $\underline{s}$ (or κ). The common landing point is called the* dynamic root *of E_κ.* □

Remark 1. In particular, the dynamic root is the unique boundary point of the characteristic Fatou component U_1 which is fixed under E_κ^n and which is the landing point of at least two dynamic rays.

Remark 2. We will later give an algorithm (Algorithm B.2) for determining $\langle \underline{s}^-, \underline{s}^+ \rangle$, given $\underline{s}$.

Remark 3. The case of parabolic parameters was not formally treated in [28]. However, the proof is the same as that given there for attracting parameters. (Alternatively, the parabolic case follows from the attracting case by using Theorem 3.6 below.)

Itineraries and kneading sequences. Recall that, given an attracting or parabolic exponential map E_κ, one can connect the singular value to ∞ in the characteristic Fatou component U_1 by a curve γ with external address $\mathrm{addr}(\kappa)$. The preimage $E_\kappa^{-1}(\gamma)$ consists of countably many curves in U_0, congruent under translation by $2\pi n$ ($n \in Z$), and these produce a partition of the dynamical plane into congruent strips. (The curve γ is unique up to homotopy within U_1, so the partition is natural except for points within U_0.) To any point $z \in J(E_\kappa)$, one can now associate an *itinerary*, which records through which strips of this partition the orbit of z passes. For more details, see [32, Section 4].

In this article, we will use the following combinatorial analog of this notion. If $\underline{s} \in \mathcal{S}$, then $\sigma^{-1}(\underline{s})$ produces a partition of $\mathcal{S}$, and the itinerary of any $\underline{t} \in \mathcal{S}$ will record where the orbit of $\underline{t}$ under σ maps with respect to this partition.

3.5. Definition (Itineraries and Kneading Sequences).
Let $\underline{s} \in \mathcal{S}$ and $\underline{r} \in \overline{\mathcal{S}}$. Then the itinerary of $\underline{r}$ with respect to $\underline{s}$ is the sequence $\mathrm{itin}_{\underline{s}}(\underline{r}) := u_1 u_2 \ldots$, where

$$\begin{cases} u_k = j & \text{if } j\underline{s} < \sigma^{k-1}(\underline{r}) < (j+1)\underline{s} \\ u_k = {}^{j}_{j-1} & \text{if } \sigma^{k-1}(\underline{r}) = j\underline{s} \\ u_k = * & \text{if } \sigma^{k-1}(\underline{r}) = \infty. \end{cases}$$

(Note that $\mathrm{itin}_{\underline{s}}(\underline{r})$ is is a finite sequence if and only if $\underline{r}$ is an intermediate external address.) We also define $\mathrm{itin}_{\underline{s}}^+(\underline{r})$ and $\mathrm{itin}_{\underline{s}}^-(\underline{r})$ to be the sequence obtained by replacing each boundary symbol ${}^{j}_{j-1}$ by j or $j-1$, respectively. When κ is a fixed attracting or parabolic parameter, we usually abbreviate $\mathrm{itin}(\underline{r}) := \mathrm{itin}_{\mathrm{addr}(\kappa)}(\underline{r})$.

We also define the kneading sequence of $\underline{s}$ to be $\mathbb{K}(\underline{s}) := \mathrm{itin}_{\underline{s}}(\underline{s})$. Similarly, the upper and lower kneading sequences of $\underline{s}$ are $\mathbb{K}^+(\underline{s}) := \mathrm{itin}_{\underline{s}}^+(\underline{s})$ and $\mathbb{K}^-(\underline{s}) := \mathrm{itin}_{\underline{s}}^-(\underline{s})$.

Remark 1. One should think of $\underline{s}$ as lying in the "combinatorial parameter plane", whereas $\underline{r}$ lies in the "combinatorial dynamical plane" associated with $\underline{s}$.

Remark 2. In the case $\underline{s} = \infty$, we can define itineraries analogously, but the addresses $j\underline{s}$ and $(j+1)\underline{s}$ in the definition will have to be replaced by $(j - \frac{1}{2})\infty$ and $(j + \frac{1}{2})\infty$. With this definition, $\mathrm{itin}_\infty(\underline{r}) = \underline{r}$ for all infinite external addresses $\underline{r}$.

Remark 3. The definition of itineraries involves a noncanonical choice of an offset for the labelling of the partition strips. Our choice was made so that the external addresses in the interval $(\overline{0}, \overline{1})$ are exactly those whose kneading sequences start with 0.

The significance of itineraries lies in the fact that they can be used to determine which periodic rays land together, as shown in [32, Theorems 3.2 and 5.4, Proposition 4.5].

3.6. Theorem (Dynamic Rays and Itineraries).
Let κ be an attracting or parabolic parameter. Then every periodic dynamic ray of E_κ lands at a periodic point, and conversely every repelling or parabolic periodic point is the landing point of such a ray.

Two periodic rays $g_{\underline{r}}$ and $g_{\underline{\tilde{r}}}$ land at the same point if and only if they have the same itinerary. □

The n-th itinerary entry u_n is locally constant (as a function of $\underline{r}$) wherever it is defined and an integer. If the n-th itinerary entry at $\underline{r}$ is a boundary symbol $u_n = {}^{j}_{j-1}$, then it is $j-1$ slightly below $\underline{r}$ and j slightly above $\underline{r}$. In other words,

$$\lim_{\underline{t}\nearrow\underline{r}} \mathrm{itin}_{\underline{s}}(\underline{t}) = \mathrm{itin}_{\underline{s}}^{-}(\underline{r}) \quad \text{and}$$

$$\lim_{\underline{t}\searrow\underline{r}} \mathrm{itin}_{\underline{s}}(\underline{t}) = \mathrm{itin}_{\underline{s}}^{+}(\underline{r})$$

for all infinite external addresses $\underline{r}$.

If the n-th itinerary entry of $\underline{r}$ is $*$, then $\underline{r}$ is an intermediate external address of length n. In this case, the n-th itinerary entries of addresses tend to $+\infty$ when approaching $\underline{r}$ from below and to $-\infty$ when approaching $\underline{r}$ from above.

We will frequently be in a situation where we compare the itineraries of an address $\underline{t}$ with respect to two different addresses $\underline{s}^1, \underline{s}^2 \in \mathcal{S}$. Therefore, let us state the following simple fact for further reference.

3.7. Observation (Itineraries and Change of Partition).
Let $\underline{s}^1, \underline{s}^2, \underline{t} \in \mathcal{S}$ with $\underline{s}^1 < \underline{s}^2$ and let $j \geq 1$ such that $\sigma^j(\underline{t})$ is defined. Then the j-th entries of the itineraries of $\underline{t}$ with respect to $\underline{s}^1$ and $\underline{s}^2$ coincide if and only if $\sigma^j(\underline{t}) \notin [\underline{s}^1, \underline{s}^2]$.

In particular, $\mathrm{itin}_{\underline{s}^1}^{-}(\underline{t}) = \mathrm{itin}_{\underline{s}^2}^{+}(\underline{t})$ if and only if $\sigma^k(\underline{t}) \notin (\underline{s}^1, \underline{s}^2)$ for all $k \geq 1$.

Proof. By definition, the j-th itinerary entry of $\underline{t}$ as a function of $\underline{s} \in \mathcal{S}$ is locally constant on $\mathcal{S} \setminus \{\sigma^j(\underline{t})\}$, which proves the "if" part. On the other hand, this function jumps by 1 as $\underline{s}$ passes $\sigma^j(\underline{t})$, proving the "only if" part. ∎

Finally, we shall require the following fact on the existence of addresses with prescribed itineraries.

3.8. Lemma (Existence of Itineraries).
Let $\underline{s} \in \overline{\mathcal{S}}$. Let $\underline{u}$ be either an infinite sequence of integers or a finite sequence of integers followed by $*$, and suppose that $\sigma^k(\underline{u}) \notin \{\mathbb{K}^+(\underline{s}), \mathbb{K}^-(\underline{s})\}$ for all $k \geq 1$. Then there exists an external address $\underline{r} \in \overline{\mathcal{S}}$ with $\mathrm{itin}_{\underline{s}}(\underline{r}) = \underline{u}$; if $\underline{u}$ is periodic then every such $\underline{r}$ is also periodic.

Furthermore, if $\underline{t} \in \overline{\mathcal{S}}$ and $k \geq 0$, then no two elements of $\sigma^{-k}(\underline{t})$ have the same itinerary with respect to $\underline{s}$. (In particular, no two intermediate external addresses have the same itinerary with respect to $\underline{s}$.)

Remark. The condition $\sigma^k(\underline{u}) \notin \{\mathbb{K}^+(\underline{s}), \mathbb{K}^-(\underline{s})\}$ is necessary. There exist periodic addresses $\underline{s}$ (for example $\underline{s} = \overline{0}$) such that both $\mathbb{K}^+(\underline{s})$ and $\mathbb{K}^-(\underline{s})$ are not realized as the itinerary of any external address. Similarly, as we will see in Lemma 7.6 (d), there exist nonperiodic addresses $\underline{s}$ with periodic kneading sequences. In this case, $\mathrm{itin}_{\underline{s}}(\underline{r}) \neq \mathbb{K}(\underline{s})$ for all periodic addresses $\underline{r}$.

Proof. The set R_k of all external addresses $\underline{r} \in \overline{\mathcal{S}}$ for which at least one of the itineraries $\mathrm{itin}_{\underline{s}}^{+}(\underline{r})$ and $\mathrm{itin}_{\underline{s}}^{-}(\underline{r})$ agrees with $\underline{u}$ in the first k entries is easily seen to be compact and nonempty for all k. Thus $R := \bigcap_k R_k \neq \emptyset$. Let $\underline{r} \in R$; then $\underline{u} \in \{\text{itin}_{\underline{s}}^+(\underline{r}), \text{itin}_{\underline{s}}^-(\underline{r})\}$.

We claim that $\text{itin}_{\underline{s}}(\underline{r})$ contains no boundary symbols. Indeed, otherwise $\sigma^k(\underline{r}) = \underline{s}$ for some $k \geq 1$, and thus $\mathbb{K}(\underline{s}) = \sigma^k(\text{itin}_{\underline{s}}(\underline{r}))$. But by the definition of R this would mean that $\mathbb{K}^+(\underline{s}) = \sigma^k(\underline{u})$ or $\mathbb{K}^-(\underline{s}) = \sigma^k(\underline{u})$, which contradicts the assumption. Consequently, $\text{itin}_{\underline{s}}(\underline{r}) = \text{itin}_{\underline{s}}^+(\underline{r}) = \text{itin}_{\underline{s}}^-(\underline{r}) = \underline{u}$ as required.

The fact that $\underline{r}$ can be chosen to be periodic if $\underline{u}$ is periodic is [32, Lemma 5.2]. The proof that no aperiodic address can have the same itinerary as a periodic address is analogous to [20, Lemma 18.12].

The final statement follows by induction from the trivial fact that changing the first entry of an address $\underline{t}$ by some integer m will also change the first entry of $\text{itin}_{\underline{s}}(\underline{t})$ by m. ∎

Properties of characteristic ray pairs. As a first application of the concept of itineraries, let us deduce two basic properties of characteristic ray pairs.

3.9. Lemma (Characteristic Ray Pairs).
Let $\langle \underline{r}^-, \underline{r}^+ \rangle$ be a characteristic ray pair of period n. Then there are numbers $\mathsf{u}_1, \ldots \mathsf{u}_n \in \mathbb{Z}$ such that

(2) $$\mathbb{K}^-(\underline{r}^-) = \mathbb{K}^+(\underline{r}^+) = \text{itin}_{\underline{r}^-}^-(\underline{r}^+) = \text{itin}_{\underline{r}^+}^+(\underline{r}^-) = \overline{\mathsf{u}_1 \ldots \mathsf{u}_n}.$$

Furthermore, if $\underline{s} \in \mathcal{S}$, then the following are equivalent.

(a) $\underline{s} \in (\underline{r}^-, \underline{r}^+)$;
(b) $\text{itin}_{\underline{s}}(\underline{r}^-) = \text{itin}_{\underline{s}}(\underline{r}^+) = \overline{\mathsf{u}_1 \ldots \mathsf{u}_n}$;
(c) $\text{itin}_{\underline{s}}(\underline{r}^-) = \text{itin}_{\underline{s}}(\underline{r}^+)$.

Proof. It follows from the definition of characteristic addresses that
$$\text{itin}_{\underline{r}^-}^-(\underline{r}^-) = \text{itin}_{\underline{r}^-}^-(\underline{r}^+).$$
By definition, the former itinerary is $\mathbb{K}^-(\underline{r}^-)$, so we have
$$\mathbb{K}^-(\underline{r}^-) = \text{itin}_{\underline{r}^i}^-(\underline{r}^+) = \overline{\mathsf{u}_1 \ldots \mathsf{u}_n}$$
for some $\mathsf{u}_1, \ldots, \mathsf{u}_n \in \mathbb{Z}$. Furthermore, no forward image of $\underline{r}^-$ or $\underline{r}^+$ belongs to $(\underline{r}^-, \underline{r}^+)$. Thus Observation 3.7 implies that
$$\text{itin}_{\underline{s}}(\underline{r}^-) = \text{itin}_{\underline{s}}(\underline{r}^+) = \overline{\mathsf{u}_1 \ldots \mathsf{u}_n}$$
for all $\underline{s} \in (\underline{r}^-, \underline{r}^+)$. This proves (2) as well as "(a)⇒(b)". Clearly (b) implies (c).

To prove that (a) follows from (c), let $\underline{s} \in \mathcal{S} \setminus (\underline{r}^-, \underline{r}^+)$. Since the interval $[\sigma^{n-1}(\underline{r}^+), \sigma^{n-1}(\underline{r}^-)]$ is mapped bijectively to $\overline{\mathcal{S}} \setminus (\underline{r}^-, \underline{r}^+)$ by the shift, there is an element of $\sigma^{-1}(\underline{s})$ between $\sigma^{n-1}(\underline{r}^+)$ and $\sigma^{n-1}(\underline{r}^-)$. So the n-th itinerary entries of $\underline{r}^-$ and $\underline{r}^+$ with respect to $\underline{s}$ are different, as required. ∎

3.10. Lemma (Unique Intermediate Addresses).
Let $\langle \underline{r}^-, \underline{r}^+ \rangle$ be a characteristic ray pair of period n, and let $\mathsf{u}_1, \ldots \mathsf{u}_n \in \mathbb{Z}$ be as in the previous lemma. Moreover, let $\underline{s}$ be any intermediate external address of length n. Then the following are equivalent:

(a) *$\underline{r}^-$ and $\underline{r}^+$ are the characteristic addresses of $\underline{s}$;*
(b) *$\text{itin}_{\underline{r}^-}(\underline{s}) = \text{itin}_{\underline{r}^+}(\underline{s}) = \mathbb{K}(\underline{s}) = \mathsf{u}_1 \ldots \mathsf{u}_{n-1}*$;*
(c) *$\underline{s} \in (\underline{r}^-, \underline{r}^+)$ and $\mathbb{K}(\underline{s}) = \mathsf{u}_1 \ldots \mathsf{u}_{n-1}*$.*

Furthermore, there is at most one $\underline{s}$ with one (and thus all) of these properties.

Remark. Note that we do not claim here that such an address $\underline{s}$ always exists. While it is not very difficult to show that this is indeed the case, we will not require this fact until Section 7, and it will be proved there (Proposition 7.4).

Proof. If $\underline{r}^-$ and $\underline{r}^+$ are the characteristic adresses of $\underline{s}$, then no forward image of $\underline{s}$ lies in $(\underline{r}^-, \underline{r}^+)$. Indeed, let κ be a parameter with $\text{addr}(\kappa) = \underline{s}$. Then the characteristic rays $g_{\underline{r}^-}$ and $g_{\underline{r}^+}$ separate the characteristic Fatou component U_1 from the other periodic Fatou components, and thus from the remainder of the singular orbit. So (a) implies (b) by Observation 3.7.

Suppose that (b) holds. Then, for each $k \in \{1, \ldots, n-1\}$, the three addresses $\sigma^{k-1}(\underline{r}^-)$, $\sigma^{k-1}(\underline{r}^+)$ and $\sigma^{k-1}(\underline{s})$ lie in the interval $(\mathsf{u}_k \underline{r}^-, (\mathsf{u}_k+1)\underline{r}^-]$. By Observation 2.2, the map σ^{n-1} thus preserves the circular order of $\underline{r}^-$, $\underline{r}^+$ and $\underline{s}$. Recall that $\sigma^{n-1}(\underline{r}^+) < \sigma^{n-1}(\underline{r}^-)$ by Lemma 3.3. Since $\sigma^{n-1}(\underline{s}) = \infty$, it follows that $\underline{r}^- < \underline{s} < \underline{r}^+$. By Lemma 3.9, it follows that (c) holds.

Now let us assume that $\underline{s}$ satisfies (c), and let $\kappa \in \mathbb{C}$ with $\text{addr}(\kappa) = \underline{s}$. Then the dynamic rays $g_{\underline{r}^-}$ and $g_{\underline{r}^+}$ have a common landing point w by Lemma 3.9 and Theorem 3.6. It easily follows that $\underline{r}^-$ and $\underline{r}^+$ are the characteristic addresses of the orbit portrait of w.

Now let $\underline{s}^-$ and $\underline{s}^+$ be the characteristic addresses of $\underline{s}$. Then we have

$$\underline{r}^- \leq \underline{s}^- < \underline{s} < \underline{s}^+ \leq \underline{r}^+,$$

and the first $n - 1$ itinerary entries of all these addresses with respect to $\underline{s}$ coincide. As above, the cyclic order of this configuration is preserved under σ^{n-1}. Since $\sigma^{n-1}(\underline{s}) = \infty$, this means that

(3) $$\sigma^{n-1}(\underline{s}^+) \leq \sigma^{n-1}(\underline{r}^+) < \sigma^{n-1}(\underline{r}^-) \leq \sigma^{n-1}(\underline{s}^-).$$

Since the outer two addresses in (3) have the same itinerary, this means that the n-th itinerary entries of $\underline{s}^-, \underline{s}^+, \underline{r}^-$ and $\underline{r}^+$ also agree. Thus all these addresses belong to the orbit portrait of w. Since both pairs $\langle \underline{s}^-, \underline{s}^+ \rangle$ and $\langle \underline{r}^-, \underline{r}^+ \rangle$ are characteristic ray pairs of this portrait, they must be equal. This proves (a).

Finally, there is at most one $\underline{s} \in \mathcal{S}$ with $\text{itin}_{\underline{r}^-}(\underline{s}) = \mathsf{u}_1 \ldots \mathsf{u}_{n-1}*$ by Lemma 3.8. This completes the proof. ∎

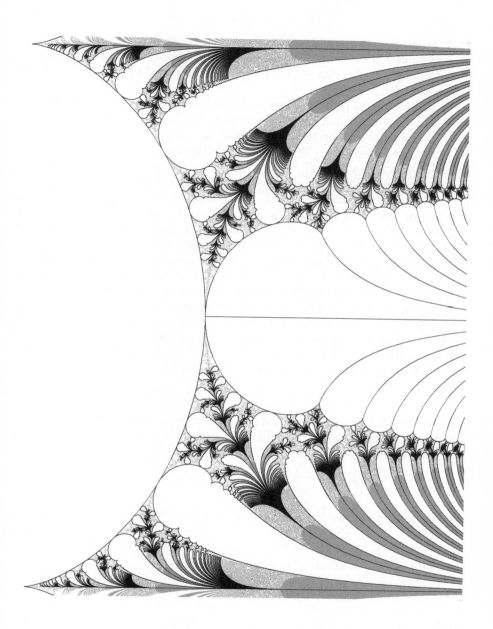

FIGURE 3. Several hyperbolic components in the strip $\{\operatorname{Im}\kappa \in [0, 2\pi]\}$. Within the period two component in the center right of the picture, internal rays at integer heights are drawn in.

4. Hyperbolic components

A *hyperbolic component* W is a maximal connected open subset of parameter space in which each parameter has an attracting periodic orbit. This (unique) orbit depends holomorphically on κ, and its period is constant throughout W.

It was shown in [1] that every hyperbolic component W is simply connected and unbounded. Furthermore, the multiplier map $\mu : W \to \mathbb{D}^*$, which maps every parameter to the multiplier of its attracting cycle, is a universal covering. Since $\exp : \mathbb{H} \to \mathbb{D}^*$ is also a universal covering, there exists a conformal isomorphism $\Psi_W : \mathbb{H} \to W$ with $\mu \circ \Psi_W = \exp$. Note that this defines Ψ_W uniquely only up to precomposition by a deck transformation of $\exp$ (i.e., an additive translation by $2\pi i k$, $k \in \mathbb{Z}$). However, in [28, Theorem 7.1], it was shown that, when W has period at least two, there is a unique choice of Ψ_W such that, for any $\kappa \in \Psi_W((-\infty, 0))$, the dynamic root lies on the distinguished boundary orbit. (This also follows from Theorem 4.4 below.) In the following, we will always fix Ψ_W to be this *preferred parametrization*.

It is well-known [1] that there exists a unique hyperbolic component W_0 of period 1. This hyperbolic component contains a left half plane and is invariant under $z \mapsto z + 2\pi i$. Since different choices for the parametrization of this component only correspond to such a translation, and thus to a relabeling of the map, there is no canonical choice for a preferred parametrization. By definition, we choose the *preferred parametrization* of the period 1 component to be the unique map respecting the real symmetry, i.e.

$$\Psi_W : \mathbb{H} \to W_0, z \mapsto z - \exp(z).$$

Internal rays. We can now foliate W by curves, called *internal rays*, along which the argument of μ is constant. More precisely, the *internal ray at height* $h \in \mathbb{R}$ is the curve

$$\Gamma_{W,h} : (-\infty, 0) \to \mathbb{C}, t \mapsto \Psi_W(t + 2\pi i h).$$

It is straightforward to see that, if W is a hyperbolic component of period ≥ 2, then $\Gamma_{W,h}(t) \to +\infty$ as $t \to -\infty$, and the homotopy class of this curve as $t \to \infty$ is independent of h (see e.g. [28, Lemma 2.1]). We say that an internal ray $\Gamma_{W,h}$ *lands* at a point $\kappa \in \hat{\mathbb{C}}$ if $\kappa = \lim_{t \nearrow 0} \Gamma_{W,h}(t)$.

4.1. Lemma (Landing of Internal Rays).
Let W be a hyperbolic component of period n. Then every internal ray $\Gamma_{W,h}$ lands at some point in $\hat{\mathbb{C}}$, which we denote by $\Psi_W(2\pi i h)$. The set of h for which this landing point belongs to $\mathbb{C}$ is open and dense. Conversely, suppose that $\kappa_0 \in \partial W \cap \mathbb{C}$. Then κ_0 is the landing point of a unique internal ray $\Gamma_{W,h}$. Furthermore, κ_0 is an indifferent parameter of period dividing n. If a is a point on this indifferent periodic orbit, then $(E_{\kappa_0}^n)'(a) = \exp(2\pi i h)$.

Remark 1. It is not difficult to see that the extended map Ψ_W is continuous on $\overline{\mathbb{H}}$. However, we shall not require this result here.

Remark 2. This lemma leaves open the possibility that some internal rays land at ∞, disconnecting the boundary of $\mathbb{C}$. Proving that this does not happen is much more difficult. The proof of this fact, in [26], uses the results of the present article. (An outline of the argument was given in [29].) This result can therefore not be used in the following sections.

Sketch of proof. It is straightforward to see that every point of $\partial W \cap \mathbb{C}$ has an indifferent periodic orbit of period dividing n. The multiplier map μ extends to a holomorphic function on a neighborhood of κ_0 (or on a finite-sheeted covering of κ_0 when $\mu(\kappa_0) = 1$, compare [19, Proof of Lemma 4.2]). Hence there exists some internal ray which lands at κ_0. If there were two internal rays of W landing at κ_0, these could be connected to form a simple closed curve $\gamma \subset \overline{W}$ which separates some part of ∂W from infinity. It is a standard fact that this is not possible: for example, it is well-known that every point of $\partial W \cap \mathbb{C}$ is structurally unstable, and thus can be approximated by attracting parameters with arbitrarily high periods (compare [22, Lemma 5.1.6]). The hyperbolic components containing these parameters thus will be separated from ∞ by γ, which is impossible since every hyperbolic component is unbounded.

The fact that the set of h for which $\Gamma_{W,h}(0)$ lands in $\mathbb{C}$ is open follows easily from the above statement about the multiplier. That this set is also dense (and in fact has full measure) follows from the F. and M. Riesz theorem [20, Theorem A.3].

Finally, it is straightforward to see that every finite limit point κ_0 of an internal ray $\Gamma_{W,h}$ has a periodic point a with $(E^n_{\kappa_0})'(a) = \exp(2\pi i h)$. The set of such parameters is easily seen to be discrete in $\mathbb{C}$, proving that $\Gamma_{W,h}$ lands at a point of $\hat{\mathbb{C}}$.

For a more detailed (self-contained!) proof, compare [26, Section 2]. ∎

When W is of period at least 2, the internal ray at height 0 is called the *central internal ray* of W. If this ray lands at a point in $\mathbb{C}$, then its landing point $\Psi_W(0)$ is called the *root* of W. The points of $\Psi_W(2\pi i \mathbb{Z} \setminus \{0\}) \cap \mathbb{C}$ are called *co-roots* of W. For the period 1 component, all points of $\Psi_W(2\pi i \mathbb{Z})$ are called co-roots. Recall that we will prove in [26] that every component has a root (and infinitely many co-roots). Without this knowledge, which we may not use at this point, we cannot be sure that bifurcations actually exist.

Classification of hyperbolic components. It is easy to see that $\text{addr}(\kappa)$ depends only on the hyperbolic component W which contains κ; this address will therefore also be denoted by $\text{addr}(W)$. Similarly, we will talk about the kneading sequence, characteristic rays etc. of W. The following theorem, which is the main result of [28], states that hyperbolic components can be completely classified terms of their combinatorics. (Note that the existence part of this result was already cited as Theorem 2.3.)

4.2. Proposition (Classification of Hyperbolic Components [28]).
For every intermediate external address $\underline{s}$, there exists exactly one hyperbolic component W with $\mathrm{addr}(W) = \underline{s}$. *We denote this component by* $\mathrm{Hyp}(\underline{s})$. *The vertical order of hyperbolic components coincides with the lexicographic order of their external addresses.* □

To explain the last statement, recall that, when W is a hyperbolic component of period ≥ 2, any internal ray $\Gamma_{W,h}$ satisfies $\Gamma_{W,h}(t) \to +\infty$ as $t \to -\infty$. Thus the family of central internal rays of hyperbolic components has a natural vertical order[2] as described in Section 2, and this is the order referred to in the Proposition. (Note that taking the central rays is not essential, as there is only one homotopy class of curves in W along which the multiplier tends to 0.)

Sectors. Since $\mu : W \to \mathbb{D}^*$ is a universal covering, parameters in W are not (as in the quadratic family) uniquely determined by their multiplier. Rather, the set $\mu^{-1}(\mathbb{D}^* \setminus [0,1))$ consists of countably many components, called *sectors* of W. If $\kappa = \Psi_W(t + 2\pi i h)$ with $h \notin \mathbb{Z}$, we denote the sector containing κ by

$$\mathrm{Sec}(\kappa) := \mathrm{Sec}(W, h) := \Psi_W\Big(\{a + 2\pi i b : a < 0 \text{ and } b \in (\lfloor h \rfloor, \lceil h \rceil)\}\Big).$$

4.3. Definition (Sector Labels).
Let $W = \mathrm{Hyp}(\underline{s})$ be a hyperbolic component and let $\kappa \in W$ be a parameter with $\mu := \mu(\kappa) \notin (0,1)$. Let γ be the principal attracting ray of E_κ, and let γ' be the component of $E_\kappa^{-1}(\gamma)$ which starts at a_0. Then $\mathrm{addr}(\gamma')$ is of the form $s_*\underline{s}$ with $s_* \in \mathbb{Z}$ (resp. $s_* \in \mathbb{Z} + \frac{1}{2}$ if $\underline{s} = \infty$). The entry $s_* = s_*(\kappa)$ is called the sector label of κ.

Remark. There are two more ways to label sectors which will appear in this article: *sector numbers* and *kneading entries*; both will be introduced in Section 7. We should warn the reader that our terminology is somewhat different from that of [17], where the term *sector label* is used to refer to what we call *kneading entries*.

The following results justify the term "sector label"; compare Figure 6(a).

4.4. Theorem (Behavior of Sector Labels).
The map $\kappa \mapsto s_*(\kappa)$ is constant on sectors of W. When κ crosses a sector boundary so that μ passes through $(0,1)$ in positive orientation, then $s_*(\kappa)$

---

[2]We should stress that we only use the "negative" ends of internal rays to define this order. A priori some internal rays could also tend to infinity as $t \to +\infty$. Using these directions might result in a different order, which we are not interested in.

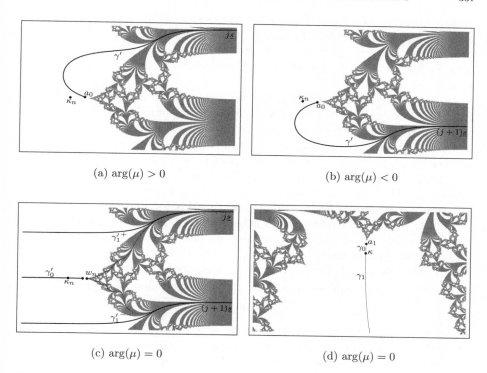

FIGURE 4. Illustration of Definition 4.3 and Theorem 4.4. Shown are parameters above, below and on the central internal ray of a period 3 hyperbolic component. (Here $\kappa_n = E_\kappa^{n-1}(\kappa)$.) For the latter parameter, the principal attracting ray in the Fatou component U_1 is also shown.

increases exactly by 1. In particular the induced map from sectors to indices is bijective. The unique sector with a given sector label s_* will be denoted by $\text{Sec}(W, s_*)$.

If the period of W is at least 2 and κ is a parameter on the internal ray between $\text{Sec}(W, j)$ and $\text{Sec}(W, j+1)$, then the distinguished boundary fixed point of E_κ^n on ∂U_1 has itinerary $\overline{\mathbf{u}_1 \ldots \mathbf{u}_{n-1} j}$ (where $\mathbb{K}(W) = \mathbf{u}_1 \ldots \mathbf{u}_{n-1}*$). In particular, the central internal ray of W is the boundary between $\text{Sec}(W, j)$ and $\text{Sec}(W, j+1)$, where j and $j+1$ are the n-th entries of the characteristic addresses $\underline{s}^+$ and $\underline{s}^-$, respectively.

If W is the unique period one component, a similar statement holds: the boundary between $\text{Sec}(W, j - \frac{1}{2})$ and $\text{Sec}(W, j + \frac{1}{2})$ is given by the internal ray $\{t + 2\pi i j : t \leq -1\}$. For parameters on this ray, the distinguished boundary fixed point has constant itinerary $\overline{j}$.

Sketch of proof. The linearizing coordinate used to define attracting dynamic rays depends holomorphically on κ. It easily follows that, as long as the principal attracting ray γ does not pass through κ, its preimage $\gamma' = \gamma'(\kappa)$ from Definition 4.3 varies continuously, which shows that s_* is constant on sectors.

In the following, let us restrict to the case where the period of W is at least two; the case of the period one component is handled analogously. Let κ_0 be a parameter with positive real multiplier, and let us set $\underline{s} := \mathrm{addr}(W)$. Then the principal attracting ray γ contains the singular value. Denote the piece of γ which connects a_1 to κ by γ_0 and the piece which connects κ to $+\infty$ by γ_1. We can define a branch φ of E_κ^{-1} on $U_1 \setminus \gamma_1$ which takes a_1 to a_0. The range of φ is then a strip S of U_0 bounded by two consecutive preimages $\gamma_1'^{-}$ and $\gamma_1'^{+}$, at external addresses $j\underline{s}$ and $(j+1)\underline{s}$ for some $j \in \mathbb{Z}$.

Let $\alpha \subset U_1 \setminus \gamma_1$ be an unbroken attracting dynamic ray, connecting a_1 to the distinguished boundary fixed point $w \in \partial U_1$. Then the image of α under φ is a curve connecting a_0 to $w_n := E_\kappa^{n-1}(w)$. Thus w_n lies between $\gamma_1'^{-}$ and $\gamma_1'^{+}$, so j is the n-th itinerary entry of w as claimed.

Denote the preimage of γ_0 in S by γ_0'. We then define two curves (in $\hat{\mathbb{C}}$),
$$\gamma'^{\pm} := \gamma_0' \cup \{-\infty\} \cup \gamma_1'^{\pm}.$$

By continuity of the linearizing coordinate, it then follows that, as $\kappa \to \kappa_0$ through parameters at positive (resp. negative) multiplier angles, the curves $\gamma'(\kappa)$ converge uniformly to γ'^{+} (resp. γ'^{-}), which completes the proof. (Compare [22, Theorem 5.5.3] for more details.) ∎

5. Local Bifurcation Results

Throughout this section, let κ_0 be a parabolic parameter of period n and intermediate external address $\underline{s}$. If the parabolic orbit portrait of E_{κ_0} is essential, then we call κ_0 a *satellite* or a *primitive* parameter, depending on the type of this orbit portrait. Similarly, we will refer to the ray period of this orbit portrait as the ray period of κ_0. Note that this ray period is also the period of the repelling (or attracting) petals of the parabolic orbit.

In this section, we will study what happens when κ_0 is perturbed into an adjacent hyperbolic component. For this purpose, we will use the following well-known statement about the analytic structure near κ_0. This result is beautifully explained, and proved using elementary complex analysis, by Milnor in [19, Section 4]. All that this local analysis requires is that there is only one singular value, and thus only one single cycle of petals at a parabolic periodic point.

5.1. Proposition (Perturbation of Parabolic Orbits).
Let κ_0 be a parabolic parameter of period n, with ray period qn.

- **(*Primitive and Co-root case*)** *If $q = 1$ (so the multiplier of the parabolic orbit is 1), then, under perturbation, the parabolic orbit splits* up into two orbits of period n that can be defined as holomorphic functions of a two-sheeted cover around κ_0.

Any hyperbolic component whose boundary contains κ_0 corresponds to one of these orbits becoming attracting (and therefore has period n).

- **(Satellite Case)** If $q \geq 2$, then, under perturbation, the parabolic orbit splits into one orbit of period n and one of period qn. The period n orbit can be defined as a holomorphic function in a neighborhood of κ_0, as can the multiplier of the period qn-orbit. The qn-orbit itself can be defined on a q-sheeted covering around κ_0.

Any hyperbolic component whose boundary contains κ_0 corresponds to one of these orbits becoming attracting (and therefore has period n or qn). □

Any hyperbolic component of period qn that touches κ_0 is called a *child component*; note that at least one such component always exists. In the satellite case, any period n component touching κ_0 is called a *parent component*. Note that any satellite parameter has at least one child and at least one parent component; a primitive parameter has at least one child component but no parent components. (We will show in Theorems 5.3 and 5.4 below that "at least one" can be replaced by "exactly one".)

We will also require the following statement on the landing behavior of periodic rays as κ_0 is perturbed. The proof is analogous to that in the case of quadratic polynomials, which can also be found in [19, Section 4]. (Recall that, by the previous proposition, the parabolic orbit of κ_0 breaks up into two orbits under perturbation. If we perturb κ_0 into an adjacent hyperbolic component, one of these orbits becomes attracting, so there is a unique repelling orbit created in the bifurcation.)

5.2. Proposition (Orbit Stability under Perturbation).
Under perturbation of a parabolic parameter κ_0 into a child or parent component, all repelling periodic points retain the same orbit portraits.

Furthermore, under perturbation into a child component, the repelling periodic orbit created in the bifurcation has the same orbit portrait as the parabolic orbit of E_{κ_0}. Under perturbation into a parent component, the rays landing at the parabolic orbit are split up, landing at distinct points of the newly created repelling orbit. □

We are now ready to describe the combinatorics of child and parent components of κ_0 (and, in particular, show that there is at most one of each, see Corollary 5.5).

5.3. Theorem (Combinatorics in a Child Component).
Let κ_0 be a parabolic parameter of period n and ray period qn, and let W be a child component of κ_0. Then $\mathrm{addr}(W) = \mathrm{addr}(\kappa_0)$; i.e., W is the unique component at address $\mathrm{addr}(\kappa_0)$.

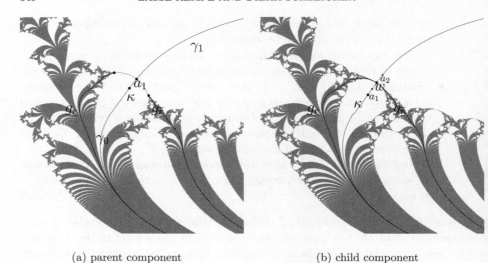

(a) parent component (b) child component

FIGURE 5. The dynamical plane just before and after a bifurcation, illustrating Theorems 5.3 and 5.4. In both pictures, unbroken attracting rays are dotted lines, broken attracting rays are solid lines, and dynamic rays are strong solid lines.

Furthermore, for points on the internal ray of W landing at κ_0, the repelling point created in the bifurcation is the distinguished boundary fixed point. Therefore κ_0 is the root point of W if and only if its parabolic orbit portrait is essential; otherwise, κ_0 is a co-root of W.

Proof. If $qn = 1$, then $W = \text{Hyp}(\infty)$ is the unique component of period 1. Now suppose that $qn > 1$. Let $\Gamma_{W,h}$ be the unique internal ray landing at κ_0 (compare Lemma 4.1); this ray has integer height $h \in \mathbb{Z}$. Let $\kappa := \Gamma_{W,h}(t)$ be a parameter on this ray. By Proposition 5.2, κ_0 and κ have the same orbit portraits and thus they have the same characteristic addresses. Lemma 3.10 then yields $\text{addr}(\kappa_0) = \text{addr}(\kappa) = \text{addr}(W)$.

Now let w be the newly created repelling point and let w' be the distinguished boundary fixed point of κ. Let α be the piece of the principal attracting ray of E_κ which connects κ to ∞. Recall from Lemma 2.4 that a_0 and $E_\kappa^{n-1}(w')$ can be connected by an unbroken dynamic ray of E_κ and thus belong to the same component of $\mathbb{C} \setminus E_\kappa^{-1}(\alpha)$. We will show that a_0 and $E_\kappa^{n-1}(w)$ are also not separated by $E_\kappa^{-1}(\alpha)$. This implies that w and w' have the same itinerary and are therefore equal by Theorem 3.6, as required.

Let $\Phi : U_1 \to \mathbb{C}$ be the linearizing coordinate for E_κ, normalized so that $\Phi(\kappa) = 1$, and let $V \subset U_1$ be the component of the preimage of $\Phi^{-1}(\mathbb{D}(0, \frac{1}{\mu}))$ which contains a_1. (Note that $\mathbb{D}(0, \frac{1}{\mu})$ is the largest disk on which Φ^{-1} exists, and that $\mu = \exp(t)$.) Then V contains the curve $\gamma := E_\kappa^n(\alpha)$, and by definition $\Phi(\gamma) = (\mu, 1)$. Since $\Phi|_V : V \to \mathbb{D}(0, \frac{1}{\mu})$ is a conformal isomorphism, the hyperbolic length of γ in V is

$$\ell_V(\gamma) = \int_\mu^1 \frac{\mu |dz|}{1 - (\mu |z|)^2} \leq \frac{\mu(1-\mu)}{1 - \mu^2} = \frac{\mu}{1 + \mu} \leq 2.$$

In other words, the hyperbolic length of γ within U_1 stays bounded as $t \to 0$. Since the euclidean length of γ (which connects κ and $E_\kappa^n(\kappa)$) is bounded below as $t \to 0$, it follows from standard estimates on the hyperbolic metric, using the fact that $w \in \partial U_1$, that the euclidean distance $\text{dist}(w, \gamma)$ is also bounded below as $t \to 0$.

Since E_κ^{n+1} is continuous in z and κ,

$$\liminf_{t \to 0} \text{dist}(E_\kappa^{n-1}(w), E_\kappa^{-1}(\alpha)) > 0$$

(recall that $E_\kappa^n(w) = w$). On the other hand, the attracting point a_0 and the repelling point $E_\kappa^{n-1}(w)$ are created from the same parabolic point, so the distance between them tends to 0 as $t \to 0$. Thus, for small enough t, these two points are not separated by $E_\kappa^{-1}(\alpha)$. Therefore we have shown $w = w'$.

To prove the final statement, observe that by Proposition 5.2, the parabolic orbit portrait of E_{κ_0} is essential if and only if the orbit portrait of $w = w'$ is essential for E_κ. By definition, the latter is the case if and only if κ belongs to the central internal ray; i.e., if κ_0 is the root point of W. ∎

Let us now turn to determining the parent component of a satellite parameter.

5.4. Theorem (Combinatorics in a Parent Component).
Let κ_0 be a satellite parabolic parameter of period n and multiplier $e^{2\pi i \frac{p}{q}}$; let W be a parent component of κ_0. If κ is a parameter on the internal ray in W which lands at κ_0, then the distinguished boundary orbit for κ is the repelling period qn orbit created from the parabolic point.

Furthermore, for such a parameter κ, let $\underline{s}'$ be the address of the attracting dynamic ray which contains the singular value. Then $\text{addr}(\kappa_0) = \underline{s}'$.

Proof. To prove the first statement, choose some small wedge-shaped repelling petals for the parabolic orbit of E_{κ_0}. If κ is close enough to κ_0 on the internal ray landing at κ_0, then, for κ, these petals (after a translation moving them to the attracting period n orbit (a_k) created in the bifurcation) are still backward invariant and contain the newly created repelling period qn orbit (w_k) (see [19, Section 4]). Choose any attracting dynamic ray which approaches a_1 through one of these petals. Recall that this ray is periodic under E_κ^n, of period q. Pulling back, it must land at the unique fixed point of the return map of this petal. That is, the points of the repelling orbit (w_k) are landing points of a cycle of attracting dynamic rays; thus (w_k) is the distinguished boundary orbit.

To prove the second part of the theorem, let $\underline{r}, \underline{\tilde{r}}$ be the characteristic addresses of κ_0. We will now find the combinatorial features of κ_0 within the attracting dynamics of κ, using attracting dynamic rays (compare Figure 5(a).) Let γ_0 be the attracting dynamic ray containing the singular value (recall that $\underline{s}' = \text{addr}(\gamma_0)$). Then the cycle of attracting rays $\gamma_j := E_\kappa^{nj}(\gamma_0)$, $j = 0, \ldots, q-1$, completely contain the singular orbit in U_1. By Proposition 5.2 and the first part of the theorem, the rays $g_{\underline{r}}$ and $g_{\underline{\tilde{r}}}$ do not land together for E_κ, but rather land separately on two points of the distinguished boundary cycle on ∂U_1. These landing points can be connected to a_1 by two attracting dynamic rays. Let $h_{\underline{r}}$ and $h_{\underline{\tilde{r}}}$ denote the curves obtained by extending $g_{\underline{r}}$ and $g_{\underline{\tilde{r}}}$ to a_1 by these attracting rays.

Now consider the preimages of that part of γ_0 which connects κ to ∞; these curves are straight lines in the linearizing coordinate of U_0, and connect $-\infty$ to $+\infty$ with external addresses of the form $m\underline{s}'$. The images of $h_{\underline{r}} \cup \gamma_0 \cup h_{\underline{\tilde{r}}}$ do not intersect these curves. Consequently for every $j \geq 0$, the j-th iterated images of $g_{\underline{r}}$, $g_{\underline{\tilde{r}}}$ and γ_0 belong to a common strip of this partition.

Thus, $\text{itin}_{\underline{s}'}(\underline{r}) = \text{itin}_{\underline{s}'}(\underline{\tilde{r}})$; so in particular $\underline{s} \in (\underline{r}, \underline{\tilde{r}})$ by Lemma 3.9. Furthermore, this itinerary agrees with $\mathbb{K}(\underline{s}')$ on its first $qn - 1$ entries, which implies by Lemma 3.10 that $\underline{r}$ and $\underline{\tilde{r}}$ are the characteristic addresses of $\underline{s}'$. Since $\underline{r}$ and $\underline{\tilde{r}}$ are also the characteristic addresses of κ_0, we conclude that $\underline{s}' = \text{addr}(\kappa_0)$. ∎

5.5. Corollary (Bifurcation Structure at a Parabolic Point).
Suppose that κ_0 is a parabolic parameter of period n and ray period qn. Let $\underline{s} := \text{addr}(\kappa)$ and $\mathbb{K}(\underline{s}) = \mathtt{u}_1 \ldots \mathtt{u}_{qn-1}*$. Then $\partial \text{Hyp}(\underline{s})$ is the unique child component of κ_0; if κ_0 is of satellite type, then $\text{Hyp}(\sigma^{(q-1)n}(\underline{s}))$ is the unique parent component of κ_0. No other hyperbolic component contains κ_0 on its boundary.

Furthermore, κ_0 is the root point of $\text{Hyp}(\underline{s})$ if and only if the parabolic orbit of E_{κ_0} has essential orbit portrait (which is always the case if $q > 1$). Otherwise, κ_0 is a co-root of $\text{Hyp}(\underline{s})$.

Proof. Let W be a child component of κ_0 (at least one such component always exists). By Theorem 5.3, $\text{addr}(W) = \underline{s}$. By the classification of hyperbolic components (Proposition 4.2), this component is unique. Similarly, suppose that $q > 1$; i.e., that κ_0 is a satellite parameter. By Theorem 5.4, $\text{Hyp}(\sigma^{(q-1)n}(\underline{s}))$ is the unique parent component of κ_0. By Proposition 5.1, no other hyperbolic component contains κ on its boundary. The final statement was already proved in Theorem 5.3. ∎

6. Bifurcation from a Hyperbolic Component

Suppose that κ_0 is a satellite parabolic parameter with parent component W and child component V. It follows from Theorem 5.4 that the address of the child component V is determined by W and the height $h \in \mathbb{Q} \setminus \mathbb{Z}$ of the internal ray $\Gamma_{W,h}$ landing at κ_0: it is the intermediate external address of a curve situated in the dynamical plane of $\kappa \in \Gamma_{W,h}$. However, this address is defined for *every* $h \in \mathbb{Q} \setminus \mathbb{Z}$, regardless of whether we know that the corresponding internal ray has a landing point!

We now use this idea to prove a combinatorial analog of the description of the structure of components bifurcating from a hyperbolic component W, as outlined in the introduction. More precisely, for every $h \in \mathbb{Q} \setminus \mathbb{Z}$, we define the address addr(W, h) of the hyperbolic component which "would" bifurcate from W at height h if the corresponding bifurcation parameter existed. Since these addresses are naturally defined in terms of curves in the dynamical plane of E_κ for $\kappa \in W$, it is straightforward to compute their itineraries with respect to $\underline{s} := \text{addr}(W)$.

Using this information, we can show that addr(W, h) behaves as we expect from pictures of parameter space; recall Figure 1 and the related discussion in the introduction. Note that, although this information is eventually about the position of hyperbolic components in *parameter* space, it will be obtained by working within the combinatorial *dynamical* plane associated to $\underline{s}$.

Combinatorial Bifurcation. As discussed above, Theorem 5.4 suggests the following definition.

6.1. Definition (Combinatorial Bifurcation).
Let $W = \text{Hyp}(\underline{s})$ be a hyperbolic component of period n, and let κ be a parameter on the internal ray $\Gamma_{W,h}$ for some $h \in \mathbb{Q} \setminus \mathbb{Z}$. Let $\underline{s}'$ be the external address of the attracting dynamic ray of E_κ which contains the singular value. Then we say that $\text{Hyp}(\underline{s}')$ bifurcates combinatorially from W (at height h). We denote the address of this component by
$$\text{addr}(W, h) := \text{addr}(A, p/q) := \text{addr}(\underline{s}, s_*, p/q) := \underline{s}'$$
where $A := \text{Sec}(W, s_*)$ is the sector containing κ and $p/q = h - \lfloor h \rfloor$ is the fractional part of h. The component $\text{Hyp}(\underline{s}')$ is called a (combinatorial) child component of W and denoted by $\text{Bif}(W, h)$.

6.2. Proposition (Child Components).
Let W be a hyperbolic component of period n, and let $h \in \mathbb{Q} \setminus \mathbb{Z}$. Then the following hold.
 (a) $\kappa_0 := \Psi_W(2\pi i h) = \Psi_{\text{Bif}(W,h)}(0)$ $(\in \hat{\mathbb{C}})$.
 (b) If $\kappa_0 \in \mathbb{C}$, then W and $\text{Bif}(W, h)$ are the only hyperbolic components containing κ_0 on their boundaries.
 (c) If two hyperbolic components have a common parabolic boundary point, then one of these components is a child component of the other.
 (d) If $\kappa \in \Gamma_{W,h}$, then the dynamic rays of E_κ at the characteristic addresses of $\text{Bif}(W, h)$ land on the distinguished boundary cycle of E_κ. For parameters in $\text{Bif}(W, h)$, the dynamic root has (exact) period n.
 (e) No two hyperbolic components have a common child component.

Proof. If $\kappa_0 := \Psi_W(2\pi i h) \in \mathbb{C}$, then κ_0 is a parabolic parameter of period n and rotation number $p/q := h - \lfloor h \rfloor$; in particular, the ray period of κ_0 is qn. Thus W is a parent component of κ_0. By Corollary 5.5, κ_0 is the root of $\mathrm{Bif}(W, h)$, and no other hyperbolic components contain κ_0 on their boundaries. This proves (b); it also proves (a) provided that $\kappa_0 \in \mathbb{C}$. Part (c) was proved in Corollary 5.5.

To prove (d), let $\kappa \in \Gamma_{W,h}$; also choose some parameter $\kappa_1 \in \mathrm{Bif}(W, h)$. Let $\underline{r}$ be the address of a periodic ray landing at a point $z_0 \in \partial U_1$ of the distinguished boundary orbit of E_κ. Recall from Theorem 3.4 that the dynamic root of E_κ is the only periodic point on ∂U_1 which has an essential orbit portrait. Thus the period of $\underline{r}$ is the same as the period of its landing point, which is qn. For $1 \le j \le n$, define

$$A_j := \{\sigma^{mn+j-1}(\underline{r}) : 0 \le m < q\}.$$

Then the landing points of the dynamic rays at the addresses in A_j are those points of the orbit of z_0 which belong to ∂U_j. Similarly as in the proof of Theorem 5.4, we can connect these landing points to the periodic point a_j using attracting dynamic rays such that the resulting "extended rays" $\widetilde{g}_{\sigma^j(\underline{r})}$ do not intersect except in their endpoints. Again, it follows that all addresses in A_1 have the same itinerary under $\widetilde{\underline{s}} := \mathrm{addr}(W, h)$, and this itinerary agrees with that of $\widetilde{\underline{s}}$ in the first $qn - 1$ entries. It is easy to see that in fact $\mathcal{O} := \{A_1, \ldots, A_n\}$ is an orbit portrait for E_{κ_1}, and that the characteristic rays of this orbit portrait both belong to A_1. (All that needs to be checked is that rays at addresses in different A_j cannot have the same landing point, which follows readily from the way that these rays are permuted by σ^n.) It now follows from Lemma 3.10 that these characteristic rays are in fact the characteristic rays of $\widetilde{\underline{s}}$, and (d) is proved.

To complete the proof of (a), suppose now that $\kappa_0' := \Psi_{\mathrm{Bif}(W,h)}(0) \in \mathbb{C}$. Then by (d) and Theorem 5.3, this parabolic parameter has period n and ray period qn. Thus Corollary 5.5 implies that $\kappa_0' = \kappa_0$.

Finally, let us prove (e). Suppose that $W' = \mathrm{Hyp}(\underline{s}')$, of period m, is a child component of W. By (d), the period n of W is uniquely determined by W'. By the definition of child components,

$$\mathrm{addr}(W) = \sigma^{m-n}(\underline{s}');$$

so $\mathrm{addr}(W)$, and thus W, is uniquely determined by W'. ∎

Analysis of Itineraries. We can now describe the addresses of child components of a given hyperbolic component $W = \mathrm{Hyp}(\underline{s})$ in terms of their itineraries under $\underline{s}$. This will be the key to understanding their behavior.

6.3. Lemma (Itineraries of Bifurcation Addresses).
Let $\underline{s}$ be an intermediate external address of length n with kneading sequence $\mathtt{u}_1 \ldots \mathtt{u}_{n-1}*$. Furthermore, let $s_* \in \mathbb{Z}$ (resp. $s_* \in \mathbb{Z} + \frac{1}{2}$ if $\underline{s} = \infty$) and

$\alpha = p/q \in \mathbb{Q} \cap (0,1)$. If $\underline{s} \neq \infty$, then $\underline{s}' := \mathrm{addr}(\underline{s}, s_*, p/q)$ is the unique intermediate address which satisfies

$$\mathrm{itin}_{\underline{s}}(\underline{s}') = \mathtt{u}_1 \ldots \mathtt{u}_{n-1}\mathtt{m}_1\mathtt{u}_1 \ldots \mathtt{u}_{n-1}\mathtt{m}_2 \ldots \mathtt{u}_1 \ldots \mathtt{u}_{n-1}\mathtt{m}_{q-1}\mathtt{u}_1 \ldots \mathtt{u}_{n-1}\mathtt{m}_q,$$

where $\mathtt{m}_q = *$, $\mathtt{m}_{q-1} = \frac{s_*}{s_*-1}$ and

$$\mathtt{m}_j = \begin{cases} s_* & \text{if } j\alpha \in [1-\alpha, 0] \pmod 1 \\ s_* - 1 & \text{otherwise} \end{cases}$$

for $j = 1, \ldots, q-2$.

If $\underline{s} = \infty$, then the above is still true, except that $\underline{s}_*$ is replaced by $\underline{s}_* + 1/2$ in the definition of the m_j.

Proof. As in Definition 6.1, let κ be a parameter on the internal ray at angle α in the sector $\mathrm{Sec}(\mathrm{Hyp}(\underline{s}), s_*)$ and let γ be the attracting dynamic ray of E_κ which contains the singular value. Then $\underline{s}' = \mathrm{addr}(\gamma)$. Set $\tilde{\underline{u}} := \mathrm{itin}_{\underline{s}}(\underline{s}')$.

Since $E_\kappa^{j-1}(\kappa)$ belongs to the Fatou component U_j, it follows immediately that $\tilde{\mathtt{u}}_{kn+j} = \mathtt{u}_j$ for $k \in \{0 \ldots q-1\}$ and $j \in \{1 \ldots n-1\}$. So we only need to show that the values $\mathtt{m}_j := \tilde{\mathtt{u}}_{jn}$ have the stated form.

For $j = 1, \ldots, q$, let us set $\gamma_j := E_\kappa^{jn-1}(\gamma)$. Note that γ_q is the attracting dynamic ray to $-\infty$ in U_0. Thus $\mathtt{m}_q = *$ and $E_\kappa(\gamma_{q-1})$ is the principal attracting ray. Thus $\mathrm{addr}(\gamma_{q-1}) = s_*\underline{s}$ by definition of s_*; so $\mathtt{m}_{q-1} = \frac{s_*}{s_*-1}$. Attracting dynamic rays do not intersect each other or their $2\pi i$-translates; thus any other entry $\mathtt{m}_j$ is either s_* or $s_* - 1$, depending on whether γ_j is above or below γ_{q-1}. Since E_κ^n permutes the curves γ_j cyclically with rotation number α, γ_j is above γ_{q-1} if and only if $j\alpha \in [1-\alpha, 1] \pmod 1$. ∎

In this and the following section, we will frequently be concerned with the question when two addresses whose itineraries coincide must in fact be the same. Let us therefore state the following simple fact for further reference.

6.4. Observation (Agreeing Itineraries).
Let $\underline{s} \in \overline{\mathcal{S}}$ and $k > 0$. Let $\underline{r}^1, \underline{r}^2$ be addresses with $\sigma^k(\underline{r}^1) \leq \sigma^k(\underline{r}^2)$ whose itineraries (under $\underline{s}$) agree in the first k entries. Suppose furthermore that, for $j = 0, \ldots, k-1$, $\sigma^j(\underline{s}) \notin [\sigma^k(\underline{r}^1), \sigma^k(\underline{r}^2)]$.

Then $\underline{r}^1 \leq \underline{r}^2$. Furthermore, σ^k maps the interval $[\underline{r}^1, \underline{r}^2]$ bijectively onto $[\sigma^k(\underline{r}^1), \sigma^k(\underline{r}^2)]$; i.e., the addresses $\underline{r}^1$ and $\underline{r}^2$ agree in the first k entries.

Proof. Note that it is sufficient to deal with the case $k = 1$; the general case follows by induction.

So suppose that $\underline{r}^1$ and $\underline{r}^2$ are as in the statement, with $k = 1$, and let I denote the interval in $\mathcal{S}$ bounded by $\underline{r}^1$ and $\underline{r}^2$. Since both addresses have the same first itinerary entry, σ maps I bijectively either to $[\sigma(\underline{r}^1), \sigma(\underline{r}^2)]$ or to $\overline{\mathcal{S}} \setminus (\sigma(\underline{r}^1), \sigma(\underline{r}^2))$. Since $\underline{s} \notin \sigma(I)$, it follows from the hypotheses that the former must be the case, as required. ∎

We can now use this observation to relate the behavior of the itineraries of the addresses addr(W, h) to that of the addresses themselves.

6.5. Proposition (Monotonicity of Itineraries).
Let $W = \mathrm{Hyp}(\underline{s})$ be a hyperbolic component of period $n \geq 2$, with kneading sequence $\mathbb{K}(W) = \mathtt{u}_1 \ldots \mathtt{u}_{n-1}*$.

(a) Let $I \subset \mathcal{S}$ denote either the interval $[\underline{s}^-, \underline{s}]$ or the interval $[\underline{s}, \underline{s}^+]$ (where $\underline{s}^- < \underline{s}^+$ are, as usual, the characteristic addresses of $\underline{s}$). Suppose that $\underline{r}^1, \underline{r}^2 \in I \setminus \{\underline{s}\}$ have the following property: if $\ell \in \{1, 2\}$ and $j \geq 0$ are such that $\sigma^{jn}(\underline{r}^\ell)$ is defined, then $\sigma^{jn}(\underline{r}^\ell) \in I$ and $\mathrm{itin}_{\underline{s}}(\sigma^{jn}(\underline{r}^\ell))$ starts with $\mathtt{u}_1 \ldots \mathtt{u}_{n-1}$. Then
$$\underline{r}^1 \leq \underline{r}^2 \iff \mathrm{itin}_{\underline{s}}(\underline{r}^1) \leq \mathrm{itin}_{\underline{s}}(\underline{r}^2).$$

(b) Let $h = \frac{p}{q} \in \mathbb{Q} \setminus \mathbb{Z}$ and $\underline{r} := \mathrm{addr}(W, h)$. If $h > 0$ (resp. if $h < 0$), then $\sigma^{jn}(\underline{r}) \in (\underline{s}^-, \underline{s}]$ (resp. $\sigma^{jn}(\underline{r}) \in [\underline{s}, \underline{s}^+)$) for all $0 \leq j < q$.

Remark. We should remark on the lexicographic order of itineraries referred to in (a). There is a natural order between integer itinerary entries and boundary symbols: $\mathtt{m} < {}^{\mathtt{j}}_{\mathtt{j}-1}$ if and only if $\mathtt{m} \leq \mathtt{j} - 1$. However, it is not clear how the symbol $*$ should fit into this order. We will fix the convention that the symbol $*$ is incomparable to any other itinerary entry, which gives our claim the strongest possible meaning.

(In fact, this is not relevant for our considerations: the itineraries of any two addresses $\underline{r}^1, \underline{r}^2 \in \mathcal{S}$ will be comparable unless at least one of them is an intermediate external address which is not a preimage of $\underline{s}$. Clearly this cannot happen in our case.)

Proof. To prove item (a); let us fix our ideas by supposing that $I = [\underline{s}^-, \underline{s}]$. As already remarked above, $\mathrm{itin}_{\underline{s}}(\underline{r}^1)$ and $\mathrm{itin}_{\underline{s}}(\underline{r}^2)$ are comparable. Note also that the orbit of $\underline{s}$ does not enter the interval I by the definition of characteristic addresses (this will enable us to apply the previous observation.)

Suppose first that $\mathrm{itin}_{\underline{s}}(\underline{r}^1) = \mathrm{itin}_{\underline{s}}(\underline{r}^2)$. If $\underline{r}^1$ and $\underline{r}^2$ are intermediate, then $\underline{r}^1 = \underline{r}^2$ by Lemma 3.8. On the other hand, if $\underline{r}^1$ and $\underline{r}^2$ are infinite, then for any $\ell \geq 0$, the hypotheses of Observation 6.4 are satisfied with $k = n\ell$. Thus the first $n\ell$ entries of $\underline{r}^1$ and $\underline{r}^2$ agree; since ℓ is arbitrary, this means that $\underline{r}^1 = \underline{r}^2$.

So now suppose that $\mathrm{itin}_{\underline{s}}(\underline{r}^1) \neq \mathrm{itin}_{\underline{s}}(\underline{r}^2)$, say $\mathrm{itin}_{\underline{s}}(\underline{r}^1) < \mathrm{itin}_{\underline{s}}(\underline{r}^2)$. Let $\ell \geq 1$ be such that these itineraries first differ in the $n\ell$-th entry. Then $\underline{r}^1$ and $\underline{r}^2$ agree in the first $(\ell - 1)n$ entries by Observation 6.4. Thus we may suppose, by passing to the $(\ell - 1)n$-th iterates, that $\ell = 1$.

Then $\sigma^{n-1}(\underline{r}^1) < \sigma^{n-1}(\underline{r}^2)$. Since σ^{n-1} preserves the circular order of $\underline{s}, \underline{r}^1$ and $\underline{r}^2$, and since $\underline{r}^1, \underline{r}^2 \leq \underline{s}$, it follows that $\underline{r}^1 < \underline{r}^2$, as required.

Now let us prove (b). We again fix our ideas by supposing that $h = p/q > 0$. Let $\mathtt{u}_n$ denote the common n-th itinerary entry of $\underline{s}^-$ and $\underline{s}^+$ with respect to

$\underline{s}$. Then the n-th entries of the addresses $\underline{s}^-$ and $\underline{s}^+$ are $\mathtt{u}_n + 1$ and $\mathtt{u}_n$, respectively.

Since $h > 0$, we thus have $s_* \geq \mathtt{u}_n + 1$ by Theorem 4.4. So for each j between 1 and $q - 1$, the nj-th itinerary entries $\mathtt{m}_j$ of $\underline{r}$ satisfy $\mathtt{m}_j \geq \mathtt{u}_n$ by Lemma 6.3. We claim that, for every $j = 0, \ldots, q - 2$,
$$\sigma^{(j+1)n}(\underline{r}) \in (\underline{s}^-, \underline{s}] \implies \sigma^{jn}(\underline{r}) \in (\underline{s}^-, \underline{s}].$$
Since $\sigma^{(q-1)n}(\underline{r}) = \underline{s}$, part (b) then follows by induction.

To prove the claim, suppose that $\sigma^{(j+1)n}(\underline{r}) \in (\underline{s}^-, \underline{s}]$. The first $n-1$ itinerary entries of $\underline{s}^-, \sigma^{jn}(\underline{r})$ and $\underline{s}$ are the same, so σ^{n-1} preserves the circular order of these addresses by Observation 2.2, Thus it is sufficient to show that $\sigma^{(j+1)n-1}(\underline{r}) > \sigma^{n-1}(\underline{s}^-)$. This is trivial if $\mathtt{m}_j > \mathtt{u}_n$, and follows from the fact that $\sigma^{(j+1)n}(\underline{r}) \in (\underline{s}^-, \underline{s}]$ if $\mathtt{m}_j = \mathtt{u}_n$. ∎

6.6. Corollary (Monotonicity of $\mathrm{addr}(W, h)$).
Let W be a hyperbolic component. Then the function $h \mapsto \mathrm{addr}(W, h)$ is strictly increasing on each of the intervals $\{h > 0\}$ and $\{h < 0\}$.

Proof. Consider the function $p/q \mapsto \mathtt{m}_1 \ldots \mathtt{m}_{q-1}$, where $\mathtt{m}_j$ are the numbers from Lemma 6.3. It is an easy exercise to check that this function is strictly increasing (with respect to lexicographic order).

If $W \neq \mathrm{Hyp}(\infty)$, the claim follows from Proposition 6.5 (a). Otherwise, the claim follows directly since itineraries and external addresses coincide in this case (compare Remark 2 after Definition 3.5). ∎

6.7. Proposition (Continuity Properties).
Let $W = \mathrm{Hyp}(\underline{s})$ be a hyperbolic component. Then $\lim_{h \to \pm\infty} \mathrm{addr}(W, h) = \underline{s}$. Furthermore, if $h_0 \in \mathbb{R}$, then the behavior of $\mathrm{addr}(W, h)$ for $h \to h_0$ is as follows.

(a) *If $h_0 \in \mathbb{Q} \setminus \mathbb{Z}$, then*
$$\lim_{h \nearrow h_0} \mathrm{addr}(W, h) = \underline{r}^- \text{ and }$$
$$\lim_{h \searrow h_0} \mathrm{addr}(W, h) = \underline{r}^+,$$
where $\underline{r} := \mathrm{addr}(W, h_0)$.

(b) *If $h_0 = 0$ and $\underline{s} \neq \infty$, then*
$$\lim_{h \nearrow 0} \mathrm{addr}(W, h) = \underline{s}^+ \text{ and }$$
$$\lim_{h \searrow 0} \mathrm{addr}(W, h) = \underline{s}^-.$$

(c) *Otherwise, the limit $\mathrm{addr}(W, h_0) := \lim_{h \to h_0} \mathrm{addr}(W, h)$ exists.*

Proof. Let $\mathtt{u}_1 \ldots \mathtt{u}_{n-1}*$ be the kneading sequence of W. Let $h_0 = p/q \in \mathbb{Q} \setminus \mathbb{Z}$, and $\underline{r} = \mathrm{addr}(W, h_0)$. Recall that, for parameters $\kappa \in \Gamma_{W, h_0}$, the landing point of the dynamic ray $g_{\underline{r}^+}$ lies on the distinguished boundary orbit of E_κ.

Since the rays landing on this cycle are permuted cyclically with rotation number p/q, it follows easily (as in Lemma 6.3) that

$$\text{itin}_{\underline{s}}(\underline{r}^+) = \overline{\mathtt{u}_1 \ldots \mathtt{u}_{n-1}\mathtt{m}_1 \ldots \mathtt{u}_1 \ldots \mathtt{u}_{n-1}\mathtt{m}_{q-2}\mathtt{u}_1 \ldots \mathtt{u}_{n-1}s_*\mathtt{u}_1 \ldots \mathtt{u}_{n-1}(s_* - 1)},$$

where the $\mathtt{m}_j$ are as in Lemma 6.3. It follows from Lemma 6.3 that the limit address $\lim_{h \nearrow h_0} \text{addr}(W, h)$ has the same itinerary under $\underline{s}$ as $\underline{r}^+$. If $n = 1$, then the two addresses must trivially be equal. Otherwise, this follows easily from Observation 6.4.

All other parts of the proposition are proved analogously. For each part, we need to prove the equality of two external addresses, at least one of which is given as a monotone limit of addresses $\text{addr}(W, h)$. In each case, it is easy to verify that the corresponding itineraries are equal, which, as above, implies that the same is true for the external addresses. ∎

We are now ready to state and prove our main theorem on the structure of the child components of a given hyperbolic component. If $\underline{s} := \text{addr}(W) \neq \infty$, then the *wake* of W is the set $\mathcal{W}(W) := (\underline{s}^-, \underline{s}^+)$; if $\underline{s} = \infty$, $\mathcal{W}(W)$ is defined to be all of $\overline{\mathcal{S}}$. In the following theorem, we use this only for simpler notation, but wakes will play an important role in the next section.

6.8. Theorem (Bifurcation Structure).
Let W be a hyperbolic component and $\underline{s} := \text{addr}(W)$. If $n \geq 2$, then the map $\text{addr}(W, \cdot) : \mathbb{Q} \setminus \mathbb{Z} \to \mathcal{S}$ has the following properties.

(a) $\text{addr}(W, \cdot)$ is strictly increasing for $h > 0$ and (separately) for $h < 0$;
(b) $\text{addr}(W, \frac{p}{q})$ is an intermediate external address of length qn (for $\frac{p}{q}$ in lowest terms);
(c) if $h \in \mathbb{Q} \setminus \mathbb{Z}$ such that $\Psi_W(2\pi i h) \in \mathbb{C}$, then the parameter $\Psi_W(2\pi i h)$ lies on the $\overline{\text{boundary of}} \text{Bif}(W, h) = \text{Hyp}(\text{addr}(W, h))$;
(d) $\overline{\mathcal{W}(W)} = \bigcup_{h \in \mathbb{Q} \setminus \mathbb{Z}} \mathcal{W}(\text{Bif}(W, h))$.
(e) $\lim_{h \to +\infty} \text{addr}(W, h) = \lim_{h \to -\infty} \text{addr}(W, h) = \underline{s}$;
(f) $\lim_{h \nearrow 0} \text{addr}(W, h) = \underline{s}^+$ and $\lim_{h \searrow 0} \text{addr}(W, h) = \underline{s}^-$.

These properties uniquely determine the map $\text{addr}(W, \cdot)$, and no such map exists if the preferred parametrization Ψ_W is replaced by some other conformal parametrization $\Psi : \mathbb{H} \to W$ with $\mu \circ \Psi = \exp$.

If $n = 1$, then the map $\text{addr}(W, \cdot)$ is strictly increasing on all of $\mathbb{Q} \setminus \mathbb{Z}$ and satisfies properties (b) to (d) above, and no other map has these properties.

Proof. Let us assume for simplicity that $n > 1$; the proofs for the case $n = 1$ are completely analogous. Property (a) is just the statement of Corollary 6.6, and (b) holds by definition. Property (c) is Proposition 6.2 (a). Properties (e) and (f) were proved in Proposition 6.7.

To establish (d), note first that the inclusion "⊃" is clear. To prove "⊂", let $\underline{r} \in \overline{\mathcal{W}(W)}$. If $\underline{r} \in \{\underline{s}^-, \underline{s}^+, \underline{s}\}$, then we are done by (e) and (f). Otherwise, there exists $h_0 \in \mathbb{R} \setminus \{0\}$ such that

$$\underline{t}^- := \lim_{h \nearrow h_0} \mathrm{addr}(W, h) \leq \underline{r} \leq \lim_{h \searrow h_0} \mathrm{addr}(W, h) =: \underline{t}^+.$$

If $h \notin \mathbb{Q} \setminus \mathbb{Z}$, then Proposition 6.7 (c) shows that $\underline{t}^- = \underline{r} = \underline{t}^+$, and we are done. Otherwise, Proposition 6.7 (a) implies that $\underline{r} \in \overline{\mathcal{W}(\mathrm{addr}(W, h))}$, which completes the proof of (d).

Let us now prove the uniqueness statements. Suppose that $a : \mathbb{Q} \setminus \mathbb{Z} \to \mathcal{S}$ also satisfies properties (a) to (d). Then, by item (c) and Corollary 5.5, we know that $a(h) = \mathrm{addr}(W, h)$ whenever $\Psi_W(ih) \in \mathbb{C}$. By Lemma 4.1, the set of such $h \in \mathbb{R}$ is open and dense, so there is a dense set of rationals on which a and $\mathrm{addr}(W, \cdot)$ agree. Properties (a) and (d) then easily imply that Proposition 6.7 is also true for the map a, which shows that both maps must be equal everywhere.

Finally, if the preferred parametrization Ψ_W is replaced by some other parametrization of W, then clearly no map a which satisfies (c) can also satisfy (a). Indeed, such a map must agree with $\mathrm{addr}(W, m + \cdot)$ on a dense set for some $m \in \mathbb{Z} \setminus \{0\}$. Therefore a is not monotone near m by (f). ■

We note the following consequence of the previous theorem for reference.

6.9. Corollary (Subwakes Fill Wake)**.**
Suppose that $\underline{s} \in \mathcal{W}(W) \setminus \{\mathrm{addr}(W)\}$. Then there exists a unique $h \in \mathbb{R}$ such that one (and only one) of the following hold.

(a) $h \in \mathbb{Q} \setminus \mathbb{Z}$ and $\underline{s} \in \mathcal{W}(\mathrm{Bif}(W, h))$,
(b) $h \in \mathbb{Q} \setminus \mathbb{Z}$ and $\underline{s}$ is a characteristic address of $\mathrm{Bif}(W, h)$, or
(c) $h \in (\mathbb{R} \setminus \mathbb{Q}) \cup (\mathbb{Z} \setminus \{0\})$ and $\underline{s} = \mathrm{addr}(W, h)$.

In particular, if $\underline{s}$ is unbounded, then Property (a) holds. ■

7. Internal Addresses

In this section, we describe the global bifurcation structure of hyperbolic components. The basic question we are now interested in is as follows: given two hyperbolic components V and W, one of which is contained in the wake of the other, how can we determine which bifurcations occur "between" V and W?

We will begin by dividing up the wake of a hyperbolic component W into *sector wakes* $\mathcal{W}(A)$ (one for every sector A of W) such that every child component of W belongs to the wake of the sector from which it bifurcates.

The wakes of two adjacent sectors are separated by a *sector boundary*, i.e. a periodic address which one should think of as the address of the parameter ray landing at the corresponding co-root of W (except that, for now, we do not know whether this co-root parameter actually exists). Every sector A has a natural associated *sector kneading sequence* $\mathbb{K}(A)$ (Definition 7.3). We give a simple description of how kneading sequences depend on the bifurcation structure of parameter space (Theorem 7.7). Finally, we will introduce *internal addresses*, which organize the information encoded in kneading sequences in a "human-readable" way.

Wakes of sectors and combinatorial arcs. Let W be a hyperbolic component, let $h_0 \in \mathbb{Z}$ and consider the sector
$$A = Sec(W, h_0 + frac12) = \Psi_W\big(\{z \in \mathbb{H} : \operatorname{Im} z \in (2\pi h_0, 2\pi(h_0 + 1))\}\big)$$
of W. The *sector boundaries* of A are defined to be
$$\underline{r}^- := \lim_{h \searrow h_0} \operatorname{addr}(W, h) \quad \text{and} \quad \underline{r}^+ := \lim_{h \nearrow h_0+1} \operatorname{addr}(W, h).$$
(Note that $\underline{r}^- = \operatorname{addr}(W, h_0)$ unless $h_0 = 0$, in which case $\underline{r}^-$ is the lower characteristic address of W, and similarly for $\underline{r}^+$.) The *wake* of A is denoted by $\mathcal{W}(A) := (\underline{r}^-, \underline{r}^+)$. If $\underline{r}$ is a sector boundary of a sector of A, we also say that $\underline{r}$ is a sector boundary of W.

Armed with this concept, we can introduce a natural (partial) order on sectors and hyperbolic components.

7.1. Definition (Combinatorial Arcs).
We define a partial ordering on the set of all hyperbolic components and sectors by writing $A \prec B$ if $\mathcal{W}(A) \supset \mathcal{W}(B)$. The combinatorial arc *$[A, B]$ is the set of all hyperbolic components or sectors C such that $A \prec C \prec B$.*

Similarly, if $\underline{s} \in \overline{\mathcal{W}(A)}$, then the combinatorial arc $[A, \underline{s}]$ is the set of all C with $A \prec C$ and $\underline{s} \in \overline{\mathcal{W}(C)}$. Note that $[A, \underline{s}]$ is linearly ordered by $\prec$.

Remark. We will often also consider open or half-open combinatorial arcs (A, B), $[A, B)$ or $(A, B]$, in which one or both of the endpoints are excluded.

Sector Boundaries and Kneading Sequences.

7.2. Lemma (Sector Boundaries and Itineraries).
Let W be a hyperbolic component of period n and with kneading sequence $\mathbb{K}(W) = \mathtt{u}_1 \ldots \mathtt{u}_{n-1}*$. Set $\underline{s} := \operatorname{addr}(W)$ and let $\underline{r} \in \mathcal{S}$. Then the following are equivalent.

   (a) *$\underline{r}$ is a sector boundary of W;*
   (b) *$\operatorname{itin}_{\underline{r}}(\underline{s}) = \mathbb{K}(W)$ and $\operatorname{itin}_{\underline{s}}(\underline{r}) = \overline{\mathtt{u}_1 \ldots \mathtt{u}_{n-1}\mathtt{m}}$ for some $\mathtt{m} \in \mathbb{Z}$.*

Furthermore, every sector boundary $\underline{r}$ satisfies
$$\sigma^j(\underline{r}) \notin \overline{\mathcal{W}(W)} \tag{4}$$
for $j = 1, \ldots, n-1$; in particular, $\underline{r}$ has (exact) period n.

Proof. It follows easily from Lemma 6.3 that for every $\mathtt{m} \in \mathbb{Z}$ there is a sector boundary $\underline{r}$ with $\operatorname{itin}_{\underline{s}}(\underline{r}) = \overline{\mathtt{u}_1 \ldots \mathtt{u}_{n-1}\mathtt{m}}$, and every sector boundary $\underline{r}$ has an itinerary of this form. Furthermore, every sector boundary $\underline{r}$ belongs to

$\overline{\mathcal{W}(W)}$, and thus satisfies $\text{itin}_{\underline{r}}(\underline{s}) = \mathbb{K}(W)$ by Observation 3.7 (recall that the iterates of $\underline{s}$ do not enter $\mathcal{W}(W)$). By Theorem 3.4, the characteristic addresses of W have period n. In particular, (a) implies (b).

Now let $\kappa \in W$ and let $\underline{r}$ be a sector boundary of W which is not a characteristic address. A hyperbolic expansion argument shows that the dynamic ray $g_{\underline{r}}$ lands on the boundary of the characteristic Fatou component U_1 of E_κ (compare [28, Proof of Theorem 6.2] or [22, Theorem 4.2.4]). By the definition of characteristic addresses, this landing point w is separated from the rest of its orbit by the characteristic rays of E_κ, and its orbit portrait is not essential. Hence, (4) holds (and $\underline{r}$ and w both have exact period n).

To prove that (b) implies (a), suppose that $\underline{r}'$ is an address which is not a sector boundary and has itinerary $\text{itin}_{\underline{s}}(\underline{r}') = \overline{u_1 \ldots u_{n-1} \mathtt{m}}$. Note that $\underline{r}'$ is necessarily periodic by Lemma 3.8. There is a sector boundary $\underline{r}$ which has the same itinerary as $\underline{r}'$; by Theorem 3.6, the dynamic rays $g_{\underline{r}}$ and $g_{\underline{r}'}$ have a common landing point. By the above, this implies that $\underline{r}$ is a characteristic address of W, and $\underline{r}' \notin \overline{\mathcal{W}(W)}$. By replacing $\underline{r}$ with the other characteristic address of W, if necessary, we may suppose that $\underline{r}$ and $\underline{r}'$ do not enclose $\underline{s}$. It then follows from Observation 6.4 that they must enclose a forward iterate of $\underline{s}$. Thus $\text{itin}_{\underline{r}'}(\underline{s}) \neq \text{itin}_{\underline{r}}(\underline{s}) = \mathbb{K}(\underline{s})$ by Observation 3.7, as required. ∎

7.3. Definition and Lemma (Kneading Entries).
Let W be a hyperbolic component of period n and $\mathbb{K}(W) = u_1 \ldots u_{n-1}*$.

(a) *Let A be a sector of W with sector boundaries $\underline{r}^- < \underline{r}^+$. Then there exists a number $\mathtt{u}(A) \in \mathbb{Z}$ (the* kneading entry *of A) such that*

$$\mathbb{K}^+(\underline{r}^-) = \mathbb{K}^-(\underline{r}^+) = \overline{u_1 \ldots u_{n-1} \mathtt{u}(A)} =: \mathbb{K}(A).$$

The sequence $\mathbb{K}(A)$ is called the kneading sequence *of the sector A.*

(b) *If $n \geq 2$, then there exists a number $\mathtt{u}(W) \in \{u_1, \ldots, u_{n-1}\}$ (the* forbidden kneading entry *of W) such that*

$$\mathbb{K}^-(\underline{s}^-) = \mathbb{K}^+(\underline{s}^+) = \overline{u_1 \ldots u_{n-1} \mathtt{u}(W)} =: \mathbb{K}^*(W)$$

(where $\underline{s}^-$ and $\underline{s}^+$ are the characteristic addresses of W). The sequence $\mathbb{K}^*(W)$ is called the forbidden kneading sequence *of W.*

The kneading entries of the sectors directly above and below the central internal ray of W are $\mathtt{u}(W) - 1$ and $\mathtt{u}(W) + 1$, respectively. If A and B are any other two adjacent sectors, with A above B, the kneading entries satisfy $\mathtt{u}(A) = \mathtt{u}(B) + 1$ (compare Figure 6(b)).

In particular, no two sectors have the same kneading entry and $\mathtt{u}(W)$ is the unique integer which is not assumed as the kneading entry of some sector of W.

(c) *In the period one case, every integer $\mathtt{u}$ is realized as the kneading entry of a sector of W, namely the sector at imaginary parts between $2\pi\mathtt{u}$ and $2\pi(\mathtt{u}+1)$). The period one component thus has no forbidden kneading sequence.*

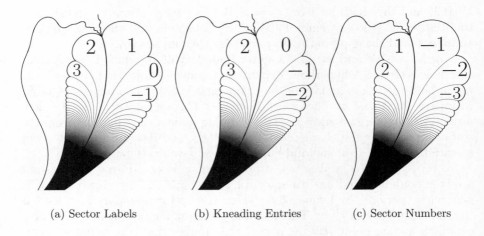

(a) Sector Labels   (b) Kneading Entries   (c) Sector Numbers

FIGURE 6. Three ways to label sectors, illustrated for the component at address $\underline{s} = 0110\frac{1}{2}\infty$, with $\underline{s}^- = \overline{011002}$ and $\underline{s}^+ = \overline{020101}$. The central internal ray of $\operatorname{Hyp}(\underline{s})$ is emphasized.

For $\mathtt{m} \neq \mathtt{u}(W)$, we denote the unique sector A satisfying $\mathtt{u}(A) = \mathtt{m}$ by $\operatorname{Sec}(W, \mathtt{m})$.

Proof. In the period one case, the sector boundaries are the addresses $\overline{m}$ with $m \in \mathbb{Z}$, and the claims are trivial. So suppose that $n \geq 2$; to prove (a), let us fix our ideas by supposing that $\underline{r}^- < \underline{r}^+ < \underline{s} := \operatorname{addr}(W)$. It follows from Lemma 6.3 that

$$\operatorname{itin}_{\underline{s}}(\underline{r}^-) = \overline{\mathtt{u}_1 \ldots \mathtt{u}_{n-1}(s_* - 1)} \quad \text{and} \quad \operatorname{itin}_{\underline{s}}(\underline{r}^+) = \overline{\mathtt{u}_1 \ldots \mathtt{u}_{n-1} s_*},$$

where $s_* = s_*(A)$ is the sector label of A. By Observation 3.7 and Lemma 7.2, we see that $\mathbb{K}^+(\underline{r}^-) = \operatorname{itin}_{\underline{s}}(\underline{r}^-)$ and $\mathbb{K}^+(\underline{r}^+) = \operatorname{itin}_{\underline{s}}(\underline{r}^+)$. Since $\underline{r}^+$ is periodic of period n, we thus have

$$\mathbb{K}^+(\underline{r}^-) = \overline{\mathtt{u}_1 \ldots \mathtt{u}_{n-1}(s_* - 1)} = \mathbb{K}^-(\underline{r}^+),$$

so $\mathtt{u}(A) = s_* - 1$ is the desired kneading entry.

We already know from Lemma 3.9 that a number $\mathtt{u}(W) \in \mathbb{Z}$ with the required property exists. By Observation 6.4, the $(n-1)$-th iterates of $\underline{s}^-$ and $\underline{s}^+$ must inclose an iterate of $\underline{s}$. Thus the entry $\mathtt{u}(W)$ must occur in $\mathbb{K}(\underline{s})$; i.e. $\mathtt{u}(W) \in \{\mathtt{u}_1, \ldots, \mathtt{u}_{n-1}\}$. The remainder of the statement follows from the fact that, for any address $\underline{r}$ of period n, the n-th entries of $\mathbb{K}^-(\underline{r})$ and $\mathbb{K}^+(\underline{r})$ differ exactly by one. ∎

Periodic addresses and hyperbolic components. The last ingredient we require for our analysis of bifurcation structure is the fact that every periodic address is the sector boundary of some hyperbolic component.

7.4. Proposition (Periodic Rays and Intermediate Addresses).
Let $\underline{r} \in \mathcal{S}$ be periodic of period n. Then there exists an intermediate external address $\underline{s}$ of length n such that $\underline{r}$ is a sector boundary of $W := \mathrm{Hyp}(\underline{s})$. If there exists $\underline{\widetilde{r}}$ such that $\langle \underline{r}, \underline{\widetilde{r}} \rangle$ is a characteristic ray pair, then $\underline{r}$ and $\underline{\widetilde{r}}$ are the characteristic addresses of W.

Proof. Let $\overline{\mathtt{u}_1 \ldots \mathtt{u}_{n-1} \mathtt{u}_n^{\mathtt{u}_n+1}}$ be the kneading sequence of r and choose a parameter κ for which the singular value lies on the dynamic ray $g_{\underline{r}}$. (It is well-known that such parameters exist for every periodic, and in fact every exponentially bounded, external address; see e.g. [3] or [15, Proposition 3.2].[3] In fact, the set of such parameters is an injective curve to infinity, called the *parameter ray* at address $\underline{s}$; see [13, 15, 14].) Consider the piece of $g_{\underline{r}}$ which connects the singular value with $+\infty$. The dynamic ray $g_{\sigma^{n-1}(\underline{r})} : (0, \infty) \to \mathbb{C}$ is a preimage component of this piece, and thus tends to $+\infty$ for $t \to \infty$, and to $-\infty$ for $t \to 0$. The ray $g_{\sigma^{n-1}(\underline{r})}$ and its translates thus cut the dynamic plane into countably many domains which we call "strips" (similarly as in the definition of itineraries for attracting parameters).

Since $g_{\sigma^{n-1}(\underline{r})} = E_\kappa^{n-1}(g_{\underline{r}})$ tends to $-\infty$ as $t \to 0$, the curve $g_{\underline{r}}$ has an intermediate external address $\underline{s}$ of length n as $t \to 0$ (as well as the usual address $\underline{r}$ as $t \to \infty$). Dynamic rays do not intersect, so both ends of the ray $g_{\underline{r}}$ tend to ∞ within the same strip. In other words, the itineraries of $\underline{r}$ and $\underline{s}$ (with respect to $\underline{r}$) coincide in the first entry. We can apply the same argument to $g_{\sigma(\underline{r})}, \ldots, g_{\sigma^{n-2}(\underline{r})}$, and conclude that $\mathrm{itin}_{\underline{r}}(\underline{r})$ and $\mathrm{itin}_{\underline{r}}(\underline{s})$ coincide in the first $n-1$ entries; i.e., $\mathrm{itin}_{\underline{r}}(\underline{s}) = \mathtt{u}_1 \ldots \mathtt{u}_{n-1}*$. To fix our ideas, let us suppose that $\underline{r} < \underline{s}$.

Note that for $j = 0, \ldots, n-1$, we have $\sigma^j(\underline{s}), \sigma^j(\underline{r}) \notin (\underline{r}, \underline{s})$. Otherwise, we could choose a minimal such j, and Observation 6.4 would imply that σ^j maps $[\underline{r}, \underline{s}]$ to $[\sigma^j(\underline{r}), \sigma^j(\underline{s})] \subset [\underline{r}, \underline{s}]$, which is clearly impossible since no interval is invariant under the shift. By Observation 3.7, it follows that

$$\mathbb{K}(\underline{s}) = \mathrm{itin}_{\underline{r}}(\underline{s}) = \mathtt{u}_1 \ldots \mathtt{u}_{n-1} * \quad \text{and} \quad \mathrm{itin}_{\underline{s}}(\underline{r}) = \mathbb{K}^-(\underline{r}) = \overline{\mathtt{u}_1 \ldots \mathtt{u}_n}.$$

By Lemma 7.2, $\underline{r}$ is a sector boundary of $W := \mathrm{Hyp}(\underline{s})$.

Finally, suppose that $\underline{\widetilde{r}} \in \mathcal{S}$ is another address of period n such that $\langle \underline{r}, \underline{\widetilde{r}} \rangle$ is a characteristic ray pair. Then $\mathrm{itin}_{\underline{r}}^-(\underline{\widetilde{r}}) = \overline{\mathtt{u}_1 \ldots \mathtt{u}_n}$ by Lemma 3.9 (2), and it follows that the cyclic order of $\underline{r}, \underline{\widetilde{r}}$ and $\underline{s}$ is preserved by σ^{n-1}. Since $\sigma^{n-1}(\underline{\widetilde{r}}) < \sigma^{n-1}((\underline{r}))$ and $\sigma^{n-1}(\underline{s}) = \infty$, this means that $\underline{r} < \underline{s} < \underline{\widetilde{r}}$. By Lemma 3.10, $\underline{r}$ and $\underline{\widetilde{r}}$ are the characteristic addresses of $\underline{s}$. ∎

Evolution of kneading sequences. We are now in a position to describe completely how kneading sequences change as we move along a combinatorial arc.

---

[3]It would also be possible to formulate the following argument completely combinatorially, without appeal to the existence of such parameters.

7.5. Lemma (Nested Components Have Different Kneading Sequences).
Let V and W be two different hyperbolic components with $W \prec V$. Then $\mathbb{K}(V) \neq \mathbb{K}(W)$.

Proof. This is a direct corollary of Lemma 3.10, "(c) $\Rightarrow$ (a)". ∎

7.6. Lemma (Kneading Sequences in Sector Wakes).
Let W be a hyperbolic component, and let A be a sector of W.
  (a) *Let $\underline{s} \in \mathcal{W}(A)$ and suppose that $m \geq 1$ is such that $\mathbb{K}(A)$ and $\mathbb{K}(\underline{s})$ have different m-th entries. Then there exists a hyperbolic component $V \in [A, \underline{s}]$ of period at most m.*
  (b) *Suppose that $V \succ A$ is a hyperbolic component such that there are no components of periods up to m in $[A, V)$. Then $\mathbb{K}^*(V)$ agrees with $\mathbb{K}(A)$ in the first m entries.*
  (c) *Suppose that $h \in \mathbb{Q} \setminus \mathbb{Z}$. Then $\mathbb{K}^*(\mathrm{Bif}(W, h)) = \mathbb{K}(\mathrm{Sec}(W, h))$.*
  (d) *Suppose that $h \in \mathbb{R} \setminus \mathbb{Q}$. Then $\mathbb{K}(\mathrm{addr}(W, h)) = \mathbb{K}(\mathrm{Sec}(W, h))$.*

Remark. Note that the addresses in (d) are non-periodic with periodic kneading sequences; compare the remark after Lemma 3.8. (It follows from Theorem 7.7 below that, conversely, these are the only external addresses with this property.)

Proof. Let K consist of all intermediate external addresses of length $\leq m$ in $\mathcal{W}(A)$. Then K is a closed, and hence compact, subset of $\overline{\mathcal{S}}$. The set $\mathcal{U} := \{\mathcal{W}(\underline{r}) : \underline{r} \in K\}$ is an open cover of K, and hence has a finite subcover. Since wakes of hyperbolic components are either nested or disjoint, it follows that $U := \bigcup \mathcal{U}$ has finitely many connected components, each of which is an element of $\mathcal{U}$.

By Proposition 7.4, the set $M := \mathcal{W}(A) \setminus \overline{U}$ contains no periodic addresses of period $\leq m$; thus each of the first m entries of $\mathbb{K}(\underline{s})$ is locally constant when considered as a function of $\underline{s} \in M$. On the other hand, if $(\underline{r}^-, \underline{r}^+) \in \mathcal{U}$, then $\mathbb{K}^-(\underline{r}^-) = \mathbb{K}^+(\underline{r}^+)$. Thus the first m kneading sequence entries remain constant throughout M and agree with those of $\mathbb{K}(A)$.

This proves (a). Items (b) through (d) are direct corollaries. ∎

We are now ready to prove the main result of this section.

7.7. Theorem (Determining Components on a Combinatorial Arc).
Suppose that A is a sector of a hyperbolic component W, and let $\underline{s} \in \mathcal{W}(A)$. Let j be the index of the first entry at which $\underline{u} := \mathbb{K}(A)$ and $\underline{\tilde{u}} := \mathbb{K}(\underline{s})$ differ (or $j = \infty$ if no such entry exists).
  (a) *Then there are no hyperbolic components of period less than j on the combinatorial arc $[A, \underline{s}]$. If $j < \infty$, then there exists a unique period j component $V \in [A, \underline{s}]$. This component has forbidden kneading sequence $\mathbb{K}^*(V) = \overline{u_1 \ldots u_j}$; if furthermore $\tilde{u}_j \in \mathbb{Z}$, then $\underline{s}$ belongs to the wake of the sector $\mathrm{Sec}(V, \tilde{u}_j)$.*

(b) *These statements remain true if $\underline{s}$ and $\underline{\tilde{u}}$ are replaced by a hyperbolic component W' and its forbidden kneading sequence $\mathbb{K}^*(W')$.*

Proof. To prove (a), let m be the minimal period of a hyperbolic component on $[A, \underline{s})$, and let $V \in [A, \underline{s})$ be a component of period m. (If there are no hyperbolic components in $[A, \underline{s})$, then $j = \infty$ by Lemma 7.6 (a), and there is nothing further to prove.) By Lemma 7.6 (a) and (b), we have $m \leq j$ and $\mathbb{K}^*(V) = \overline{\mathsf{u}_1 \ldots \mathsf{u}_m}$. In particular, V is unique by Lemma 7.5.

If $\tilde{\mathsf{u}}_m \notin \mathbb{Z}$, then $\tilde{\mathsf{u}}_m \neq \mathsf{u}_m$, so $m = j$ and we are done. Otherwise, $\underline{s} \in \mathcal{W}(B)$ for some sector B of V. By choice of m and uniqueness of V, there are no components of period $\leq m$ on $[B, \underline{s})$, and it follows from Lemma 7.6 (a) that $\tilde{\mathsf{u}}_m = \mathsf{u}(B) \neq \mathsf{u}(V) = \mathsf{u}_m$; in particular, $m = j$.

Part (b) can be reduced to (a) by choosing $\underline{s}$ to be an address just outside $\mathcal{W}(W)$. ∎

Internal addresses. Repeated applications of the preceding theorem enable us to determine the periods and combinatorial order of all hyperbolic components on the combinatorial arc between $\mathrm{Hyp}(\infty)$ and $\underline{s}$ solely from $\mathbb{K}(\underline{s})$. *Internal addresses*, introduced for Mandel- and Multibrot sets in [17], organize this information in a more convenient way.

Since neither sector labels nor kneading entries can be easily identified in a picture of parameter space, let us introduce a third labelling method for the sectors of a hyperbolic component. The *sector number* of a given sector is the nonzero integer obtained by counting sectors in counterclockwise orientation, starting at the central internal ray. In other words, the sector number of a sector A is the integer $\mathsf{u}(A) - \mathsf{u}(W) \in \mathbb{Z} \setminus \{0\}$ and can be found in parameter space by counting from the root of W (compare Figure 6(c)).

7.8. Definition (Internal Addresses).
Let $\underline{s} \in \overline{\mathcal{S}}$. Consider the sequence $W_1 = \mathrm{Hyp}(\infty) \prec W_2 \prec \ldots$ of all hyperbolic components $W \in [\mathrm{Hyp}(\infty), \underline{s}]$ which have the property that all components on $(W, \underline{s}]$ have higher period than W. Denote by n_j the period of W_j, and by m_j the sector number of the sector of W_j containing $\underline{s}$. (We adopt the convention that $m_j = \infty$ if $\underline{s} = \mathrm{addr}(W_j)$ and $m_j = k + \frac{1}{2}$ if $\underline{s}$ is the sector boundary with kneading sequence $\mathbb{K}(\underline{s}) = \mathsf{u}_1 \ldots \mathsf{u}_{n_j - 1} \overline{\mathsf{u}(W_j) + k + 1 \atop \mathsf{u}(W_j) + k}$.)

The internal address of $\underline{s}$ is defined as the (finite or infinite) sequence
$$(n_1, m_1) \mapsto (n_2, m_2) \mapsto (n_3, m_3) \mapsto \ldots$$

The internal address of an intermediate external address $\underline{s}$ is also called the internal address of the associated hyperbolic component $\mathrm{Hyp}(\underline{s})$.

Remark 1. By Theorem 7.7, we could alternatively define W_{j+1} as the unique period of lowest period on the combinatorial arc $(W_j, \underline{s}]$. In particular, the components are indeed ordered as stated and the sequence n_j is strictly increasing. (These facts also follow easily directly from the definition.)

Remark 2. Internal addresses do not label hyperbolic components uniquely, reflecting certain symmetries of parameter space: two child components of a given sector with the same denominator but differing numerators of the bifurcation angle have the same internal address. This is the only ambiguity and thus uniqueness can be achieved by specifying bifurcation angles in the internal address, see Theorem A.7.

7.9. Corollary (Computing Internal Addresses).
Two external addresses have the same internal address if and only if they have the same kneading sequence.

Furthermore, the internal address $(1, m_1) \mapsto (n_2, m_2) \mapsto (n_3, m_3) \mapsto \ldots$ *of any* $\underline{s} \in \mathcal{S}$ *can be determined inductively from* $\underline{u} := \mathbb{K}(\underline{s})$ *by the following procedure:*

Set $m_1 := u_1$. *To compute* (n_{i+1}, m_{i+1}) *from* n_i, *continue the first* n_i *entries of* $\underline{u}$ *periodically to a periodic sequence* $\underline{u}^i$. *Then* n_{i+1} *is the position of the first difference between* $\underline{u}$ *and* $\underline{u}^i$. *Furthermore,*

$$m_{i+1} = \begin{cases} u_{n_{i+1}} - u^i_{n_{i+1}} & \text{if } u_{n_{i+1}} \in \mathbb{Z} \\ \infty & \text{if } u_{n_{i+1}} = * \\ k - u^i_{n_{i+1}} + \frac{1}{2} & \text{if } u_{n_{i+1}} = \frac{k+1}{k}. \end{cases}$$

(If $\mathbb{K}(\underline{s})$ *is periodic of period* n_{i+1}, *or if* $\underline{s}$ *is intermediate of length* n_{i+1}, *then the algorithm terminates, and the internal address is finite.)*

Remark 1. $\underline{u}^i$ is the kneading sequence of the sector of W_i containing $\underline{s}$. The forbidden kneading sequence $\mathbb{K}^*(W_i)$ can be obtained by repeating the first n_i entries of $\underline{u}^{i-1}$ periodically. In particular, if $\underline{s}$ is an intermediate external address, then the forbidden kneading sequence of $\underline{s}$ consists of the first n_{k-1} entries of $\mathbb{K}(\underline{s})$ repeated periodically, where k is the length of the internal address of $\underline{s}$.

Remark 2. As an example, let us consider $\underline{s} = 030\frac{1}{2}\infty$. This address has kneading sequence $\mathbb{K}(\underline{s}) = 0200*$. Applying the above procedure, we obtain that $\underline{u}^1 = \overline{0}$, $\underline{u}^2 = \overline{02}$ and $\underline{u}^3 = \overline{0200}$, resulting in the internal address

$$(1, 0) \mapsto (2, 2) \mapsto (4, -2) \mapsto (5, \infty).$$

Remark 3. There is an obvious converse algorithm: given the internal address of $\underline{s}$, we can determine the kneading sequence $\mathbb{K}(\underline{s})$ by inductively defining $\underline{u}^i$.

Proof. The correctness of the algorithm is an immediate corollary of Theorem 7.7. In particular, the internal address of $\underline{s}$ depends only on $\mathbb{K}(\underline{s})$. Conversely, applying this procedure to two different kneading sequences will produce different internal addresses. ∎

We note the following consequences of Theorem 7.7 for further reference.

7.10. Corollary (Combinatorics of Nested Wakes).
Let W and V be hyperbolic components with $W \prec V$. Then all entries of $\mathbb{K}(W)$ also occur in $\mathbb{K}(V)$.

Proof. Let p be the period of W, and let $\mathbf{u}_1 \ldots \mathbf{u}_{p-1}*$ be the kneading sequence of W. The proof proceeds by induction on the number n of hyperbolic components on the combinatorial arc (W, V) which have period less than p. If $n = 0$, then it follows from Theorem 7.7 that $\mathbb{K}^*(V)$ begins with $\mathbf{u}_1 \ldots \mathbf{u}_{p-1}$, and we are done. (Recall from Lemma 7.3 that all entries of $\mathbb{K}^*(V)$ occur in $\mathbb{K}(V)$.)

If $n > 0$, then let $V' \in (W, V)$ be a component of period $\leq p$. We can now apply the induction hypothesis first to W and V', and then to V' and V. ∎

7.11. Corollary (Components on the Combinatorial Arc).
Let $\underline{s} \in \mathcal{S}$ and $\underline{u} := \mathbb{K}(\underline{s})$. Suppose that $n \geq 1$ is an index such that $\mathbf{u}_n \in \mathbb{Z}$ and $\mathbf{u}_j \in \mathbb{Z} \setminus \{\mathbf{u}_n\}$ for all $j < n$. Then there exists a hyperbolic component W with $\mathbb{K}^*(W) = \overline{\mathbf{u}_1 \ldots \mathbf{u}_n}$ and $\underline{s} \in \mathcal{W}(W)$.

Proof. By the internal address algorithm, n appears in the internal address of $\underline{s}$; let V be the associated period n component. The child component of V containing $\underline{s}$ has the required property. ∎

Infinitely many essential periodic orbits. To conclude this section, we will give a simple necessary and sufficient criterion for an attracting exponential map to have infinitely many essential periodic orbits. (A non-necessary sufficient condition under which this occurs was the main result of [2].)

7.12. Proposition (Infinitely Many Essential Orbits).
Let W be a hyperbolic component and $\kappa \in W$. Then the characteristic ray pairs of essential periodic orbits of E_κ are exactly the characteristic ray pairs of hyperbolic components V with $V \prec W$.

In particular, the number of essential periodic orbits of E_κ is finite if and only if the internal address of W is of the form

$$(5) \qquad (1, m_1) \mapsto (n_2, m_2) \mapsto (n_3, m_3) \mapsto \ldots (n_k, \infty),$$

with $n_j | n_{j+1}$ for all $j < k$. In this case, the number of essential periodic orbits is exactly $k - 1$.

Remark. Using the internal address algorithm, it is simple to convert (5) to a (somewhat more complicated) statement about the kneading sequence of W.

Proof. The first statement follows immediately from Lemma 3.9 and Proposition 7.4. In particular, E_κ has only finitely many essential periodic orbits if and only if W is contained in only finitely many wakes. This is the case if and only if W can be reached by finitely many bifurcations from the period one component $\mathrm{Hyp}(\infty)$, which is exactly what the statement about internal addresses means. ∎

Appendix A. Further Topics

In this appendix, we treat some further developments which are naturally related to the discussion in this article. We first prove a theorem which associates external addresses to unbounded connected subsets of $\mathbb{C}$ which intersect closures of hyperbolic components in at most one point and which will be needed in [14]. After this, we outline a number of related results which might eventually lead to a proof of the fact that any non-hyperbolic component in parameter space would be bounded.

Addresses of Connected Sets. An important application of the results in this article is to obtain control over the combinatorial position of curves and, more generally, connected sets within exponential parameter space.

To make this precise, suppose that $A \subset \mathbb{C}$ is an unbounded connected set containing at most finitely many attracting or indifferent parameters. Let $\underline{s} \in \mathcal{S}$ and suppose that there exist two hyperbolic components W_1, W_2 with $\mathrm{addr}(W_1) < \underline{s} < \mathrm{addr}(W_2)$ and the following property: for sufficiently large $R > 0$, the (unique) component U of

$$\{\mathrm{Re}\, z > R\} \setminus \left(\Gamma_{W_1,0}((-\infty, -1]) \cup \Gamma_{W_2,0}((-\infty, -1]) \right)$$

which is unbounded but has bounded imaginary parts satisfies $U \cap A = \emptyset$. In this case, we say that *A is separated from $\underline{s}$*. We define

$$\mathrm{Addr}(A) := \{\underline{s} \in \mathcal{S} : A \text{ is not separated from } \underline{s}\}.$$

Note that $\mathrm{Addr}(A)$ is a closed subset of $\mathcal{S}$. Furthermore, $\mathrm{Addr}(A)$ is nonempty since A is unbounded.

Remark 1. If $\gamma : [0, \infty) \to \mathbb{C}$ is a curve to infinity which contains no indifferent parameters, then $\mathrm{Addr}(\gamma)$ consists of a single external address; compare also [26, Section 2]. In particular, if $G_{\underline{s}}$ is a *parameter ray tail* as defined in [13, 15], then $\mathrm{Addr}(\gamma) = \{\underline{s}\}$.

Remark 2. $\mathrm{Addr}(A)$ may contain several addresses, for instance if A consists of several parameter rays together with a common landing point. As we will see below, such cases are rather restricted.

A.1. Theorem (Addresses of Connected Sets).
Let $A \subset \mathbb{C}$ be an unbounded connected set containing at most finitely many attracting or indifferent parameters. If $\# \mathrm{Addr}(A) \geq 2$, then either

(a) *all addresses in $\mathrm{Addr}(A)$ share the same infinite internal address (and thus the same non-periodic kneading sequence), or*

(b) *$\mathrm{Addr}(A)$ consists of the two characteristic addresses of some hyperbolic component.*

COMBINATORIAL BIFURCATIONS OF EXPONENTIAL MAPS   359

Proof. Suppose that $\#\operatorname{Addr}(A) \geq 2$. We claim that, for any hyperbolic component W, either $\operatorname{Addr}(A) \subset \mathcal{W}(W)$ or $\operatorname{Addr}(A) \cap \mathcal{W}(W) = \emptyset$. In fact, we prove the following stronger fact.

Claim. If $\operatorname{Addr}(A) \cap \mathcal{W}(W) \neq \emptyset$, then there exists $h_0 \in \mathbb{R} \cup \{\infty\}$ such that $\operatorname{Addr}(A) = \{\operatorname{addr}(W, h_0)\}$ if $h \notin \mathbb{Q} \setminus \mathbb{Z}$ and $\operatorname{Addr}(A) \subset \{\mathcal{W}(\operatorname{Bif}(W, h_0))\}$ otherwise. (We adopt the convention that $\operatorname{addr}(W, \infty) = \operatorname{addr}(W)$.)

Proof. Let us suppose that $\operatorname{addr}(W) \neq \infty$ (the proof in the period one case is completely analogous), and let $\underline{s} \in \operatorname{Addr}(A) \setminus \operatorname{addr}(W)$. To fix our ideas, assume that $\underline{s} < \operatorname{addr}(W)$.

By Corollary 6.9, there exists a unique $h_0 \neq 0$ such that $\underline{s} = \operatorname{addr}(W, h_0)$ if $h_0 \notin \mathbb{Q} \setminus \mathbb{Z}$ and $\underline{s} \in \overline{\mathcal{W}(\operatorname{Bif}(W, h_0))}$ otherwise.

By Lemma 4.1, we can choose rational h^- and h^+ with $h^- < h_0 < h^+$ arbitrarily close to h_0 such that $\Psi_W(ih^\pm) \in \mathbb{C}$. Since A contains at most finitely many indifferent parameters, we may assume that
$$A \cap \left(\Psi_W(i[h^-, h^+]) \setminus \Psi_W(ih_0)\right) = \emptyset.$$

Consider the Jordan arc
$$\gamma := \Gamma_{\operatorname{Bif}(W, h^-), 0} \cup \{\Psi_W(ih^-)\} \cup \widetilde{\gamma} \cup \{\Psi_W(ih^+)\} \cup \Gamma_{\operatorname{Bif}(W, h^+), 0},$$
where $\widetilde{\gamma} \subset W$ is some curve connecting $\Psi_W(ih^-)$ and $\Psi_W(ih^+)$. Then γ does not intersect A. Let U denote the component of $\mathbb{C} \setminus \gamma$ which does not contain a left half plane. Since
$$\underline{r}^+ := \operatorname{addr}(W, h^+) > \underline{s} > \underline{r}^- := \operatorname{addr}(W, h),$$
it follows by the definition of $\operatorname{Addr}(A)$ that $A \subset U$. Therefore A is separated from every address in $\mathcal{S} \setminus [\underline{r}^-, \underline{r}^+]$, and so $\operatorname{Addr}(A) \subset [\underline{r}^-, \underline{r}^+]$. Letting h_1^+ and h_1^- tend to h_0, we have
$$\operatorname{Addr}(A) \subset \left[\lim_{h \nearrow h_0} \operatorname{addr}(W, h), \lim_{h \searrow h_0} \operatorname{addr}(W, h)\right],$$
as required, proving the claim. △

Let us now distinguish two cases.
Case 1: Some address in $\operatorname{Addr}(A)$ *has an infinite internal address.* It then follows from the claim that all addresses have the same internal address, and by Corollary 7.9, they also all share the same kneading sequence.
Case 2: All addresses in $\operatorname{Addr}(A)$ *have a bounded internal address.* It follows by the definition of internal addresses that there exists some hyperbolic component W such that $\operatorname{Addr}(A) \subset \overline{\mathcal{W}(W)}$ while $\operatorname{Addr}(A)$ is not contained in the closure of the wake of any child component of W. Since $\#\operatorname{Addr}(A) \geq 2$, it follows from the claim that $\operatorname{Addr}(A)$ must consist of the two characteristic addresses of W. ∎

We record the following special case for use in [14] (see there or in [13, 15] for the definition of a parameter ray).

A.2. Corollary (Parameter Rays Accumulating at a Common Point).
Suppose that $G_{\underline{s}^1}$ and $G_{\underline{s}^2}$ are parameter rays which have a common accumulation point $\kappa_0 \in \mathbb{C}$. Then $|s_j^1 - s_j^2| \leq 1$ for all $j \geq 1$.

Proof. The set $A := G_{\underline{s}^1} \cup G_{\underline{s}^2} \cup \{\kappa_0\}$ is unbounded and connected. Since $\mathrm{Addr}(A)$ contains $\underline{s}^1$ and $\underline{s}^2$, by the previous theorem either $\mathbb{K}(\underline{s}^1) = \mathbb{K}(\underline{s}^2)$ or $\underline{s}^1$ and $\underline{s}^2$ are the characteristic addresses of some hyperbolic component. In either case, the claim follows. ■

In fact, we can sharpen Theorem A.1 to the following statement.

A.3. Theorem (Addresses of Connected Sets II).
Let $A \subset \mathbb{C}$ be an unbounded connected set which contains at most one attracting or indifferent parameter. Then exactly one of the following holds.
  (a) $\mathrm{Addr}(A)$ *consists of a single external address (bounded or unbounded).*
  (b) $\mathrm{Addr}(A)$ *consists of two bounded external addresses, both of which have the same infinite internal address (and thus the same non-periodic kneading sequence).*
  (c) $\mathrm{Addr}(A)$ *consists of the two characteristic addresses of some hyperbolic component.*
  (d) $\mathrm{Addr}(A)$ *consists of at least three but finitely many preperiodic addresses. Moreover, there is a postsingularly finite parameter $\kappa_0 \in \mathbb{C}$ such that, for every $\underline{s} \in \mathrm{Addr}(A)$, the parameter ray $G_{\underline{s}}$ lands at κ_0.*

Remark. The *Squeezing Lemma*, proved in [26], also shows that $\mathrm{Addr}(A)$ cannot contain intermediate or exponentially unbounded addresses.

Sketch of proof. By Theorem A.1, it only remains to consider the case where all addresses in $\mathrm{Addr}(A)$ share the same infinite internal address a.

First suppose that $\mathrm{Addr}(A)$ contains some unbounded infinite external address $\underline{s}$. We need to show that $\mathrm{Addr}(A) = \{\underline{s}\}$. Let $(W_i)_{i \geq 1}$ be the hyperbolic components appearing in the internal address of $\underline{s}$, and set

$$I := \bigcap_i \overline{\mathcal{W}(W_i)}.$$

Then I is a closed connected subset of $\mathcal{S}$, and $\mathrm{Addr}(A) \subset I$. We claim that I contains no intermediate external addresses (and thus consists of a single point).

Indeed, if $\underline{r} \in I$ was an intermediate external address, then by Corollary 7.10, every entry of $\mathbb{K}(W_i)$ is one of the finitely many symbols of $\mathbb{K}(\underline{r})$. However, $\mathbb{K}(W_i) \to \mathbb{K}(\underline{s})$ by Corollary 7.9, and $\mathbb{K}(\underline{s})$ is unbounded. This is a contradiction.

Finally, it remains to consider the case where $\mathrm{Addr}(A)$ consists of at least three external addresses. By the previous step, all addresses in $\mathrm{Addr}(A)$ are bounded; let M be an upper bound on the size of the entries in their common kneading sequence, and let $d := 2M + 2$. Then the map

$$\underline{s} \mapsto \sum_{j \geq 1} \frac{s_k}{d^j} \pmod{1}$$

takes $\mathrm{Addr}(A)$ injectively to a set $\widetilde{A} \subset \mathbb{R}/\mathbb{Z}$ which has the property that, for any wake of a hyperbolic component W *in the Multibrot set* $\mathcal{M}_d$ *of degree* d, either $\widetilde{A} \subset \mathcal{W}(W)$ or $\widetilde{A} \cap \mathcal{W}(W) = \emptyset$. It follows from the Branch Theorem for Multibrot sets (see [17, Theorem 9.1] or [30]) that $\widetilde{A}$ (and $\mathrm{Addr}(A)$) consists of finitely many preperiodic addresses.

It is easy to see that, in this case, for any $\underline{s}, \underline{r} \in \mathrm{Addr}(A)$, $\mathrm{itin}_{\underline{s}}(\underline{r}) = \mathbb{K}(\underline{s})$. By the main result of [16], there exists a parameter κ_0 for which the dynamic ray $g_{\underline{s}}$ lands at the singular value (and thus the singular orbit is finite for this parameter). It follows from [32, Proposition 4.4] that all rays $g_{\underline{r}}$ with $\underline{r} \in \mathrm{Addr}(A)$ also land at the singular value. It follows easily from Hurwitz's theorem and the stability of orbit portraits (compare Proposition 5.2) that all parameter rays $G_{\underline{r}}$ with $\underline{r} \in \mathrm{Addr}(A)$ land at κ_0 (compare [22, Theorem 5.14.5] or [27, Theorem IV.6.1] for details). ∎

We believe that the ideas of [26] can be extended to show that all "queer", i.e. nonhyperbolic, stable components (which conjecturally do not exist) must be bounded; compare the discussion in [26, Section 8]. The following corollary, which states that such a component could be unbounded in at most two directions, is a first step in this direction.

A.4. Corollary (Nonhyperbolic Components).
Suppose that U *is a nonhyperbolic stable component in exponential parameter space (or more generally, any connected subset of parameter space which contains no attracting, indifferent or escaping parameters). Then* $\mathrm{Addr}(U)$ *consists of at most two external addresses.*

Proof. This follows directly from the previous theorem except in the case of item (d). In the latter case, it follows since U cannot intersect any of the parameter rays landing at the associated postsingularly finite parameters, and thus is separated from all but at most two of these addresses. ∎

The *bifurcation locus* $\mathcal{B}$ in exponential locus is the set of parameters which are not structurally stable (compare [12, Section 8]). It is known (compare [22, Theorem 5.1.4 and Lemma 5.1.5]) that $\mathcal{B}$ is exactly the closure of the set of escaping parameters.

For quadratic polynomials, a famous theorem by Douady and Hubbard from [8] asserts that local connectivity of the quadratic bifurcation locus (i.e., the boundary of the Mandelbrot set) implies density of hyperbolicity in the space of quadratic polynomials (see [30, Corollary 3.6]). Local connectivity is not the right property to investigate in the exponential family, since

$\mathcal{B}$ is clearly not locally connected. However, one can still formulate other topological questions in this spirit.

For example, it seems reasonable to expect that every parameter in the bifurcation locus is either on a parameter ray or the landing point of a ray. The following corollary shows that this, again, would imply density of hyperbolicity.

A.5. Corollary.
Suppose that every parameter $\kappa \in \mathcal{B}$ is either on a parameter ray or the landing point of a parameter ray.

Then every structurally stable parameter κ is hyperbolic, and in particular hyperbolic parameters are dense in $\mathbb{C}$.

Proof. By Theorem A.3, at most finitely many parameter rays can accumulate on a given nonhyperbolic component U. Since ∂U is contained in the bifurcation locus, this means by assumption that ∂U is finite, which is impossible. So every structurally stable parameter is hyperbolic. Structural stability is dense by [12, Theorem 10], so we are done. ∎

Angled Internal Addresses. Internal addresses do not label hyperbolic components uniquely. For completeness, we will now discuss a way of decorating internal addresses with bifurcation angles to restore uniqueness. As explained in [17], these angles explain certain (at least combinatorial) symmetries of Mandel- and Multibrot sets, and, in the limit, also of exponential parameter space.

A.6. Definition (Angled Internal Address)**.**
Let $\underline{s} \in \mathcal{S}$, and let W_j be the components in the internal address of W_j. Then the angled internal address of $\underline{s}$ is

$$(1, h_1) \mapsto (n_2, h_2) \mapsto (n_3, h_3) \mapsto \ldots,$$

where $W_{j+1} \subset \mathcal{W}(\operatorname{Bif}(W_j, h_j))$.

A.7. Theorem (Uniqueness of Angled Internal Addresses)**.**
No two hyperbolic components share the same angled internal address.

Sketch of proof. Suppose that $W_1 \neq W_2$ have the same angled internal address a. Let M be an upper bound for the entries of $\mathbb{K}(W_1) = \mathbb{K}(W_2)$, and set $d := 2M + 2$. Similarly as in the proof of Theorem A.3, it then follows that the Multibrot Set $\mathcal{M}_d$ contains two different hyperbolic components $\widetilde{W}_1$ and $\widetilde{W}_2$ which both have the same angled internal address. This contradicts uniqueness of angled internal addresses for Multibrot sets [17, Theorem 9.2]. ∎

A Combinatorial Tuning Formula. Let $\underline{s}$ be an intermediate external address of length at least 2. We will give an analog of the concept of tuning for polynomials, on a combinatorial level. For every i, let us denote by r_{i-1}^i the first n entries of the sector boundary $\mathrm{Bdy}(\underline{s},{}_{i-1}^i)$.

A map $\tau : \overline{\mathcal{S}} \to \mathcal{W}(\underline{s})$ is called a tuning map for $\underline{s}$, if $\tau(-\infty) = \underline{s}$ and

$$\tau(k\underline{r}) = \begin{cases} r_{\mathrm{u}_n+k-1}^{\mathrm{u}_n+k}\tau(\underline{r}) & \tau(\underline{r}) > \underline{s} \\ r_{\mathrm{u}_n+k}^{\mathrm{u}_n+k+1}\tau(\underline{r}) & \tau(\underline{r}) < \underline{s} \\ r_{\mathrm{u}_n+k-\frac{1}{2}}^{\mathrm{u}_n+k+\frac{1}{2}}\tau(\underline{r}) & \tau(\underline{r}) = \underline{s}. \end{cases}$$

(In other words, $\tau(\underline{r})$ is obtained from $\underline{r}$ by replacing each symbol with a suitable sequence r_{i-1}^i of n symbols.) There are exactly two such maps, which are uniquely defined by choosing $\tau(\overline{0})$ to be either $\overline{r_{\mathrm{u}_n}^{\mathrm{u}_n+1}}$ or $\overline{r_{\mathrm{u}_n-1}^{\mathrm{u}_n}}$. (Note that under a tuning map, some addresses which are not exponentially bounded will be mapped to addresses which are exponentially bounded. This is related to the fact that topological renormalization fails for exponential maps; see [24] or [22, Section 4.3].)

A.8. Theorem (Tuning Theorem).
If the internal address of $\underline{r}$ is $(1, m_1) \mapsto (n_2, m_2) \mapsto (n_3, m_3) \mapsto \ldots$, and the internal address of $\underline{s}$ is $(1, \widetilde{m}_1) \mapsto (\widetilde{n}_2, \widetilde{m}_2) \mapsto \ldots \mapsto (n, \infty)$, then the internal address of $\tau(\underline{r})$ is

$$(1, \widetilde{m}_1) \mapsto (\widetilde{n}_2), \widetilde{m}_2) \mapsto \ldots \mapsto (n, m_1') \mapsto (n * \widetilde{n}_2, m_2) \mapsto (n * \widetilde{n}_3, m_3) \mapsto \ldots,$$

where m_1' is $m_1 + 1$ or m_1, depending on whether $m_1 \geq 0$ or $m_1 < 0$.

Sketch of proof. This can be easily inferred from the well-known tuning formula for Multibrot sets (see e.g. [19, Theorem 8.2] or [17, Proposition 6.7]).

Alternatively, it is not difficult to give a direct combinatorial proof of this fact; see [22, 5.11.2]. □

Appendix B. Combinatorial Algorithms

In this article, we have seen several ways to describe hyperbolic components, and our results allow us to compute any of these from any other. Since these algorithms have not always been made explicit in the previous treatment, we collect them here.

First note the kneading sequence of a hyperbolic component W can be easily computed from its intermediate external address according to the definition (Definition 3.5). How to compute the internal address of a hyperbolic component from its kneading sequence was shown in Corollary 7.9, which also describes how to obtain the forbidden kneading sequence $\mathbb{K}^*(W)$ from $\mathbb{K}(W)$. Furthermore, $\mathbb{K}^*(W)$ can easily be computed from its characteristic ray pair $\langle \underline{s}^-, \underline{s}^+ \rangle$.

Algorithm B.1 (Computing $\mathrm{addr}(W)$ given $\langle \underline{s}^-, \underline{s}^+ \rangle$).
Given: *Kneading sequence* $\mathbb{K}(W) = \mathrm{u}_1 \ldots \mathrm{u}_{n-1}*$ *and some* $\underline{r} \in \overline{\mathcal{W}(W)}$.

Aim: *Compute* addr(W).

Algorithm: *Let* $\underline{s}^n := \infty$, *and compute* $\underline{s}^{n-1}, \ldots, \underline{s}^1$ *inductively by choosing* $\underline{s}^{j-1}$ *to be the unique preimage of* $\underline{s}^j$ *in* $(\mathbf{u}_{j-1}\underline{r}, (\mathbf{u}_{j-1}+1)\underline{r})$. *Then* addr($W$) = $\underline{s}^1$.

Proof. By Observation 3.7 and Lemma 3.8, addr(W) is the unique address $\underline{s}$ with itin$_{\underline{r}}(\underline{s}) = \mathbb{K}(W)$. This is exactly the address $\underline{s}^1$ computed by this algorithm. ∎

Algorithm B.2 (Computing $\langle \underline{s}^-, \underline{s}^+ \rangle$ given addr(W)).
Given: *Kneading sequence* $\mathbb{K}(W) = \mathbf{u}_1 \ldots \mathbf{u}_{n-1}*$, *some* $\underline{r} \in \overline{\mathcal{W}(W)}$ *and some* $s_* \in \mathbb{Z}$.

Aim: *Compute* Bdy$\left(W, {}^{s_*}_{s_*-1}\right)$.

Algorithm: *Compute the unique preimage* $\widetilde{\underline{r}}$ *of* $s_*\underline{r}$ *whose itinerary (with respect to* $\underline{r}$*) begins with* $\mathbf{u}_1 \ldots \mathbf{u}_{n-1}$, *as in Algorithm B.1. The sought address is obtained by continuing the first n entries of* $\widetilde{\underline{r}}$ *periodically.*

Remark. To actually compute the characteristic addresses of W, first compute the forbidden kneading entry $\mathbf{u}_n = \mathbf{u}(W)$, and then apply the algorithm to $s_* = \mathbf{u}_n$ and $s_* = \mathbf{u}_n + 1$.

Proof. Using Observation 6.4 (similarly to Proposition 6.5 (a), one shows that the interval between $\widetilde{\underline{r}}$ and the required address is mapped bijectively by σ^n. ∎

Algorithm B.3 (Compute the Angled Internal Address of W).
Given: *An intermediate external address* $\underline{s}$.

Aim: *Compute the angled internal address of* $\underline{s}$.

Algorithm: *Let n be the length of* $\underline{s}$, *and calculate the kneading sequence* $\underline{\mathbf{u}} = \mathbf{u}_1 \ldots \mathbf{u}_{n-1}* := \mathbb{K}(\underline{s})$. *Set* $n_1 := 1$ *and* $m_1 := \mathbf{u}_1$.

Given, for some $j \geq 1$, two numbers $n_j < n$ and $p^j \in \mathbb{Z}$, we calculate three numbers $n_{j+1} > n_j$, $m_{j+1} \in \mathbb{Z}$ and $h_j \in \mathbb{Q}$. The algorithm terminates when n_{j+1} is equal to n; at this point the angled internal address of $\underline{s}$ *will be given by*

$$(n_1, h_1) \mapsto (n_2, h_2) \mapsto \ldots \mapsto (n_j, h_j) \mapsto (n, \infty).$$

Step 1: Calculation of n_{j+1} and the corresponding sector number. *Define n_{j+1} to be the first index at which* $\underline{\mathbf{u}}$ *and* $\underline{\mathbf{u}}^j := \overline{\mathbf{u}_1 \ldots \mathbf{u}_{n_j}}$ *differ. If $n_{j+1} \neq n$, then set* $m_{j+1} := \mathbf{u}_{n_{j+1}} - \mathbf{u}^j_{n_{j+1}}$.

Step 2: Determining the denominator. *Let us inductively define a finite sequence $\ell_1 \leq \ell_2 \leq \cdots \leq \ell_r$ as follows. Set $\ell_1 := n_{j+1}$; if n_j does not divide ℓ_k, let ℓ_{k+1} be the first index at which* $\underline{\mathbf{u}}^j$ *and* $\overline{\mathbf{u}_1 \ldots \mathbf{u}_{\ell_k}}$ *differ. Otherwise, we terminate, setting $r := k$. Set $q_j := \ell_r/n_j$.*

Step 3: Determining the numerator. If $q_j = 2$, set $p_j := 1$. Otherwise, calculate

$$x := \#\{k \in \{2, \ldots, q_j - 2\} : \sigma^{kn_j}(\underline{s}) \text{ lies between } \underline{s} \text{ and } \sigma(\underline{s})\}.$$

Set $p_n := x + 1$ if $\sigma(\underline{s}) > \underline{s}$, and $p_n := n_j - x - 1$ otherwise.

Step 4: Determining h_j. h_j is defined to be $m_j + p_n/q_n$ if $j = 1$ or $m_j < 0$, and $m_j - 1 + p_n/q_n$ otherwise.

Proof. Let W_j denote the j-th component in the internal address of $\underline{s}$. By Corollary 7.9, the value of n_j computed by our algorithm will be the period of W_j, and if $n_j \neq n$, then m_j is the number of the sector containing $\underline{s}$.

If $n_j \neq n$, let V_j be the child component of W containing $\underline{s}$. By Theorem 7.7, this component has period $\ell_r = q_j n_j$. By Proposition 6.5 (a), the interval $\mathcal{W}(V_j)$ is mapped bijectively by $\sigma^{(q_j-2)n_j}$. (Note that this shows, in particular, that $\ell_r < 2n_j + n$, and thus limits the iterations necessary in Step 2.) Therefore, the order of the iterates of $\underline{s}$ and those of $\mathrm{addr}(V_j)$ is the same. It follows that $V_j = \mathrm{Bif}(W_j, h_j)$, where h_j is defined as indicated. ∎

Remark. In Step 3, we could have instead first calculated the intermediate external address of the bifurcating component in question (using Algorithm B.4 below), and then calculated its rotation number. This introduces an extra step in the algorithm, but makes the proof somewhat simpler.

It remains to indicate how an intermediate external address can be recovered from its angled internal address. Let us first note how to handle the special case of computing a combinatorial bifurcation.

Algorithm B.4 (Computing Combinatorial Bifurcations).
Given: $\underline{s} := \mathrm{addr}(W)$, $s_* \in \mathbb{Z}$ and $\alpha = \frac{p}{q} \in (0,1) \cap \mathbb{Q}$.
Aim: *Compute* $\mathrm{addr}(W, s_*, \alpha)$.

Algorithm: *Compute the unique address whose itinerary under $\underline{s}$ is as given by Lemma 6.3. (Alternatively, compute the sector boundaries of the given sector and apply the Combinatorial Tuning Formula.)* ∎

Algorithm B.5 (Computing $\mathrm{addr}(W)$ from an angled internal address).
Given: *The angled internal address* $a = (n_1, h_1) \mapsto \ldots \mapsto (n, \infty)$ *of some hyperbolic component W.*
Aim: *Compute the unique intermediate external address $\underline{s}$ whose angled internal address is a.*

Algorithm: *Let W_i denote the component represented by the i-th entry of a. We will compute $\underline{s}^i := \mathrm{addr}(W_i)$ inductively as follows; note that $\underline{s}^1 = \infty$.*

Compute the upper characteristic address $\underline{r}^+$ of the hyperbolic component $V_i := \mathrm{Bif}(W_i, h_i,)$ (by applying Algorithms B.4 and B.2, or by using the combinatorial tuning formula).

Let $\underline{t}^1 \leq \underline{r}^+$ be maximal such that $\underline{t}^1$ is periodic of period at most n_{i+1}. If the period of $\underline{t}^1$ is strictly less than n_{i+1}, then $\underline{t}^1$ is the upper characteristic address of some hyperbolic component. Compute the lower characteristic address $\underline{t}^{1-}$ of this component and let $\underline{t}^2 \leq \underline{t}^{1-}$ be maximal such that $\underline{t}^2$ is periodic of period at most n_{i+1}.

Continue until an address $\underline{t}^k$ is computed which is periodic of period n_{i+1}. This is the upper characteristic addresses of $\underline{s}^{i+1}$, and we can compute $\underline{s}^i$ using Algorithm B.1.

Proof. Recall that $\underline{s}^{i+1}$ is not contained in the wake of any component $U \prec V_i$ which has smaller period than n_{i+1}, and by the uniqueness of angled internal addresses (Theorem A.7), there are only finitely many periodic addresses $\underline{t}^j$ which are encountered in each step. Thus, the algorithm will indeed terminate and compute an address which is periodic of period n_{i+1}. The associated hyperbolic component then has angled internal address

$$(n_1, h_1) \mapsto \ldots \mapsto (n_j, h_j) \mapsto (n_{j+1}, \infty)$$

and the claim follows by Theorem A.7. ∎

Finally, let us mention that it is very simple to decide whether a given hyperbolic component is a child component, and to calculate the address of the parent component in this case. In particular, it is easy to decide whether two given hyperbolic components have a common parabolic boundary point (or, in fact, *any* common boundary point, compare [26, Proposition 8.1]).

Algorithm B.6 (Primitive and Satellite Components).
Let $\underline{s} \in \mathcal{S}$ be an intermediate external address, with $\mathbb{K}(\underline{s}) = u_1 \ldots u_{n-1}*$. If there exists some $u_n \in \mathbb{Z}$ such that $\overline{u_1 \ldots u_{n-1} u_n}$ is periodic with period $j < n$, then $\mathrm{Hyp}(\underline{s})$ is a child component of $\mathrm{Hyp}(\sigma^{n-j}(\underline{s}))$.

Otherwise, $\mathrm{Hyp}(\underline{s})$ is a primitive component.

Proof. This is an immediate consequence of Corollary 5.5 and Theorem 6.8. ∎

List of Symbols

$[A, \underline{s})$	(combinatorial arc)	350
$[A, B]$	(combinatorial arc)	350
$\prec$	(combinatorial order)	350
U_1	(characteristic Fatou component)	325
$\mathrm{Addr}(A)$	(addresses associated to a connected set A)	358
$\mathrm{addr}(A, p/q)$	(combinatorial bifurcation in Sector A at angle p/q)	343
$\mathrm{addr}(\gamma)$	(external address of γ)	323
$\mathrm{addr}(\kappa)$	(intermediate external address of κ)	325
$\mathrm{addr}(\underline{s}, s_*, p/q)$	(combinat. bifurc. in $\mathrm{Sec}(\mathrm{Hyp}(\underline{s}), s_*)$ at angle p/q)	343
$\mathrm{addr}(W)$	(external address of W)	335
$\mathrm{addr}(W, h)$, $h \in \mathbb{Q} \setminus \mathbb{Z}$	(combinatorial bifurcation at height h)	343
$\mathrm{addr}(W, h)$, $h \neq 0$	(combinatorial bifurcation at height h)	347
$\mathrm{Bif}(W, h)$	(child component of W at height h)	343
$F(t)$	(model for exponential growth)	323
$\Gamma_{W,h}$	(internal ray at height h)	334
$g_{\underline{s}}$	(dynamic ray)	323
$\mathrm{Hyp}(\underline{s})$	(hyperbolic component at address $\underline{s}$)	336
$I(E_\kappa)$	(set of escaping points)	323
$\mathrm{itin}_{\underline{s}}(\underline{r})$	(itinerary of $\underline{r}$)	329
$J(E_\kappa)$	(Julia set)	323
$\mathbb{K}(\underline{s})$	(kneading sequence of $\underline{s}$)	329
$\mathbb{K}^+(\underline{s})$, $\mathbb{K}^-(\underline{s})$	(upper and lower kneading sequences of $\underline{s}$)	329
$\mathbb{K}^*(W)$	(forbidden kneading sequence)	351
μ	(multiplier map)	334
Ψ_W	(preferred parametrization of W)	334
$\mathcal{S}, \overline{\mathcal{S}}$	(sequence spaces)	324
$s_*(\kappa)$	(sector label)	336
$\underline{s}^-, \underline{s}^+$	(characteristic addresses of $\underline{s}$)	328
$\mathrm{Sec}(\kappa)$	(sector of κ)	336
$\mathrm{Sec}(W, h)$	(sector of W)	336
$\mathrm{Sec}(W, \mathrm{m})$	(sector with kneading entry m)	352
$\mathrm{Sec}(W, s_*)$	(sector with sector label s_*)	337
σ	(shift map)	324
τ	(tuning map)	363
$\mathrm{u}(A)$	(kneading entry of A)	351
$\mathrm{u}(W)$	(forbidden kneading entry)	351
$\mathcal{W}(A)$	(wake of a sector A)	350
$\mathcal{W}(W)$	(wake of W)	348

Index

attracting dynamic ray, 325
  broken, 326
  principal, 325

characteristic addresses/rays
  of an attracting or parabolic map, 328
  of an orbit portrait, 327
characteristic Fatou component, 325
characteristic ray pair, 328
characteristic sector, 327
child component
  of a hyperbolic component, 343
  of a parabolic parameter, 339
co-roots of a hyperbolic component, 335
combinatorial arc, 350
combinatorial bifurcation, 343
curve to $\pm\infty$, 322

distinguished boundary orbit, 326
dynamic ray, 323
dynamic root, 328

external address, 323
  infinite, 323
  intermediate, 324
  of a curve γ, 323, 324
  of a hyperbolic component, 335
  of an attracting or parabolic map, 325

forbidden kneading sequence/entry, 351

hyperbolic component, 334

infinite external address, 323
intermediate external address, 324
  of a curve γ, 324
  of a hyperbolic component, 335
  of an attracting or parabolic map, 325
internal address, 355
  angled, 362
internal ray, 334
  central, 335
itinerary, 328

kneading entry, 351
  forbidden, 351
kneading sequence, 329
  forbidden, 351
  of a sector, 351

nonhyperbolic components, 361

orbit portrait, 327
  primitive, 327
  satellite, 327

parent component, 339
preferred parametrization, 334
primitive parameter, 338
principal attracting ray, 325

queer components, 361

ray period
  of a parabolic parameter, 338
  of an orbit portrait, 327
root of a hyperbolic component, 335

satellite parameter, 338
sector, 336
sector boundary, 350
sector label, 336
sector number, 355

tuning map, 363

vertical order, 325

wake
  of a hyperbolic component, 348
  of a sector, 350

References

[1] I. Noel Baker and Philip J. Rippon, *Iteration of exponential functions*, Ann. Acad. Sci. Fenn. Ser. A I Math. **9** (1984), 49–77.

[2] Ranjit Bhattacharjee and Robert L. Devaney, *Tying hairs for structurally stable exponentials*, Ergodic Theory Dynam. Systems **20** (2000), no. 6, 1603–1617.

[3] Clara Bodelón, Robert L. Devaney, Michael Hayes, Gareth Roberts, Lisa R. Goldberg, and John H. Hubbard, *Hairs for the complex exponential family*, Internat. J. Bifur. Chaos Appl. Sci. Engrg. **9** (1999), no. 8, 1517–1534.

[4] ———, *Dynamical convergence of polynomials to the exponential*, J. Differ. Equations Appl. **6** (2000), no. 3, 275–307.

[5] Robert L. Devaney, *Julia sets and bifurcation diagrams for exponential maps*, Bull. Amer. Math. Soc. (N.S.) **11** (1984), no. 1, 167–171.

[6] Robert L. Devaney, Núria Fagella, and Xavier Jarque, *Hyperbolic components of the complex exponential family*, Fund. Math. **174** (2002), no. 3, 193–215.

[7] Robert L. Devaney, Lisa R. Goldberg, and John H. Hubbard, *A dynamical approximation to the exponential map by polynomials*, Preprint, MSRI Berkeley, 1986, published as [3, 4].

[8] Adrien Douady and John Hubbard, *Etude dynamique des polynômes complexes*, Prépublications mathémathiques d'Orsay (1984 / 1985), no. 2/4.

[9] Dominik Eberlein and Dierk Schleicher, *Rational parameter rays of Multibrot sets*, in preparation, compare Thesis 1999-02 on the Stony Brook Thesis Server.

[10] Alexandre È. Eremenko, *On the iteration of entire functions*, Dynamical systems and ergodic theory (Warsaw, 1986), Banach Center Publ., vol. 23, PWN, Warsaw, 1989, pp. 339–345.

[11] Alexandre È. Eremenko and Mikhail Yu. Lyubich, *Iterates of entire functions*, Preprint, Physico-Technical Institute of Low-Temperature Physics Kharkov, 1984.

[12] ———, *Dynamical properties of some classes of entire functions*, Ann. Inst. Fourier (Grenoble) **42** (1992), no. 4, 989–1020.

[13] Markus Förster, *Parameter rays for the exponential family*, Diplomarbeit, Techn. Univ. München, 2003, Available as Thesis 2003-03 on the Stony Brook Thesis Server.

[14] Markus Förster, Lasse Rempe, and Dierk Schleicher, *Classification of escaping exponential maps*, Preprint, 2004, arXiv:math.DS/0311427, to appear in Proc. Amer. Math. Soc.

[15] Markus Förster and Dierk Schleicher, *Parameter rays for the exponential family*, Preprint, 2005, arXiv:math.DS/0505097, submitted for publication.

[16] John H. Hubbard, Dierk Schleicher, and Mitsuhiro Shishikura, *A topological characterization of postsingularly finite exponential maps and limits of quadratic differentials*, Manuscript, 2004, submitted for publication.

[17] Eike Lau and Dierk Schleicher, *Internal addresses in the Mandelbrot set and irreducibility of polynomials*, Preprint #1994/19, Institute for Mathematical Sciences, SUNY Stony Brook, 1994, arXiv:math.DS/9411238.

[18] John Milnor, *On rational maps with two critical points*, Experiment. Math. **9** (2000), no. 4, 481–522, arXiv:math.DS/9709226.

[19] ———, *Periodic orbits, externals rays and the Mandelbrot set: an expository account*, Astérisque (2000), no. 261, xiii, 277–333, arXiv:math.DS/9905169, Géométrie complexe et systèmes dynamiques (Orsay, 1995).

[20] ———, *Dynamics in one complex variable*, third ed., Annals of Mathematics Studies, vol. 160, Princeton University Press, Princeton, NJ, 2006.

[21] Carsten L. Petersen and Gustav Ryd, *Convergence of rational rays in parameter spaces*, The Mandelbrot set, theme and variations, London Math. Soc. Lecture Note Ser., vol. 274, Cambridge Univ. Press, Cambridge, 2000, pp. 161–172.
[22] Lasse Rempe, *Dynamics of exponential maps*, doctoral thesis, Christian-Albrechts-Universität Kiel, 2003, http://e-diss.uni-kiel.de/diss_781/.
[23] _____, *A landing theorem for periodic rays of exponential maps*, Proc. Amer. Math. Soc **134** (2006), no. 9, 2639–2648, arXiv:math.DS/0307371.
[24] _____, *Topological dynamics of exponential maps on their escaping sets*, Ergodic Theory Dynam. Systems **26** (2006), no. 6, 1939–1975, arXiv:math.DS/0309107.
[25] _____, *Nonlanding dynamic rays of exponential maps*, Ann. Acad. Sci. Fenn. **32** (2007), 353–369, arXiv:math.DS/0511588.
[26] Lasse Rempe and Dierk Schleicher, *Bifurcations in the space of exponential maps*, Preprint #2004/03, Institute for Mathematical Sciences, SUNY Stony Brook, 2004, arXiv:math.DS/0311480, submitted for publication.
[27] Dierk Schleicher, *On the dynamics of iterated exponential maps*, Habilitation thesis, TU München, 1999.
[28] _____, *Attracting dynamics of exponential maps*, Ann. Acad. Sci. Fenn. Math. **28** (2003), 3–34.
[29] _____, *Hyperbolic components in exponential parameter space*, C. R. Math. Acad. Sci. Paris **339** (2004), no. 3, 223–228.
[30] _____, *On fibers and local connectivity of Mandelbrot and Multibrot sets*, Fractal geometry and applications: a jubilee of Benoît Mandelbrot. Part 1, Proc. Sympos. Pure Math., vol. 72, Amer. Math. Soc., Providence, RI, 2004, pp. 477–517.
[31] Dierk Schleicher and Johannes Zimmer, *Escaping points of exponential maps*, J. London Math. Soc. (2) **67** (2003), no. 2, 380–400.
[32] _____, *Periodic points and dynamic rays of exponential maps*, Ann. Acad. Sci. Fenn. Math. **28** (2003), 327–354.

DEPARTMENT FOR MATHEMATICAL SCIENCES, UNIVERSITY OF LIVERPOOL, LIVERPOOL, L69 7ZL, UNITED KINGDOM
E-mail address: l.rempe@liverpool.ac.uk

INTERNATIONAL UNIVERSITY BREMEN, P.O. BOX 750 561, 28725 BREMEN, GERMANY
E-mail address: dierk@iu-bremen.de

BAKER DOMAINS

P.J. RIPPON

ABSTRACT. This paper surveys the current state of knowledge about Baker domains, including: the rate of growth of iterates in a Baker domain, the connection with singular values, and the various different types of Baker domain. Also, many examples of functions with Baker domains are given and it is proved that transcendental meromorphic functions of the form $f(z) = z + R(z)e^{Q(z)}$, where R is rational and Q is a polynomial, have infinitely many invariant Baker domains.

Dedicated to the memory of Professor Noel Baker.

1. INTRODUCTION

The extension of the Fatou-Julia theory of complex dynamics from rational functions to transcendental meromorphic functions leads to the occurrence of two new types of dynamical phenomena, known as Baker domains and wandering domains. Noel Baker made extensive contributions to our understanding of both these phenomena, and Baker domains are named after him. The aim of this paper is to survey the current state of knowledge about Baker domains.

Throughout this paper $f : \mathbb{C} \to \hat{\mathbb{C}}$ is a transcendental meromorphic function and we denote by f^n, $n = 0, 1, 2, \ldots$, the nth iterate of f. The *Fatou set* $F(f)$ is defined to be the set of points $z \in \mathbb{C}$ such that $(f^n)_{n \in \mathbb{N}}$ is well-defined, meromorphic and forms a normal family in some neighborhood of z. The complement of $F(f)$ is called the *Julia set* $J(f)$ of f. An introduction to the properties of these sets can be found in [19], and also [48] for the case of entire functions.

The set $F(f)$ is completely invariant, so for any component U of $F(f)$ there exists, for each $n = 0, 1, 2, \ldots$, a component of $F(f)$, which we call U_n, such that $f^n(U) \subset U_n$. If, for some $p \geq 1$, we have $U_p = U_0 = U$, then we say that U is a *periodic* component of *period* p (or *p-periodic* component), assuming p to be minimal, and $U_0, \ldots, U_{p-1}$ is a *p-cycle*. There are then five possible types of periodic components, as follows:

1991 *Mathematics Subject Classification.* Primary: 30D40.

- U contains an attracting periodic point z_0 of period p. In this case, $f^{np}(z) \to z_0$ as $n \to \infty$ for $z \in U$, and U is called an *attracting component*.
- ∂U contains a periodic point z_0 of period p and $f^{np}(z) \to z_0$ as $n \to \infty$ for $z \in U$. In this case, U is called a *parabolic component* or *Leau domain*.
- $f^p : U \to U$ is analytically conjugate to a Euclidean rotation (through an angle that is an irrational multiple of π) of the unit disc onto itself. In this case, U is called a *Siegel disc*.
- $f^p : U \to U$ is analytically conjugate to a Euclidean rotation (through an angle that is an irrational multiple of π) of an annulus onto itself. In this case, U is called a *Herman ring*.
- There exists $z_0 \in \partial U$ such that $f^{np}(z) \to z_0$ for $z \in U$ as $n \to \infty$ but $f^p(z_0)$ is not defined. In this case, U is called a *Baker domain*.

See [19, Theorem 6] for this result, including detailed historical references. This classification can be extended to various other cases such as analytic self-maps of the punctured plane $\mathbb{C}^* = \mathbb{C}\setminus\{0\}$ (see [8]) and functions meromorphic outside a small exceptional set (see [10] and [28]).

If U_n is not eventually periodic, then we say that U is a *wandering component* of $F(f)$, or a *wandering domain*.

The name Baker domain was introduced by Eremenko and Lyubich; see their fundamental paper [35]. Before that, names such as *domain of attraction of* ∞ [37] and *essentially parabolic domain* [13] were used. Both [19] and [48] include an introduction to Baker domains and this survey aims to build on these accounts.

If f is a transcendental *entire* function with a Baker domain U, then $f^n(z) \to \infty$ as $n \to \infty$, for $z \in U$. If f is a meromorphic function with a p-cycle of Baker domains $U_0, \ldots, U_{p-1}$, then there is at least one component U_k, $0 \le k \le p-1$, with the property that $f^{np}(z) \to \infty$ as $n \to \infty$, for $z \in U_k$.

In [4, Theorem 3.1], Baker showed that if f is a transcendental entire function, then any multiply connected component of $F(f)$ must be a wandering domain. Thus, if U is a Baker domain of an *entire* function f, then U is simply connected. This is not true in general for meromorphic functions, even those with only finitely many poles, as we see in the next section. Also, we note that any multiply connected Baker domain must be infinitely connected, by a result of Baker, Kotus and Lü [13].

The study of Baker domains has revealed that they are in a sense more complicated than the other types of invariant Fatou components. In particular, there are different types of Baker domains and they are related to singular values in various different ways.

The plan of the paper is as follows. In Section 2 we give many examples of Baker domains. In Section 3 we describe the rate of growth of iterates in a Baker domain and in Section 4 we discuss the relationship between Baker domains and singular values (this section has considerable overlap with the survey [24, Section 2.2]). In Section 5 we describe results which classify Baker domains into different types and discuss the boundaries of Baker domains. Finally, in Section 6 we show that certain types of meromorphic functions have infinitely many Baker domains.

2. Examples of Baker domains

The first example of a Baker domain was given by Fatou [37, Example I].

Example 2.1. Fatou observed that for the function
$$f(z) = z + 1 + e^{-z}$$
we have $\Re(f(z)) > \Re(z)$ and $f^n(z) \to \infty$ whenever $\Re(z) > 0$. Hence the right half-plane $\{z : \Re(z) > 0\}$ is contained in an invariant Baker domain U. Fatou showed that for this function U is the only component of $F(f)$.

This example illustrates a common way to prove the existence of a Baker domain for a meromorphic function f, namely, to use analytic estimates for f in order to find a domain V which is invariant under f^p, for some $p \in \mathbb{N}$, and which contains at least one point z such that $f^{np}(z) \to z_0$ but $f^p(z_0)$ is not defined. Then, by the classification of periodic components, V is contained in a periodic component U of $F(f)$ (with period some divisor of p) and U must be a Baker domain. The following examples were all constructed this way.

Example 2.2. Baker [3] showed that, for sufficiently large positive a, the function
$$f(z) = z + \frac{\sin \sqrt{z}}{\sqrt{z}} + a$$
has an invariant Baker domain which contains an invariant parabola-shaped domain $V = \{x + iy : y^2 < 4(x+1), x > a^2\}$. Figure 1 shows an approximate picture of this Baker domain in the case $a = 6$ (the white points are slowly escaping), kindly provided by Dominique Fleischmann, who has recently shown that this function has an invariant Baker domain for all $a > 0$ [38].

Example 2.3. Baker, Kotus and Lü [13, page 606] showed that
$$f(z) = \frac{1}{z} - e^z$$
has a 2-cycle U_0, U_1 of Baker domains such that
$$f^{2n}(z) \to \infty \text{ for } z \in U_0, \quad f^{2n}(z) \to 0 \text{ for } z \in U_1,$$
and U_0 contains a set of the form $V = \{z : 3\pi/4 < \arg z < 5\pi/4, |z| > R\}$, where $R > 0$, which is invariant under f^2.

Figure 1

Example 2.4. König [44, Example 3] showed that, for certain values of the real constants a and b,
$$f(z) = \frac{1}{z} + ae^{-z} + b$$
has a 3-cycle U_0, U_1, U_2 of Baker domains such that
$$f^{3n}(z) \to \infty \text{ for } z \in U_0, \quad f^{3n}(z) \to b \text{ for } z \in U_1, \quad f^{3n}(z) \to 0 \text{ for } z \in U_2,$$
and U_0 contains a right half-plane which is invariant under f^3.

Example 2.5. Rippon and Stallard [53, Theorem 3] showed that
$$f(z) = az(1 + e^{-z^p}), \quad \text{where } a > 1, \, p \in \mathbb{N},$$
has p invariant Baker domains, one in each sector $\{z : |\arg z - 2k\pi/p| < \pi/p\}$, $k = 0, 1, \ldots, p-1$, and that $g(z) = e^{2\pi i/p} f(z)$ has a p-cycle of Baker domains. A similar example was given by Morosawa [47].

Example 2.6. Domínguez [33, Example 1] showed that
$$f(z) = z + 2 + e^{-z} + \frac{1/100}{z - (1 + \pi i)},$$
has an invariant infinitely connected Baker domain, which contains an invariant right half-plane.

Another way to prove the existence of certain Baker domains is to start with an analytic self-map g of $\mathbb{C}^* = \mathbb{C} \setminus \{0\}$, with known behaviour near 0 or ∞, and lift the dynamics to $\mathbb{C}$ as follows.

Let $\pi(z) = e^{az}$, where $a \neq 0$, be a projection map from $\mathbb{C}$ to $\mathbb{C}^*$. Since $g(\pi(z)) \neq 0$, for $z \in \mathbb{C}$, there is an entire function f such that $\pi(f(z)) = g(\pi(z))$, for $z \in \mathbb{C}$. It was proved by Bergweiler [22] that in this situation we have
$$J(f) = \pi^{-1}(J(g)).$$
In particular, the set $J(f)$ is periodic with period $2\pi i/a$.

This process is often called a *logarithmic change of variable* or a *logarithmic lift*. We now show how this technique can be used to give several examples of Baker domains.

Example 2.7. On applying the above process to
$$g(w) = cwe^{-w}, \quad \text{where } c \neq 0,$$
with $\pi(z) = e^{-z}$, we obtain
$$f(z) = z - \log c + e^{-z}.$$
If $0 < |c| < 1$, then g has an attracting fixed point at 0 with attracting basin V, so $U = \pi^{-1}(V)$ is a single component of $F(f)$ which contains a right half-plane in which $f^n \to \infty$. Such a Baker domain is a generalisation of the type given by Fatou (Example 2.1).

Example 2.8. In [40], Herman applied this process with $g(w) = e^{2\pi i\theta}we^w$. For suitable irrational $\theta \in (0, 1)$, the function g has an invariant Siegel disc V with centre 0, on which g is conjugate to an irrational rotation. Using $\pi(z) = e^z$, we obtain
$$f(z) = z + 2\pi i\theta + e^z,$$
and $U = \pi^{-1}(V)$ is a single component of $F(f)$ which contains a left half-plane. In this case, the dynamics of g around 0 lifts to give $\Im(f^n) \to \infty$ in U, so this component of $F(f)$ must be a Baker domain and f is univalent in U. Moreover, for some choices of θ the Siegel disc V is bounded by a quasicircle, so the boundary of the corresponding Baker domain U is a Jordan curve in $\hat{\mathbb{C}}$; see [15].

Example 2.9. Starting with
$$g(w) = cw^2 e^{-w}, \quad \text{where } c \neq 0,$$
we obtain, with $\pi(z) = e^{-z}$ again, the entire function
$$f(z) = 2z - \log c + e^{-z}.$$
In this case, g has a super-attracting fixed point at 0, for all $c \neq 0$. If V is the corresponding basin of attraction, then $U = \pi^{-1}(V)$ is a Baker domain which contains a right half-plane. In [21, Theorem 1] Bergweiler considered the function $f(z) = 2z + 2 - \log 2 - e^z$, which is of this type (apart from a trivial change of variable), and showed (using the theory of polynomial-like maps) that ∂U is a Jordan curve in $\hat{\mathbb{C}}$ and f is univalent on U.

Example 2.10. Baker and Domínguez [8, Theorem 5.1] took
$$g(w) = we^{-w},$$
which, with $\pi(z) = e^{-z}$, gives
$$f(z) = z + e^{-z}.$$
The function g has a parabolic fixed point at $0 \in J(g)$, with parabolic component $V \subset F(g)$ containing $\mathbb{R}^+ = \{x \in \mathbb{R} : x > 0\}$; see [48, page 17], for example. Also, $\mathbb{R}^- = \{x \in \mathbb{R} : x < 0\} \subset J(g)$ since $g(\mathbb{R}^-) \subset \mathbb{R}^-$, $g^n \to \infty$ on $\mathbb{R}^-$ and g belongs to the class $\mathcal{B}$; see Theorem 4.1. Lifting these properties of g by π, we deduce that $\pi^{-1}(V)$ has infinitely many components $U_k, k \in \mathbb{Z}$, all congruent under translation by integer multiples of $2\pi i$, such that $U_k \subset \{z : (2k-1)\pi < \Im(z) < (2k+1)\pi\}$ and ∂U_k is asymptotic to the lines $y = (2k \pm 1)\pi$. For each $k \in \mathbb{Z}$, we have $f^n \to \infty$ and $\Re(f^n) \to \infty$ in U_k, and the lines $y = (2k \pm 1)\pi$ lie in $J(f)$. Therefore the U_k are distinct invariant Baker domains.

In Section 6 we describe various types of meromorphic functions with infinitely many invariant Baker domains; these can be considered as generalisations of Example 2.10.

Example 2.11. Barański and Fagella [16, page 422] started with
$$g(w) = e^{2\pi i\theta} w e^{\frac{1}{2}\beta(w-1/w)}, \quad \text{where } 0 < \theta < 1, \ 0 < \beta < 1,$$
which is a self-map of $\mathbb{C}^*$ that does not extend analytically to $\mathbb{C}$. With $\pi(z) = e^{iz}$, this gives
$$f(z) = z + 2\pi\theta + \beta \sin z.$$
It is known that for suitable values of θ and β, the map g has an invariant Herman ring V, symmetric with respect to the unit circle, on which g is conjugate to an irrational rotation; see [40] or [6]. Thus $U = \pi^{-1}(V)$ is an invariant Baker domain, symmetric with respect to the real axis.

Finally, we mention a method for obtaining Baker domains due to Eremenko and Lyubich [34], which uses results from approximation theory. The idea of the method is to construct:

- a function g that is analytic on one or more unbounded closed sets V_j, with the property that, for each j, $f(V_j) \subset V_j$ and $f^n \to \infty$ in V_j, where f is any function that is uniformly close to g on $\bigcup_j V_j$;
- a function h that is analytic on one or more unbounded closed sets W_j, with the property that, for each j, $f^n(z) = O(1)$ for $z \in W_j$, where f is any function that is uniformly close to h on $\bigcup_j W_j$.

Then, we use Arakelyan's theorem [39] to construct an entire function f which is uniformly close to g on each V_j and to h on each W_j. We deduce that each V_j is contained in a Baker domain of f which does not meet any of the sets W_j. Moreover, if g is chosen to be univalent in a large enough neighbourhood of

V_j, then we can arrange that f is univalent in V_j. Using other approximation results, such as Nersejan's theorem [39], we can also apply this method to obtain meromorphic examples. The following examples of Baker domains were all constructed using this approximation method.

Example 2.12. Eremenko and Lyubich [34, Example 3] constructed an entire function f with a Baker domain containing a right half-plane in which f is univalent.

Example 2.13. König [44, Example 1] constructed a meromorphic function f with a finite number of poles and an invariant infinitely connected Baker domain containing a half-plane.

Example 2.14. Rippon and Stallard [53, Theorem 4] constructed an entire function f with a p-cycle of Baker domains, each shaped like a sector, in each of which f is univalent.

Example 2.15. Barański and Fagella [16, page 424] constructed an entire function f with a spiral-shaped Baker domain in which f is univalent.

3. Growth in Baker domains

For many of the examples described in Section 2 it is clear that, at most points z of the Baker domain, $|f(z)|$ is no greater than a multiple of $|z|$ and the iterates $f^n(z)$ do not tend to ∞ rapidly. The following result [51, Theorem 1] shows that this property always holds in a Baker domain.

Theorem 3.1. *Let U be a p-periodic Baker domain of a meromorphic function f in which $f^{np} \to \infty$. Then, for any compact subset K of U, there exist constants $C > 0$ and $n_0 \in \mathbb{N}$ such that*

$$(3.1) \qquad |f^{np}(z')| \leq C|f^{np}(z)|, \quad \text{for } z, z' \in K, n \geq n_0;$$

for all $z \in U$,

$$(3.2) \qquad \ln|f^{np}(z)| = O(n) \quad \text{as } n \to \infty;$$

for any $z_0 \in U$ and any path $\Gamma = \bigcup_{n=0}^{\infty} f^{np}(\Gamma_0)$, where Γ_0 joins z_0 to $f^p(z_0)$ in U and $0 \notin \Gamma$, there is a constant $C > 0$ such that

$$(3.3) \qquad \frac{1}{C}|z| \leq |f(z)| \leq C|z|, \quad \text{for } z \in \Gamma.$$

Note that the path Γ in this result lies in U and tends to ∞.

The rate at which iterates tend to infinity in a Baker domain was first studied by Baker [3], in connection with the question of whether a transcendental entire function of small growth has no unbounded components of $F(f)$. In [7], he proved Theorem 3.1 for an entire function f in the case $p = 1$ by using classical estimates for the density of the hyperbolic metric in a simply connected domain; recall that a Baker domain of an entire function is simply connected.

In any hyperbolic domain D (that is, a domain which omits at least 2 points of $\mathbb{C}$), we can define the hyperbolic metric $[z, z']_D$, for $z, z' \in D$; see [31], for example. If D is also simply connected, then the density ρ_D of this metric satisfies:

(3.4) $$\frac{1}{2\text{dist}(z, \partial D)} \leq \rho_D(z) \leq \frac{2}{\text{dist}(z, \partial D)}.$$

In the case of a simply connected Baker domain, the result (3.1) is proved using the contracting property of the hyperbolic metric and the lower bound in (3.4), and then (3.2) and (3.3) follow by applying (3.1) to suitable compact subsets of U; see [19, Lemma 7].

In a multiply connected hyperbolic domain, however, we have a weaker lower bound for the density of the hyperbolic metric. Using this, correspondingly weaker growth estimates than those in Theorem 3.1 can be obtained in a multiply connected Baker domain; see [13, Lemma 4.1] and [19, Lemma 7].

The proof of Theorem 3.1 for a general meromorphic function was based on work of Bonfert [29]. Briefly, Bonfert used results from the theory of Fuchsian groups to show that if g is an analytic self-map of an unbounded hyperbolic domain D in which $g^n \to \infty$, and g does not extend analytically to ∞, then

- a lower bound for the hyperbolic density of the form (3.4) holds at points on and near any orbit $g^n(z_0)$, $n = 0, 1, 2, \ldots$, where $z_0 \in D$;
- whether $[g^{n+1}(z_0), g^n(z_0)]_D \to 0$ does not depend on the choice of $z_0 \in D$.

Note that for $z_0 \in D$, the quantity $[g^{n+1}(z_0), g^n(z_0)]_D$ is non-increasing by the contraction property of the hyperbolic metric.

Bonfert's results can also be used to relate the geometric properties of a p-periodic Baker domain U of a meromorphic function f to the proximity of f^p to the identity function in U, as stated below; see [51, Theorem 2].

Theorem 3.2. *Let U be a p-periodic Baker domain of a meromorphic function f in which $f^{np} \to \infty$ and for $z_0 \in U$ let $z_n = f^n(z_0)$ and $\Gamma = \bigcup_{n=0}^{\infty} f^{np}(\Gamma_0)$, where Γ_0 is a path in U joining z_0 to z_p.*

(a) *There exists $C_1 > 0$ such that*

(3.5) $$\text{dist}(z, \partial U) \leq C_1 \frac{|f^p(z) - z|}{[f^p(z), z]_U}, \quad \text{for } z \in \Gamma.$$

If we also have $[z_{(n+1)p}, z_{np}]_U \not\to 0$ as $n \to \infty$, then there exists $C_2 > 0$ such that

(3.6) $$\text{dist}(z, \partial U) \leq C_2 |f^p(z) - z|, \quad \text{for } z \in \Gamma.$$

(b) *There exist $c_1 > 0$ and $c_2 > 0$ such that*

(3.7) $$\text{dist}(z, \partial U) \geq c_1 \frac{|f^p(z) - z|}{[f^p(z), z]_U} \geq c_2 |f^p(z) - z|, \quad \text{for } z \in \Gamma.$$

It can be seen from this result that the closer f^p is to the identity in a p-periodic Baker domain the thinner is the Baker domain, and vice versa.

A less general result of this type was proved in [53, Theorem 3] and used there to show the *non-existence* of Baker domains under certain circumstances; see [53, Theorem 3]. Also, a version of this result was used in [16, Lemma 3.3] to classify univalent Baker domains of entire functions; see Theorem 5.3.

4. Baker domains and singular values

Let f be a transcendental meromorphic function. For $p \in \mathbb{N}$, we denote by $\text{sing}(f^{-p})$ the set of finite *inverse function singularities* of f^p; that is, the set of points $w \in \mathbb{C}$ such that some branch of $(f^p)^{-1}$ can be analytically continued (along some path) up to but not through w. We also call the points of $\text{sing}(f^{-p})$ the *singular values* of f^p. The set $\text{sing}(f^{-1})$ of singular values of f consists of the *critical values* of f (those values $f(z_0)$, where $f'(z_0) = 0$) and the finite *asymptotic values* of f (those values α such that $f(z) \to \alpha$ as $z \to \infty$ along a path Γ).

For $p \in \mathbb{N}$, we can relate $\text{sing}(f^{-p})$ to $\text{sing}(f^{-1})$ as follows:

$$(4.1) \quad f^{p-1}\left(\text{sing}(f^{-1}) \setminus A_{p-1}\right) \subset \text{sing}(f^{-p}) \subset \bigcup_{j=0}^{p-1} f^j\left(\text{sing}(f^{-1}) \setminus A_j\right),$$

where $A_j = A_j(f) = \{z : f^j \text{ is not analytic at } z\}$; see [41, Theorem 7.1.2] and also [1, Lemma 2] for the case of a transcendental entire function.

The *postsingular set* $P(f)$ is defined as

$$P(f) = \bigcup_{n=0}^{\infty} f^n(\text{sing}(f^{-1}) \setminus A_n),$$

or, in some texts, the closure of this set. By (4.1), the set $P(f)$ is the union of all the sets $\text{sing}(f^{-p})$, $p \in \mathbb{N}$.

Attracting components, parabolic components, Siegel discs and Herman rings of a meromorphic function f all have close connections with the singular values of f; see [19, Theorem 7]. For Baker domains there are also connections with singular values, but they are less precise. In [35] Eremenko and Lyubich introduced the class

$$\mathcal{B} = \{f : \text{sing}(f^{-1}) \text{ is bounded}\}$$

and proved the following fundamental result.

Theorem 4.1. *If f is a transcendental entire function and $f \in \mathcal{B}$, then there is no component of $F(f)$ in which $f^n \to \infty$. In particular, the function f has no Baker domains.*

The idea of the proof of Theorem 4.1 is as follows. Let $f \in \mathcal{B}$ and choose R so large that $\text{sing}(f^{-1}) \subset \{z : |z| < R/2\}$ and $|f(0)| < R$. Now put $D_R = \{z : |z| > R\}$. If Δ_0 is an open disc in $F(f)$ in which $f^n \to \infty$, then we can assume that $|f^n| > R$ in Δ_0, for $n \geq 0$, so each $\Delta_n = f^n(\Delta_0)$ lies in a component V_n of $f^{-1}(D_R)$. Now each V_n is simply connected and does not contain 0, so we can consider a component W_n of $\log V_n$ and take T_n to be the component of $\log \Delta_n$ in W_n. Then for some branch of the logarithm $F_n(s) = \log f(e^s)$ maps W_n univalently onto the half-plane $\{t : \Re(t) > \log R\}$ and T_n univalently onto T_{n+1}. By Koebe's $\frac{1}{4}$-theorem, F_n is expanding on T_n for all large values of n (since W_n contains no disc of radius greater than π and $\inf_{s \in T_n} \Re(s) \to \infty$ as $n \to \infty$), but $\text{diam}\, T_n \leq 2\pi$, for $n \geq 0$, which gives a contradiction.

The following result extends Theorem 4.1 to p-cycles of Baker domains.

Theorem 4.2. *If f is a transcendental meromorphic function and $\text{sing}(f^{-p})$ is bounded, then there is no component of $F(f)$ in which $f^{np} \to \infty$. In particular, the function f has no p-periodic Baker domains.*

This was proved in [54, Theorem A], under the slightly stronger assumption that
$$\bigcup_{j=0}^{p-1} f^j \left(\text{sing}(f^{-1}) \setminus A_j \right) \text{ is bounded,}$$
but the proof in [54] also gives Theorem 4.2. This proof builds on Eremenko and Lyubich's method of proof of Theorem 4.1, and on the partial proof given in [19, Theorem 16]. The extra ingredient used in [54] is that, for R large enough, each component of $f^{-p}(\{z : |z| > R\} \cup \{\infty\})$ is simply connected; for a meromorphic function such components can be bounded or unbounded.

As a corollary of Theorem 4.2, we note that if f belongs to the Speiser class $\mathcal{S}$ of transcendental meromorphic functions with only finitely many singular values, then $\text{sing}(f^{-p})$ is bounded for all $p \in \mathbb{N}$, by (4.1), so f has no Baker domains. Such functions are also known to have no wandering domains; see [14].

Bargmann [17, Theorem 4] used a fundamental result of König (see Theorem 5.1) and results on normal families of covering maps to give a stronger version of Theorem 4.1 for Baker domains. This shows that if f has a Baker domain, then the (unbounded) set of singular values of f cannot be too sparse.

Theorem 4.3. *If f is an entire function with an invariant Baker domain, then there exist constants $C > 1$ and $r_0 > 0$ such that every annulus of the form $\{z : r/C < |z| < Cr\}$ with $r \geq r_0$ meets $\text{sing}(f^{-1})$.*

As part of the proof of Theorem 4.3, Bargmann showed that an invariant Baker domain of a transcendental entire function contains a path Γ tending to ∞ on which f' is bounded and also (3.3) holds.

Theorem 4.3 has been generalised to meromorphic functions with a finite number of poles [57], by adapting the argument used by Eremenko and Lyubich to prove Theorem 4.1.

Theorem 4.4. *If f is a meromorphic function with a finite number of poles and with a p-periodic Baker domain, then there exist constants $C > 1$ and $r_0 > 0$ such that every annulus $\{z : r/C < |z| < Cr\}$ with $r \geq r_0$ meets $sing(f^{-p})$.*

The paper [57] also gives examples to show that this generalisation does not extend to meromorphic functions with infinitely many poles.

However, it should be noted that Baker domains need not contain singular values. For example it is straightfoward to check that this is the case in each of Examples 2.8 and 2.11. Moreover, the example considered by Bergweiler $f(z) = 2z + 2 - \log 2 - e^z$ has an invariant Baker domain U in which f is univalent and $\text{dist}(P(f), U) > 0$, where $\text{dist}(., .)$ denotes the Euclidean distance; see [21, Theorem 1]. In the same paper, Bergweiler [21, Theorem 3] showed that a Baker domain cannot lie too far from the set $P(f)$ in the following sense.

Theorem 4.5. *Let f be a transcendental entire function with an invariant Baker domain U. If $U \cap sing(f^{-1}) = \emptyset$, then there exists a sequence (p_n) such that $p_n \in P(f)$, $|p_n| \to \infty$, $|p_{n+1}/p_n| \to 1$ and $\text{dist}(p_n, U) = o(|p_n|)$ as $n \to \infty$.*

For some classes of meromorphic functions, we can assert that any cycle of Baker domains must contain a singular value. For example, Bergweiler used an observation of Herman [40, page 609] to show that this is the case for certain functions arising from Newton's method (see [20]) and pointed out that similar reasoning applies to functions of the form $f(z) = z + p(z)e^{q(z)}$, where p and q are polynomials; see also [26] for other examples.

We end this section by listing several other particular types of meromorphic function for which we can assert the existence of Baker domains containing singular values. The first is due to Hinkkanen [42].

Theorem 4.6. *Let f be a transcendental meromorphic function such that*
$$f(z) = z + az^{-m} + O(z^{-m-\delta}) \text{ as } z \to \infty, z \in V,$$
where $a \in \mathbb{C}^*$, $\delta > 0$ and $V = \{z : |\arg z - t_k| < \pi/(m+1)\}$ for some $k \in \mathbb{Z}$ with $t_k = (2\pi k + \arg a)/(m+1)$. Then f has an invariant Baker domain U, which contains an unbounded Jordan domain in V whose boundary is tangent to ∂V at ∞, and $U \cap sing(f^{-1})$ is non-empty.

Sometimes it is possible to prove that there is a Baker domain containing infinitely many singular values, as in the following result of Rippon and Stallard [53, Theorem 2].

Theorem 4.7. *Let f be a transcendental meromorphic function such that*
$$(4.2) \qquad f(z) = az + bz^k e^{-z}(1 + o(1)) \quad as \; \Re(z) \to \infty,$$
where $k \in \mathbb{N}$, $a > 1$ and $b > 0$. Then f has an invariant Baker domain U, which contains $\{z : |z^k e^{-z}| < \rho, |z| > R\}$ for each $\rho > 0$ and large enough values of $R > 0$, and $U \cap sing(f^{-1})$ is unbounded.

Recently the author showed that if (4.2) holds, then the sets of critical points of f in U and critical values of f in U are unbounded; see [51, Theorem 3].

Note that if f is of the form (4.2) with $k \in \mathbb{Z} \setminus \mathbb{N}$, then f has an invariant Baker domain U, which contains an invariant set of the form $\{z : \Re(z) > R\}$ for large enough values of R. However, for such k we cannot deduce that $U \cap sing(f^{-1})$ is non-empty. For example, the function $f(z) = 2z + 2e^{-2}e^{-z}$ has such an invariant Baker domain U and f is univalent in U. This follows from the corresponding properties of the function $f(z) = 2z + 2 - \log 2 - e^z$ mentioned in Example 2.9, after a suitable linear change of variable.

Finally we give another result of this type, due to Bergweiler [23, Theorem 3].

Theorem 4.8. *Let f be a transcendental entire function of finite order such that*
$$(4.3) \qquad f(z) = z + a + o(1) \quad as \; z \to \infty, \; |\arg z| \leq \eta.$$
where $a, \eta > 0$. Then f has an invariant Baker domain U, which contains $\{z : |\arg z| \leq \eta, \Re(z) > R\}$ for some $R > 0$, and $U \cap sing(f^{-1})$ is unbounded.

A detailed discussion of the background to this result can also be found in the survey article [24].

Recently, Lauber [45] has studied the Baker domains of various families of entire functions that satisfy the hypotheses of Theorem 4.8 (after a trivial change of variable), in particular the functions $g_b(z) = z - 1 + bze^z$, where $b \in \mathbb{C}$. For the functions g_b, there is a single invariant Baker domain U_b which contains all the critical points of g_b with at most one exception; also, the set of parameters b for which there is such an exceptional critical point contains a copy of the Mandelbrot set.

5. TYPES OF BAKER DOMAINS

Let U be a simply connected Baker domain of f and let ψ be a Riemann map from the open unit disc $\mathbb{D}$ onto U. Then $g = \psi^{-1} \circ f \circ \psi$ is an analytic self-map of $\mathbb{D}$. Hence by the Denjoy-Wolff theorem (see [31], for example), there is a point $\zeta \in \overline{\mathbb{D}}$ such that $g^n \to \zeta$ locally uniformly in $\mathbb{D}$. The fact that $f^n \to \infty$ in U implies that $\zeta \in \partial \mathbb{D}$. In [32], Cowen introduced a remarkable classification of such self-maps g of $\mathbb{D}$ whose Denjoy-Wolff point lies in $\partial \mathbb{D}$, proving that g must be semi-conjugate in $\mathbb{D}$ to a Möbius transformation in one of three different ways. Cowen's result leads to a corresponding classification of the original simply connected Baker domain. This classification is expressed in terms of the following notation, introduced by König in [44].

Let G be a domain and let $f : G \to G$ be analytic. Then a domain $V \subset G$ is *absorbing* (or *fundamental*) for f if: V is simply connected, $f(V) \subset V$ and for each compact set $K \subset G$ there exists $N = N_K$ such that $f^N(K) \subset V$.

Now let $\mathbb{H} = \{z : \Re(z) > 0\}$. The triple (V, ϕ, T) is called a *conformal conjugacy* (or *eventual conjugacy*) of f in G if:

(a) V is *absorbing* for f;
(b) $\phi : G \to \Omega \in \{\mathbb{H}, \mathbb{C}\}$ is analytic and univalent in V;
(c) $T : \Omega \to \Omega$ is a bijection, and $\phi(V)$ is absorbing for T;
(d) $\phi(f(z)) = T(\phi(z))$, for $z \in G$.

In this situation we write $f \sim T$. We note that properties (b) and (d) imply that f is univalent in V.

Cowen proved that any self-map g of $\mathbb{D}$ whose Denjoy-Wolff point lies in $\partial \mathbb{D}$ has a conformal conjugacy in $\mathbb{D}$, and the Riemann map enables us to transfer this to any meromorphic function with a simply connected Baker domain. In [44, Theorem 1], König succeeded in generalising this result to some multiply connected Baker domains.

Theorem 5.1. *Let f be a meromorphic function with a finite number of poles, and with a p-periodic Baker domain U. Then there exists a conformal conjugacy of f^p in U, which is one of the following (mutually exclusive) types:*

(a) $T(z) = z + 1$ *and* $\Omega = \mathbb{C}$;
(b) $T(z) = z \pm i$ *and* $\Omega = \mathbb{H}$;
(c) $T(z) = az$, *for* $a > 1$, *and* $\Omega = \mathbb{H}$.

Case (a) is called *parabolic of type I*, case (b) is called *parabolic of type II*, and case (c) is called *hyperbolic*.

Moreover, König gave geometric criteria for these three cases, as follows [44, Theorem 3].

Theorem 5.2. *Let U be a p-periodic Baker domain of a meromorphic function f in which $f^{np} \to \infty$ and on which f^p has a conformal conjugacy. For $z_0 \in U$, put*

$$c_n = c_n(z_0) = \frac{|f^{(n+1)p}(z_0) - f^{np}(z_0)|}{\operatorname{dist}(f^{np}(z_0), \partial U)}.$$

Then exactly one of the following cases holds:

(a) $f^p \sim z + 1$, *which is equivalent to*

$$\lim_{n \to \infty} c_n = 0, \quad \text{for } z_0 \in U;$$

(b) $f^p \sim z \pm i$, which is equivalent to
$$\liminf_{n\to\infty} c_n > 0 \quad \text{for } z_0 \in U, \text{ but } \inf_{z_0 \in U} \limsup_{n\to\infty} c_n = 0;$$

(c) $f^p \sim az$ for some $a > 1$, which is equivalent to
$$c_n > c \quad \text{for } z_0 \in U, n \in \mathbb{N}, \text{where } c = c(f) > 0.$$

In case (a), there is at least one singular value of f in U.

These geometric criteria enable us to classify some of the Baker domains in Section 2. For example, it is easy to check that the function $f(z) = z+1+e^{-z}$, studied by Fatou, satisfies
$$\frac{f^{n+1}(z_0) - f^n(z_0)}{\operatorname{dist}(f^n(z_0), \partial U)} = O\left(\frac{1}{n}\right).$$

So this example is parabolic of type I, as are Examples 2.2, 2.6 and 2.10. On the other hand, Example 2.8 is parabolic of type II, and Examples 2.4 and 2.9 are hyperbolic; see [46] for a discussion of the stability of these three types.

In [16], Barański and Fagella considered invariant *univalent* Baker domains U; that is, invariant Baker domains of entire functions in which f is univalent. In this situation, the restriction of f to U is conjugate to a conformal mapping from $\mathbb{D}$ onto $\mathbb{D}$, and this implies that parabolic Baker domains of type I cannot occur. The paper [16] studies the remaining two cases and divides the hyperbolic univalent Baker domains into two types, depending on whether backward iterates of $f|_U$ approach a finite point of ∂U, called *hyperbolic of type I*, or approach ∞, called *hyperbolic of type II*. The Bergweiler function $f(z) = 2z + 2 - \log 2 - e^z$ is hyperbolic of type I, whereas Example 2.11 is hyperbolic of type II and Example 2.8 is parabolic (of type II).

The following necessary and sufficient condition from [16] is closely related to Theorems 3.2 and 5.2.

Theorem 5.3. *Let U be an invariant univalent Baker domain of an entire function f. Then U is hyperbolic if and only if*
$$\sup_{z \in U} \frac{\operatorname{dist}(z, \partial U)}{|f(z) - z|} < \infty.$$

Bergweiler [23] and independently Fagella and Henriksen [36] have studied the Baker domains of functions for which the conjugate self-map of $\mathbb{D}$ extends analytically to a neighbourhood of the Denjoy-Wolff point on $\partial \mathbb{D}$. Such Baker domains include those for which the restriction of f to U is a proper map (see [59] for results about proper maps), in which case f is conjugate to a finite Blaschke product on $\mathbb{D}$. A classification of such Baker domains expressed in terms of the hyperbolic metric is given in [23], and one in terms of the nature of the 'invariant petals' which exist in U is given in [36].

Finally in this section, we discuss some of the work that has been done to describe the geometric properties of the boundaries of Baker domains. This work began with the paper of Baker and Weinreich [15], which studied the nature of the boundary of *any* unbounded invariant Fatou component of a transcendental entire function f. They showed that in most cases such boundaries are highly irregular, in the sense that ∞ belongs to the impression of every prime end of U. In particular, if ∂U is a Jordan curve in $\hat{\mathbb{C}}$, then U must be a Baker domain and f must be univalent in U. Examples 2.8 and 2.9 show that the latter possibility can occur.

The key tool in this study of boundaries is the fact that if ψ is a conformal map from the unit disc $\mathbb{D}$ onto U, then $g = \psi^{-1} \circ f \circ \psi$ is an *inner function*, that is, an analytic self-map of $\mathbb{D}$ whose angular limits have modulus 1 almost everywhere on $\partial \mathbb{D}$. The paper [15] initiates a version of the Fatou-Julia theory for inner functions, a topic taken further in [18].

More precise information about such unbounded invariant components U can be obtained by studying the set

$$\Theta = \{e^{i\theta} : \psi(re^{i\theta}) \to \infty \text{ as } r \to 1\},$$

introduced by Kisaka [43]. The idea here is that each point in Θ corresponds to a different *access* to ∞ from within U. For a Baker domain U the set $\Theta \neq \emptyset$, by Theorem 3.1. In [9], Baker and Domínguez obtained the following result.

Theorem 5.4. *Let U be an invariant Baker domain of an entire function f such that f is not univalent in U. Then $\overline{\Theta}$ contains a perfect set.*

Thus, in this situation U has infinitely many accesses to ∞, so ∂U has infinitely many components. In [18] Bargmann further develops the Fatou-Julia theory for inner functions and proves that if U is a Baker domain of an entire function f, which is parabolic of type I, then $\overline{\Theta} = \partial \mathbb{D}$.

In [16] Barański and Fagella applied their classification of univalent Baker domains to show that univalent Baker domains which are hyperbolic of type II must have disconnected boundaries. However, it is an open question whether such Baker domains can have infinitely many accesses to ∞.

6. Functions with infinitely many Baker domains

In [52] various families of meromorphic functions were given, each function having infinitely many invariant Baker domains in which the dynamical behaviour near ∞ is analogous to that in Leau domains near a parabolic fixed point. Here we describe the results of [52] in a slightly different way which makes it easier to deduce similar results for further families of functions. First we describe the underlying dynamics near ∞ of functions analytic in a right half-plane with a certain asymptotic behaviour there. Roughly speaking, for these functions there are infinitely many strip-like invariant domains, which play the role of attracting petals with respect to ∞, and they are 'separated' by strip-like sets which play the role of repelling petals.

First we define
$$R(t,s) = \{z : \Re(z) \geq t, |\Im(z)| \leq s\}, \quad t > 0, \ s > 0,$$
$$T_k(\theta) = \{x + iy : x > t_k(\theta), |y - 2k\pi| < \min\{\theta, \pi/4 + (x - t_k(\theta))\cot\theta/2\}\},$$
where $0 < \theta < \pi$ and $t_k(\theta) > 0$ is to be chosen, and
$$S_k = \{x + iy : x \geq s_k, |y - (2k+1)\pi| \leq \pi/4\},$$
where $s_k > 0$ is to be chosen.

Theorem 6.1. *Let $f(z) = z + \phi(z)e^{-z}$ be analytic in $H = \{z : \Re(z) > x_0\}$, where $x_0 > 0$, and suppose that, for each $s > 0$,*

(6.1) $$\sup\{|\arg\phi(z)| : z \in R(t,s)\} \to 0 \text{ as } t \to \infty.$$

Then, for each $k \in \mathbb{Z}$ and $3\pi/4 < \theta < \pi$, and for $t_k(\theta) > 0$, $s_k > 0$ sufficiently large, we have

(a) $f(T_k(\theta)) \subset T_k(\theta)$;
(b) *if $z \in T_k(7\pi/8)$, then*
$$\Re(f^n(z)) \to \infty, \quad \Im(f^n(z)) - 2k\pi \to 0, \quad f^{n+1}(z) - f^n(z) \to 0$$
as $n \to \infty$;
(c) *there exist paths γ_k and Γ_k in S_k such that γ_k is bounded, Γ_k joins γ_k to ∞ and*
$$\text{if } z \in \Gamma_k, \text{ then } f^n(z) \in \gamma_k, \text{ for some } n \geq 0;$$
(d) *if $z \in S_{k-1} \cup S_k$, then $f^n(z) \notin S_{k-1} \cup S_k$, for some $n > 0$.*

We refer to the sets $T_k(\theta)$ with properties (a) and (b) as *attracting petals* and the sets S_k with properties (c) and (d) as *repelling petals*.

Remark If $f(z) = z + \phi(z)e^{bz}$, $b \neq 0$, is analytic in the half-plane $\rho(H)$, where $\rho(z) = -z/b$, and ϕ satisfies

(6.2) $$\sup\{|\arg\phi(z) - \theta| : z \in \rho(R(t,s))\} \to 0 \text{ as } t \to \infty,$$

then the change of variables $w = -bz - i(\theta + \arg(-b))$, gives a function of the form in Theorem 6.1.

The proofs of the various parts of Theorem 6.1 are all contained in the proof of [52, Theorem 1], which is similar to Theorem 6.1 but with a slightly less general formulation. In [52] the function f is assumed to be meromorphic in $\mathbb{C}$ but in Theorem 6.1 it only needs to be analytic in a half-plane, so the notion of a Baker domain of f may not be defined. Also, parts (c) and (d) were not included in the statement of [52, Theorem 1] but were part of the proof; see [52, Lemma 2.4]. These parts of Theorem 6.1 are expressed in a way that is preserved under a univalent change of variable. This allows us to apply Theorem 6.1 to various types of meromorphic functions, after a change of variable if necessary, and conclude that they have infinitely many attracting and repelling petals. We can then deduce that each attracting petal lies in an invariant Baker domain by using the following lemma.

Lemma 6.1. *Let V_n, $n = 0, 1, \ldots$, be unbounded simply connected domains, each bounded by a simple curve tending to ∞ in both directions, such that*

(6.3) $$V_n \subset V_0, \quad \text{for } n = 1, 2, \ldots, \quad \text{and} \quad \bigcap_{n=0}^{\infty} \overline{V_n} = \emptyset,$$

and let f be analytic in V_0 such that

(6.4) $$f(V_n) \subset V_n, \quad \text{for } n = 0, 1, \ldots.$$

Then $f^n(z) \to \infty$ as $n \to \infty$, for $z \in V_0$. Thus, if f is also a transcendental meromorphic function, then V_0 is a subset of an invariant Baker domain of f.

Proof. Let ϕ be a Riemann map from the open unit disc $\mathbb{D}$ onto V_0. Since V_0 is a Jordan domain in $\hat{\mathbb{C}}$, the map ϕ extends to a homeomorphism from $\overline{\mathbb{D}}$ onto $\overline{V_0} \cup \{\infty\}$; see [50, Theorem 2.6]. Now put $g = \phi^{-1} \circ f \circ \phi$. Then g is a self-map of $\mathbb{D}$, and so has a Denjoy-Wolff point $\zeta \in \overline{\mathbb{D}}$, such that $g^n(t) \to \zeta$ for $t \in \mathbb{D}$; see [31], for example. Hence $f^n(z) \to \phi(\zeta) \in \partial V_0 \cup \{\infty\}$ for $z \in V_0$. If $\phi(\zeta) \neq \infty$, then $\phi(\zeta) \in \overline{V_n}$ for all n, by (6.4), so we obtain a contradiction to (6.3). □

The question then arises as to whether each different attracting petal lies in a different Baker domain. To illustrate how this can be deduced from Theorem 6.1, we prove the following result from [52].

Theorem 6.2. *Let f be a meromorphic function such that*

(6.5) $$f(z) = z + az^k e^{-z}(1 + o(1)) \quad \text{as } \Re(z) \to \infty,$$

where $a > 0$. Then f has infinitely many distinct invariant Baker domains U_k, $k \in \mathbb{Z}$, such that, for each $k \in \mathbb{Z}$ and $3\pi/4 < \theta < \pi$, and for $t_k(\theta)$ sufficiently large:

(a) $U_k \supset T_k(\theta)$;
(b) *if $z \in U_k$, then $f^n(z)$ eventually lies in $T_k(7\pi/8)$, so*
$$\Re(f^n(z)) \to \infty, \quad \Im(f^n(z)) - 2k\pi \to 0, \quad f^{n+1}(z) - f^n(z) \to 0$$
as $n \to \infty$;
(c) U_k *contains at least one singularity of f^{-1}.*

Proof. It is clear that f satisfies (6.1), so we can apply Theorem 6.1 immediately, without any change of variable. By Theorem 6.1(a) and Lemma 6.1, each set $T_k(\theta)$, $3\pi/4 < \theta < \pi$, with $t_k(\theta)$ large enough, lies in an invariant Baker domain U_k.

We now prove part (b), which also shows that the Baker domains U_k, $k \in \mathbb{Z}$, are distinct. Suppose that $z \in U_k$. Choose $z' \in T_k = T_k(7\pi/8)$ and join z to z' by a path σ in U_k. Then $f^n \to \infty$ uniformly on σ. Thus if R is so large that

(6.6) $$\gamma_{k-1} \cup L_k \cup \gamma_k \subset \{z : |z| \leq R\},$$

where γ_k are the paths defined in Theorem 6.1(c) and

$$L_k = \partial T_k \cap \{x + iy : |y - 2k\pi| < 7\pi/8\},$$

then there exists $N \in \mathbb{N}$ such that

(6.7) $$f^n(\sigma) \subset \{z : |z| > R\}, \quad \text{for } n \geq N.$$

Now suppose that for some $n \geq N$, we have

$$f^n(z) \notin T_k \cup S_{k-1} \cup S_k.$$

Since $f^n(z') \in T_k(7\pi/8)$, we deduce by (6.6) and (6.7) that there is a point $w \in f^n(\sigma) \cap (\Gamma_{k-1} \cup \Gamma_k)$, where Γ_k are the paths defined in Theorem 6.1. Thus $f^m(w) \in \gamma_{k-1} \cup \gamma_k$ for some $m \geq 0$, by Theorem 6.1(c), which contradicts (6.7). Hence $f^n(z) \in T_k \cup S_{k-1} \cup S_k$, for $n \geq N$, and it follows from Theorem 6.1(d) that $f^n(z)$ lies eventually in T_k, as required.

Finally, to prove part (c) we note that if U_k contains no singularity of f^{-1}, then $f : U_k \longrightarrow U_k$ is a univalent map. Thus, for $z \in U_k$ the hyperbolic distance between $f^{n+1}(z)$ and $f^n(z)$ remains constant, and this is contradicted by the fact that $f^{n+1}(z) - f^n(z) \to 0$ as $n \to \infty$ and $\text{dist}(f^n(z), \partial U_k) > \pi/2$ for large n.

Alternatively, we can apply König's criterion, Theorem 5.2(a), to deduce that U_k is parabolic of type I and so U_k contains a singular value of f. $\square$

Remark If

$$f(z) = z + az^k e^{bz}(1 + o(1)) \quad \text{as } \Re(-bz) \to \infty,$$

where $a, b \in \mathbb{C} \setminus \{0\}$, then the change of variable $w = -bz - i\arg(a/(-b)^{k-1})$, gives a function satisfying (6.5) with $a > 0$.

Next, we consider meromorphic functions of the form

$$f(z) = z + R(z)e^{Q(z)},$$

where R is rational and Q is a non-constant polynomial. Functions of this form were studied by Stallard [58], who showed that they do not have wandering domains. A key step in the proof of this result is to establish that all but a finite number of critical values of f lie in invariant domains of f. These invariant domains are constructed, after the change of variable $w = Q(z)$, in a similar way to the sets $T_k(\theta)$ in Theorem 6.1, and it is natural to expect therefore that f has infinitely many invariant Baker domains. Indeed, using Theorem 6.1 we can obtain the following result. The proof is closely related to that of [58, Theorem 2.1], but here we also need to prove that the invariant domains found there correspond to infinitely many *distinct* invariant Baker domains.

Theorem 6.3. *Let*
$$f(z) = z + R(z)e^{Q(z)},$$
where R is rational and Q is a polynomial with $\deg Q = m \geq 1$. Then f has m infinite families of invariant Baker domains, each family lying for large values of z in a sector of angle $2\pi/m$. Each such Baker domain is parabolic of type I and contains at least one critical value of f.

Proof. On substituting $z = cz'$, where c is a complex constant, we can assume that
$$f(z) = z + \lambda \frac{S(z)}{T(z)} e^{Q(z)},$$
where S, T and Q are monic polynomials and $\lambda \neq 0$. Write $w = Q(z)$ and introduce m inverse branches of Q, defined in $A = \{w : |w| > M, w \notin \mathbb{R}^+\}$ for M large enough, of the form
$$z = Q_j^{-1}(w) = w^{1/m} + c_0 + c_1 w^{-1/m} + \cdots, \quad 1 \leq j \leq m.$$
In the definition of $Q_j^{-1}(w)$ we take $\arg w^{1/m} \in (2\pi(j-1)/m, 2\pi j/m)$.

For $j = 1, \ldots, m$, we put $A_j = Q_j^{-1}(A)$, $B_j = Q_j^{-1}(A \cap \{w : \Re(w) < 0\})$. Note that ∂A_j is a simple curve which tends to ∞ in the two directions $\arg z = (2j-1)\pi/m \pm \pi/m$ and ∂B_j is a simple curve which tends to ∞ in the two directions $\arg z = (2j-1)\pi/m \pm \frac{1}{2}\pi/m$.

Now put $F_j = Q \circ f \circ Q_j^{-1}|_A$, $n = \deg S - \deg T$ and $k = m + n - 1$. Then

(6.8)
$$\begin{aligned}f(Q_j^{-1}(w)) &= Q_j^{-1}(w) + \lambda w^{n/m} e^w (1 + O(w^{-1/m})) \quad \text{as } w \to \infty \\ &= Q_j^{-1}(w) + o(1) \quad \text{as } \Re(w) \to -\infty, \, \Im(w) \text{ bounded.}\end{aligned}$$

Hence
$$\begin{aligned}F_j(w) &= Q\left(Q_j^{-1}(w) + \lambda w^{n/m} e^w (1 + O(w^{-1/m}))\right) \\ &= w + Q'(Q_j^{-1}(w)) \lambda w^{n/m} e^w (1 + O(w^{-1/m}))(1 + o(1)) \\ &= w + m w^{(m-1)/m} \lambda w^{n/m} e^w (1 + o(1)) \\ &= w + \lambda m w^{k/m} e^w (1 + o(1)) \quad \text{as } \Re(w) \to -\infty, \, \Im(w) \text{ bounded.}\end{aligned}$$

To justify the second line above we can either use the Taylor series for Q about $Q_j^{-1}(w)$ and apply Cauchy's estimate to the coefficients or use the fact that Q is univalent on the set A_j so, by the Koebe distortion theorem,
$$\frac{Q(\zeta) - Q(z)}{Q'(z)(\zeta - z)} = 1 + o(1),$$

uniformly as $\zeta \to z$, for $z \in B_j$; see [30, proof of Lemma 6.3] for a similar application.

Now the function $\phi_j(w) = \lambda m w^{k/m}$ is analytic in the left half-plane and satisfies (6.2) there, with $\theta = \arg \lambda + (2j-1)k\pi/m$. Thus we can apply Theorem 6.1 and the remark following it (with $b = 1$) to F_j. We deduce that F_j has infinite families of attracting and repelling petals in the left half-plane, of the form $T'_k(\theta) = \tau(T_k(\theta))$ and $S'_k = \tau(S_k)$, where $T_k(\theta)$ and S_k are attracting and repelling petals with the properties given in Theorem 6.1, and

$$\tau(w) = -w - i(\arg \lambda + (2j-1)k\pi/m + \pi).$$

For $j = 1, \ldots, m$ and $k \in \mathbb{Z}$, we put $V_{j,k}(\theta) = Q_j^{-1}(T'_k(\theta))$ and $W_{j,k} = Q_j^{-1}(S'_k)$. Then

(6.9) $V_{j,k}(\theta) \cup W_{j,k} \subset B_j \subset A_j$,   for $j = 1, \ldots, m$, $k \in \mathbb{Z}$, $3\pi/4 < \theta < \pi$.

By (6.8) and (6.9), we can assume that $t_k(\theta)$ and s_k are chosen so large when applying Theorem 6.1 that

(6.10) \quad if $w \in \bigcup_{k \in \mathbb{Z}}(T'_k(\theta) \cup S'_k)$,   then $f(Q_j^{-1}(w)) \in A_j$.

We now claim that

$$f(V_{j,k}(\theta)) \subset V_{j,k}(\theta), \quad \text{for } j = 1, \ldots, m, \ k \in \mathbb{Z}, \ 3\pi/4 < \theta < \pi.$$

For if $z \in V_{j,k}(\theta)$, then $z = Q_j^{-1}(w)$, where $w \in T'_k(\theta)$, so $Q(f(z)) = F_j(w)$ is in $T'_k(\theta)$. Also, $f(z) \in A_j$ by (6.10), so $f(z) = Q_j^{-1}(Q(f(z)))$ is in $V_{j,k}(\theta)$, as required. Each such invariant domain $V_{j,k}(\theta)$ lies in a component $U_{j,k}$ of $F(f)$ and, by Theorem 6.1(a) and Lemma 6.1, each $U_{j,k}$ is an invariant Baker domain.

We can now prove that these Baker domains are distinct and that all orbits $f^n(z)$, $z \in U_{j,k}$, eventually land in $V_{j,k}(7\pi/8)$. For the function F_j, there are paths γ_k and Γ_k in the sets S_k with the properties stated in Theorem 6.1(c) and (d). The images under the univalent map $Q_j^{-1} \circ \tau$ of these paths and sets have similar properties, since $f(z) = Q_j^{-1}(Q(f(z)))$ for $z \in W_{j,k} = Q_j^{-1}(S'_k)$, by (6.10), and these properties enable us to use a similar argument to that in the proof of Theorem 6.2(b); we omit the details.

Finally, we note that any orbit of F_j in one of the invariant domains $T'_j(7\pi/8)$ satisfies

$$\frac{F_j^{n+1}(w) - F_j^n(w)}{\mathrm{dist}(F_j^n(w), \partial T'_k(7\pi/8))} \to 0 \quad \text{as } n \to \infty.$$

Since the mapping $z = Q_j^{-1}(w)$ is univalent on A, it has uniformly bounded distortion on any square of side 4π in $\{w : \Re(w) < 0, |w| > 2M\}$. Thus we

deduce that, for $z \in V_{j,k}(7\pi/8)$ and hence for $z \in U_{j,k}$, we have
$$\frac{f^{n+1}(z) - f^n(z)}{\operatorname{dist}(f^n(z), \partial U_{j,k})} \to 0 \text{ as } n \to \infty,$$
so $U_{j,k}$ is parabolic of type I, by König's criterion. Hence $U_{j,k}$ must contain a critical value of f, because f has no finite asymptotic values; see [58, page 218]. □

Remark It follows from the above proof that if Q is a monic polynomial, then for each j, $1 \le j \le m$, the Baker domains $U_{j,k}$, $k \in \mathbb{Z}$, are all asymptotic at ∞ to the ray $\{z : \arg z = (2j-1)\pi/m\}$, $j = 1, \ldots, m$.

The following special case of Theorem 6.3 was also given in [52] (apart from a trivial change of variable).

Corollary 6.1. *For $p \in \mathbb{N}$, let $f(z) = z(1 + e^{z^p})$. Then*

(a) *the function f has infinitely many invariant Baker domains in each sector $\{z : 2\pi(j-1)/p < \arg z < 2\pi j/p\}$, $j = 1, \ldots, p$;*
(b) *the function $g(z) = e^{2\pi i/p} f(z)$ has infinitely many p-cycles of Baker domains.*

Proof. Part (a) holds by Theorem 6.3 and the above remark, together with the fact that no Baker domain can meet $\{z : \arg z = 2\pi(j-1)/p\}$, $j = 1, \ldots, p$, since these rays are invariant and $f^n(z)$ tends to infinity too rapidly on them, by Theorem 3.1. (In fact, these rays lie in $J(f)$, as can be seen by considering the fast rate at which the orbits of nearby pairs of points on these rays separate; see [55, Theorem 4].)

To obtain part (b), we use the fact that $g^p = f^p$, so $J(g) = J(f)$. □

Theorem 6.3 is related to an interesting question of Xavier Buff, namely, is it possible to find a transcendental entire function with only finitely many fixed points and no Baker domains? This question is related to results in the paper [30], where it is shown that if a transcendental entire function f has a logarithmic singularity over 0, then the Newton map $N_f(z) = z - f(z)/f'(z)$ has a Baker domain associated with that singularity; see [27] for related results.

Since any transcendental entire function of finite order with only finitely many fixed points must be of the form $f(z) = z + P(z)e^{Q(z)}$, where P and Q are polynomials with $\deg Q \ge 1$, Theorem 6.3 gives the answer 'yes' at least in this case. The general question, however, concerns a function of the form $f(z) = z + P(z)e^{g(z)}$, where g is any entire function. This question remains open. However, there are many examples of functions of this form with infinite order which have infinitely many Baker domains. The following example was mentioned in [52].

Corollary 6.2. Let $f(z) = z + \exp(-e^z)$. Then the function f has infinitely many invariant Baker domains in each strip $\{z : |\Im(z) - 2\pi j| < \pi\}$, $j \in \mathbb{Z}$.

Proof. We follow the approach in the proof of Theorem 6.3, but the details are simpler. Write $w = e^z$ and introduce inverse branches

$$z = L_j(w) = \log|w| + i\arg_j(w), \quad \text{where } (2j-1)\pi < \arg_j(w) < (2j+1)\pi,$$

defined on $A = \{w : |\arg w| < \pi\}$. Then consider $F_j(w) = \exp(f(L_j(w))) = w \exp(e^{-w})$. By Theorem 6.1, the function F_j has infinitely many attracting and repelling petals, $T_k(\theta)$ and S_k, $k \in \mathbb{Z}$, in the right half-plane (moreover, by Theorem 6.2, the function F_j has infinitely many invariant Baker domains). Now put $V_{j,k}(\theta) = L_j(T_k(\theta))$. Then the sets $V_{j,k}(\theta)$, $k \in \mathbb{Z}$, are asymptotic in the right half-plane to the line $\Im(z) = 2\pi j$. Since $T_k(\theta)$ is invariant under F_j and

$$f(L_j(w)) = L_j(w) + e^{-w} = L_j(w) + o(1) \text{ as } \Re(w) \to \infty,$$

the set $V_{j,k}(\theta)$ is invariant under f. Hence, for each $k, j \in \mathbb{Z}$, the function f has an invariant Baker domain $U_{j,k}$ containing $V_{j,k}(\theta)$. We can then use a similar argument to that in the proof of Theorem 6.2(b), but here involving the sets $L_j(S_k)$ and their associated paths, to show that these Baker domains are distinct. Also, these Baker domains do not meet the lines $\Im(z) = (2j\pm1)\pi$, $j \in \mathbb{Z}$, since these lines are invariant and $f^n(z)$ tends to infinity too rapidly on them. This completes the proof. □

Similar reasoning (but using more techniques from the proof of Theorem 6.3) can be applied to other functions of the form $f(z) = z + R(z)e^{g(z)}$, whenever g has infinitely many inverse branches g_j^{-1}, $j \in \mathbb{Z}$, defined on a left half-plane $B = \{w : \Re(w) < -M\}$, where $M > 0$, and the inverse image sets $g_j^{-1}(B)$ are sufficiently large that the following property holds:

if $z \in g_j^{-1}(B)$, then $f(z) \notin g_i^{-1}(B)$, where $i \neq j$.

For example, the following result can be obtained by using this approach; we omit the details.

Corollary 6.3. Let $f(z) = z + e^{\sin z}$. Then, for every $j \in \mathbb{Z}$, the function f has infinitely many invariant Baker domains asymptotic in the upper half-plane to the line $\Re(z) = 2\pi j - \pi/2$.

Remark Bergweiler [25] has recently given an example of a transcendental meromorphic function which has no fixed points and no invariant Baker domains, thus providing a partial answer to Buff's question.

References

[1] I.N. Baker, Limit functions and sets of non-normality in iteration theory, *Ann. Acad. Sci. Fenn. Ser. A I Math.*, 467 (1970) 11 pp.

[2] I.N. Baker, The domains of normality of an entire function, *Ann. Acad. Sci. Fenn. Ser. A I Math.*, 1 (1975) 277–283.

[3] I.N. Baker, The iteration of polynomials and transcendental entire functions, *J. Austral. Math. Soc. Ser. A*, 30 (1980/81) 483–495.

[4] I.N. Baker, Wandering domains in the iteration of entire functions, *Proc. London Math. Soc.*, (3) 49 (1984) 563–576.

[5] I.N. Baker, Some entire functions with multiply-connected wandering domains, *Ergodic Theory Dynam. Systems*, 5 (1985) 163–169.

[6] I.N. Baker, Wandering domains for maps of the punctured plane, *Ann. Acad. Sci. Fenn. Ser. A I Math.*, 12 (1987) 191–198.

[7] I.N. Baker, Infinite limits in the iteration of entire functions, *Ergodic Theory Dyn. Systems*, 8 (1988) 503–507.

[8] I.N. Baker and P. Domínguez, Analytic self-maps of the punctured plane, *Complex Var. Theory Appl.*, 37 (1998) 67–91.

[9] I.N. Baker and P. Domínguez, Boundaries of unbounded Fatou components of entire functions, *Ann. Acad. Sci. Fenn. Math.*, 24 (1999) 437–464.

[10] I.N. Baker, P. Domínguez and M. Herring, Dynamics of functions meromorphic outside a small set, *Ergodic Theory Dynam. Systems*, 21 (2001) 647–672.

[11] I.N. Baker, J. Kotus and Lü Yinian, Iterates of meromorphic functions II: Examples of wandering domains, *J. London Math. Soc.*, 42 (1990) 267–278.

[12] I.N. Baker, J. Kotus and Lü Yinian, Iterates of meromorphic functions I, *Ergodic Theory Dynam. Systems*, 11 (1991) 241–248.

[13] I.N. Baker, J. Kotus and Lü Yinian, Iterates of meromorphic functions III: Preperiodic domains, *Ergodic Theory Dynam. Systems*, 11 (1991) 603–618.

[14] I.N. Baker, J. Kotus and Lü Yinian, Iterates of meromorphic functions IV: Critically finite functions, *Results Math.*, 22 (1992) 651–656.

[15] I.N. Baker and J. Weinreich, Boundaries which arise in the dynamics of entire functions, Proceedings of the International Colloquium on Complex Analysis, Bucharest, *Revue Roumaine Math. Pures Appl.*, 36 (1991) 413–420.

[16] K. Barański and N. Fagella, Univalent Baker domains, *Nonlinearity*, 14 (2001) 411–429.

[17] D. Bargmann, Normal families of covering maps, *J. Analyse. Math.*, 85 (2001) 291–306.

[18] D. Bargmann, Iteration of inner functions and boundaries of components of the Fatou set, *Transcendental dynamics and complex analysis*, Cambridge University Press, 2008.

[19] W. Bergweiler, Iteration of meromorphic functions, *Bull. Amer. Math. Soc.*, 29 (1993) 151–188.

[20] W. Bergweiler, Newton's method and a class of meromorphic functions without wandering domains, *Ergodic Theory Dynam. Systems*, 13 (1993) 231–247.

[21] W. Bergweiler, Invariant domains and singularities, *Math. Proc. Camb. Phil. Soc.*, 117 (1995) 525–532.

[22] W. Bergweiler, On the Julia set of analytic self-maps of the punctured plane, *Analysis*, 15 (1995) 251–256.

[23] W. Bergweiler, Singularities in Baker domains, *Comput. Methods Funct. Theory*, 1 (2001) 41–49.

[24] W. Bergweiler, Periodic Fatou components and singularities of the inverse function, *Topics in analysis and its applications, NATO Sci. Ser. II Math. Phys. Chem.*, 147 (2004) 47–59, Kluwer Acad. Publ., Dordrecht, 2004.

[25] W. Bergweiler, Newton's method and Baker domains, Preprint.
[26] W. Bergweiler and N. Terglane, On the zeros of solutions of differential equations of the second order, *J. London Math. Soc.*, 58 (1998) 311–330.
[27] W. Bergweiler, F. von Haeseler, H. Kriete, H. -G. Meier and N. Terglane, Newton's method for meromorphic functions, *Complex analysis and its applications (Hong Kong 1993)*, Pitman Res. Notes Math. Ser., 305 Longman Scientific & Technical, Harlow, 1994, 147–158.
[28] A. Bolsch, Periodic Fatou components of meromorphic functions, *Bull. London. Math. Soc.*, 31 (1999) 543–555.
[29] P. Bonfert, On iteration in planar domains, *Michigan Math. J.*, 44 (1997) 47–68.
[30] X. Buff and J. Rückert, Virtual immediate basins of Newton maps and asymptotic values, *International Math. Res. Notes*, Article ID 65498 (2006) 1–18.
[31] L. Carleson and T.W. Gamelin, *Complex dynamics*, Springer, 1993.
[32] C. Cowen, Iteration and solutions of functional equations for functions analytic in the unit disk, *Trans. Amer. Math. Soc.*, 265 (1981) 69–95.
[33] P. Domínguez, Dynamics of transcendental meromorphic functions, *Ann. Acad. Sci. Fenn. Math.*, 23 (1998) 225–250.
[34] A.E. Erememko and M.Yu Lyubich, Examples of entire functions with pathological dynamics, *J. London Math. Soc.* (2), 36 (1987) 458–468.
[35] A.E. Erememko and M.Yu Lyubich, Dynamical properties of some classes of entire functions, *Ann. Inst. Fourier, Grenoble*, 42 (1992) 989–1020.
[36] N. Fagella and Ch. Henriksen, Deformation of entire functions with Baker domains, *Discrete Contin. Dyn. Syst.*, 15 (2006) 379–394.
[37] P. Fatou, Sur l'itération des fonctions transcendantes entières, *Acta Math. France*, 47 (1926) 337-360.
[38] D.S. Fleischmann, On a family of Baker domains, Preprint.
[39] D. Gaier, *Lectures on complex approximation*, Birkhäuser, 1985.
[40] M. Herman, Are there critical points on the boundary of singular domains?, *Comm. Math. Phys.*, 99 (1985) 593–612.
[41] M. Herring, *An extension of the Julia–Fatou theory of iteration*, Ph.D. Thesis, University of London, 1994.
[42] A. Hinkkanen, Iteration and the zeros of the second derivative of a meromorphic function, *Proc. London Math. Soc.*, 65 (1992) 629–650.
[43] M. Kisaka, On the connectivity of Julia sets of transcendental entire functions, *Ergodic Theory Dynam. Systems*, 18 (1998) 189–205.
[44] H. König, Conformal conjugacies in Baker domains, *J. London Math. Soc.*, 59 (1999) 153–170.
[45] A. Lauber, On the stability of Julia sets of functions having Baker domains, Dissertation, University of Göttingen, 2004.
[46] A. Lauber, Bifurcations of Baker domains, *Nonlinearity*, 20 (2007), 1535-45,
[47] S. Morosawa, An example of cyclic Baker domains, *Mem. Fac. Sci. Kochi Univ. (Math.)*, 20 (1999) 123–126.
[48] S. Morosawa, Y. Nishimura, M. Taniguchi and T. Ueda, *Holomorphic dynamics*, Cambridge University Press, 2000.
[49] R. Nevanlinna, *Analytic functions*, Springer, 1970.
[50] Ch. Pommerenke, *Boundary behaviour of conformal maps*, Springer-Verlag, 1992.
[51] P.J. Rippon, Baker domains of meromorphic functions, *Ergodic Theory Dynam. Systems*, 26 (2006) 1225–1233.
[52] P.J. Rippon and G.M. Stallard, Families of Baker domains I, *Nonlinearity*, 12 (1999) 1005–1012.

[53] P.J. Rippon and G.M. Stallard, Families of Baker domains II, *Conform. Geom. Dyn.*, 3 (1999) 67–78.

[54] P.J. Rippon and G.M. Stallard, Iteration of a class of hyperbolic meromorphic functions. *Proc. Amer. Math. Soc.*, 127 (1999) 3251–3258.

[55] P.J. Rippon and G.M. Stallard, On sets where iterates of a meromorphic function zip towards infinity, *Bull. London Math. Soc.*, 32 (2000) 528–536.

[56] P.J. Rippon and G.M. Stallard, Escaping points of meromorphic functions with a finite number of poles, *J. Analyse Math.*, 96 (2005) 225–245.

[57] P.J. Rippon and G.M. Stallard, Singularities of meromorphic functions with Baker domains, *Math. Proc. Camb. Phil. Soc.*, 141 (2006) 371–382.

[58] G.M. Stallard, A class of meromorphic functions with no wandering domains, *Ann. Acad. Sci. Fenn. Ser. A I Math.*, 16 (1991) 211–226.

[59] N. Steinmetz, *Rational iteration*, de Gruyter, 1993.

THE OPEN UNIVERSITY, DEPARTMENT OF MATHEMATICS, WALTON HALL, MILTON KEYNES MK7 6AA, UK

E-mail address: p.j.rippon@open.ac.uk

ESCAPING POINTS OF THE COSINE FAMILY

GÜNTER ROTTENFUSSER AND DIERK SCHLEICHER

Dedicated to the memory of Noel Baker

ABSTRACT. We study the dynamics of iterated cosine maps $E \colon z \mapsto ae^z + be^{-z}$, with $a, b \in \mathbb{C} \setminus \{0\}$. We show that the points which converge to ∞ under iteration are organized in the form of rays and, as in the exponential family, every escaping point is either on one of these rays or the landing point of a unique ray. Thus we get a complete classification of the escaping points of the cosine family, confirming a conjecture of Eremenko in this case. We also get a particularly strong version of the "dimension paradox": the set of rays has Hausdorff dimension 1, while the set of points these rays land at has not only Hausdorff dimension 2 but infinite and sometimes full planar Lebesgue measure.

1. INTRODUCTION

The dynamics of iterated polynomials has been investigated quite successfully, particularly in the past two decades. The study begins with a description of the *escaping points:* those points which converge to ∞ under iteration. It is well known that the set of escaping points is an open neighborhood of ∞ which can be parametrized by *dynamic rays*. The Julia set can then be studied in terms of landing properties of dynamic rays.

For entire transcendental functions, the point ∞ is an essential singularity (rather than a superattracting fixed point as for polynomials). This makes the investigation of the dynamics much more difficult. In particular, there is no obvious structure of the set of escaping points. Fatou [7] observed that certain transcendental Julia sets contained curves consisting of escaping points, and asked whether this property was true more generally. Eremenko [5] showed that for every entire transcendental function, the set of escaping points is always non-empty, and he asked whether each connected component (or even every path component) was unbounded. In [2], this question was answered in the affirmative for the special case of exponential functions $z \mapsto \lambda e^z$ for every $\lambda \in (0, 1/e)$ (these maps have attracting fixed points), and in [18] for arbitrary $\lambda \in \mathbb{C}^*$: every escaping point can be connected to ∞ along a unique curve running entirely along escaping points. In this paper, we extend this result

---

Keywords: Dynamic rays, cosine, iteration, escaping point, Hausdorff dimension
2000 *Mathematics Subject Classification.* 30D05,37F10,37F20,37F35.

to the family of cosine maps $E_{a,b}\colon z \mapsto ae^z + be^{-z}$ with arbitrary $a,b \in \mathbb{C}^*$; as in [18], we also provide an explicit classification of all escaping points. The existence of curves in the escaping set was shown for many entire functions of bounded type in [3].

In many ways, the escaping points of the cosine family behave quite similarly to those of the exponential family; therefore, the present paper is very similar to [18]. Most of it is based on the Diploma thesis [12] of 2002. Based on experience with these and other families of maps, we believe that similar results should hold for much larger classes of entire transcendental functions, possibly of bounded type (which means that the set of asymptotic or critical values is bounded). This paper is thus a contribution to the program to make polynomial tools, in particular dynamic rays, available for the study of iterated entire transcendental functions.[1]

Our main result is a classification of escaping points for every map $E_{a,b}$; this classification is the same for all such maps (with natural exceptions if one or both critical orbits escape). A by-product is an affirmative answer to Eremenko's question as mentioned above: every path connected component of the set of escaping points is a curve starting at ∞. As for exponential functions, but quite unlike the polynomial case, certain of these curves land at points in $\mathbb{C}$ which are escaping points themselves. A *dynamic ray* is a connected component of the escaping set, removing the landing points (for those curves which land at escaping points). It turns out that the union of all the uncountably many dynamic rays has Hausdorff dimension 1 (in analogy to results [9, 18] for the exponential family). However, by a result of McMullen [10], the set of escaping points in the cosine family has infinite planar Lebesgue measure (this is one main difference to the exponential case, where the set of escaping points has zero measure). In certain cases, the escaping set even has full planar Lebesgue measure. Therefore, the entire measure of the escaping set sits in the landing points of those rays which land at escaping points.

In [15, 16], these results are extended even further: if both critical orbits of $E_{a,b}$ are strictly preperiodic, then the set R of dynamic rays still has Hausdorff dimension 1 and every dynamic ray lands somewhere in $\mathbb{C}$. Therefore, $\mathbb{C} \setminus R$ is "most of $\mathbb{C}$" (the complement of a one-dimensional set). It turns out that each $z \in \mathbb{C} \setminus R$ is the landing point of one or several of the rays in R: the rays in the one-dimensional set R manage to connect all the remaining points to ∞ by curves in R! This highlights another difference between the dynamics in the cosine family and in the exponential family: in the exponential family, there

---

[1]In the years since this paper was originally written, it was shown in the Ph.D. thesis of the first author [13] that Eremenko's question is true for many further entire functions of bounded type, in particular those of finite order; but there are also entire functions of finite type for which every path component of the set of escaping points is bounded; see [14].

is no case known where the Julia set is the entire complex plane and every dynamic ray lands [11, Theorem 1.6]; this is due to the enormous contraction in the asymptotic tract of the asymptotic value 0, while the singular values in the cosine family are simply two critical values with much better-behaved properties.

We start this paper in Section 2 by setting up a partition for symbolic dynamics. In Section 3, we construct *ray tails* which are curves of escaping points terminating at ∞. These ray tails are extended to entire *dynamic rays* in Section 4. In Section 5, we prove that points on the same dynamic ray move away from each other very quickly, and this leads to a complete classification of escaping points in Section 6. Finally, in Section 7 we discuss the implications in terms of Hausdorff dimension and Lebesgue measure.

ACKNOWLEDGMENT. We would like to thank Alexandra Kaffl and Johannes Rückert and in particular Markus Förster and Lasse Rempe for helpful comments on earlier drafts of this paper, and John Milnor for having asked very helpful questions. We are grateful to the Institut Henri Poincaré in Paris for its hospitality while this paper was finished, and to Adrien Douady and Hans Henrik Rugh for having organized the program which brought us there in the fall of 2003. The first author would like to thank the DAAD for its support during the stay.

NOTATION. We consider the maps

$$E_{a,b}(z) := ae^z + be^{-z} \text{ for } a, b \in \mathbb{C}^* := \mathbb{C} \setminus \{0\}$$

and their iterates $E_{a,b}^{\circ n}(z)$. Usually we will omit the parameters a and b and write $E(z)$ for $E_{a,b}(z)$. Set $c := \frac{1}{2}\ln\left(\frac{b}{a}\right)$, where the branch of the logarithm is chosen such that $|\mathrm{Im}(c)| \leq \pi/2$. The critical points of E are

$$C_{crit} = \{c + i\pi n : n \in \mathbb{Z}\}$$

and the critical values $v_{1/2} = \pm 2\sqrt{ab}$, choosing signs so that v_1 is the image of $c + 2\pi i \mathbb{Z}$, while v_2 is the image of $c + i\pi + 2\pi i\mathbb{Z}$. There are no asymptotic values in $\mathbb{C}$. We will use the notation $\mathbb{C}^* := \mathbb{C} \setminus \{0\}$ and $\overline{\mathbb{C}} := \mathbb{C} \cup \{\infty\}$. Also, we will need $F \colon \mathbb{R}_0^+ \to \mathbb{R}_0^+$, $F(t) := \exp(t) - 1$. Now let $\alpha := \ln a$ and $\beta := \ln b$, choosing branches so that $|\mathrm{Im}(\alpha)| \leq \pi$, $|\mathrm{Im}(\beta)| \leq \pi$. Furthermore, let $K := \min\{|a|, |b|\}$, $K_{\max} := \max\{|a|, |b|\}$ and $M := \max\{|\alpha|, |\beta|\}$. Given $a, b \in \mathbb{C}^*$, let

$$(1) \quad T_{a,b} \geq \max\left\{\sqrt{\left|\frac{2b}{a}\right|}(|a|+|b|), \sqrt{\left|\frac{2a}{b}\right|}(|a|+|b|), 8|ab|, 1, \frac{1}{2}\ln\left|\frac{2b}{a}\right|, \frac{1}{2}\ln\left|\frac{2a}{b}\right|, \ln\frac{4}{|a|}, \ln\frac{4}{|b|}\right\} + M + 2$$

be the least value for which $F(T_{a,b}) \geq T_{a,b} + M + 4$. Note that

(2) $$2\sqrt{|ab|} \leq \max\{8|ab|, 1\} :$$

if $|ab| < 1/16$, then $2\sqrt{|ab|} < 1/2 < 1$, while if $|ab| \geq 1/16$, then $2\sqrt{|ab|} \leq 8|ab|$. Thus $T_{a,b} \geq 2\sqrt{|ab|} + M + 2$.

We will use the following sets (their significance will be explained in Section 2 and Figure 2): if $\text{Im}(v_1) \geq \text{Im}(v_2)$, then set

(3) $$\begin{aligned}\mathcal{A} &:= \{z \in \mathbb{C}: z = \lambda v_1 + (1-\lambda)v_2; \lambda \in [0,1]\} \\ &\cup \{z \in \mathbb{C}: \text{Re}(z) = \text{Re}(v_1), \text{Im}(z) \geq \text{Im}(v_1)\};\end{aligned}$$

this is the segment between v_1 and v_2, together with the vertical ray starting at v_1 in upwards direction. If $\text{Im}(v_1) < \text{Im}(v_2)$, we reverse the last inequality in the definition of $\mathcal{A}$: the last ray is replaced by the downwards vertical ray at v_1. In both cases, set $\mathbb{C}' := \mathbb{C} \setminus \mathcal{A}$.

2. Escaping Points and Symbolic Dynamics

In this section, we set up a partition of the complex plane and define symbolic dynamics of escaping points.

Definition 2.1 (Escaping Points).

A point $z \in \mathbb{C}$ with $|E^{\circ k}(z)| \to \infty$ for $k \to \infty$ is called an escaping point *and its orbit is called an* escaping orbit.

Lemma 2.2 (Real Parts of Escaping Orbits).

If (z_k) is an orbit with $|z_k| \to \infty$ for $k \to \infty$, then $|\text{Re}(z_k)| \to \infty$.

PROOF. This follows from the standard estimate

(4) $$|z_{k+1}| \leq |a|\exp(\text{Re}(z_k)) + |b|\exp(\text{Re}(-z_k)) .$$

□

In order to introduce symbolic dynamics we need a useful partition of the complex plane. First define

$$\mathbb{Z}_S := \mathbb{Z} \times \{L, R\}, \quad \mathbb{Z}_R := \mathbb{Z} \times \{R\} \text{ and } \mathbb{Z}_L := \mathbb{Z} \times \{L\},$$

and we write $n_R := (n, R)$ and $n_L := (n, L)$ for $n \in \mathbb{Z}$. Let us first consider the "straight partition" (Figure 1): for $j \in \mathbb{Z}_R$ define the strip R'_j as follows

$$R'_j := \{z \in \mathbb{C} : \text{Im}\, c + 2\pi j < \text{Im}\, z < \text{Im}\, c + 2\pi(j+1), \text{Re}\, c < \text{Re}\, z\}$$

and for $j \in \mathbb{Z}_L$

$$R'_j := \{z \in \mathbb{C}: \text{Im}\, c + 2\pi j < \text{Im}\, z < \text{Im}\, c + 2\pi(j+1), \text{Re}\, z < \text{Re}\, c\} .$$

Then the R'_j form a rather simple partition of $\mathbb{C}$. The image of the boundary of each strip is the part of the straight line through the critical values ending in v_1. It will be convenient to modify the partition so that the real part of the image of the boundary is bounded. To introduce our partition (see Figure 2) we now define $\mathcal{A}$ as in (3) and set $\mathbb{C}' := \mathbb{C} \setminus \mathcal{A}$. Then define the strips R_j as connected components of $E^{-1}(\mathbb{C} \setminus \mathcal{A})$, so that

$$E \colon R_j \to \mathbb{C}'$$

is a conformal isomorphism for all $j \in \mathbb{Z}_S$. For $s \in \mathbb{Z}_S$ we denote the inverse mapping of $E \colon R_s \to \mathbb{C}'$ by $L_s \colon \mathbb{C}' \to R_s$. Note that the strips are open and that the union of their closures is $\mathbb{C}$. Label the strips such that $\overline{R}_{0_R}$ contains some right end of $\mathbb{R}^+$, $\overline{R}_{0_L}$ contains some left end of $\mathbb{R}^-$, and such that $z \in R_{(n,L)}$ iff $z + 2\pi i \in R_{(n+1,L)}$ and $z \in R_{(n,R)}$ iff $z + 2\pi i \in R_{(n+1,R)}$. The following property of the strips will be used throughout.

Lemma 2.3 (Height of Strips Bounded by 3π).
If $z, w \in R_{n,L} \cup R_{n,R}$, then $|\text{Im}(z) - \text{Im}(w)| < 3\pi$ and $|\text{Im}(z) - 2\pi i n| < 3\pi$.

PROOF. The straight line through v_1 and v_2 divides $\mathbb{C}$ into two half-planes, and the preimages under $E_{a,b}$ of each of them are horizontal half-strips of height π. Therefore, each R_s is contained in three of those half-strips. Since $\overline{R}_{0_L}$ and $\overline{R}_{0_R}$ intersect the real axis, the second part follows immediately. □

Lemma 2.4 (Sign of Real Part).
If $|E(z)| > |a| + |b|$ for some $z \in \mathbb{C}$, then $\text{Re}(z) > 0$ iff $z \in R_{n,R}$ for some $n \in \mathbb{Z}$, and $\text{Re}(z) < 0$ iff $z \in R_{n,L}$ for some $n \in \mathbb{Z}$.

PROOF. Suppose that $\text{Re}(c) < 0$. Then $\{z \in R_{n,L} \colon \text{Re}(z) > 0\}$ is empty for all n, while every $z \in R_{n,R}$ with $\text{Re}(z) < 0$ has real parts between $\text{Re}(c)$ and 0. The vertical line through c (containing all the critical points) is mapped under E to the segment connecting the two critical values $\pm 2\sqrt{ab}$, while the imaginary axis maps to an ellipse with major axis $|a| + |b|$. Therefore if $|E(z)| > |a| + |b|$, then $|\text{Re}(z)|$ has the same sign as the unbounded part of the strip $R_{(n,L)}$ or $R_{(n,R)}$ containing z.

The case $\text{Re}(c) \geq 0$ is analogous. □

There are a number of further conceivable partitions, such as the one in Figure 3, all with their particular advantages but with a different syntax of symbolic sequences.

ESCAPING POINTS OF THE COSINE FAMILY 401

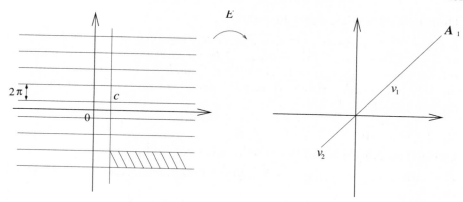

FIGURE 1. The partition formed by the R'_j: the strips have simple shapes, but the real part of $\mathcal{A}_1$ is unbounded.

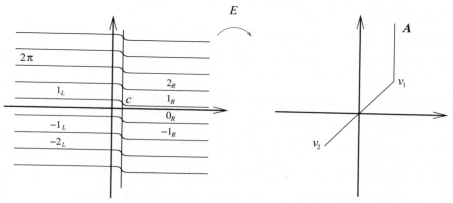

FIGURE 2. The partition we use: the set $\mathcal{A}$ has bounded real parts.

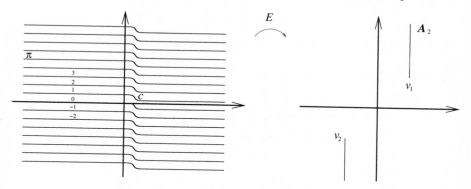

FIGURE 3. A partition with bounded real parts of $\mathcal{A}_2$ which could be used.

Definition 2.5 (External Address).

Let $\mathcal{S} := \mathbb{Z}_S^{\mathbb{N}} = \{(s_1 s_2 s_3 \ldots): s_k \in \mathbb{Z}_S\}$ be the sequence space over $\mathbb{Z}_S$ and let $\sigma \colon \mathcal{S} \to \mathcal{S}, (s_1 s_2 s_3 s_4 \ldots) \mapsto (s_2 s_3 s_4 \ldots)$ be the shift on $\mathcal{S}$. We will often use the notation $\underline{s} = (s_1 s_2 s_3 \ldots)$. For all $z \in \mathbb{C}$ with $E^{\circ n}(z) \in \mathbb{C}'$ for all $n \in \mathbb{N}$ the external address $S(z) \in \mathcal{S}$ is the sequence of the symbols of the strips containing $z, E(z), E^{\circ 2}(z), \ldots$.

Set $|(n, L)| = |(n, R)| := |n|$ for $n \in \mathbb{Z}$.

Definition 2.6 (Minimal Potential of External Addresses).

For a sequence $\underline{s} = s_1 s_2 s_3 \ldots \in \mathcal{S}$, define its minimal potential $t_{\underline{s}} \in \mathbb{R}_0^+ \cup \{\infty\}$ via

$$t_{\underline{s}} = \inf \left\{ t > 0 \colon \lim_{k \geq 1} \frac{|s_k|}{F^{\circ(k-1)}(t)} = 0 \right\}.$$

Notice that $t_{\sigma(\underline{s})} = F(t_{\underline{s}})$.

Definition 2.7 (Exponentially Bounded).

A sequence $\underline{s} \in \mathcal{S}$ is exponentially bounded if there is an $x > 0$ such that $|s_k| \leq F^{\circ(k-1)}(x)$ for all $k \geq 1$.

This condition is preserved under the shift, but the constant changes:

$$\underline{s}' = \sigma(\underline{s}) \implies |s_k'| \leq F^{\circ(k-1)}(F(x)).$$

An equivalent definition of exponential boundedness (as used in [18]) is the existence of $x, A > 0$ with $|s_k| \leq AF^{\circ(k-1)}(x)$ for all $k \geq 1$. It was shown in [18, Theorem 4.2 (1)] that a sequence $\underline{s}$ is exponentially bounded iff $t_{\underline{s}} < \infty$.

Lemma 2.8 (External Addresses are Exponentially Bounded).

For $E = E_{a,b}$ choose $\delta > 0$ with $|a| + |b| \leq e^{\delta} - (\delta + 1)$. Then every orbit (z_k) satisfies the bound

$$\max\{|\mathrm{Re}(z_k)|, |\mathrm{Im}(z_k)|\} \leq |z_k| < F^{\circ(k-1)}(|z_1| + \delta).$$

In particular, every orbit in $\mathbb{C}'$ has exponentially bounded external address.

PROOF. For all k we can estimate

$$\begin{aligned}
|z_{k+1}| + \delta &= |ae^{z_k} + be^{-z_k}| + \delta \\
&\leq |a| \exp(\mathrm{Re}(z_k)) + |b| \exp(\mathrm{Re}(-z_k)) + \delta \\
&\leq (|a| + |b|) \exp(|z_k|) + \delta \leq (e^{\delta} - (\delta + 1)) \exp(|z_k|) + \delta \\
&= e^{|z_k|+\delta} - (\delta + 1)e^{|z_k|} + \delta \leq e^{|z_k|+\delta} - (\delta + 1) + \delta \\
&= F(|z_k| + \delta) \ .
\end{aligned}$$

Induction yields

$$|z_k| + \delta < F^{\circ(k-1)}(|z_1| + \delta)$$

for all $k \geq 1$. If the orbit avoids $\mathcal{A}$, then the external address $s_1 s_2 s_3 \ldots$ is defined and we have (using Lemma 2.3 in the first inequality):

$$2\pi|s_k| \leq |\operatorname{Im}(z_k)| + 3\pi \leq |z_k| + \delta + 3\pi - \delta < F^{\circ(k-1)}(|z_1| + \delta) + 3\pi - \delta.$$

$\square$

3. Tails of Dynamic Rays

In this section, we show that the set of escaping points of E contains uncountably many curves starting at ∞; in Section 4, these curves will be extended further, and in Section 6, we will show that all escaping points are associated to these curves.

Definition 3.1 (Tail of Ray).
A ray tail with external address $\underline{s} \in \mathcal{S}$ is an injective curve $g_{\underline{s}} \colon [\tau, \infty) \to \mathbb{C}$ ($\tau > 0$) with the following properties:

- *each point on the curve escapes within $\mathbb{C}'$*
- *each point on the curve has external address $\underline{s}$*
- *either $\lim_{t \to \infty} \operatorname{Re}(g_{\underline{s}}(t)) = +\infty$ or $\lim_{t \to \infty} \operatorname{Re}(g_{\underline{s}}(t)) = -\infty$*
- *all $t \geq \tau$ satisfy $E^{\circ k}(g_{\underline{s}}(t)) = \pm F^{\circ(k-1)}(t) + 2\pi i s_k + O(1)$ as $k \to \infty$.*

The real value t is called the *potential* of $z = g_{\underline{s}}(t)$.

Given an exponentially bounded external address $\underline{s}$, let $T_{\underline{s}} \geq T_{a,b}$ be the least value for which $4\pi|s_{n+1}| < F^{\circ n}(T_{\underline{s}})$ for all $n \in \mathbb{N}$ (such a finite $T_{\underline{s}}$ exists since $\underline{s}$ is exponentially bounded). Note that $T_{\sigma(\underline{s})} \leq F(T_{\underline{s}})$.

Lemma 3.2 (Minimal Potentials).
For every exponentially bounded $\underline{s}$ and every $t > t_{\underline{s}}$, there is an $N \in \mathbb{N}$ such that for all $n \geq N$, $F^{\circ n}(t) > T_{\sigma^n(\underline{s})}$.

PROOF. Since $t > t_{\underline{s}}$, there is an $N > 0$ such that $F^{\circ n}(t) > 4\pi|s_{n+1}|$ for all $n \geq N$; enlarge N if necessary so that also $F^{\circ N}(t) > T_{a,b}$. Then $F^{\circ n}(t) = F^{\circ(n-N)}(F^{\circ N}(t)) > F^{\circ(n-N)}(T_{\sigma^N(\underline{s})}) \geq T_{\sigma^n(\underline{s})}$. $\square$

Proposition 3.3 (Existence of Tails of Rays).
For all $a, b \in \mathbb{C}^*$ and for every exponentially bounded sequence $\underline{s} \in \mathcal{S}$ there exists a ray tail $g_{\underline{s}}([T_{\underline{s}}, \infty))$ with external address $\underline{s}$. Furthermore

$$g_{\underline{s}}(t) = \begin{cases} t - \alpha + 2\pi i s_1 + r_{\underline{s}}(t) & (\text{if } s_1 \in \mathbb{Z}_R) \\ -t + \beta + 2\pi i s_1 + r_{\underline{s}}(t) & (\text{if } s_1 \in \mathbb{Z}_L) \end{cases}$$

with

$$|r_{\underline{s}}(t)| \leq (C_1 + 8\pi|s_2|)e^{-t}$$

with a constant C_1 depending only on a and b. Moreover, for $t \geq T_{\underline{s}}$,
$$E(g_{\underline{s}}(t)) = g_{\sigma(\underline{s})}(F(t)) .$$

OVERVIEW: We construct a family of maps $g_{\underline{s}}^n \colon \mathbb{R}^+ \to \mathbb{C}$ for $n \in \mathbb{N}$ as follows:

(5) $$g_{\underline{s}}^n(t) := L_{s_1} \circ \cdots \circ L_{s_n} \circ (\pm F^{\circ n}(t) + 2\pi i s_{n+1}) .$$

The $\pm$ depends on s_{n+1}: it is $+$ for $s_{n+1} \in \mathbb{Z}_R$ and $-$ for $s_{n+1} \in \mathbb{Z}_L$ (then $\pm F^n(t) + 2\pi i s_{n+1} \in \overline{R_{s_{n+1}}}$ for large t). We will show that the $g_{\underline{s}}^n$ are defined for all $t \geq T_{\underline{s}}$ independently of n and converge uniformly to the desired function $g_{\underline{s}}(t)$. For this proof we need several lemmas. The first of them gives control over the inverse branches of E.

Lemma 3.4 (Control on L_s).
Let $w \in \mathbb{C}'$ and $z = L_s(w)$ with
$$|w| > \max\left\{\sqrt{\left|\frac{2b}{a}\right|}(|a|+|b|), \sqrt{\left|\frac{2a}{b}\right|}(|a|+|b|), 8|ab|, 1\right\}$$
and $s \in \mathbb{Z}_R$. Then there is a $p \in \mathbb{Z}$ so that
$$z = \ln w - \alpha + 2\pi i p + r^*$$
with $|r^*| < |2\frac{b}{a}e^{-2z}| < 1$ and $|r^*| < 8|ab| \cdot |w|^{-2} < \frac{1}{|w|} < 1$.
Similarly, for $s \in \mathbb{Z}_L$ there is a $p \in \mathbb{Z}$ so that
$$z = -\ln w + \beta + 2\pi i p + r^*$$
with $|r^*| < |2\frac{a}{b}e^{-2z}| < 1$ and $|r^*| < 8|ab| \cdot |w|^{-2} < \frac{1}{|w|} < 1$.

Note that the branch of $\ln w$ in this lemma is immaterial because of the ambiguity in p.

PROOF. Heuristically, $E(z) \approx ae^z$ if $\mathrm{Re}(z) \gg 0$ and $E(z) \approx be^{-z}$ if $\mathrm{Re}(z) \ll 0$. We discuss the case $s \in \mathbb{Z}_R$; the other case is similar. We have
$$w = E(z) = ae^z + be^{-z} = ae^z\left(1 + \frac{b}{a}e^{-2z}\right)$$

(6) $$\implies z \in \ln\frac{w}{a} - \ln\left(1 + \frac{b}{a}e^{-2z}\right) + 2\pi i\mathbb{Z}.$$

Lemma 2.4 implies that $\mathrm{Re}(z) > 0$, so we have
$$\sqrt{\left|\frac{2b}{a}\right|}(|a|+|b|) < |w| \leq |a|e^{\mathrm{Re}(z)} + |b|e^{-\mathrm{Re}(z)} \leq (|a|+|b|)e^{\mathrm{Re}(z)} ,$$

hence $e^{\text{Re}(z)} > \sqrt{|2b/a|}$ and thus
$$\left|\frac{b}{a}e^{-2z}\right| = \left|\frac{b}{a}\right|e^{-2\text{Re}(z)} < \frac{1}{2}.$$
Since $|\ln(1+u)| < (2\ln 2)|u|$ for $|u| < \frac{1}{2}$, it follows that
$$\left|\ln\left(1+\frac{b}{a}e^{-2z}\right)\right| < 2\ln 2\left|\frac{b}{a}e^{-2z}\right| < 1$$
(here we use the principal branch of ln).

Together with (6) the first claim follows. For the estimate of $|r^*|$ in terms of $|w|$ it follows from $2\ln 2\left|\frac{b}{a}e^{-2z}\right| < 1$ that $|be^{-z}| < \frac{1}{2\ln 2}|ae^z|$ and hence by (4) we get $|w| < 2|ae^z|$. We thus obtain
$$|ab|\frac{2}{|ae^z|^2} < 8|ab|\cdot|w|^{-2} < |w|^{-1} < 1.$$
$\square$

The next lemma gives us control on $g_{\underline{s}}^1 = L_{s_1} \circ (\pm F + 2\pi i s_2)$.

Lemma 3.5 (Control on $L_s \circ F$).
Choose $a, b \in \mathbb{C}^*$ and $n \in \mathbb{Z}$. Then for $t > 0$ such that
$$F(t) > \max\left\{\sqrt{\left|\frac{2b}{a}\right|}(|a|+|b|), \sqrt{\left|\frac{2a}{b}\right|}(|a|+|b|), 8|ab|, 1\right\}$$
and $F(t) > 4\pi|n|$, we have for $s \in \mathbb{Z}_R$
$$L_s(\pm F(t) + 2\pi i n) \in t - \alpha + \pi i \mathbb{Z} + r \quad \text{with} \quad |r| < (4+8\pi|n|)e^{-t} < 4$$
and for $s \in \mathbb{Z}_L$ we have
$$L_s(\pm F(t) + 2\pi i n) \in -t + \beta + \pi i \mathbb{Z} + r \quad \text{with} \quad |r| < (4+8\pi|n|)e^{-t} < 4.$$

PROOF. For $s \in \mathbb{Z}_R$, Lemma 3.4 gives
$$L_s(\pm F(t) + 2\pi i n) \in \ln\left(\pm F(t)\left(1 \pm \frac{2\pi i n}{F(t)}\right)\right) - \alpha + 2\pi i \mathbb{Z} + r^*$$
$$= t + \ln(1 - e^{-t}) + \ln\left(1 \pm \frac{2\pi i n}{F(t)}\right)$$
$$-\alpha + 2\pi i \mathbb{Z} + r^* + (i\pi)$$
with $|r^*| < |F(t) + 2\pi i n|^{-1} \leq F(t)^{-1} = e^{-t}\frac{e^t}{e^t-1} < 2e^{-t} < 1$. Here, the last term $(+i\pi)$ only occurs in the case $L_s(-F(t)\ldots)$.
Since $2\pi|n|/F(t) < 1/2$, we get
$$\left|\ln\left(1 \pm \frac{2\pi i n}{F(t)}\right)\right| \leq (2\ln 2)\frac{2\pi|n|}{F(t)} \leq 8\pi|n|e^{-t} < 2$$

and
$$|\ln(1 - e^{-t})| < 2e^{-t} < 1.$$

Thus we get
$$L_s(\pm F(t) + 2\pi i n) \in t - \alpha + \pi i \mathbb{Z} + r$$

with $|r| < (4 + 8\pi |n|) e^{-t} < 4$.
The proof for $s \in \mathbb{Z}_L$ is analogous. $\square$

Lemma 3.6 (Bound on Real Parts).
For all $\underline{s} \in \mathcal{S}$, $n \in \mathbb{N}$ and $t > 0$ with $t \geq T_{a,b}$
$$\left|\operatorname{Re}\left(g_{\underline{s}}^n(t)\right)\right| > t - (M+2).$$

Moreover, if $|\operatorname{Re}(z)| > \max\left\{\frac{1}{2}\ln\left|\frac{2b}{a}\right|, \frac{1}{2}\ln\left|\frac{2a}{b}\right|, \ln\frac{4}{|a|}, \ln\frac{4}{|b|}\right\}$, then
$$|E'(z)| > \frac{1}{2} K e^{|\operatorname{Re}(z)|} > 2.$$

PROOF. We will prove the first part via induction simultaneously for all external addresses $\underline{s}$ and all $t \geq T_{a,b}$. For $n = 0$ clearly
$$\left|\operatorname{Re}(g_{\underline{s}}^0(t))\right| = t > t - (M+2).$$

Assume that $\left|\operatorname{Re}(g_{\underline{s}}^{n-1}(t))\right| > t - (M+2)$. Then by Lemma 3.4 and the inductive hypothesis we get

$$\begin{aligned}
\left|\operatorname{Re}(g_{\underline{s}}^n(t))\right| &= \left|\operatorname{Re}\left(\pm \ln\left(g_{\sigma(\underline{s})}^{n-1}(F(t))\right) \pm \{\alpha, \beta\} + 2\pi i p + r^*\right)\right| \\
&\geq \left|\operatorname{Re}\left(\ln\left(g_{\sigma(\underline{s})}^{n-1}(F(t))\right)\right)\right| - M - 1 \\
&= \left|\ln\left|g_{\sigma(\underline{s})}^{n-1}(F(t))\right|\right| - M - 1 \\
&\geq \left|\ln\left|F(t) - (M+2)\right|\right| - M - 1 \\
&= \ln(e^t - M - 3) - M - 1 \\
&> t - M - 2.
\end{aligned}$$

The last step needs an elementary calculation based on
$$F(t) \geq F(T_{a,b}) \geq T_{a,b} + M + 4.$$

For the second part, we write:
$$|E'(z)| = \begin{cases} |a|e^{|\operatorname{Re}(z)|} \cdot \left|1 - \frac{b}{a}e^{-2|\operatorname{Re}(z)|}\right| & \text{if } \operatorname{Re}(z) > 0 \\ |b|e^{|\operatorname{Re}(z)|} \cdot \left|1 - \frac{a}{b}e^{-2|\operatorname{Re}(z)|}\right| & \text{if } \operatorname{Re}(z) < 0 \end{cases}.$$

By hypothesis, $\left|\frac{b}{a}e^{-2|\mathrm{Re}(z)|}\right| < \frac{1}{2}$, resp. $\left|\frac{a}{b}e^{-2|\mathrm{Re}(z)|}\right| < \frac{1}{2}$, so we get

$$|E'(z)| > \frac{1}{2}\min\{|a|,|b|\}e^{|\mathrm{Re}(z)|}$$

and hence $|E'(z)| > 2$. □

We can now finish the construction of dynamic ray tails.

PROOF OF PROPOSITION 3.3. We show first that $g_{\underline{s}}^n(t)$ converges uniformly in t to a limit function $g_{\underline{s}}([T_{\underline{s}},\infty)) \to \mathbb{C}$. For $t \geq T_{\underline{s}}$, we write

$$|g_{\underline{s}}^{n+1}(t) - g_{\underline{s}}^n(t)| = \left| L_{s_1} \circ \cdots \circ L_{s_n} \circ \overbrace{L_{s_{n+1}}(\pm F(F^{\circ n}(t)) + 2\pi i s_{n+2})}^{=:\mu_n} \right.$$

$$\left. - L_{s_1} \circ \cdots \circ L_{s_n} \circ \underbrace{(\pm F^{\circ n}(t) + 2\pi i s_{n+1})}_{=:\nu_n} \right|.$$

Since $\mu_n, \nu_n \in \overline{R}_{s_{n+1}}$, we get $|\mathrm{Im}(\mu_n - \nu_n)| \leq 3\pi$ by Lemma 2.3; by Lemma 3.5, we get $|\mathrm{Re}(\mu_n - \nu_n)| < 4 + M$, hence $|\mu_n - \nu_n| < 4 + 3\pi + M$.

By construction, all $w \in \mathcal{A}$ have $|\mathrm{Re}(w)| \leq 2\sqrt{|ab|}$. However, by Lemma 3.6 and (2),

$$|\mathrm{Re}(L_{s_k} \circ \cdots \circ L_{s_n}(\nu_n))| > F^{\circ(k-1)}(t) - (M+2) \geq T_{\underline{s}} - (M+2)$$
$$> \max\{8|ab|, 1\} \geq 2\sqrt{|ab|}$$

and similarly for μ_n, so the same branch of $L_{s_{k-1}}$ applies to $L_{s_k} \circ \cdots \circ L_{s_n}(\nu_n)$ and $L_{s_k} \circ \cdots \circ L_{s_n}(\mu_n)$.

By Lemma 3.6, $L_{s_k} \circ \cdots \circ L_{s_n}(\nu_n)$ and $L_{s_k} \circ \ldots \circ L_{s_n}(\mu_n)$ are both in the domain $\{z \in \mathbb{C}: \mathrm{Re}(z) > T_{\underline{s}} - (M+2)\}$ or both in the symmetric domain $\{z \in \mathbb{C}: \mathrm{Re}(z) < -T_{\underline{s}} + (M+2)\}$; these satisfy $|E'(z)| > \frac{1}{2}Ke^{|\mathrm{Re}(z)|} > 2$. In particular,

$$|\mathrm{Re}(\nu_n)| = F^{\circ n}(t) > t + (M+4) \quad \text{and so} \quad |\mathrm{Re}(\mu_n)| > t.$$

Therefore,

$$|L_{s_n}(\mu_n) - L_{s_n}(\nu_n)| < \frac{|\mu_n - \nu_n|}{\frac{1}{2}Ke^t} < \frac{8 + 6\pi + 2M}{K}e^{-t}.$$

After repeated application, we get

(7) $$|g_{\underline{s}}^{n+1}(t) - g_{\underline{s}}^n(t)| < \frac{8 + 6\pi + 2M}{2^{n-1}K}e^{-t}.$$

Therefore, the $g_{\underline{s}}^n$ converge uniformly to a continuous limit function

$$g_{\underline{s}}\colon [T_{\underline{s}}, \infty) \to \mathbb{C}.$$

By construction, $E(g_{\underline{s}}^n(t)) = g_{\sigma(\underline{s})}^{n-1}(F(t))$, so in the limit we obtain the desired relation

$$E(g_{\underline{s}}(t)) = g_{\sigma(\underline{s})}(F(t))$$

for $t \geq T_{\underline{s}}$. In order to estimate r, we get

$$|g_{\underline{s}}(t) - g_{\underline{s}}^1(t)| \leq \sum_{n=1}^{\infty} |g_{\underline{s}}^{n+1}(t) - g_{\underline{s}}^n(t)|$$

$$< \sum_{n=1}^{\infty} \frac{8 + 6\pi + 2M}{2^{n-1}K} e^{-t} = 2\frac{8 + 6\pi + 2M}{K} e^{-t}.$$

By Lemma 3.5, we have

$$g_{\underline{s}}(t) = t - \alpha + 2\pi i s_1 + r_{\underline{s}}(t),$$

resp.

$$g_{\underline{s}}(t) = -t + \beta + 2\pi i s_1 + r_{\underline{s}}(t)$$

with

$$r_{\underline{s}}(t) \leq \left(2\frac{8 + 6\pi + 2M}{K} + 4 + 8\pi|s_2|\right)e^{-t}.$$

Finally we prove injectivity. If $g_{\underline{s}}(t_1) = g_{\underline{s}}(t_2)$ for $t_2 \geq t_1 \geq T_{\underline{s}}$, then we get

$$g_{\sigma^n(\underline{s})}(F^{\circ n}(t_1)) = g_{\sigma^n(\underline{s})}(F^{\circ n}(t_2))$$

for all $n \geq 0$, with the bounds

$$\left|g_{\sigma^n(\underline{s})}(F^{\circ n}(t_j)) - (\pm F^{\circ n}(t_j) + \{-\alpha \text{ or } +\beta\} - 2\pi i s_{n+1})\right|$$
$$= |r_{\sigma^n(\underline{s})}(F^{\circ n}(t_j))| < (C_1 + 8\pi|s_{n+2}|)e^{-F^{\circ n}(t_j)}$$
$$= 8\pi|s_{n+2}|/F^{\circ(n+1)}(t_j) + o(1) .$$

Since $t_1, t_2 > t_{\underline{s}}$, the right-hand side is bounded; this implies that $|F^{\circ n}(t_1) - F^{\circ n}(t_2)|$ must be bounded as well as $n \to \infty$. This implies that $t_1 = t_2$. □

4. Dynamic Rays

In this section we construct dynamic rays by extending ray tails to as low potentials as possible. The idea is to use the relation $E(g_{\underline{s}}(t)) = g_{\sigma(\underline{s})}(F(t))$ to pull ray tails back by the dynamics: $g_{\underline{s}}$ is a branch of $E^{-1} \circ g_{\sigma(\underline{s})} \circ F$.

Theorem 4.1 (Existence of Dynamic Rays).

(1) If neither of the two critical orbits escapes then for every exponentially bounded $\underline{s}$ there exists an injective curve $g_{\underline{s}} : (t_{\underline{s}}, \infty) \to \mathbb{C}$ consisting of escaping points such that

$$E(g_{\underline{s}}(t)) = g_{\sigma(\underline{s})}(F(t)) \text{ for all } t > t_{\underline{s}}$$

which extends the ray tail with external address $\underline{s}$ as constructed in Proposition 3.3. In particular, it inherits its asymptotics for large t.

(2) If at least one of the critical orbits escapes, then (1) is still true for every $\underline{s}$, unless $\underline{s}$ is such that there is an $n \geq 1$ and a $t_0 > F^{\circ n}(t_{\underline{s}})$ with $g_{\sigma^n(\underline{s})}(t_0) \in \{v_1, v_2\}$. For those exceptional $\underline{s}$, there is an injective curve $g_{\underline{s}} : (t_{\underline{s}}^*, \infty) \to \mathbb{C}$ with the same properties as before, where $t_{\underline{s}}^*$ is the largest potential which has an $n \geq 1$ such that $g_{\sigma^n(\underline{s})}(F^{\circ n}(t_{\underline{s}}^*)) \in \{v_1, v_2\}$.

The curve $g_{\underline{s}} : (t_{\underline{s}}, \infty) \to \mathbb{C}$ is called the *dynamic ray at external address $\underline{s}$*.

REMARK. Notice that it is no longer required that all points on the ray share the external address $\underline{s}$. Since the partition $\{R_{\underline{s}}\}$ of $\mathbb{C}$ is unnatural from the dynamical point of view, there is no reason why the rays should respect it. Only the points with large potential (those which are on ray tails) have external address $\underline{s}$. Note that the definition of $t_{\underline{s}}$ is the same as for the case of exponential functions $\lambda \exp$ in [18]: since $E_{a,b}(z) \approx ae^z$ resp. $E_{a,b}(z) \approx be^{-z}$ in far right resp. left half-planes, this coincidence of the values of $t_{\underline{s}}$ reflects the fact that the value of $t_{\underline{s}}$ for exponential functions does not depend on the complex parameter λ.

PROOF. We want to show that $g_{\underline{s}}(t)$ exists for $t > t_{\underline{s}}$. Choose $\varepsilon > 0$. By Lemma 3.2, there is an $N \in \mathbb{N}$ such that $F^{\circ N}(t_{\underline{s}} + \varepsilon) > T_{\sigma^N(\underline{s})}$.

By Proposition 3.3, the ray tail $g_{\sigma^N(\underline{s})}(t)$ exists for $t \geq T_{\sigma^N(\underline{s})}$. Since $E^{\circ N}$ maps the ray tail $g_{\underline{s}}$ to a tail of $g_{\sigma^N \underline{s}}$, it follows that there is a branch of E^{-N} which sends (a subset of) the ray tail $g_{\sigma^N} : [T_{\sigma^N(\underline{s})}, \infty) \to \mathbb{C}$ to a curve $g_{\underline{s}} : [t_{\underline{s}} + \varepsilon, \infty) \to \mathbb{C}$. In other words, the ray tail $g_{\underline{s}}$ can be extended to potentials $t_{\underline{s}} + \varepsilon$. Since ε was arbitrary, we have shown the existence of the dynamic ray $g_{\underline{s}} : (t_{\underline{s}}, \infty) \to \mathbb{C}$. Injectivity follows from Proposition 3.3. This proves the second claim.

Note that the pull-back of a ray tail is possible if and only if it contains no critical value. Therefore, this construction can be carried out for all ray tails except those mentioned in the exceptions of the third claim. This proves the theorem. □

REMARK. Note that the pull-back in the proof of the previous theorem need not respect the partition we had initially used for constructing the ray tails: the branch of every E^{-1} is determined using the ray tails and is then continued analytically. It may well happen that a dynamic ray crosses the partition boundary, but only at potentials below $T_{\underline{s}}$.

Next we investigate under which conditions an escaping point is on a ray.

Theorem 4.2 (Fast Escaping Points are on Ray).
Let (z_k) be an escaping orbit and let $\underline{s}$ be such that $z_k \in \overline{R_{s_k}}$ for all k; suppose also that there exists a $t' > t_{\underline{s}}$ with $|\operatorname{Re}(z_k)| \geq F^{\circ(k-1)}(t')$ for infinitely many k. If in addition

(8) $\qquad |\operatorname{Re}(z_k)| > 2\sqrt{|ab|} + 1 \quad \text{for all } k, \text{ as well as} \quad t' \geq T_{\underline{s}}$

then there is a $t \geq t'$ such that $z_1 = g_{\underline{s}}(t)$.

If (8) is not satisfied for all k, then there is an $N \in \mathbb{N}$ such that $z_{N+1} = g_{\sigma^N(\underline{s})}(F^{\circ N}(t))$ for some $t \geq t'$; moreover, at least if the two critical orbits do not escape, there is an external address $\underline{s}'$ which differs from $\underline{s}$ only in finitely many entries such that $z_1 = g_{\underline{s}'}(t)$ for some $t \geq t' > t_{\underline{s}'} = t_{\underline{s}}$.

PROOF. For $k \in \mathbb{N}$ pick $t_k > 0$ such that $F^{\circ(k-1)}(t_k) = |\operatorname{Re}(z_k)|$. By assumption, $t_k \geq t'$ for infinitely many k, and by Lemma 2.8, the sequence (t_k) is bounded above. Moreover, by Lemma 2.3, we have

$$|\operatorname{Im}(z_k)| \leq 2\pi(|s_k| + 3/2) \ll F^{\circ(k-1)}(t')$$

for all sufficiently large k. Therefore,

$$|\operatorname{Im}(z_k)| \ll F^{\circ(k-1)}(t') \leq |\operatorname{Re}(z_k)|$$

for infinitely many k. From now on we will only look at such k.

Suppose first that $|\operatorname{Re}(z_k)| > 2\sqrt{|ab|} + 1$ for all k. Then all $z_k \notin \mathcal{A}$, hence $z_k \in R_{s_k}$, and thus

$$z_l = L_{s_l} \circ \cdots \circ L_{s_{k-1}}(z_k)$$

for all $l \in \{k-1, \ldots, 2, 1\}$.

Consider the points $w_k := \pm F^{\circ(k-1)}(t_k) + 2\pi i s_k$, where the signs are chosen such that for all k, z_k and w_k are in the same (right or left) half-plane; then, using Lemma 2.3 again,

(9) $\qquad\qquad\qquad |z_k - w_k| \leq 3\pi.$

Therefore, given $\varepsilon \in (0,1)$, the derivative bound in Lemma 3.6 implies $|z_{k-1} - L_{s_{k-1}}(w_k)| < \varepsilon$ for sufficiently large k. But this implies that the same branch of $L_{s_{k-2}}$ applies to both points, and repeated application of this argument shows that if k is sufficiently large, then for all $l \in \{k-1, \ldots, 1\}$ we have

$$|z_l - L_{s_l} \circ \cdots \circ L_{s_{k-1}}(w_k)| < \varepsilon \ .$$

If in addition $t_k \geq T_{\underline{s}}$, then $g_{\underline{s}}^{k-1}(t_k) = L_{s_1} \circ \cdots \circ L_{s_{k-1}}(w_k)$, and we have $|z_1 - g_{\underline{s}}^{k-1}(t_k)| < \varepsilon$. Now suppose $t' \geq T_{\underline{s}}$. Let t be a limit point of the sequence (t_k) (restricted to such k as mentioned above). Obviously $t \geq t' \geq T_{\underline{s}}$. By uniform convergence of $g_{\underline{s}}^{k-1}$ to $g_{\underline{s}}$ for potentials at least $T_{\underline{s}}$, we get $|g_{\underline{s}}^{k-1}(t_k) - g_{\underline{s}}(t_k)| < \varepsilon$ (possibly by enlarging k). Finally, for t_k close enough to t (t is a limit point of (t_k)) $|g_{\underline{s}}(t_k) - g_{\underline{s}}(t)| < \varepsilon$. Combining this, it follows that

$$|z_1 - g_{\underline{s}}(t)| \leq |z_1 - g_{\underline{s}}^{k-1}(t_k)| + |g_{\underline{s}}^{k-1}(t_k) - g_{\underline{s}}(t_k)| + |g_{\underline{s}}(t_k) - g_{\underline{s}}(t)| < 3\varepsilon$$

for sufficiently large k. Hence $g_{\underline{s}}(t) = z_1$ because $\varepsilon > 0$ was arbitrary.

If the condition $|\mathrm{Re}(z_k)| > 2\sqrt{|ab|} + 1$ does not hold for all k, then there is an $N \in \mathbb{N}$ such that it holds for all $k \geq N$; similarly, by Lemma 3.2, if $t' \geq T_{\underline{s}}$ is not satisfied, then for sufficiently large N, we have $F^{\circ N}(t') \geq T_{\sigma^N(\underline{s})}$. Therefore, there is an $N \in \mathbb{N}$ with $z_{N+1} = g_{\sigma^N(\underline{s})}(F^{\circ N}(t))$ for some $t \geq t'$. Pulling back N times along the orbit from z_1 to z_{N+1}, it follows that $z_1 = g_{\underline{s}'}(t)$ for some external address $\underline{s}'$ which can differ from $\underline{s}$ only in the first N entries; thus $t_{\underline{s}'} = t_{\underline{s}}$. However, if the pull-back runs through a dynamic ray which contains an escaping critical value, then this pull-back is impossible — and only then. $\square$

Proposition 4.3 (Controlled Escape for Points on Rays).

For every exponentially bounded external address $\underline{s}$ and for every $t > t_{\underline{s}}$, the orbit of $g_{\underline{s}}(t)$ satisfies the asymptotic bound

$$E^{\circ k}(g_{\underline{s}}(t)) = \begin{cases} F^{\circ k}(t) - \alpha + 2\pi i s_{k+1} + o(1) & (\text{if } s_{k+1} \in \mathbb{Z}_R) \\ -F^{\circ k}(t) + \beta + 2\pi i s_{k+1} + o(1) & (\text{if } s_{k+1} \in \mathbb{Z}_L) \end{cases}$$

as $k \to \infty$. In particular, for every real $p > 0$ it satisfies

$$\frac{|\mathrm{Im}(E^{\circ k}(g_{\underline{s}}(t)))|^p}{\mathrm{Re}(E^{\circ k}(g_{\underline{s}}(t)))} \longrightarrow 0 .$$

PROOF. By Proposition 3.3 we have good error bounds for the dynamic rays $g_{\sigma^k(\underline{s})}$ for potentials greater than $T_{\sigma^k(\underline{s})}$. Since $t > t_{\underline{s}}$, there exists a k_0 such that $F^{\circ k}(t) > T_{\sigma^k(\underline{s})}$ for all $k \geq k_0$ (Lemma 3.2). Then

$$E^{\circ k}(g_{\underline{s}}(t)) = g_{\sigma^k(\underline{s})}(F^{\circ k}(t)) = F^{\circ k}(t) - \alpha + 2\pi i s_{k+1} + r_{\sigma^k(\underline{s})}(F^{\circ k}(t))$$

$$\text{resp.} \quad = -F^{\circ k}(t) + \beta + 2\pi i s_{k+1} + r_{\sigma^k(\underline{s})}(F^{\circ k}(t))$$

with

$$|r_{\sigma^k(\underline{s})}(F^{\circ k}(t))| < (C_1 + 8\pi|s_{k+2}|)e^{-F^{\circ k}(t)},$$

where C_1 only depends on a and b. This last expression tends to 0 as $k \to \infty$ (extremely fast). Along the orbit of $g_{\underline{s}}(t)$, the absolute values of real parts thus grow like $F^{\circ k}(t)$, while the imaginary parts are bounded in absolute value by the asymptotically much smaller quantity $AF^{\circ k}(t_{\underline{s}}+\varepsilon)$ for any $\varepsilon \in (0, t-t_{\underline{s}})$. In particular we have

$$\ln\left(\frac{|\mathrm{Im}(E^{\circ k}(g_{\underline{s}}(t)))|^p}{|\mathrm{Re}(E^{\circ k}(g_{\underline{s}}(t)))|}\right) < pF^{\circ(k-1)}(t_{\underline{s}}+\varepsilon) - F^{\circ(k-1)}(t) + O(1) \to -\infty.$$

This proves the last claim. □

5. Eventually Horizontal Escape

The main result in this section shows essentially that, for any given external address, there is at most one point with this external address which is not on a dynamic ray.

Define

$$R_h := \max\left\{\ln\left(\frac{2h+8\pi}{|a|\pi}\right), \ln\left(\frac{2h+8\pi}{|b|\pi}\right), \frac{1}{2}\ln\left|\frac{2b}{a}\right|, \frac{1}{2}\ln\left|\frac{2a}{b}\right|, |a|+|b|\right\}$$

(depending on the parameters $a, b \in \mathbb{C}^*$).

Lemma 5.1 (Exponential Separation of Orbits).

Let $R > R_h$ and let (z_k) and (w_k) be two escaping orbits with $|\mathrm{Re}(z_k)| > R$, $|\mathrm{Re}(w_k)| > R$, $\mathrm{Re}(z_k)/\mathrm{Re}(w_k) > 0$ and $|\mathrm{Im}(z_k - w_k)| < h$ for some $h > 0$ and all k. Define $d_k := \mathrm{Re}(z_k - w_k)$ for all k. If $|\mathrm{Re}(z_1)| - |\mathrm{Re}(w_1)| \geq 3$ and neither of the critical values escapes, then the following holds:

(1) *$|d_{k+1}| \geq \exp(|d_k|)$ and $|\mathrm{Re}(z_k)| = |\mathrm{Re}(w_k)| + |d_k|$ for all $k \geq 1$;*
(2) *if $\underline{s} \in \mathcal{S}$ is such that $z_k \in R_{s_k}$ for all k, then $z_1 = g_{\underline{s}'}(t')$ for some $t' > t_{\underline{s}}$ and some $\underline{s}'$ which differs from $\underline{s}$ in only finitely many entries; if $|d_1|$ is sufficiently large, then $\underline{s}' = \underline{s}$;*
(3) *if $w_1 = g_{\underline{s}''}(t'')$ for some external address $\underline{s}''$ and some $t'' > t_{\underline{s}''}$, then $t' > t''$.*

If one or both the critical values escape, then the first and third statements are still true, and there is an $N \in \mathbb{N}$ with $z_{N+1} = g_{\sigma^N(\underline{s})}(F^{\circ N}(t))$ for some $t > t_{\underline{s}}$.

PROOF. 1. Let $w_k = t_k + iu_k$ with real t_k, u_k. We write

$$|w_k| = |E(w_{k-1})| = \begin{cases} |a|e^{t_{k-1}} \cdot \left|1 + \frac{b}{a}e^{-2w_{k-1}}\right| & \text{if } t_{k-1} > 0 \\ |b|e^{-t_{k-1}} \cdot \left|1 + \frac{a}{b}e^{2w_{k-1}}\right| & \text{if } t_{k-1} < 0 \end{cases}.$$

By assumption, $\mathrm{Re}(z_k)$ and $\mathrm{Re}(w_k)$ always have the same sign. By definition of d_k, the property $|\mathrm{Re}(z_k)| = |\mathrm{Re}(w_k)| + |d_k|$ is equivalent to the fact that d_k

has the same sign as $\mathrm{Re}(z_k)$ and $\mathrm{Re}(w_k)$, i.e. $\frac{t_k}{d_k} > 0$. Since $|\mathrm{Re}(w_{k-1})| > R$, we have $e^{-2|\mathrm{Re}(w_{k-1})|} < \min\left\{\left|\frac{a}{2b}\right|, \left|\frac{b}{2a}\right|\right\}$ and thus

(10) $$\frac{3}{2}|a|e^{|t_{k-1}|} \geq |w_k| \geq \frac{1}{2}|a|e^{|t_{k-1}|} \quad \text{if } t_{k-1} > 0,$$

and

(11) $$\frac{3}{2}|b|e^{|t_{k-1}|} \geq |w_k| \geq \frac{1}{2}|b|e^{|t_{k-1}|} \quad \text{if } t_{k-1} < 0.$$

Since $\mathrm{Re}(w_k)$ and $\mathrm{Re}(z_k)$ always have the same signs, there exists a $K' \in \{|a|, |b|\}$ (depending on s_k) such that

(12) $$\frac{3K'}{2}e^{|\mathrm{Re}(z_{k-1})|} \geq |z_k| \geq \frac{K'}{2}e^{|\mathrm{Re}(z_{k-1})|}$$

and

(13) $$\frac{3K'}{2}e^{|\mathrm{Re}(w_{k-1})|} \geq |w_k| \geq \frac{K'}{2}e^{|\mathrm{Re}(w_{k-1})|}.$$

Now let $|\mathrm{Re}(z_{k-1})| - |t_{k-1}| \geq 3$ be the inductive hypothesis; then $|d_{k-1}| \geq 3$ and $\frac{t_{k-1}}{d_{k-1}} > 0$. Since $|\mathrm{Im}(z_k - w_k)| < h$, Pythagoras' Theorem implies

$$(\mathrm{Re}(z_k))^2 = |z_k|^2 - (\mathrm{Im}(z_k))^2 \geq |z_k|^2 - (|w_k| + h)^2$$

(14) $$\geq \left(\frac{K'}{2}\exp|t_{k-1} + d_{k-1}|\right)^2 - \left(\frac{3}{2}K'\exp|t_{k-1}| + h\right)^2.$$

This difference is positive: from

$$|t_{k-1}| > R_h > \max\left\{\ln\left(\frac{2h}{|a|\pi}\right), \ln\left(\frac{2h}{|b|\pi}\right)\right\}$$

it follows that $\frac{2h}{K'e^{|t_{k-1}|}} < \pi$; since $|d_{k-1}| \geq 3$, we get

$$e^{|d_{k-1}|} \geq e^3 > 3 + \pi > 3 + \frac{2h}{K'e^{|t_{k-1}|}}$$

and therefore

$$\frac{K'}{2}e^{|t_{k-1}|}e^{|d_{k-1}|} \geq \frac{3K'}{2}e^{|t_{k-1}|} + h.$$

Hence we can extract the root in (14) and obtain

(15) $$|t_k + d_k| \geq \frac{K'}{2}\exp|t_{k-1} + d_{k-1}|\sqrt{1 - \left(\frac{3\exp|t_{k-1}| + 2h/K'}{\exp|t_{k-1} + d_{k-1}|}\right)^2}.$$

Since $|d_{k-1}| \geq 3$ and $\frac{t_{k-1}}{d_{k-1}} > 0$, we get as above

$$0 < \frac{3\exp|t_{k-1}| + 2h/K'}{\exp|t_{k-1} + d_{k-1}|} = \frac{3 + (2h/K')\exp(-|t_{k-1}|)}{\exp|d_{k-1}|}$$

(16) $$< \frac{3 + \pi}{\exp|d_{k-1}|} \leq \frac{3+\pi}{e^3} < \frac{1}{2}.$$

The radicand in (15) is thus in $\left(\frac{1}{2}, 1\right)$. Since $|t_k| = |\text{Re}(w_k)| \leq |w_k|$, we get

$$|t_k + d_k| - |t_k| \geq \frac{K'}{4} \exp|t_{k-1} + d_{k-1}| - \frac{3}{2} K' \exp|t_{k-1}|$$
$$= \frac{K'}{4} \exp|t_{k-1}| \left(\exp|d_{k-1}| - 6\right) > 0.$$

Thus $|\text{Re}(z_k)| > |\text{Re}(w_k)|$ and $\frac{t_k}{d_k} > 0$. Since $|t_{k-1}| \geq R$, we get

$$\frac{K'}{4} \exp|t_{k-1}| > \frac{K'}{4} \exp(\ln(8/K')) = 2.$$

Thus

$$|d_k| = |t_k + d_k| - |t_k| > 2(\exp|d_{k-1}| - 6)$$
$$= \exp|d_{k-1}| + (\exp|d_{k-1}| - 12) \geq \exp|d_{k-1}|.$$

This shows the first claim.

2. The second claim will be done in four steps:

   (a) $|t_k + d_k| \geq F^{\circ(k-N)}(t_N + d_N + \gamma') - \gamma'$ for some $\gamma' \in \mathbb{R}$ and $N \in \mathbb{N}$ and for all $k \geq N$ (using the estimates from the first statement),
   (b) $|t_k| + \gamma \leq F^{\circ(k-N)}(t_N + \gamma)$ for some $\gamma \in \mathbb{R}$ and all $k > N$ (from the maximal growth rate along orbits),
   (c) this will imply an upper bound on $|w_k|$, hence on $|\text{Im}(w_k)|$ and on $|s_k|$, giving $F^{\circ(N-1)}(t_{\underline{s}}) \leq |t_N| + \gamma$,
   (d) the conclusion will then follow.

(a) Choose $\gamma' < \ln\left(\frac{K}{4}\right)$ (possibly negative). By (15), (16) and since $K' \geq K$, we have

$$|t_k + d_k| = |\text{Re}(z_k)| \geq \frac{K}{4} e^{|t_{k-1}+d_{k-1}|} = e^{|t_{k-1}+d_{k-1}| + \ln\frac{K}{4}}$$
$$\geq e^{|t_{k-1}+d_{k-1}|+\gamma'} - \gamma' - 1 = F(|t_{k-1}+d_{k-1}| + \gamma') - \gamma'$$

provided $|t_{k-1} + d_{k-1}|$ is sufficiently large. There is thus an $N > 0$ such that inductively for all $k \geq N$

(17) $$|t_k + d_k| + \gamma' \geq F^{\circ(k-N)}(|t_N + d_N| + \gamma').$$

(b) Choose $\gamma > \ln\left(\frac{3K_{\max}}{2}\right)$. We get

$$|t_k| = |\text{Re}(w_k)| \leq |w_k| \leq \frac{3K_{\max}}{2} e^{|t_{k-1}|} = F\left(|t_{k-1}| + \ln\frac{3K_{\max}}{2}\right) + 1$$
$$\leq F(|t_{k-1}| + \gamma) - \gamma,$$

again provided t_{k-1} is sufficiently large. Possibly by enlarging N, we have for all $k \geq N$

$$|t_k| + \gamma \leq F^{\circ(k-N)}(|t_N| + \gamma).$$

(c) For every $\varepsilon > 0$ the definition of $t_{\underline{s}}$ shows that there are infinitely many $k \geq N$ with

$$\begin{aligned}
F^{\circ(k-N)}(F^{\circ(N-1)}(t_{\underline{s}} - \varepsilon)) &= F^{\circ(k-1)}(t_{\underline{s}} - \varepsilon) < 2\pi|s_k| \leq |\text{Im}(w_k)| + 3\pi \\
&\leq |w_k| + 3\pi \leq \frac{3K_{\max}}{2}e^{|t_{k-1}|} + 3\pi \\
&< e^{|t_{k-1}|+\gamma} + 3\pi = F(|t_{k-1}| + \gamma) + 1 + 3\pi \\
&\leq F^{\circ(k-N)}(|t_N| + \gamma) + 1 + 3\pi \ .
\end{aligned}$$

Since this holds for arbitrarily large k, we get

$$F^{\circ(N-1)}(t_{\underline{s}} - \varepsilon) \leq |t_N| + \gamma \ .$$

Since $\varepsilon > 0$ was arbitrary, it follows

$$t_{\sigma^{N-1}(\underline{s})} = F^{\circ(N-1)}(t_{\underline{s}}) \leq |t_N| + \gamma \ .$$

(d) Enlarge N if necessary so that

$$|t_N + d_N| + \gamma' > |t_N| + \gamma + 2$$

and

$$F^{\circ(N-1)}(t_{\underline{s}} + 2) > T_{\sigma^{N-1}(\underline{s})} + 1$$

(such N exists by Lemma 3.2). Then

$$\begin{aligned}
|\text{Re}(z_k)| &= |t_k + d_k| \geq F^{\circ(k-N)}\left(|t_N + d_N| + \gamma'\right) - \gamma' \\
&> F^{\circ(k-N)}\left(|t_N| + \gamma + 2\right) - \gamma' \geq F^{\circ(k-N)}\left(t_{\sigma^{N-1}(\underline{s})}\right) + 2) - \gamma' \\
&\gg F^{\circ(k-N)}\left(t_{\sigma^{N-1}(\underline{s})} + 1\right)
\end{aligned}$$

for infinitely many $k > N$. Now Theorem 4.2 implies that there is a $k \geq N$ and a $t^* > t_{\sigma^{k-1}(\underline{s})}$ such that

$$z_{k+1} = g_{\sigma^k(\underline{s})}(t^*) \ .$$

If no critical value escapes, then we can pull back k times, and it follows that there is an $\underline{s}'$ which differs from $\underline{s}$ only in finitely many entries, and a $t' > t_{\underline{s}} = t_{\underline{s}'}$ with

$$z_1 = g_{\underline{s}'}(t') \ .$$

If $|d_1|$ is large enough so that $|t_k + d_k| \geq \max\{F^{\circ(k-1)}(T_{\underline{s}}), 2\sqrt{|ab|} + 1\}$ for all k, then the conditions of Theorem 4.2 are satisfied immediately so that $z_1 = g_{\underline{s}}(t)$ for some $t \geq T_{\underline{s}}$.

3. If $w_1 = g_{\underline{s}''}(t'')$ for $t'' > t_{\underline{s}''}$, then $t > t''$ by Proposition 4.3. □

6. Classification of Escaping Points

In this section we show that all escaping points are organized in the form of dynamic rays which are associated to exponentially bounded external addresses, and we complete the classification of escaping points.

Definition 6.1 (Limit Set, Landing Point, Uniform Escape).
The limit set of the ray $g_{\underline{s}}$ is defined as the set of all possible limit points of $g_{\underline{s}}(t_k)$ for $t_k \searrow t_{\underline{s}}$. We say that the ray $g_{\underline{s}}$ lands at a point w if $\lim_{t' \searrow t_{\underline{s}}} g_{\underline{s}}(t')$ exists and is equal to w (the limit set consists of only one point). If $g_{\underline{s}}$ lands at an escaping point $w = g_{\underline{s}}(t_{\underline{s}})$, then we say that the ray and landing point escape uniformly if for every $R \in \mathbb{R}$ there exists an $N \geq 0$ such that for every $n \geq N$, we have $|\mathrm{Re}(g_{\underline{s}}^n([t_{\underline{s}}, \infty)))| > R$.

For $R > 0$ let Y_R be defined as
$$Y_R := \{z \in \mathbb{C} \colon |\mathrm{Re}(z)| < R\} \ .$$

Lemma 6.2 (Escaping Set Connected).
Let $R \geq R_{3\pi}$ with
$$\frac{K}{2} e^R \geq \frac{2 + 3\pi}{1 - 3/e^2}$$
and let (z_k) be an escaping orbit which is completely contained in $\mathbb{C} \setminus Y_{R+2}$. Then there exists a closed connected set $C \subset \overline{\mathbb{C}} \setminus Y_R$ with $\{z_1, \infty\} \subset C$ such that the orbit of every $z \in C$ is completely contained in $\mathbb{C} \setminus Y_R$ and escapes such that
$$|\mathrm{Re}(E^{\circ(k-1)}(z))| \geq |\mathrm{Re}(z_k)| - 2.$$
Let $\underline{s}$ be the external address of z_1. All points in $C \setminus \{\infty\}$ have external address $\underline{s}$ and

- *either $C \setminus \{\infty\} = g_{\underline{s}}([t, \infty))$ for some $t > t_{\underline{s}}$,*
- *or the ray $g_{\underline{s}}$ lands at z_1, and $C = \{z_1, \infty\} \cup g_{\underline{s}}((t_{\underline{s}}, \infty))$.*

In particular, all points in $C \setminus \{\infty\}$, except possibly z_1, lie on the ray $g_{\underline{s}}$.

REMARK. The external address of z_1 is defined uniquely because the orbit of z_1 never enters the vertical strip $Y_R \supset \mathcal{A}$, so it can never hit the boundary of our partition.

PROOF. By (12), for every k there is a $K' \in \{|a|, |b|\}$ (depending on k) such that
$$\frac{3}{2} K' \exp|\mathrm{Re}(z_k)| \geq |E(z_k)| = |z_{k+1}| \geq \frac{K'}{2} \exp|\mathrm{Re}(z_k)| \geq \frac{K'}{2} e^R$$
(18)
$$\geq \frac{2 + 3\pi}{1 - 3/e^2} > 0$$

and thus
$$|z_{k+1}| - 2 - 3\pi \geq \frac{3}{e^2}|z_{k+1}|.$$

For $k \geq 1$ define
$$S_k := \left\{ z \in R_{s_k} : \begin{cases} \text{Re}(z) > \text{Re}(z_k) - 2, \text{ if } \text{Re}(z_k) > 0 \\ \text{Re}(z) < \text{Re}(z_k) + 2, \text{ if } \text{Re}(z_k) < 0 \end{cases} \right\}.$$

Our first claim is $E(S_k) \supset S_{k+1}$ for all k. Every $z \in R_{s_k}$ with
$$|\text{Re}(z)| \leq |\text{Re}(z_k)| - 2$$
satisfies (as in (18))
$$|E(z)| \leq \frac{3}{2}K' \exp|\text{Re}(z)| \leq \frac{3}{2}K' \exp(|\text{Re}(z_k)| - 2) \leq \frac{3}{e^2}|z_{k+1}|.$$

Since $E : \overline{R_{s_k}} \to \mathbb{C}$ is surjective, it follows that $E(S_k)$ contains every point $w \in S_{k+1}$ with
$$|w| > \frac{3}{e^2}|z_{k+1}|.$$

But $w \in S_{k+1}$ implies
$$|\text{Re}(w)| \geq |\text{Re}(z_{k+1})| - 2 \quad \text{and} \quad |\text{Im}(w) - \text{Im}(z_{k+1})| < 3\pi,$$
and thus
$$|w| \geq |z_{k+1}| - 2 - 3\pi \geq \frac{3}{e^2}|z_{k+1}|.$$

Hence we get $w \in E(S_k)$ and the first claim is proved.

Since $\mathcal{A} \cap S_{k+1} = \emptyset$ we obtain a connected set $C'_k \subset S_k$ such that $E: C'_k \to S_{k+1}$ is a conformal isomorphism. For $k \geq 1$ consider the sets
$$C_k := \{ z \in S_1 : E^{\circ i}(z) \in \overline{S}_{i+1}, \text{ for } i = 0, 1, \ldots, k-1 \} \cup \{\infty\}.$$

The sets C_k are non-empty (because $\{z_1, \infty\} \subset C_k$), compact and nested: $C_{k+1} \subset C_k$. We just proved that $E^{\circ(k-1)} : C_k \to S_k \cup \{\infty\}$ is a homeomorphism for all k. Therefore, all C_k are connected. The nested intersection of non-empty connected compact sets is non-empty connected and compact; therefore
$$C := \bigcap_{k \geq 1} C_k$$
is a closed connected and compact subset of $\overline{\mathbb{C}}$ with $\{z_1, \infty\} \subset C$. Set $C' := C \setminus \{\infty\}$. For $z \in C'$ we have $E^{\circ(k-1)}(z) \in \overline{S_k}$ for all k and thus $|\text{Re}(E^{\circ(k-1)}(z))| > |\text{Re}(z_k)| - 2 \to \infty$. Hence C' only consists of escaping points with orbits in $\mathbb{C} \setminus Y_R$ which have the same external address as z_1.

It remains to show that there is at most one point in C' which is not on the dynamic ray $g_{\underline{s}}$. By Lemma 5.1, there is a $\xi > 0$ such that every $z \in C'$ with

$|\mathrm{Re}(z)| > \xi$ is on $g_{\underline{s}}$. In particular, C' contains an unbounded connected part of the tail of $g_{\underline{s}}$.

Suppose that there are two points $z_1, w_1 \in C'$ with orbits (z_k) and (w_k) such that $|\mathrm{Re}(z_k - w_k)| < 3$ for all k. By Lemma 2.3, we also have $|\mathrm{Im}(z_k - w_k)| < 3\pi$, hence $|z_k - w_k| < 3 + 3\pi$. The derivative E' is bounded below along the orbits of z_1 resp. w_1 by $|ae^R - be^{-R}| \geq \frac{K}{2}e^R > 4$. By pulling back, choosing the branch $E^{-1}: S_{k+1} \to C_k$, we get

$$|z_{k-1} - w_{k-1}| < (3\pi + 3)/4$$

and thus inductively $|z_{k-j} - w_{k-j}| < (3\pi + 3)/4^j$. Hence

$$|z_1 - w_1| < (3\pi + 3)/4^{k-1}$$

for all k and thus $z_1 = w_1$. Therefore, if $z_1 \neq w_1$, then by Lemma 5.1 at least one of these points (say z_1) satisfies $z_1 = g_{\underline{s}'}(t)$ for an external address $\underline{s}'$ which differs from $\underline{s}$ in only finitely many positions, and $t > t_{\underline{s}'} = t_{\underline{s}}$; say $\sigma^N(\underline{s}') = \sigma^N(\underline{s})$. But since $E^{\circ N}(C') \cup \{\infty\}$ is connected, a single branch of E^{-N} maps $E^{\circ N}(C')$ to C', avoiding $\mathcal{A}$ in the pull-back, so it follows even that $\underline{s}' = \underline{s}$.

Therefore, every point in C' with at most one exception is on $g_{\underline{s}}$. Since $g_{\underline{s}}: (t_{\underline{s}}, \infty) \to \mathbb{C}$ is continuous and C' is closed, the set

$$\{t \in (t_{\underline{s}}, \infty) : g_{\underline{s}}(t) \in C'\}$$

is closed in $(t_{\underline{s}}, \infty)$. We use this to show that if $g_{\underline{s}}(t) \in C'$, then $g_{\underline{s}}(t') \in C'$ for all $t' > t$: otherwise, there would be $t_2 > t_1 > t_{\underline{s}}$ with $g_{\underline{s}}(t_1), g_{\underline{s}}(t_2) \in C'$, but $g_{\underline{s}}((t_1, t_2)) \cap C' = \emptyset$. But then at least for large N, there would be a large gap between $g_{\sigma^N(\underline{s})}(F^{\circ N}(t_2))$ and $g_{\sigma^N(\underline{s})}(F^{\circ N}(t_1))$ which could not be filled by points in $g_{\sigma^N(\underline{s})}((t_{\underline{s}}, t_1))$ or $g_{\sigma^N(\underline{s})}((t_2, \infty))$ (Lemma 5.1 (3)) or by the single exceptional point, so $E^{\circ N}(C')$ could not be connected.

It follows that except for possibly a single point, C' equals either $g_{\underline{s}}([t, \infty))$ for some $t > t_{\underline{s}}$, or $g_{\underline{s}}((t_{\underline{s}}, \infty))$. But since C' is closed and connected, an extra point can (and must) occur only in the second case, and this is what we claimed. $\square$

For an external address $\underline{s}$ let $S(\underline{s})$ be the space of external addresses $\underline{s}'$ which differ from $\underline{s}$ at only finitely many entries. Clearly, all $\underline{s}' \in S(\underline{s})$ have $t_{\underline{s}'} = t_{\underline{s}}$.

Lemma 6.3 (Limit Set does Not Intersect Ray).

Let $g_{\underline{s}}$ be a ray with the property that the entire orbits of all of its points avoid Y_R for an $R > 2\sqrt{|ab|} + M + 2$ and $R > R_h$ for $h = 2\pi$. Let $L_{\underline{s}}$ be the limit set of $g_{\underline{s}}$. Then $L_{\underline{s}}$ is disjoint from $g_{\underline{s}'}$ for all $\underline{s}' \in S(\underline{s})$.

PROOF. Suppose there exists a $\underline{s}' \in S(\underline{s})$ and a $t > t_{\underline{s}'}$ with $g_{\underline{s}'}(t) \in L_{\underline{s}}$. Since all $|\mathrm{Re}(E^{\circ k}(g_{\underline{s}}(t)))| > R$ for all $t > t_{\underline{s}}$ and all $k \in \mathbb{N}$, all points $g_{\underline{s}}(t)$ have external address $\underline{s}$, and so has the limit $g_{\underline{s}'}(t)$ (Theorem 4.2).

Now choose a potential $t' \in (t_{\underline{s}}, t)$. By Proposition 4.3 (and possibly after finitely many iterations) we get $|\text{Re}(g_{\underline{s}'}(t))| - |\text{Re}(g_{\underline{s}'}(t'))| \geq 4$. Then there is a potential $t'' > t_{\underline{s}}$ arbitrarily close to $t_{\underline{s}}$ such that $g_{\underline{s}}(t'')$ is arbitrarily close to $g_{\underline{s}'}(t)$. More precisely we assume that

$$t'' < t' \quad \text{and} \quad |g_{\underline{s}}(t'') - g_{\underline{s}'}(t)| < 1,$$

hence $|\text{Re}(g_{\underline{s}}(t''))| - |\text{Re}(g_{\underline{s}'}(t'))| > 3$. Since $g_{\underline{s}}(t'')$ and $g_{\underline{s}'}(t')$ both have external address $\underline{s}$, we have $\left|\text{Im}\left(E^{\circ k}(g_{\underline{s}}(t'')) - E^{\circ k}(g_{\underline{s}'}(t'))\right)\right| \leq 2\pi$ for all k. By Lemma 5.1 this means $t'' > t'$ and we get a contradiction. Hence $L_{\underline{s}} \cap g_{\underline{s}'}((t_{\underline{s}}, \infty)) = \emptyset$ for all $\underline{s}' \in S(\underline{s})$. $\square$

Theorem 6.4 (Escaping Points are Organized in Rays).

For every escaping point w there exists a unique exponentially bounded external address $\underline{s}$ and a unique potential $t \geq t_{\underline{s}}$ such that exactly one of the following holds:

- *either $t > t_{\underline{s}}$ and $w = g_{\underline{s}}(t)$,*
- *or $t = t_{\underline{s}}$ and the dynamic ray $g_{\underline{s}}$ lands at w such that w and the ray $g_{\underline{s}}$ escape uniformly,*
- *or, if one of the singular values escapes: $v_i = g_{\underline{s}}(t)$ for some $\underline{s}$ and $t \geq t_{\underline{s}}$, and the point w maps to $g_{\underline{s}}(t')$ with $t_{\underline{s}} \leq t' < t$ after finitely many iterations.*

PROOF. Let R be as defined in Lemma 6.2. If the entire orbit of w lies in $\mathbb{C} \setminus Y_{R+2}$, then let $\underline{s}$ be the external address of w. By Lemma 6.2, either $w = g_{\underline{s}}(t)$ for some $t > t_{\underline{s}}$, or the ray $g_{\underline{s}}$ lands at w with uniform escape. In both cases, the external address $\underline{s}$ is uniquely determined by the orbit of w, and so is t along dynamic rays because each ray is an injective curve. To finish the uniqueness claim, we have to show that $g_{\underline{s}}(t) \neq g_{\underline{s}}(t_{\underline{s}})$ for all $t > t_{\underline{s}}$: this follows from Lemma 6.3.

If the entire orbit of w is not in $\mathbb{C} \setminus Y_R$, then there exists a finite iterate of w whose orbit has that property. This iterate is either on a dynamic ray or the landing point of a ray. By pulling back along the orbit of w the claim is proved for all w; if a singular value lies on a dynamic ray along this pull-back, then an exception may occur as stated in the third case of the claim. $\square$

Now we want to show under which conditions a landing point escapes. We need slow and fast external addresses, as in [18].

Definition 6.5 (Slow and Fast External Addresses).

We say that an external address $\underline{s}$ is slow if there are $A, x > 0$ and infinitely many n for which $|s_{n+k}| \leq AF^{\circ(k-1)}(x)$ for all $k \geq 1$. Otherwise we call $\underline{s}$ fast.

Note that every external address $\underline{s}$ with $t_{\underline{s}} > 0$ is fast, but the converse is not true: the two external addresses $1\,2\,1\,3\,1\,4\,1\,5\ldots$ and $1\,2\,1\,2\,3\,1\,2\,3\,4\,1\,2\,3\,4\,5\ldots$ (with arbitrary entries L or R) are both unbounded with $t_{\underline{s}} = 0$, but the first one is fast while the second one is slow.

Now the following result holds in complete analogy to [18, Proposition 6.8], so we omit the proof.

Proposition 6.6 (Uniform Escape for Fast Addresses).

An external address $\underline{s}$ is fast if and only if the ray $g_{\underline{s}}$ lands at an escaping point so that the ray and landing point escape uniformly. □

Corollary 6.7 (Uniform Escape of Ray and Landing Point).

If a dynamic ray lands at an escaping point, then the ray and its landing point escape uniformly.

PROOF. Let $g_{\underline{s}}$ be a dynamic ray and w be its escaping landing point. If $\underline{s}$ is fast (in particular, if $t_{\underline{s}} > 0$), then $g_{\underline{s}}$ lands at w with uniform escape by Proposition 6.6. Therefore, $\underline{s}$ is slow and $t_{\underline{s}} = 0$.

The point w cannot be on any other dynamic ray $g_{\underline{s}'}$ by the same argument as in [18, Corollary 6.9]: every point $g_{\underline{s}'}(t)$ with $t > t_{\underline{s}'}$ can be approximated arbitrarily closely by pieces of other rays almost parallel to $g_{\underline{s}'}$ so that $g_{\underline{s}'}(t)$ is not accessible by any ray. Therefore, by our classification, w is the escaping landing point of some other dynamic rays $g_{\underline{s}'}$, and thus $\underline{s}'$ is necessarily fast. This means that the dynamic rays $g_{\underline{s}'}$ and $g_{\underline{s}}$ land together at w, so for every $N \in \mathbb{N}$ there is a $k \in \mathbb{N}$ such that the first entries of $\sigma^k(\underline{s}')$ and $\sigma^k(\underline{s})$ differ by at least N, while $g_{\sigma^k(\underline{s}')}$ and $g_{\sigma^k(\underline{s})}$ land together at $E^{\circ k}(w)$. But the $2\pi i n$-translates of the same rays must land at the $2\pi i n$-translates of $E^{\circ k}(w)$, and this is a topological impossibility if $N > 1$. □

As a result, we obtain the following classification of escaping points:

Theorem 6.8 (Classification of Escaping Points).

If no critical value escapes, then the set of escaping points is classified by external addresses $\underline{s} \in \mathcal{S}$ so that

- *for $\underline{s}$ with $t_{\underline{s}} < \infty$ so that $\underline{s}$ is slow, the associated set of escaping points is the ray $g_{\underline{s}}((t_{\underline{s}}, \infty))$;*
- *for $\underline{s}$ with $t_{\underline{s}} < \infty$ so that $\underline{s}$ is fast, the associated set of escaping points is $g_{\underline{s}}([t_{\underline{s}}, \infty))$, i.e. the ray including its escaping landing point with uniform escape;*
- *for $\underline{s}$ with $t_{\underline{s}} = \infty$, there is no associated set of escaping points.*

The sets of escaping points to all $\underline{s} \in \mathcal{S}$ are disjoint, and their union is the entire set of escaping points of E. □

REMARK. Using methods as in [8], one can show that every path component of the set of escaping points is exactly the set of escaping points associated to any particular exponentially bounded external address.

REMARK. If one or both critical values escape, the necessary modifications are straightforward as in [18].

7. Epilogue: Hausdorff Dimension

Theorem 7.1 (Dimension Paradox).
The union of all dynamic rays has Hausdorff dimension 1, while the set of the escaping landing points has Hausdorff dimension 2 and even infinite planar Lebesgue-measure.

SKETCH OF PROOF. McMullen [10] has shown that the set of escaping points for every $E_{a,b}$ has infinite planar Lebesgue measure. By our classification, every escaping point is either on a dynamic ray or the landing point of a dynamic ray. It thus suffices to prove that the union of all dynamic rays has Hausdorff dimension 1. We sketch a lemma of Karpińska [9] which generalizes to show that the set of rays has dimension less than $1 + 1/p$ for every $p > 0$. Given $p > 0$ and $\xi > 0$, let

$$P_{p,\xi} := \{(x+iy) \in \mathbb{C}: |x| > \xi, |y| < |x|^{1/p}\}.$$

We will show that $S_{p,\xi} := \{z \in \mathbb{C}: E^{\circ k}(z) \in P_{p,\xi} \text{ for all } k\}$ has Hausdorff dimension at most $1 + 1/p$ for large ξ.

Let $Q \subset P_{p,\xi}$ be a square of side length 2π with boundaries parallel to the coordinate axes, and with real parts in $[x, x + 2\pi]$. If $|x| > \xi$ is large, then $E(Q)$ is almost an annulus between radii e^x and $e^{2\pi}e^x$ (up to a factor $|a|$ or $|b|$ which will not matter). Then $E(Q) \cap P_{p,\xi}$ is contained in two approximate rectangles of width $e^x(e^{2\pi} - 1)$ and height $2e^{x/p}e^{2\pi/p}$. Filling with squares of side length 2π, we need approximately $Ae^{x(1+1/p)}$ such squares (where $A > 0$ is some bounded factor). Pulling those back into the original square Q, we obtain the same number of squares, with diameters approximately $e^{-x}/2\pi$. Now

$$\frac{\log\left(Ae^{x(1+1/p)}\right)}{-\log\left(e^{-x}/2\pi\right)} = \frac{x(1+1/p) + \log(A)}{x + \log(2\pi)} = 1 + \frac{1}{p} + O(1/\xi).$$

We can refine the tiling of $S_{p,\xi}$ by iterating this procedure: the squares with side length 2π in $E(Q) \cap P_{p,\xi}$ can themselves be replaced by smaller squares. In the limit, we obtain a covering of $S_{p,\xi}$ with small tiles, which gives an upper bound of $1 + 1/p$ for the Hausdorff dimension of $S_{p,\xi}$.

In Proposition 4.3 we have shown that every point z on a ray satisfies the parabola condition $|\text{Im}(E^{\circ n}(z))|^p < |\text{Re}(E^{\circ n}(z))|$ for any $p > 0$ for all but finitely many n. Therefore, the union of all rays is contained in the countable union

$$\bigcup_{n\geq 0} E^{-N}(S_{p,\xi})$$

and still has Hausdorff dimension at most $1+1/p$ for every $p>0$. Therefore, the union of all rays has Hausdorff dimension 1. □

Recently, Karpińska's ideas about Hausdorff dimension of escaping points have been extended to large families of transcendental functions of bounded type; see [17].

Theorem 7.2 (Almost Every Orbit Escapes).
If each critical value either lands on a repelling cycle or escapes to infinity on a dynamic ray, then the set $\mathbb{C} \setminus I$ has Lebesgue measure zero. In fact, Lebesgue-almost every point in $\mathbb{C}$ is the escaping endpoint of a dynamic ray.

PROOF. By McMullen [10], the set I has positive Lebesgue measure for all maps $z \mapsto E(z) = ae^z + be^{-z}$. By a result of Bock [1], at least one of the following two condition holds: (i) almost every orbit converges to ∞ or to the postcritical set, and (ii) almost every orbit intersects every positive measure set infinitely often. Since I has positive measure, condition (i) must hold. If a critical value is on a repelling cycle, an orbit can converge to the corresponding postcritical set only if it falls exactly onto it, which happens only for countably many points. This proves the claim if both critical values land on repelling cycles.

If both critical orbits escape on a ray, or one escapes on a ray and the other is preperiodic, then we employ ideas of Schubert [19] (compare also [16, Sec. 8]): it is easy to check that on a disk D of radius π centered at a point with large real part ξ, the set I has large density that converges to 1 very quickly: most points $z \in D$ have $|\text{Re}(E(z))| > e^{\xi/2} \gg \xi$; the remaining points have density of roughly $e^{-\xi/2}$. Iterating this argument shows that the density of I in D is roughly $e^{-\xi/2}$. It follows that for every $\eta > 0$ and every $r > 0$ there is a $\xi > 0$ so that for any disk D' of radius r at real part at least ξ, the density of I on D' is at least $1 - \eta$.

Let P be the postcritical set of E. This set is locally finite, and by Proposition 4.3, there are only finitely many pairs of points in P that have mutual distance less than any given $r > 0$. Let C be the set of points in $\mathbb{C}$ that converge to $P \cup \{\infty\}$. If $C \setminus I$ has positive measure, let w be a Lebesgue density point which is not one of the countably many that eventually fall onto P. For any $\xi > 0$, there is an $N \in \mathbb{N}$ and a $p \in P$ with $|\text{Re}\, p| > \xi$

and a radius $\rho > 0$ so that $D_\rho(p) \cap P = \{p\}$ and $|E^N(w) - p| = \rho/2$. Consider the disk $D' := D_{\rho/2}(E^N(w))$; then there is a neighborhood U' of w so that $E^N\colon U' \to D'$ is a conformal isomorphism: since $D' \cap P = \emptyset$, the set U' can be obtained by pulling back D' along the orbit of w. We may even assume that $\rho > \pi$: otherwise, $E\colon D_\rho(p) \to E(D_\rho(p))$ is a conformal isomorphism and the Koebe 1/4-theorem gives us a much larger value of ρ around the point $E(p)$ provided ξ is large enough (it may happen that $E(D_\rho(p))$ intersects P in more points, but this is no problem for the necessary pullback argument); repeating this argument finitely often, we eventually get $\rho > \pi$.

The density of I on D' is close to 1 by Schubert's argument above, so by the Koebe distortion theorem I must have large density on U' near w, but this is a contradiction and shows that $C \setminus I$ must have measure zero. □

As mentioned in the introduction, the dynamics is particularly easy to understand when both critical orbits are strictly preperiodic: see [15, 16]. In this case, *every* dynamic ray lands somewhere in $\mathbb{C}$ and, conversely, every point in $\mathbb{C}$ which is not on a dynamic ray is the landing point of a ray. The set of dynamic rays still has Hausdorff dimension 1, so the set of landing points of this 1-dimensional set of rays is the entire complex plane minus this 1-dimensional set of rays. In that case, almost every $z \in \mathbb{C}$ (with respect to planar Lebesgue measure) is an escaping point.

References

[1] Heinrich Bock: *On the dynamics of entire functions on the Julia set.* Results Math. **30** (1996), no. 1-2, 16–20.
[2] Robert L. Devaney and Michał Krych, *Dynamics of* $\exp(z)$, Ergodic Theory Dynam. Systems **4** (1984), no. 1, 35–52.
[3] Robert L. Devaney and Folkert Tangerman, *Dynamics of entire functions near the essential singularity*, Ergodic Theory Dynam. Systems **6** (1986), no. 4, 489–503.
[4] Adrien Douady and John Hubbard: *Etude dynamique des polynômes complexes.* Prépublications mathématiques d'Orsay **2** (1984) and **4** (1985).
[5] Alexandre Eremenko: *On the iteration of entire functions.* In: *Dynamical systems and ergodic theory*, Banach Center publications, Polish Scientific Publishers **23** (1989), 339-345.
[6] Alexandre Eremenko and Mikhail Lyubich: *Dynamical properties of some classes of entire functions.* Annales de l'Institut Fourier, Grenoble **42** 4 (1992), 989–1020.
[7] Pierre Fatou, *Sur l'itération des fonctions transcendantes entières*, Acta Math. **47** (1926), 337–370.
[8] Markus Förster, Lasse Rempe, Dierk Schleicher, *Classification of escaping exponential maps.* Proceedings of the American Mathematical Society, to appear.
[9] Bogusława Karpińska: *Hausdorff dimension of the hairs without endpoints for* $\lambda\exp(z)$. Comptes Rendus Acad. Sci. Paris **328** Série I (1999), 1039–1044.
[10] Curt McMullen: *Area and Hausdorff dimension of Julia sets of entire functions.* Transactions of the American Mathematical Society **300** 1 (1987), 329–342.

[11] Lasse Rempe: *Dynamics of exponential maps*. Ph.D. thesis, Christian-Albrechts-Universität Kiel (2003); http://e-diss.uni-kiel.de/diss_781.

[12] Günter Rottenfußer: *Dynamic rays for a family of entire transcendental maps*. Diploma thesis, Technische Universität München (2002).

[13] Günter Rottenfußer, *On the dynamical fine structure of entire transcendental functions*, doctoral thesis, International University Bremen (2005).

[14] Günter Rottenfußer, Johannes Rückert, Lasse Rempe, Dierk Schleicher, *Dynamic rays of bounded-type entire functions*, submitted.

[15] Dierk Schleicher: *The dynamical fine structure of iterated cosine maps and a dimension paradox*. Duke Mathematics Journal **136** 2 (2007), 343–356.

[16] Dierk Schleicher: *Hausdorff dimension, its properties, and its surprises*. American Mathematical Monthly **114** 6 (2007), 509–528. ArXiv math.DS/0505099.

[17] Dierk Schleicher and Michael Thon, *Hausdorff dimension of dynamic rays for certain entire functions of bounded type*, manuscript.

[18] Dierk Schleicher and Johannes Zimmer: *Escaping points of exponential maps*. Journal of the London Mathematical Society **67** 2 (2003), 380–400.

[19] Hendrik Schubert: *Area of Fatou sets of trigonometric functions*. Proceedings of the American Mathematical Society, to appear.

GÖRRESSTR. 20, 80798 MÜNCHEN, GERMANY

E-mail address: guenter.rottenfusser@gmx.net

SCHOOL OF ENGINEERING AND SCIENCE, RESEARCH I, JACOBS UNIVERSITY (FORMERLY INTERNATIONAL UNIVERSITY), POSTFACH 750 561, D-28725 BREMEN, GERMANY

E-mail address: dierk@iu-bremen.de

DIMENSIONS OF JULIA SETS OF TRANSCENDENTAL MEROMORPHIC FUNCTIONS

GWYNETH M. STALLARD

Dedicated to the memory of Professor Noel Baker

Abstract. We look at the relationships between the different dimensions that can be used to describe the size of the Julia set of a transcendental meromorphic function and also look at the different values that these dimensions can take. We summarise the main results in this area and indicate some of the techniques that are used.

1. INTRODUCTION

Let f be a meromorphic function which is not rational of degree one and denote by $f^n, n \in \mathbb{N}$, the nth iterate of f. The Fatou set, $F(f)$, is defined to be the set of points, $z \in \mathbb{C}$, such that $(f^n)_{n \in \mathbb{N}}$ is well-defined, meromorphic and forms a normal family in some neighbourhood of z. The complement, $J(f)$, of $F(f)$ is called the Julia set of f. An introduction to the properties of these sets can be found in, for example, [2] for rational functions and in [3] for transcendental meromorphic functions.

Most Julia sets are very intricate and their size cannot be described well in terms of classical geometry. There are a number of definitions of dimension that are well suited to describing the size of fractals such as Julia sets and in Section 2 we discuss the advantages and disadvantages of these different dimensions.

In Section 3 we look at the relationships between these different dimensions — when they give the same value for the size of a Julia set and when they give different values.

In Section 4 we look at the possible values for the dimension of the Julia set of a transcendental entire function and summarise some of the main results concerning exponential functions.

In Section 5 we look at the possible values for the dimension of the Julia set of a transcendental meromorphic function with poles.

We use the following notation:

$$S(f) = \{z : z \text{ is a finite singularity of } f^{-1}\},$$

$$P(f) = \{z : z \text{ is a finite singularity of } f^{-n} \text{ for some } n \in \mathbb{N}\},$$

$$B = \{\text{transcendental meromorphic } f : S(f) \text{ is bounded}\},$$

$$I(f) = \{z : f^n(z) \neq \infty, f^n(z) \to \infty \text{ as } n \to \infty\}.$$

2. Definitions of dimensions

In this section we define the main definitions of dimension that are used to describe the size of Julia sets. We discuss the advantages and disadvantages of the different dimensions and look at the relationships between them. This material is covered in more detail in [10].

2.1. Box dimensions.
First we define the box dimension of a bounded set $A \subseteq \mathbb{C}$ — this is probably the simplest 'respectable' definition of a dimension and is widely used. Suppose that we have a grid composed of lines parallel to the axes so that each box in the grid has sides of length d. We count the number $N_d(A)$ of boxes in this grid that meet the set A. The box dimension of A is defined to be

$$\dim_B A = \lim_{d \to 0} \frac{\log N_d(A)}{-\log d},$$

if this limit exists. There are several equivalent definitions of the box dimension of a set.

It is important to note that the box dimension of a set is only well defined if the limit in the definition exists — this is not always the case. Because of this, we often look at the *upper* and *lower* box dimensions of a set. The upper box dimension of A is defined to be

$$\overline{\dim}_B A = \limsup_{d \to 0} \frac{\log N_d(A)}{-\log d},$$

and the lower box dimension of A is defined to be

$$\underline{\dim}_B A = \liminf_{d \to 0} \frac{\log N_d(A)}{-\log d}.$$

The box dimension of a set exists precisely when the upper and lower box dimensions are equal.

2.2. Hausdorff dimension.
Another definition of dimension that is widely used is that of Hausdorff dimension. Here, instead of covering A with sets that all have the same size, we look at coverings of A with sets of diameter at most equal to d. For each $t > 0$, we put

$$H_t(A) = \liminf_{d \to 0} \sum_i r_i^t,$$

where the infimum is taken over all possible covers of A with sets of diameter $r_i < d$. Thus

(2.1) $$H_t(A) \leq \lim_{d \to 0} N_{d(A)}(\sqrt{2}d)^t.$$

The Hausdorff dimension $s = \dim_H A$ of the set A is defined to be the unique value satisfying

(2.2) $$H_t(A) = \begin{cases} \infty & \text{for } t < s \\ 0 & \text{for } t > s. \end{cases}$$

It can be shown that each of the functions H_t is a measure and this enables us to use results from measure theory to obtain estimates of the Hausdorff dimension of a set. The following result which is part of Frostman's Lemma is often used to obtain a lower bound for the Hausdorff dimension of a set.

Lemma 2.1. *Let A be a compact subset of $\mathbb{R}^n$ and let $s > 0$. Suppose that there exist a measure μ supported on A and constants $C, d > 0$ such that every ball B of radius $r < d$ satisfies*

$$\mu(B) \leq Cr^s.$$

Then

$$dim_H A \geq s.$$

Note that if the box dimension of the set A exists then

(2.3) $$\lim_{d \to 0} N_{d(A)} d^t = \begin{cases} \infty & \text{for } t < \dim_B A \\ 0 & \text{for } t > \dim_B A. \end{cases}$$

It follows from (2.1), (2.2) and (2.3) that

(2.4) $$\dim_H A \leq \dim_B A.$$

A simple example of a set which has different Hausdorff and box dimensions is the set $A = \{1/n : n \in \mathbb{N}\}$. This is just a countable collection of points, each of which has zero dimension. As we would expect, $\dim_H A = 0$. However, by covering A with intervals of length $1/n^2$, it is easy to check that $\dim_B A = 1/2$. More generally, Hausdorff dimension is stable under countable unions; that is,

$$\dim_H \bigcup_{i=1}^{\infty} A_i = \sup_{1 \leq i \leq \infty} \dim_H A_i,$$

whereas, as this example shows, box dimension is not stable under countable unions — a significant drawback.

2.3. Modified box dimensions. In order to overcome this drawback, a more recent definition of dimension has been introduced — the *modified* box dimension. The modified upper box dimension of a set A is defined as

$$\overline{\dim}_{MB} A = \inf\{\sup_{i\in\mathbb{N}} \overline{\dim}_B A_i : A \subset \bigcup_{i=1}^{\infty} A_i\},$$

where the infimum is over all possible countable covers of A with bounded sets A_i, $i \in \mathbb{N}$. The modified lower box dimension is defined similarly. It follows from the definition that

$$\overline{\dim}_{MB} A \leq \overline{\dim}_B A.$$

By using the fact that Hausdorff dimension is stable under countable unions, together with (2.4), it can also be shown that

$$\dim_H A \leq \overline{\dim}_{MB} A.$$

The modified box dimensions are stable under countable unions and are defined for unbounded sets as well as bounded ones. Unlike Hausdorff dimension, they are not defined using measures. The modified upper box dimension can, however, be shown to be equal to the packing dimension which *is* defined by measures. (This is shown in [10, Proposition 3.8] for a bounded set A. This proof applies equally well when A is unbounded.)

2.4. Packing dimension. When defining the Hausdorff dimension of a set A, we considered coverings of A with sets of small diameter and tried to make these coverings as efficient as possible. When defining the packing dimension of a set A we consider packings of A with disjoint balls of small radius and try to make these packings as dense as possible.

For each $t > 0$, we put

$$P_t^0(A) = \limsup_{d\to 0} \sum_i r_i^t,$$

where the supremum is taken over all possible collections of disjoint sets of diameter $r_i < d$ with centres in A. Unfortunately, the functions P_t^0 are not measures and so we modify them as follows to produce measures:

$$P_t(A) = \inf\{\sum_{i=1}^{\infty} P_t^0(A_i) : A \subset \bigcup_{i=1}^{\infty} A_i\}.$$

The packing dimension $s = \dim_P A$ of the set A is defined to be the unique value satisfying
$$P_t(A) = \begin{cases} \infty & \text{for } t < s \\ 0 & \text{for } t > s. \end{cases}$$

3. Relationships between dimensions of Julia sets

We have now given the main definitions of dimension that are used to describe the size of Julia sets and seen that they are related as follows.

$$\dim_H A \le \overline{\dim}_{MB} A = \dim_P A \le \overline{\dim}_B A.$$

The lower box dimension can be less than or greater than the packing dimension.

The Hausdorff and packing dimensions have several nice properties — they are defined for every set, they are stable under countable unions and are defined using measures — and so these are the dimensions that we are particularly interested in. Although the box dimensions do not have these properties, they can sometimes be easy to estimate and so it is useful to have conditions which ensure that these different dimensions are equal.

3.1. **Box and packing dimensions.** For a transcendental meromorphic function, f, the Julia set is unbounded since, by Picard's Theorem, all but at most two points in the Julia set must have pre-images accumulating at the essential singularity at infinity. Thus, strictly speaking, the box dimensions of $J(f)$ are not well-defined. There are, however, well-defined 'local' upper and lower box dimensions of the Julia set. In [19, Theorem 2] we proved that, apart from at the exceptional set $E(f)$ which contains at most two points, these local box dimensions are *constant*. Moreover, this enabled us to show that the local upper box dimension of the Julia set (away from $E(f)$) is equal to the packing dimension of the Julia set — this result was proved for rational functions in [31, Theorem 7.1].

Theorem 3.1. *Let f be a meromorphic function. Then there exist $\overline{d} > 0$ and $\underline{d} > 0$ such that, for any bounded open connected set U satisfying $U \cap J(f) \ne \emptyset$ and $\overline{U} \cap E(f) = \emptyset$,*

$$\overline{dim}_B(U \cap J(f)) = \overline{d},$$

and

$$\underline{dim}_B(U \cap J(f)) = \underline{d}.$$

Further,

$$dim_P J(f) = \overline{d}.$$

In Theorem 3.1,
$$E(f) = \{z : O^-(z) \text{ is finite}\},$$
where
$$O^-(z) = \{w : f^n(w) = z, \text{ for some } n \in \mathbb{N}\}.$$

Theorem 3.1 enables us to use simple box dimension calculations to estimate the size of the packing dimension of the Julia set.

3.2. **Hyperbolic functions.** There is a large class of *rational* functions for which all the definitions of dimension that we have described give the same value for the size of the Julia set. Indeed, the following result follows fairly easily from results of Sullivan [30, Theorems 3 and 4].

Theorem 3.2. *Let f be a hyperbolic rational function. Then*
$$dim_H J(f) = dim_P J(f) = dim_B J(f).$$

An obvious question to ask is whether this result is true for transcendental meromorphic functions. Before answering this question, we must establish what is meant by the word hyperbolic. A rational function f is hyperbolic if f satisfies the following equivalent conditions (see [2, Section 9.7] and [4, Section 5.2]):

- $\overline{P}(f) \cap J(f) = \emptyset$;
- $\overline{P}(f)$ is a compact subset of $F(f)$;
- f is expanding, in the sense that there exist $K > 1$ and $c > 0$ such that $|(f^n)'(z)| > cK^n$ for each $z \in J(f)$, $n \in \mathbb{N}$.

These conditions are not equivalent for a transcendental meromorphic function f, since the Julia set of such a function is unbounded, but in [18] we showed that, if f is in the class
$$\hat{B} = \{f : f \in B \text{ and } J(f) \cap \bar{P}(f) = \emptyset\},$$
then

- $P(f)$ is bounded;
- f is expanding in the sense that there exist $K > 1$ and $c > 0$ such that $|(f^n)'(z)| > cK^n(|f^n(z)| + 1)/(|z| + 1)$ for each $n \in \mathbb{N}$ and each $z \in J(f)$ such that f^n is analytic at z.

Thus it makes sense to say that functions in the class $\hat{B}$ are hyperbolic. The result of Theorem 3.2, however, does not hold for functions in $\hat{B}$. In [28], [29] and [19] we showed that there are examples of transcendental meromorphic functions in the class $\hat{B}$ (both with and without poles) whose Julia sets have packing dimensions that are strictly greater than their Hausdorff dimensions.

We discuss these functions later in the paper.

The Hausdorff and packing dimensions of the Julia set must be equal when the Hausdorff dimension of the Julia set is equal to two. The only other example of a transcendental meromorphic function that we are aware of for which the Hausdorff and packing dimensions of the Julia set are known to be equal is $\tan z$ — in this case the Julia set is simply equal to the real line and so has dimension one.

Question 1 Does there exist a transcendental meromorphic function f for which $\dim_H J(f) = \dim_P J(f) = d$, where d is not an integer?

4. Transcendental entire functions

In this section we restrict our attention to transcendental *entire* functions. For such a function, the Julia set must contain continua since any multiply connected component of the Fatou set is a bounded wandering domain (of the type sometimes described as a Baker wandering domain) — this was proved by Baker [1]. Thus we have the following result.

Theorem 4.1. *Let f be a transcendental entire function. Then*
$$1 \leq \dim_H J(f) \leq \dim_P J(f) \leq 2.$$

Later in this section we will give examples which show that, for each $d \in (1, 2]$, there is a transcendental entire function f for which $\dim_H J(f) = d$.

Question 2 Does there exist a transcendental entire function f for which $\dim_H J(f) = 1$?

We also give examples of transcendental entire functions for which the packing dimension of the Julia set is equal to two even though the Hausdorff dimension of the Julia set is arbitrarily close to one.

4.1. Exponential functions.
It was proved by Misiurewicz [16] that the Julia set of the exponential function e^z is equal to the whole plane and so, trivially, has Hausdorff dimension equal to two. More interesting examples, from the point of view of dimension, were given by McMullen who proved the following result [15, Theorems 1.2 and 1.3].

Theorem 4.2. *Let $f_\lambda(z) = \lambda e^z$, where $\lambda \neq 0$. Then*
$$\dim_H J(f_\lambda) = 2.$$
Further, if $0 < \lambda < 1/e$, then the area of $J(f_\lambda)$ is zero.

In order to prove this result, McMullen first proved the following general result which is useful for obtaining a lower bound for the Hausdorff dimension of a set. The proof of this result uses Lemma 2.1.

Lemma 4.3. *For each $n \in \mathbb{N}$, let $\mathcal{A}_n$ be a finite collection of disjoint compact subsets of $\mathbb{R}^d$, each of which has positive d-dimensional measure, and define*

$$\mathcal{U}_n = \bigcup_{A_n \in \mathcal{A}_n} A_n, \quad A = \bigcap_{n=1}^{\infty} \mathcal{U}_n.$$

Suppose also that, for each $A_n \in \mathcal{A}_n$, there exists $A_{n+1} \in \mathcal{A}_{n+1}$ and a unique $A_{n-1} \in \mathcal{A}_{n_1}$ such that

$$A_{n+1} \subseteq A_n \subseteq A_{n-1}.$$

If Δ_n, d_n are such that, for each $A_n \in \mathcal{A}_n$,

$$vol(\mathcal{U}_{n+1} \cap A_n)/vol(A_n) \geq \Delta_n$$

and

$$diam\, A_n \leq d_n, \ d_n \to 0 \text{ as } n \to \infty,$$

then

$$dim_H A \geq d - \limsup_{n \to \infty} \sum_{i=1}^{n} \frac{|\log \Delta_i|}{|\log d_n|}.$$

Functions of the form $f_\lambda(z) = \lambda e^z$ were studied by Devaney and Krych [6]. They showed that there is a unique attracting fixed point p_λ and a unique repelling fixed point q_λ with $0 < p_\lambda < 1 < q_\lambda$. The half plane $\{z : \Re(z) < q_\lambda\}$ is contained in the attracting basin of p_λ and

(4.1) $\qquad J(f_\lambda) = \{z : \Re(f_\lambda^n(z)) \geq q_\lambda, \text{ for all } n \in \mathbb{N}\}.$

About half of the points in the right half-plane

$$H = \{z : \Re(z) \geq q_\lambda\}$$

remain there after applying f — this enabled McMullen to use Lemma 4.3. To do this, take the collection of strips

$$S_k = \{z \in H : |\Im(z) - 2k\pi| \leq \pi/4\}, \ k \in \mathbb{Z}$$

and pack them with boxes

$$B_k^j = \{z \in S_k : |\Re(z) - q_\lambda - j\pi| \leq \pi/4\}, \ j \in \mathbb{N}.$$

Lemma 4.3 can then be applied to the collection of sets given by

$$\mathcal{A}_n = \{A_n : f_\lambda^n(A_n) = B_k^j \text{ for some } k \in \mathbb{Z}, j \in \mathbb{N}, \text{ and } A_n \subset A_{n-1}$$
$$\text{for some } A_{n-1} \in \mathcal{A}_{n-1}\},$$

for $n \geq 2$, with $\mathcal{A}_1$ being one of the boxes B_k^j. The set A defined in Lemma 4.3 is then clearly a subset of $J(f_\lambda)$ by (4.1). It follows from Koebe's distortion theorem (see, for example, [8]), that there exists $C > 0$ such that, for each $A_n \in \mathcal{A}_n$ and each $n \in \mathbb{N}$,

$$\mathrm{vol}(\mathcal{U}_{n+1} \cap A_n)/\mathrm{vol}(A_n) \geq C \sum_{B_k^j \cap f_\lambda^{n+1}(A_n) \neq \emptyset} \mathrm{vol}(B_k^j)/\mathrm{vol} f_\lambda^{n+1}(A_n).$$

This implies that the quantities Δ_n in Lemma 4.3 are all greater than some constant, for each $n \in \mathbb{N}$. (This is where we use the fact that roughly a half of all points in any right half-plane remain there after applying f_λ.) It is then straightforward to check that Lemma 4.3 implies that the set A (which is a subset of $J(f_\lambda)$) has Hausdorff dimension two.

In fact $J(f_\lambda)$ is actually a 'Cantor bouquet' of curves whose endpoints are the only accessible points from the basin of attraction (see [5], [6] and [7]). Karpińska ([11] and [12]) proved the following remarkable result.

Theorem 4.4. *Let $f_\lambda(z) = \lambda e^z$, where $0 < \lambda < 1/e$, and let $\mathcal{C}_\lambda$ denote the set of endpoints of the curves that make up $J(f_\lambda)$. Then*

$$dim_H \mathcal{C}_\lambda = 2$$

while

$$dim_H(J(f_\lambda) \setminus \mathcal{C}_\lambda) = 1.$$

To prove the first part of this result, Karpińska modified McMullen's proof of Theorem 4.2. The key change was to restrict the sets A_n such that, for a given $A_n \in \mathcal{A}_n$, the sets $A_{n+1} \subset A_n$ are roughly those for which $f_\lambda^{n+1}(A_{n+1})$ is a box in the upper and lower quarters of the set $f^{n+1}(A_n)$. This still produces a set A which has Hausdorff dimension two. Karpińska showed that each point in this set A is accessible from the basin of attraction and is therefore contained in $\mathcal{C}_\lambda$. Note that this subset of $\mathcal{C}_\lambda$ consists of points that tend to infinity very quickly under iteration.

By using methods of thermodynamic formalism, Karpińska [11, Theorem 2] showed that the set

$$J_{bd}(f_\lambda) = \{z \in J(f_\lambda) : \{f_\lambda^n(z)\} \text{ is bounded}\}$$

has Hausdorff dimension which tends to 1 from above as λ tends to zero. This set is also a subset of the set of endpoints $\mathcal{C}_\lambda$.

Theorem 4.5. *Let $f_\lambda(z) = \lambda e^z$, where $0 < \lambda < 1/e$. Then*

$$dim_H J_{bd}(f_\lambda) \geq 1$$

and, if λ is sufficiently small, then

$$1 + \frac{1}{\log(\log 1/\lambda)} < dim_H J_{bd}(f_\lambda) < 1 + \frac{1}{\log(\log(\log 1/\lambda))}.$$

Urbański and Zdunik [32] used conformal measures to show that the set
$$J_r(f_\lambda) = J(f_\lambda) \setminus I(f_\lambda),$$
consisting of those points in the Julia set that do not tend to infinity under iteration, has Hausdorff dimension less than two. This set contains $J_{bd}(f_\lambda)$ and is also a subset of the set of endpoints $\mathcal{C}_\lambda$. (They also give a direct proof of this result using Karpińska's Theorem 4.5 together with other results.) By using the methods of thermodynamic formalism, Urbański and Zdunik [33] show that the Hausdorff dimension of $J_r(f_\lambda)$ is actually a real analytic function of λ for $\lambda \in (0, 1/e)$.

Theorem 4.6. *Let $f_\lambda(z) = \lambda e^z$, where $0 < \lambda < 1/e$. Then*
$$dim_H J_r(f_\lambda) < 2$$
and $dim_H J_r(f_\lambda)$ is a real analytic function of λ.

These results show that, for the exponential family, the main contribution to the size of the Julia set is from those points whose iterates tend to infinity under iteration. Kotus and Urbański [14] showed that this is not the case for elliptic functions.

Theorem 4.7. *Let f be an elliptic function and let q be the maximal multiplicity of the poles of f. Then*
$$dim_H J(f) > \frac{2q}{1+q} \geq dim_H I(f).$$

4.2. Julia sets with small Hausdorff dimension. Recall that, for a transcendental entire function f, we have $1 \leq dim_H J(f) \leq 2$. We now give examples which show that 1 is the best possible lower bound here. Recall that, in McMullen's proof of Theorem 4.2, the key observation is that for an exponential function f, the Julia set consists of those points that remain in a right half plane under iteration and approximately *half* of the points in a right half plane remain there after applying f. This suggests that in order to obtain a function whose Julia set has Hausdorff dimension less than two it would be a good idea to construct a function whose Julia set lies in a relatively small part of the plane. This was the motivation in [21] for looking at the following function which is described in [17]. The transcendental entire function E is defined by
$$E(z) = \frac{1}{2\pi i} \int_L \frac{\exp(e^t)}{t-z} dt,$$
where L is the boundary of the region
$$G = \{z : \Re(z) > 0, |\Im(z)| < \pi\}$$
described in a clockwise direction, for $z \in \mathbb{C} \setminus \overline{G}$, and by analytic continuation elsewhere. It is shown in [17] that $E(z)$ is close to $\exp(e^z)$ for $z \in G$ and is close to $-1/z$ elsewhere.

In [21], we studied the family of functions defined by
$$f_K(z) = E(z) - K, \text{ where } K \in \mathbb{R}.$$
For large K, $F(f_K)$ is the attracting basin of a fixed point in the left half plane and $J(f_K)$ is a Cantor bouquet of curves in the half-strip
$$G_K = G \cap \{z : \Re(z) > K'\},$$
where $K' > \log \log K/2$. Clearly $J(f_K)$ is contained in a relatively small part of the plane which becomes progressively smaller as K gets larger. We have
$$J(f_K) = \{z : f_K^n(z) \in G_K \text{ for all } n \in \mathbb{N}\}$$
and the proportion of points in a vertical cross-section of G_K that remain in G_K after applying f_K tends to zero as we move across to the right of G_K. This is the key difference between these functions and the exponential functions that allowed us to prove the following result in [21].

Theorem 4.8. *The functions f_K defined above satisfy*
$$1 \leq dim_H J(f_K) \leq 1 + \delta(K),$$
where $\delta(K) \to 0$ as $K \to \infty$.

In order to obtain an upper bound for the Hausdorff dimension of a set it is sufficient to consider just one sequence of covers with sets of progressively smaller diameter. A 'natural' way to do this for a Julia set is to begin with a cover $\mathcal{C}$ and then take successive pre-images of the sets in $\mathcal{C}$ to give a sequence of covers. This method was used in the proof of Theorem 4.8 and also by Karpińska to prove the second part of Theorem 4.4.

In Theorem 4.8, the 'natural' sequence of covers is defined as follows. We begin by covering the strip G_K with the collection of boxes B_n, $n \in \mathbb{N}$, defined by
$$B_n = \{z : |\Re(z - z_n)| < 3C/2, |\Im(z - z_n)| < 3C/2\},$$
where $z_n = K' + nC$ and C is some constant with $3C/2 > \pi$. Now take an arbitrary set B_N and define the following sequence of covers of $J(f_K)$.
$$\mathcal{A}_k = \{A_k : A_k \text{ is a component of } f_K^{-k}(B_n) \text{ for some } n \in \mathbb{N}, A_k \cap B_N \neq \emptyset\}.$$
It is not difficult to show that $\max\{\text{diam } A_k : A_k \in \mathcal{A}_k\} \to 0$ as $k \to \infty$. Thus, to prove Theorem 4.8, it is sufficient to show that, for each $A_k \in \mathcal{A}_k$ and each $k \in \mathbb{N}$,
$$\sum_{A_{k+1} \in \mathcal{A}_{k+1},\ A_{k+1} \cap A_k \neq \emptyset} (\text{diam } A_{k+1})^{1+\delta(K)} \leq (\text{diam } A_k)^{1+\delta(K)}.$$

The set $P(f_K)$ lies in the left half-plane and so we can use Koebe's distortion theorem to show that there exists a constant $c > 0$ such that, for each $A_k \in \mathcal{A}_k$ and each $k \in \mathbb{N}$,

$$\sum_{A_{k+1} \in \mathcal{A}_{k+1},\ A_{k+1} \cap A_k \neq \emptyset} \frac{(\operatorname{diam} A_{k+1})^{1+\delta(K)}}{(\operatorname{diam} A_k)^{1+\delta(K)}}$$

$$\leq \sum_{A_{k+1} \in \mathcal{A}_{k+1},\ A_{k+1} \cap A_k \neq \emptyset} \frac{c(\operatorname{diam} f_K^k(A_{k+1}))^{1+\delta(K)}}{(\operatorname{diam} f_K^k(A_k))^{1+\delta(K)}} = c \sum_{U \in H_s} \frac{(\operatorname{diam} U)^{1+\delta(K)}}{(\operatorname{diam} B_s)^{1+\delta(K)}},$$

where $B_s = f_K^k(A_k)$ and

$$H_s = \{U : U \text{ is a component of } f_K^{-1}(B_n) \text{ for some } n \in \mathbb{N},\ U \cap B_s \neq \emptyset\}.$$

The right hand quantity can be shown to be less than one thus proving Theorem 4.8.

4.3. Lower bounds for the Hausdorff dimension of a Julia set.

We have seen how a 'natural' sequence of covers of a Julia set, $J(f)$, can be used to obtain an upper bound for $\dim_H J(f)$. We now show how a lower bound for $\dim_H J(f)$ can be obtained by showing that a natural choice of covers is the best possible choice of covers. We prove the following general result.

Theorem 4.9. *Let E be a compact set and, for each $n \in \mathbb{N}$, let $\mathcal{A}_n = \{A_n\}$ be a collection of disjoint sets such that, for some $t > 0$,*

(1) $A_n \cap E \neq \emptyset$ for each $A_n \in \mathcal{A}_n$, $n \in \mathbb{N}$;
(2) there exists $M \in \mathbb{N}$ such that

$$\sum_{A_{n+M} \subset A_n,\ A_{n+M} \in \mathcal{A}_{n+M}} (\operatorname{diam} A_{n+M})^t > (\operatorname{diam} A_n)^t,$$

for each $A_n \in \mathcal{A}_n$, $n \in \mathbb{N}$;
(3) there exist $r_0 > 0$, $C > 0$, $m \in \mathbb{N}$ such that for all $r < r_0$, $z \in E$ there exists $P \in \mathbb{N} \cup \{0\}$ such that

$$\sum_{A_{PM+m} \cap B(z,r) \neq \emptyset,\ A_{PM+m} \in \mathcal{A}_{PM+m}} (\operatorname{diam} A_{PM+m})^t < C(\operatorname{diam} B(z,r))^t.$$

Then $\dim_H E \geq t$.

Proof. Suppose that $\dim_H E < t$. Then there exists a cover of E with sets $B_i = B(z_i, r_i)$, $i \in I$, such that, for each $i \in I$, we have $z_i \in E$, $r_i < r_0$ and

(4.2) $$\sum_{i \in I} (\operatorname{diam} B_i)^t < \sum_{A_m \in \mathcal{A}_m} (\operatorname{diam} A_m)^t / C.$$

It follows from condition (3) that, for each $i \in I$, there exists $P(i) \in \mathbb{N} \cup \{0\}$ such that

(4.3) $$\sum_{A_{P(i)M+m} \cap B_i \neq \emptyset,\, A_{P(i)M+m} \in \mathcal{A}_{P(i)M+m}} (\operatorname{diam} A_{P(i)M+m})^t < C(\operatorname{diam} B_i)^t.$$

Since E is compact, we can cover E with a finite number of the sets B_i. We label these sets $B_1, B_2, \ldots, B_n$ in such a way that $P(i) \leq P(i+1)$ for $1 \leq i < n$. It follows from (4.3) and condition (2) that

$$\sum_{A_m \in \mathcal{A}_m} (\operatorname{diam} A_m)^t$$

$$\leq \sum_{A_{P(1)M+m}} (\operatorname{diam} A_{P(1)M+m})^t$$

$$= \sum_{A_{P(1)M+m} \cap B_1 = \emptyset} (\operatorname{diam} A_{P(1)M+m})^t + \sum_{A_{P(1)M+m} \cap B_1 \neq \emptyset} (\operatorname{diam} A_{P(1)M+m})^t$$

$$\leq \sum_{A_{P(2)M+m} \cap B_1 = \emptyset} (\operatorname{diam} A_{P(2)M+m})^t + C(\operatorname{diam} B_1)^t$$

(4.4) $$\leq \sum_{A_{P(2)M+m} \cap (B_1 \cup B_2) = \emptyset} (\operatorname{diam} A_{P(2)M+m})^t$$

$$+ \sum_{A_{P(2)M+m} \cap B_2 \neq \emptyset} (\operatorname{diam} A_{P(2)M+m})^t + C(\operatorname{diam} B_1)^t$$

$$\leq \sum_{A_{P(3)M+m} \cap (B_1 \cup B_2) = \emptyset} (\operatorname{diam} A_{P(3)M+m})^t + C[(\operatorname{diam} B_1)^t + (\operatorname{diam} B_2)^t]$$

$$\leq \vdots$$

$$\leq \sum_{A_{P(n)M+m} \cap \bigcup_{i=1}^n B_i = \emptyset} (\operatorname{diam} A_{P(n)M+m})^t + C \sum_{i=1}^n (\operatorname{diam} B_i)^t,$$

where $A_{P(i)M+m}$ denotes a set in $\mathcal{A}_{P(i)M+m}$ wherever it appears in the above equation. Since the sets B_i, $1 \leq i \leq n$, form a cover of E, it follows from condition (1) that

$$\{A_{P(n)M+m} \in \mathcal{A}_{P(n)M+m} : A_{P(n)M+m} \cap \bigcup_{i=1}^n B_i = \emptyset\} = \emptyset.$$

Thus, from (4.2) and (4.4) we have

$$\sum_{i \in I} (\operatorname{diam} B_i)^t < \sum_{A_m \in \mathcal{A}_m} (\operatorname{diam} A_m)^t / C \leq \sum_{i=1}^n (\operatorname{diam} B_i)^t$$

which is a contradiction. This completes the proof.

We used this method in [27] to obtain exact values for the Hausdorff dimensions of the Julia sets of a family of transcendental entire functions, thus showing that every value in $(1,2]$ is attained by the Hausdorff dimension of such a Julia set. The examples considered in [27] are similar to those described in Section 4.2 but have their Julia sets in a larger area of the plane — larger than a strip but smaller than an angle.

For each $p \in (0, \infty)$, the transcendental entire function E_p is defined by

$$E_p(z) = \frac{1}{2\pi i} \int_{L_p} \frac{\exp(e^{(\log t)^{1+p}})}{t-z} dt,$$

where L_p is the boundary of the region

$$G_p = \{z = x + iy : |y| < \frac{\pi x}{(1+p)(\log x)^p}, x \geq 3\}$$

described in a clockwise direction, for $z \in \mathbb{C} \setminus \overline{G_p}$, and by analytic continuation elsewhere. The function E_p is close to $\exp(e^{(\log z)^{1+p}})$ for $z \in \overline{G_p}$ and is bounded elsewhere. (Note that many other functions can be defined in this way by taking the boundary to be a curve on which the integrand is small.)

In [27], we studied the family of functions defined by

$$f_{p,K}(z) = E_p(z) - K, \text{ where } K \in \mathbb{R}.$$

Here $(\log t)^{1+p}$ and $(\log x)^{1+p}$ are taken to be the values which are real for $t > e$. For large K, $F(f_{p,K})$ is the attracting basin of a fixed point in the left half plane and $J(f_{p,K})$ is a Cantor bouquet of curves in $G_{p,K} = G \cap \{z : \Re(z) \geq K'_p\}$, where $K'_p \to \infty$ as $K \to \infty$. We have

$$J(f_{p,K}) = \{z : f_{p,K}^n(z) \in G_{p,K} \text{ for all } n \in \mathbb{N}\}.$$

We proved the following result.

Theorem 4.10. *For each $p \in (0, \infty)$, there exists $C(p) > 0$ such that, for each $K > C(p)$, the functions $f_{p,K}$ defined above satisfy*

$$dim_H J(f_{p,K}) = 1 + \frac{1}{1+p}.$$

For a function $f_{p,K}$, a 'natural' sequence of covers of $J(f_{p,K})$ is defined as follows. We begin by covering $G_{p,K}$ with the collection of boxes B_n, $n \in \mathbb{N}$, defined by

$$B_n = \{z = x + iy : x_n \leq x \leq x_n + R_n, |y| \leq R_n\},$$

where
$$x_1 = K'_p, \quad x_{n+1} = x_n + R_n, \quad R_n = \frac{\pi x_n}{(1+p)(\log x_n)^p}.$$

Now define the following collection of sets inductively:
$$\mathcal{A}_1 = \{A_1 : A_1 \text{ is a component of } f^{-1}(B_n) \text{ for some } n \in \mathbb{N}, A_1 \subset \bigcup_{n \in \mathbb{N}} B_n\},$$

$$\mathcal{A}_{k+1} = \{A_{k+1} : A_{k+1} \text{ is a component of } f^{-(k+1)}(B_n) \text{ for some } n \in \mathbb{N}$$
$$\text{and } A_{k+1} \subset \bigcup_{A_k \in \mathcal{A}_k} A_k\}.$$

In [27], we show that, if K is sufficiently large, then the following is true.
$$\lim_{k \to \infty} \sum_{A_k \in \mathcal{A}_k} (\text{diam } A_k)^t < \infty \quad \Leftrightarrow \quad t > 1 + \frac{1}{1+p}.$$

This shows that $1 + \frac{1}{1+p}$ is an *upper* bound for $\dim_H J(f_{p,K})$. To show that it is also a *lower* bound, we apply Theorem 4.9 to the collections $\mathcal{A}_n$ defined above but with
$$B_n = \{z = x + iy : x_n \leq x \leq x_n + R_n/\pi, |y| \leq R_n/\pi\},$$
to ensure that sets in $\mathcal{A}_n$ are disjoint. This shows that, in some sense, the natural sequence of covers is also the most efficient sequence of covers.

It is very unusual to be able to obtain an exact value for the Hausdorff dimension of a Julia set. For these functions, the sum $\sum_{A_k \in \mathcal{A}_k} (\text{diam } A_k)^t$ is approximately equal to

$$\sum_{n=1}^{\infty} (\text{diam } B_n)^t \sum_{f(b)=x_n} \frac{1}{|(f'(b)|^t} \sum_{f^{k-1}(z)=b} \frac{1}{|(f^{k-1})'(z)|^t},$$

by Koebe's distortion theorem. The expression $\sum_{n=1}^{\infty} (\text{diam } B_n)^t \sum_{f(b)=x_n} \frac{1}{|(f'(b)|^t}$ can be approximated by the integral

$$\int_{x_1}^{\infty} \frac{1}{x(\log x)^{(p+1)(t-1)}} dx$$

which converges precisely when $t > 1 + \frac{1}{1+p}$.

We also used the method of Theorem 4.9 in [23] to obtain the following result.

Theorem 4.11. *If f is an entire function in the class B then*
$$\dim_H J(f) > 1.$$

Let f be an entire function in the class B. To prove Theorem 4.11 we took $R > 0$ such that $S(f) \subset B(0,R)$ and a point $z_0 \in J(f) \cap I(f)$ such that $|f^n(z_0)| \geq 8R$. We then took a set
$$I_n \subset \{z : f^n(z) = z_0\}$$
chosen in such a way that the points in I_n lie in a compact part of the plane and
$$z \in I_n \Rightarrow |f^k(z)| \geq 2R, \text{ for } 0 \leq k \leq n.$$
Thus
$$I = \bigcup_{n=1}^\infty I_n \subset J(f) \cap I(f).$$
We defined
$$s = \inf\left\{t : \sum_{n \in \mathbb{N}} \sum_{z \in I_n} \left(\frac{1 + |z_0|^2}{|(f^n)'(z)|(1 + |z|^2)}\right)^t < \infty\right\}.$$
The proof of Theorem 4.11 in [23] is then split into two parts. Firstly, we showed that $s > 1$ and secondly we showed that $\dim_H J(f) > s$. The proof of this second part was based on the method described in Theorem 4.9 applied to the following sets:
$$\mathcal{A}_n = \{g(A) : g \text{ is a branch of } f^{-n} \text{ with } g(z_0) \in I_n\},$$
where A is a disc centered at z_0.

To prove that $s > 1$, we defined measures μ_t by
$$\mu_t(B) = c_t \sum_{n \in \mathbb{N}} \sum_{z \in I \cap B} \left(\frac{1 + |z_0|^2}{|(f^n)'(z)|(1 + |z|^2)}\right)^t, \ t > s,$$
such that $\mu_t(\mathbb{C}) = 1$ and then let μ_s be a weak limit of the measures μ_t as $t \to s$. The measure μ_s is supported on a bounded part of the plane E. We use bounded distortion to show that there exists $C_1 > 0$ such that

(4.5) $\qquad \mu_s(A_n) \geq C_1(\operatorname{diam} A_n)^s, \text{ for each } A_n \in \mathcal{A}_n, n \in \mathbb{N}.$

We then look at pre-images of the curve $\gamma = \{z : |z| = |z_0|\}$ and show that there exists $m > 0$ such that, for each $n \in \mathbb{N}$, there are 2^n disjoint unbounded curves $\gamma_{n,i}$, $1 \leq i \leq n$, crossing E with $f^{mn+1}(\gamma_{n,i}) \subset \gamma$. The sets A_{mn+1} cover a certain proportion of the curves $\gamma_{n,i} \cap E$ and so there exists $C_2 > 0$ such that
$$\sum_{A_{mn+1} \in \mathcal{A}_{mn+1}, \ A_{mn+1} \cap E \neq \emptyset} \operatorname{diam} A_n \geq 2^n C_2, \text{ for each } n \in \mathbb{N}.$$

Together with (4.5), this shows that $s > 1$ thus proving Theorem 4.11.

4.4. Packing dimensions of Julia sets of transcendental entire functions.

In Section 4.2, we introduced a family of transcendental entire functions f_K for which $\dim_H J(f_K) \to 1$ as $K \to \infty$. In [29] we showed that $\dim_P J(f_K) = \dim_B J(f_K) = 2$, for large K, and so the Hausdorff and packing dimensions of the Julia sets are as different as they could be for a transcendental entire function. The functions f_K are certainly hyperbolic since $\overline{P}(f_K)$ is bounded and disjoint from the Julia set — this shows that the result of Theorem 3.2 does not hold for transcendental entire functions.

Recall that $J(f_K)$ is contained in a half-strip

$$G_K = \{z : \Re(z) > K', \ |\Im(z)| < \pi\}.$$

In order to obtain an estimate for $\dim_B J(f_K)$, we use the fact that $f_K^{-1}(\mathbb{R} \cap G_K) \subset J(f_K)$. The set $f_K^{-1}(\mathbb{R} \cap G_K)$ comprises of unbounded curves which, near a point $x \in \mathbb{R}$, are separated by a vertical distance of approximately $1/e^x$. So, there exists some $c_1 > 0$ such that every box of side c_1/e^x within a distance 1 of x must contain a point in $J(f_K)$.

If we fix $X \in \mathbb{R} \cap J(f_K)$, then this result is true for $f^n(X)$, for each $n \in \mathbb{N}$. Pulling back by f_K^n and using bounded distortion, we show that, close to X, there are at least $c_2 e^{2f_K^n(X)}$ disjoint boxes of side $c_3/(e^{f_K^n(X)}(f_K^n)'(X))$ which meet $J(f_K)$, for some $c_2, c_3 > 0$. The result then follows from the fact that

$$\lim_{n \to \infty} \frac{\log c_2 e^{2f_K^n(X)}}{\log e^{f_K^n(X)}(f_K^n)'(X)/c_3} = 2.$$

In [19] we developed these ideas further to prove the following result.

Theorem 4.12. *If f is an entire function in the class B then*

$$dim_P J(f) = 2.$$

The proof of this result uses the fact that, if f is an entire function in the class B, then $J(f)$ contains an unbounded continuum Γ. We use a logarithmic change of variables, as introduced by Eremenko and Lyubich in [9], and note that the set $\exp^{-1}(\Gamma)$ comprises of unbounded continua which are separated by a vertical distance of 2π. We pull these back under $f^n \circ \exp$ and, using bounded distortion, we show that there exist $c_4, c_5 > 0$ such that, close to a point $z_0 \in J(f)$, there are at least $c_4 |\log f^n(z_0)|^2$ boxes of side $c_5 |f^n(z_0)|/|(f^n)'(z_0)|$ which meet $f^{-n} \circ \exp(\exp^{-1}(\Gamma)) = f^{-n}(\Gamma) \subset J(f)$, for each $n \in \mathbb{N}$. Thus

$$\overline{\dim}_B J(f) \geq 2 \lim_{n \to \infty} \frac{\log \log |f^n(z_0)|}{\log |(f^n)'(z_0)/f^n(z_0)|}.$$

By using Wiman-Valiron theory, we are able to show that, for each $\epsilon > 0$, there exists a point $z_0 \in J(f)$ for which this limit is at least $1 - \epsilon$. This is sufficient to prove Theorem 4.12.

(Note that in [19] we actually prove the stronger result that Theorem 4.12 holds for the set $A(f)$ which consists of those points that escape to infinity as quickly as posible — this is a subset of $J(f)$ for functions in the class B.)

Question 3 Does there exist a transcendental entire function f for which $\dim_P J(f) < 2$?

5. MEROMORPHIC FUNCTIONS WITH POLES

In this section we consider those transcendental meromorphic functions that have at least one pole. If such a function has only *finitely* many poles, then it follows from [20, Theorem 4] that the Julia set contains a continuum and so we have the following result.

Theorem 5.1. *Let f be a transcendental meromorphic function with finitely many poles. Then*
$$1 \leq \dim_H J(f) \leq \dim_P J(f) \leq 2.$$

We also showed in [20] that if f is a transcendental meromorphic function with finitely many poles then $J(f)$ contains no free Jordan arc.

If a function has *infinitely* many poles, however, then the situation is rather different. For example, if $f(z) = \tan z$, then the Julia set is the real axis, the upper and lower half planes are invariant components of the Fatou set and 0 is a parabolic fixed point. If $f_\lambda(z) = \lambda \tan z$, with $0 < \lambda < 1$, then 0 is an attracting fixed point and the Julia set is a Cantor set in the real line.

In [22] we proved the following result.

Theorem 5.2. *Let f be a transcendental meromorphic function. Then*
$$0 < \dim_H J(f) \leq 2.$$

This result can be proved by using Ahlfors' Five Islands Theorem to construct a disc D containing two disjoint simply connected domains G_1 and G_2 such that f^2 is a univalent map of G_i onto D for $i = 1, 2$. This sets up an iterated function scheme for which the invariant set (or attractor) is a subset of $J(f)$ with positive Hausdorff dimension.

It is in fact possible to obtain transcendental meromorphic functions whose Julia sets have Hausdorff dimension arbitrarily close to 0. Consider again the functions $f_\lambda(z) = \lambda \tan z$, with $0 < \lambda < 1$. For such a function, the upper and lower half planes are invariant and are so contained in the Fatou set. Thus the Julia set is contained in the real line. Since 0 is attracting, an interval containing 0, and all the pre-images of this interval must also be in the Fatou set. Thus the Julia set is a Cantor set in the real line.

As $\lambda \to 0$, the size of the interval around 0 increases. The size of the pre-images also increase and it might be expected that the Hausdorff dimension of $J(f_\lambda)$ would tend to zero. This is not, however, the case as is shown by the following result of Kotus [13, Example 1].

Theorem 5.3. *Let $f_\lambda(z) = \lambda \tan z$ with $\lambda \in \mathbb{C}$. Then*

$$dim_H J(f_\lambda) \geq 1/2.$$

Kotus actually proves a more general result than this in her paper. She considers transcendental meromorphic functions with infinitely many poles with some regularity conditions on the location of the poles and the behaviour of the function near the poles. She obtains a lower bound for the Hausdorff dimension of the Julia set — this lower bound depends on the positions of the poles. To do this she uses a natural sequence of covers of the Julia set (as described in Section 4) and Lemma 2.1.

In [22] we proved the following result.

Theorem 5.4. *For each $m \in \mathbb{N}$ with $m \geq 4$, the function f_m defined by*

$$f_m(z) = \lambda_m \tan z / \prod_{r=1}^{m}(z - r\pi),$$

where $\lambda_m = (2m)^{-m}$, satisfies

$$dim_H J(f_m) \leq 4/m.$$

This shows that 0 is indeed the best possible lower bound for the Hausdorff dimension of the Julia set of a transcendental meromorphic function. The reason for dividing $\lambda \tan z$ by a polynomial was to ensure that the first pre-images of the interval around the attracting fixed point at 0 become larger the further away from 0 they are — this makes the Julia set more sparse in the real line. The polynomial was chosen so as not to introduce any further poles. The proof of Theorem 5.4 involves looking at a natural sequence of covers of the Julia set as before. The initial cover here is a collection of intervals containing the poles of f.

In [28] we proved the following result.

Theorem 5.5. *For each $m \in \mathbb{N}$ with $m \geq 4$, the function f_m defined by*

$$f_m(z) = \lambda_m \tan z / \prod_{r=1}^{m} (z - r\pi),$$

where $\lambda_m = (2m)^{-m}$, is hyperbolic and satisfies

$$dim_P J(f_m) \geq 1/2.$$

This gives another class of functions whose Julia sets have different Hausdorff and packing dimensions and shows that the result of Theorem 3.2 does not hold for transcendental meromorphic functions with poles.

The proof of Theorem 5.5 involves identifying a subset A of $J(f_m)$ with box dimension equal to half and then applying Theorem 3.1. The set A consists of those first pre-images of the poles of f_m that are just to the left of the pole of f_m at $(2m + 1/2)\pi$. The points in A are very close to the points $1/(c_m(n + 1/2)\pi)$ for each $n \in \mathbb{N}$ and some constant c_m. Thus

$$\dim_B A = \dim_B \{1/n : n \in \mathbb{N}\} = 1/2.$$

In [25] we considered the functions defined by

$$f_{p,\lambda}(z) = \lambda \sum_{n=p^2}^{\infty} \frac{2z}{n^{2p} - z^2}, \quad p \in \mathbb{N}, \ \lambda > 0.$$

These functions are similar to the functions defined by $f_\lambda(z) = \lambda \tan z$ except that the poles are at $\pm n^p$ instead of at $\pm(n + 1/2)\pi$. As before, we have $J(f_{p,\lambda}) \subset \mathbb{R}$. Spreading the poles out by increasing the values of p gives an alternative way of obtaining Julia sets of small Hausdorff dimension and in [25] we proved the following result.

Theorem 5.6. *Let $f_{p,\lambda}$ be as defined above. If $p \geq 6$ and $0 < \lambda < 1/6^p$, then*

$$dim_H J(f_{p,\lambda}) \leq 1/p.$$

For a given value of p, we were also able to show that taking λ to be large causes the Hausdorff dimension of the Julia set to be close to 1. More precisely, we proved the following.

Theorem 5.7. *Let $f_{p,\lambda}$ be as defined above. If $p \geq 6$ and $p^{4p-1}/10^5 \log p < \lambda < p^{4p-1}/10^4 \log p$, then*

$$dim_H J(f_{p,\lambda}) \geq 1 - (30 \log \log p / \log p).$$

In order to prove Theorems 5.6 and 5.7, we used the following result which we proved in [24].

Theorem 5.8. *Let f be a hyperbolic function in the class B. Then there exists $s(f)$ such that*

$$\lim_{n\to\infty} \sum_{f^n(z)=a} \left(\frac{1+|f^n(z)|^2}{|(f^n)'(z)|(1+|z|^2)}\right)^t = \begin{cases} \infty & \text{for } t < s(f) \\ 0 & \text{for } t > s(f), \end{cases}$$

for each $a \in J(f)$.

For the functions $f_{p,\lambda}$ defined above, we showed in [24] that $\dim_H J(f_{p,\lambda}) = s(f_{p,\lambda})$ and then used this to prove Theorems 5.6 and 5.7. Further, in [26], we showed that, for a fixed $p \geq 6$, $s(f_{p,\lambda})$ is a continuous function of λ for $0 < \lambda < \frac{p^{4p-1}}{10^4 \log p}$. Together with Theorems 5.6 and 5.7, this shows that each value in $(0,1)$ is attained by $\dim_H J(f_{p,\lambda})$ for some p, λ. (Note that the results in [26] are proved for a class of families of functions depending on a parameter — this class includes the functions $f_{p,\lambda}$ that we have just discussed.)

Finally, we note that it was shown by Sullivan [30] that for a hyperbolic rational function f, the Hausdorff dimension of the Julia set is always equal to $s(f)$. In [24] we showed that this is not the case for transcendental meromorphic functions — for the hyperbolic functions $f_{p,K}$ described in Section 4.3, $s(f_{p,K})$ is strictly less than $\dim_H J(f_{p,K})$.

References

[1] I.N. Baker. The domains of normality of an entire function. *Ann. Acad. Sci. Fenn. Ser. A I Math.*, 1, (1975), 277-283.

[2] A.F. Beardon. *Iteration of Rational Functions* (Springer, 1991).

[3] W. Bergweiler. Iteration of meromorphic functions. *Bull. Amer. Math. Soc.*, 29 (1993), 151-188.

[4] L. Carleson and T.W. Gamelin. *Complex Dynamics* (Springer, 1993).

[5] R.L. Devaney and L. Goldberg. Uniformization of attracting basins for exponential maps. *Duke Math. J.* 2 (1987), 253-266.

[6] R. Devaney and M. Krych. Dynamics of exp(z). *Ergod. Th. Dynam. Sys.*, 4 (1984), 35-52.

[7] R.L. Devaney and F. Tangerman. Dynamics of entire functions near the essential singularity. *Ergod. Th. and Dynam. Sys.*, 6 (1986), 489-503.

[8] P.L. Duren. *Univalent Functions* (Springer, 1953).

[9] A.E. Eremenko and M.Yu Lyubich. Dynamical properties of some classes of entire functions. *Ann. Inst. Fourier (Grenoble)*, 42 (1992), 989-1020.

[10] K.J. Falconer. *Fractal Geometry: Mathematical Foundations and Applications* (Wiley, 1990).

[11] B. Karpińska. Area and Hausdorff dimension of the set of accessible points of the Julia sets of λe^z and $\lambda \sin z$. *Fund. Math.*, 159 (1999), 269-287.

[12] B. Karpińska. Hausdorff dimension of the hairs without endpoints for $\lambda \exp z$. *C. R. Acad. Sci. Paris*, 328 (1) (1999), 1039-1044.

[13] J. Kotus. On the Hausdorff dimension of Julia sets of meromorphic functions II. *Bull. Soc. Math. France*, 123 (1995), 33-46.

[14] J. Kotus and M. Urbański. Hausdorff dimension and Hausdorff measures of Julia sets of elliptic functions. *Bull. London Math. Soc.*, 35 (2003), 269-275.
[15] C. McMullen. Area and Hausdorff dimension of Julia sets of entire functions. *Trans. Amer. Math. Soc.*, 300 (1987), 329-342.
[16] M. Misiurewicz. On iterates of e^z. *Ergod. Th. Dynam. Sys.*, 1 (1981), 103-106.
[17] G. Pólya and G. Szegö. *Problems and Theorems in Analysis I* (Part III, problems 158-160). Springer, New York, 1972.
[18] P.J. Rippon and G.M. Stallard. Iteration of a class of hyperbolic meromorphic functions. *Proc. Amer. Math. Soc.*, 127 (1999), 3251-3258.
[19] P.J. Rippon and G.M. Stallard. Dimensions of Julia sets of meromorphic functions. *J. London Math. Soc.*, (2) 71 (2005), 669-683.
[20] P.J. Rippon and G.M. Stallard. Escaping points of meromorphic functions with a finite number of poles. *J. d'Analyse Math.*, 96 (2005), 225-245.
[21] G.M. Stallard. The Hausdorff dimension of Julia sets of entire functions. *Ergod. Th. and Dynam. Sys.*, 11 (1991), 769-777.
[22] G.M. Stallard. The Hausdorff dimension of Julia sets of meromorphic functions. *J. London Math. Soc.*, (2) 49 (1994), 281-295.
[23] G.M. Stallard. The Hausdorff dimension of Julia sets of entire functions II. *Math. Proc. Camb. Phil. Soc.*, 119 (1996), 513-536.
[24] G.M. Stallard. The Hausdorff dimension of Julia sets of hyperbolic meromorphic functions. *Math. Proc. Camb. Phil. Soc.*, 127 (1999), 271-288.
[25] G.M. Stallard. The Hausdorff dimension of Julia sets of meromorphic functions II. *J. London Math. Soc.*, (2) 60 (1999), 847-859.
[26] G.M. Stallard. The Hausdorff dimension of Julia sets of hyperbolic meromorphic functions II. *Ergod. Th. and Dynam. Sys.*, 20 (2000), 895-910.
[27] G.M. Stallard. The Hausdorff dimension of Julia sets of entire functions IV. *J. London Math. Soc.*, (2) 61 (2000), 471-488.
[28] G.M. Stallard. Dimensions of Julia sets of hyperbolic meromorphic functions. *Bull. London Math. Soc.*, 33 (2001), 689-694.
[29] G.M. Stallard. Dimensions of Julia sets of hyperbolic entire functions. *Bull. London Math. Soc.*, 36 (2004), 263-270.
[30] D. Sullivan. 'Conformal dynamical systems', *Geometric dynamics*, Lecture Notes in Math., 1007 (Springer, New York, 1983) 725-752.
[31] M. Urbański. Geometry and ergodic theory of conformal non-recurrent dynamics. *Ergod. Th. and Dynam. Sys.*, 17 (1997), 1449-1476.
[32] M. Urbański and A. Zdunik. The finer geometry and dynamics of the hyperbolic exponential family. *Michigan Math. Journal*, 51 (2003), 227-250.
[33] M. Urbański and A. Zdunik. Real analyticity of Hausdorff dimension of finer Julia sets of exponential family. *Ergod. Th. and Dynam. Sys.*, 24 (2004), 279-315.

THE OPEN UNIVERSITY, DEPARTMENT OF MATHEMATICS, WALTON HALL, MILTON KEYNES MK7 6AA, UK

E-mail address: g.m.stallard@open.ac.uk

ABEL'S FUNCTIONAL EQUATION AND ITS ROLE IN THE PROBLEM OF *CROISSANCE RÉGULIÈRE*.

GEORGE SZEKERES

Professor Szekeres completed the manuscript of this paper in April 2005, sadly just a few months before he died, aged 94.

In memory of I.N. Baker

From the earliest years of his mathematical life Noel Baker shared with me a strong interest in Abel's functional equation

$$(1) \qquad A(f(x)) = A(x) + 1$$

where f is a continuous strictly increasing function of a real variable x. Yet the nearest we ever came to a joint publication were two papers published side by side in the Journal of the Australian Mathematical Society; see [5] and [1]. Inspired by an earlier work of Noel, I showed that if f is an entire or rational function with $f'(0) = 1$ and having a Taylor expansion

$$f(z) = z + \sum_{n \geq 2} a_n z^n$$

then with the exception of $z/(a+az)$ the formal fractional iterates

$$(2) \qquad f_\sigma(z) = z + \sigma a_2 z^2 + \sum_{n > 2} a_n^{(\sigma)} z^n, \qquad \sigma \in \mathbb{R}$$

of f (the coefficients $a_n^{(\sigma)}$ being formally determined from $f \circ f_\sigma = f_\sigma \circ f$) cannot be holomorphic at $z = 0$ for all real σ. In fact, by results of Baker, Écalle and Liverpool (see, for example, [4]), the set of σ for which the series (2) converges forms a discrete one-dimensional lattice in $\mathbb{R}$. Noel immediately generalized my result to all meromorphic f. At that time he was strongly influenced by the works of the French mathematicians G. Julia and P. Fatou [2], long before their work was resurrected as part of the subject of dynamical systems.

The basic tool of fractional iterations is Abel's functional equation (1) or some of its variants. This simple but intriguing functional equation has an interesting history. Abel himself never published it in his short lifetime (he was barely 27 when he died) but his brief note, in which he clearly states the role of the equation in fractional iterations, was fortunately rescued for posterity by his teacher Holmboe and was included in the second volume of Abel's collected works under the cryptic title *Détermination d'une fonction au moyen d'une équation qui ne contient q'une seule variable*; that is, "Determination of a function by means of an equation which only contains a single variable" (not a word about iteration in the title).

In fact in order to construct a family of iterates of f, all one has to do is to find a continuous strictly increasing solution A of (1) (there are infinitely many such solutions). If $B = A^{-1}$ is the inverse of A then the functions

$$f_\sigma(x) = B(A(x) + \sigma), \qquad \sigma \in \mathbb{R}$$

form a family of fractional iterates satisfying

$$f_\sigma \circ f_\tau = f_\tau \circ f_\sigma = f_{\sigma+\tau}$$

for all $\sigma, \tau \in \mathbb{R}$, $f_1 = f$. Of course B itself satisfies

(3) $$B(y+1) = f(B(y)).$$

My own interest in Abel's equation came from an entirely different source. What is particularly interesting about equation (1) is that $A(x)$ grows more slowly to infinity than any iterate $f_\sigma(x)$, $\sigma > 0$. The ease with which (1) produces such functions strongly suggests that the equation should have a significant role in any theory of "orders of infinity" in the sense of G.H. Hardy. This fact has been almost completely ignored, with one notable exception, that of Paul Lévy [3]. What Lévy envisaged was an intrinsic property of real functions which could single out a class of "regularly growing" functions (*croissance régulière* in Lévy's terminology), which could then hopefully provide a universal comparison scale for real functions in the spirit of Hardy's orders of infinity. Such a class $\mathcal{R}$ of (preferably analytic) functions ought to have the following two basic properties:

(P1) *Comparability* : If $f, g \in \mathcal{R}$ then $f(x) - g(x)$ has the same sign (positive or negative) for all sufficiently large x.

(P2) *Completeness* : If h is any continuous strictly increasing positive function then $\exists f \in \mathcal{R}$ such that $f(x) = h(x)$ for infinitely many arbitrarily large x.

Lévy mentions other desirable properties of $\mathcal{R}$ such as closedness under differentiation, but here we postulate just one further property, namely:

(P3) : If $f \in \mathcal{R}$ and g is defined by

$$g(x) = cf(x+a) + b, \qquad c > 0, a, b \in \mathbb{R}$$

then $g \in \mathcal{R}$.

Unfortunately Lévy's proposed condition for *croissance régulière*, as envisaged by him, led nowhere and at that time it appeared almost impossible to verify (or even to guess) whether a given function does satisfy Lévy's condition or not. Some years ago I recast his condition in a more manageable form to make it at least accessible to computer examination [6]. Here I wish to present another modification which promises to get us nearer to capturing the elusive concept of regular growth.

Let us assume that $f(x)$ is analytic for $x \geq 0$ and $f'(0) = b > 1$,
$$f(x) = bx + \sum_{n>1} b_n x^n, \qquad b_n \geq 0 \text{ for all } n > 1.$$
Then there is an analytic solution A of Abel's equation (1) with an expansion
$$A(x) = \frac{1}{\log b} \log x + \sum_{j \geq 0} a_j x^j$$
where the coefficients a_j are determined recursively from (1), with the exception of a_0 which is anyhow irrelevant. Of course then $D(x) = 1/A'(x)$ is also analytic with an expansion

(4) $$D(x) = x \log b + \sum_{j>1} d_j x^j$$

where the coefficients d_j are formally determined from

(5) $$D(f(x)) = f'(x) D(x).$$

We shall be more interested in a similar function $E(x)$ satisfying
(6) $$E(f(x)) = f'(x) \big(E(x) + D'(x) \big)$$
and having an expansion
(7) $$E(x) = \sum_{j \geq 0} e_j x^j, \qquad e_0 = -\frac{b \log b}{b-1}$$
with the coefficients e_j determined formally from (6) (except for e_1 which can be arbitrarily prescribed). We normalize e_1 by setting
(8) $$e_1 = \frac{2b_2}{b(b-1)} \left(1 - \frac{b \log b}{b-1} \right)$$
for reasons explained in [6, equation (2-16)].

We introduce now the function
$$K(x) = \frac{E(x)}{D(x)}$$
which satisfies
(9) $$K(f(x)) = K(x) + \frac{D'(x)}{D(x)}.$$
From here and the assumed convexity of f it follows that
$$\psi(x) = \lim_{n \to \infty} K(f^n(x))$$
exists and of course $\psi(f(x)) = \psi(x)$. Therefore
$$\varphi(\sigma) = \psi(A^{-1}(\sigma)) \qquad \text{satisfies} \qquad \varphi(\sigma + 1) = \varphi(\sigma).$$

In fact

$$A(f(x)) = A(x) + 1, \qquad B(A(x) + 1) = f(x), \qquad B(\sigma + 1) = f(B(\sigma))$$

and so

$$\varphi(\sigma + 1) = \psi(B(\sigma + 1)) = \psi(f(B(\sigma))) = \psi(B(\sigma)) = \varphi(\sigma).$$

What is noteworthy here is that we have associated with f a periodic function φ with period 1; for convenience (and quite justifiably) we shall call φ the Lévy transform of f. In the present context the merit of the Lévy transform is that it allows us to express the desired regularity of growth condition of f by some regularity condition of the Fourier coefficients of φ. We note that the normalization condition (8) simply expresses the fact that $\int_0^1 \varphi(x) \mathrm{d}x = 0$, as shown in [6].

Let us write the Fourier coefficients of φ in the form $\widehat{\varphi_n} = \mathrm{e}^{\alpha_n + i\beta_n}$, α_n, β_n real so that $|\widehat{\varphi_n}| = \mathrm{e}^{\alpha_n}$. We shall call α_n the Lévy-Fourier coefficients of f. The most obvious regularity property of the Lévy-Fourier coefficients to consider is total monotonity (see [6] for the definition), and we postulate:

f is regularly growing (in the sense of Lévy) if the Lévy-Fourier coefficients are totally monotonic.

As a trivial illustration consider $f(x) = bx$, $b > 1$. We then have

$$A(x) = \frac{\log x}{\log b}, \qquad D(x) = x \log b, \qquad E(x) = -\frac{b}{b-1} \log b$$

because of the normalization (8). Therefore

$$K(x) = -\frac{b}{b-1} \frac{1}{x}, \qquad \psi(x) = 0$$

and the Lévy-Fourier coefficients are all 0, hence totally monotonic. Thus bx is (trivially) regularly growing.

In general, however, one cannot verify the regularity postulate because we do not have sufficient control over the Lévy-Fourier coefficients. In principle one can calculate any number of these and in many examples there are hundreds of Lévy-Fourier coefficients which are known to be totally monotonic. We say $f(x)$ is regularly growing of order k if the first k of the Lévy-Fourier coefficients are totally monotonic. Given f this can certainly be verified for any k but, apart from the trivial example of a linear f, I am not aware of any function which is regularly growing of order ∞. Until such an example is produced (or shown that it does not exist) one can hardly call the problem of regular growth fully solved by our criterion. I am afraid I must leave the problem in this somewhat perplexing state.

References

[1] I.N. Baker, *Fractional iteration near a fixpoint of multiplier 1*, J. Austral. Math. Soc. **4** (1964), 143–148.

[2] P. Fatou, *Sur l'itération des fonctions transcendantes entières*, Acta Math., **47** (1926), 337-370.

[3] P. Lévy, *Fonctions à croissance régulière et itérations d'order fractionnaire*, Ann. di Math. Pure Appl. (4) **5** (1928), 270-298.

[4] L.S.O. Liverpool, *Fractional iteration near a fix point of multiplier 1*, J. London Math. Soc. (2) **9** (1974/75), 599–609.

[5] G. Szekeres, *Fractional iteration of entire and rational functions*, J. Austral. Math. Soc. **4** (1964), 129–142.

[6] G. Szekeres, *Abel's equation and regular growth: variations on a theme by Abel*, Experimental Mathematics **7** (1998), 85-100.

Department of Mathematics, University of Adelaide, Adelaide, South Australia